T0222795

Lecture Notes in Computer Science   10672

Commenced Publication in 1973
Founding and Former Series Editors:
Gerhard Goos, Juris Hartmanis, and Jan van Leeuwen

More information about this series at http://www.springer.com/series/7407

Roberto Moreno-Díaz · Franz Pichler
Alexis Quesada-Arencibia (Eds.)

Computer Aided Systems Theory – EUROCAST 2017

16th International Conference
Las Palmas de Gran Canaria, Spain, February 19–24, 2017
Revised Selected Papers, Part II

 Springer

Editors
Roberto Moreno-Díaz
University of Las Palmas de Gran Canaria
Las Palmas de Gran Canaria
Spain

Alexis Quesada-Arencibia
University of Las Palmas de Gran Canaria
Las Palmas de Gran Canaria
Spain

Franz Pichler
Johannes Kepler University Linz
Linz
Austria

ISSN 0302-9743   ISSN 1611-3349   (electronic)
Lecture Notes in Computer Science
ISBN 978-3-319-74726-2   ISBN 978-3-319-74727-9   (eBook)
https://doi.org/10.1007/978-3-319-74727-9

Library of Congress Control Number: 2018930745

LNCS Sublibrary: SL1 – Theoretical Computer Science and General Issues

Printed on acid-free paper

This Springer imprint is published by Springer Nature
The registered company is Springer International Publishing AG
The registered company address is: Gewerbestrasse 11, 6330 Cham, Switzerland

Preface

The concept of CAST as a computer-aided systems theory was introduced by Franz Pichler in the late 1980s to refer to computer theoretical and practical development as tools for solving problems in system science. It was thought of as the third component (the other two being CAD and CAM) required to complete the path from computer and systems sciences to practical developments in science and engineering.

Franz Pichler, of the University of Linz, organized the first CAST workshop in April 1988, which demonstrated the acceptance of the concepts by the scientific and technical community. Next, Roberto Moreno-Díaz, of the University of Las Palmas de Gran Canaria, joined Franz Pichler, motivated and encouraged by Werner Schimanovich, of the University of Vienna (present Honorary Chair of Eurocast), and they organized the first international meeting on CAST, (Las Palmas February 1989), under the name EUROCAST 1989. The event again proved to be a very successful gathering of systems theorists, computer scientists, and engineers from most European countries, North America, and Japan.

It was agreed that EUROCAST international conferences would be organized every two years, alternating between Las Palmas de Gran Canaria and a continental European location. Since 2001 the conference has been held exclusively in Las Palmas. Thus, successive EUROCAST meetings took place in Krems (1991), Las Palmas (1993), Innsbruck (1995), Las Palmas (1997), and Vienna (1999), before being held exclusively in Las Palmas in 2001, 2003, 2005, 2007, 2009, 2011, 2013, and 2015, in addition to an extra-European CAST conference in Ottawa in 1994. Selected papers from these meetings were published as Springer *Lecture Notes in Computer Science* volumes 410, 585, 763, 1030, 1333, 1798, 2178, 2809, 3643, 4739, 5717, 6927, 6928, 8111, 8112, and 9520, respectively, and in several special issues of *Cybernetics* and *Systems: An International Journal*. EUROCAST and CAST meetings are definitely consolidated, as shown by the number and quality of the contributions over the years.

EUROCAST 2017 took place in the Elder Museum of Science and Technology of Las Palmas, during February 19–24, and it continued with the approach tested at previous conferences as an international computer-related conference with a true interdisciplinary character. The participants profiles are presently extended to include fields that are in the frontier of science and engineering of computers, of information and communication technologies, and the fields of social and human sciences. The best paradigm is the Web, with its associate systems engineering, CAD-CAST tools, and professional application products (Apps) for services in the social, public, and private domains.

There were specialized workshops, which, on this occasion, were devoted to the following topics:

1. Pioneers and Landmarks in the Development of Information and Communication Technologies, chaired by Pichler (Linz), Stankovic (Nis), Kreuzer Felisa and Kreuzer James (USA)
2. Systems Theory and Applications, chaired by Pichler (Linz) and Moreno-Díaz (Las Palmas)

3. Stochastic Models and Applications to Natural, Social, and Technological Systems, chaired by Nobile and Di Crescenzo (Salerno)
4. Theory and Applications of Metaheuristic Algorithms, chaired by Affenzeller and Jacak (Hagenberg) and Raidl (Vienna)
5. Embedded Systems Security, chaired by Mayrhofer (Linz) and Schmitzberger (Linz)
6. Model-Based System Design, Verification, and Simulation, chaired by Nikodem (Wroclaw), Ceska (Brno), and Ito (Utsunomiya)
7. Systems in Industrial Robotics, Automation and IoT, chaired by Stetter (Munich), Markl (Vienna), and Jacob (Kempten)
8. Applications of Signal Processing Technology, chaired by Huemer (Linz), Zagar (Linz), Lunglmayr (Linz), and Haselmayr (Linz)
9. Algebraic and Combinatorial Methods in Signal and Pattern Analysis, chaired by Astola (Tampere), Moraga (Dortmund), and Stankovic (Nis)
10. Computer Vision, Deep Learning, and Applications, chaired by Penedo (A Coruña) and Radeva (Barcelona)
11. Computer- and Systems-Based Methods and Electronic Technologies in Medicine, chaired by Rozenblit (Tucson), Hagelauer (Linz), Maynar (Las Palmas), and Klempous (Wroclaw)
12. Cyber-Medical Systems, chaired by Rudas (Budapest), Kovács (Budapest), and Fujita (Iwate)
13. Socioeconomic and Biological Systems: Formal Models and Computer Tools, chaired by Schwaninger (St. Gallen), Schoenenberger (Basel), Tretter (Munich), Cull (Corvallis USA), and Suárez-Araújo (Las Palmas)
14. Intelligent Transportation Systems and Smart Mobility, chaired by Sanchez-Medina (Las Palmas), Celikoglu (Istanbul), Olaverri-Monreal (Wien), Garcia-Fernandez (Madrid), and Acosta-Sanchez (La Laguna)

In this conference, as in previous ones, most of the credit for the success is due to the workshop chairs. They and the sessions chairs, with the counseling of the international Advisory Committee, selected from 160 presented papers, after oral presentations and subsequent corrections, the 117 revised papers included in this volume.

The event and this volume were possible thanks to the efforts of the chairs of the workshops in the diffusion and promotion of the conference, as well as in the selection and organization of all the material. The editors would like to express their thanks to all the contributors, many of whom are already Eurocast participants for years, and particularly to the considerable interaction of young and senior researchers, as well as to the invited speakers, Prof. Christian Müller-Scholer from Hamburg, Prof. Manuel Maynar, from Las Palmas, and Prof. Jaakko Astola from Tampere, for their readiness to collaborate. We would also like to thank the director of the Elder Museum of Science and Technology, D. José Gilberto Moreno, and the museum staff. Special thanks are due to the staff of Springer in Heidelberg for their valuable support.

September 2017   Roberto Moreno-Díaz
                                                                           Franz Pichler
                                                             Alexis Quesada-Arencibia

Organization

Organized by

Instituto Universitario de Ciencias y Tecnologías Cibernéticas
Universidad de Las Palmas de Gran Canaria, Spain

Johannes Kepler University Linz,
Linz, Austria

Museo Elder de la Ciencia y la Tecnología
Las Palmas de Gran Canaria, Spain

Conference Chair

Roberto Moreno-Díaz, Las Palmas

Program Chair

Franz Pichler, Linz

Honorary Chair

Werner Schimanovich, Austrian Society for Automation and Robotics

Organizing Committee Chair

Alexis Quesada Arencibia
Instituto Universitario de Ciencias y Tecnologías Cibernéticas
Universidad de Las Palmas de Gran Canaria
Campus de Tafira
35017 Las Palmas de Gran Canaria, Spain
Phone: +34-928-457108
Fax: +34-928-457099
e-mail: alexis.quesada@ulpgc.es

Contents – Part II

Model-Based System Design, Verification and Simulation

Applications of Signal Processing Technology

Algebraic and Combinatorial Methods in Signal and Pattern Analysis

Computer Vision, Deep learning and Applications

**Computer and Systems Based Methods
and Electronic Technologies in Medicine**

Intelligent Transportation Systems and Smart Mobility

Contents – Part I

Systems Theory, Socio-economic Systems and Applications

Stochastic Models and Applications to Natural, Social and Technical Systems

On Fractional Stochastic Modeling of Neuronal Activity Including Memory Effects

Giacomo Ascione and Enrica Pirozzi[✉]

Dipartimento di Matematica e Applicazioni, Università di Napoli Federico II,
Via Cintia, 80126 Napoli, Italy
{giacomo.ascione,enrica.pirozzi}@unina.it

Abstract. In order to model the memory and to describe the memory effects in the firing activity of a single neuron subject to a time-dependent input current, a fractional stochastic Langevin-type equation is considered. Two different discretization formulas are derived and the corresponding algorithms are implemented by means of R-codes for several values of the parameters. Reset mechanisms after successive spike times are suitably imposed to compare simulation results. The firing rates and some neuronal statistical estimates obtained by means the two algorithms are provided and discussed.

1 Introduction

The behavior of the neuronal membrane potential and its firing activity can be described by the well-known Leaky Integrate-and-Fire (LIF) model ([4–6,9] and references therein), even though this model is not able to include any memory effects such as those due to the history of the firing itself or those related to correlated inputs. Indeed, several correlated causes can develop features of the neuronal dynamics. For instance, the dynamics of membrane neuronal gates or the action of ionic currents, as those related to voltage-dependent potassium channels [7,10], can be exploit to explain the spike-frequency adaptation. Moreover, different time scales are involved in the neuronal dynamics as, for instance, the slower one of calcium dependent channels and the faster one typical of the membrane voltage variation. By considering all involving dynamics, the value of membrane voltage cannot be represented by a Markovian process such as that of LIF model, but models with correlations are mandatory in order to include memory effects.

An attempt to explain adaptation in the firing activity is that to insert in the LIF model the inputs with no-zero correlation; in [11] the usual white noise (an uncorrelated Gaussian process) involved in the stochastic LIF dynamics is replaced by the temporally correlated (colored) inputs, i.e. a zero mean Gaussian stochastic process with covariance $c(s,t) = \frac{1}{2\tau} \exp\left\{-\frac{|t-s|}{\tau}\right\}$, with $t \geq s \geq 0$ and where τ is the so-called correlation time. In such kind of models, the focus is

This paper is partially supported by G.N.C.S.- INdAM.

R. Moreno-Díaz et al. (Eds.): EUROCAST 2017, Part II, LNCS 10672, pp. 3–11, 2018.
https://doi.org/10.1007/978-3-319-74727-9_1

centered on the rule and on the effects of the correlation time τ on the variation of the firing rate. Anyway, although extremely interesting from both the mathematical and physiological points of view, the stochastic evolution of the above model can be interpreted as a double integrated [15] Gaussian process over time. Hence, the study of the memory effects can be quite complex to be carried out. Furthermore, in order to take into account the adaptation phenomenon as a consequence of the effect of variations in the calcium concentration, in [7] a time-inhomogeneous LIF model is proposed for the neuronal activity, based on the following stochastic differential equation for $t \geq 0$:

$$dV(t) = - \left[\frac{V(t) - \rho(t)}{\theta(t)} - \mu \right] dt + \sigma dW(t), \quad V(0) = V_0, \tag{1}$$

where $\rho(t)$ and $\theta(t)$ are the time-dependent resting potential and decay time, respectively, μ a constant signal, σ the intensity of the noise and $W(t)$ the standard Brownian motion. Even though uncorrelated inputs are considered, time dependent functions are adopted to explain the action of multiple time-scale dynamics. In this case, the theory of Gauss-Markov processes and the First Passage Time problem turned out especially useful (see [4–7]). Note that, also in this case, the mathematical manipulations, applied to model phenomenological evidences in order to obtain the tractable Eq. (1), incorporate memory effects, making then difficult to identify them.

Recently, several models have been proposed to describe the no-memoryless neuronal voltage by using fractional derivatives. The fractional stochastic approach for neuronal modeling can be found in [1,3,12] in which some memory effects and correlations are involved. Due to the property to include different scales and ranges of time, the fractional stochastic models seem to be more general and powerful than models including colored noise and time-inhomogeneous LIF models.

Here, along the lines of [2,13], a stochastic model based on fractional stochastic Langevin equation including a time-dependent current is considered; for such a model the corresponding stochastic process is not Markovian and includes the memory of the process itself. Specifically, starting from [13,14], we consider the fractional Leaky Integrate-and-Fire (FLIF) model based on the following equation, for $t \geq 0$:

$$C_m \frac{d^\alpha V(t)}{dt^\alpha} = -g_L(V(t) - V_r) + I_{inj}(t), \quad V(0) = V_0, \tag{2}$$

where $V(t)$ is the neuronal membrane potential, V_r the resting potential, C_m the membrane capacitance, g_L the leak conductance, α is the order of the fractional derivate ($0 < \alpha \leq 1$) and the injecting current $I_{inj}(t)$ includes a Gaussian white noise. The above dynamics records a spike (the firing) if the voltage $V(t)$ crosses a threshold value V_{th}, i.e. if $V(t) \geq V_{th}$, and the reset $V(t^+) \to V_{reset}$ is imposed, without applying any reset to the current $I_{inj}(t)$. The time t such that $V(t) \geq V_{th}$ is the time of the firing, i.e. a spike is generated at time t.

The above fractional model is presented and investigate in [13] with a constant current I_{inj}; the same Langevin-type equation is also considered in [2], but it is investigated in a different way. Here, we compare the two study approaches from which two different simulation procedures can be derived. We point out that, even if the same stochastic fractional equation is involved in the two models, a specific reset mechanism is required for the model in [2] to make it comparable to that in [13]. We investigate the results of the two R-codes suitably made up to implement the corresponding algorithms. In particular, to put in evidence the effects of the memory of the process on the firing activity, we provide simulations including both a constant and a time-depending external current. The First Passage Time (FPT) probability density, often used to provide approximations of Inter-Spike Interval (ISI) (for instance, see [4,8]) is estimated by histograms of first crossing (through the neuronal threshold V_{th}) time of simulated paths of Eq. (2).

2   The Fractional Model

Let us consider the neuronal model based on the following fractional equation, for $t \geq 0$:

$$\frac{d^\alpha V}{dt^\alpha} = -\frac{1}{\theta}(V(t) - V_r) + I(t) + \sigma\xi(t), \quad \text{if} \quad V > V_{th}, \text{ then } V \to V_{reset}, \quad (3)$$

with $V(0) = V_0$, and where, comparing with (2), $\theta = C_m/g_L$, $I(t)$ stands for $I_{inj}(t)/C_m$ and $\xi(t)$ represents a white Gaussian noise. Following the notation used in [2], we denote the Caputo derivative as D_α^*, while the fractional integral is denoted as J_α. Hence, we have:

$$J_\alpha f(t) = \frac{1}{\Gamma(\alpha)} \int_0^t (t-s)^{\alpha-1} f(s) ds \qquad (4)$$

so that $D_\alpha^* = J_{1-\alpha}D$, where D is the usual derivative. In particular, one has $J_{1-\alpha}^{-1} = DJ_\alpha$. The Eq. (3) can be also re-written as:

$$D_\alpha^* V(t) = -\frac{1}{\theta}(V(t) - V_r) + I(t) + \sigma\xi(t). \qquad (5)$$

Along the lines of [13], a simulation algorithm can be obtained. In particular, applying the representation of [2] in [13], we have

$$D_\alpha^* V(t) = \frac{1}{\Gamma(1-\alpha)} \int_0^t \frac{V'(s)}{(t-s)^\alpha} ds.$$

One can consider a discretization of $V'(s)$ such as

$$D_\alpha^* V(t) \simeq \frac{\Delta t^{-\alpha}}{\Gamma(2-\alpha)} \sum_{k=0}^{n-1} (V_{k+1} - V_k)[(n-k)^{1-\alpha} - (n-k-1)^{1-\alpha}]$$

where Δt is fixed, n is chosen such that $n\Delta t = t$ and $V_k = V(k\Delta t)$. Then, by using Eq. (5) and isolating V_n in the sum on the left side of the above equation, we have, for $n \geq 1$,

$$V_n = V_{n-1} + (\Delta t)^\alpha \Gamma(2 - \alpha) \left(-\frac{1}{\theta}(V_{n-1} - V_r) + I_{n-1} + \sigma\xi_{n-1} \right)$$

$$- \sum_{k=0}^{n-2}(V_{k+1} - V_k)[(n - k)^{1-\alpha} - (n - k - 1)^{1-\alpha}] \tag{6}$$

where $I_{n-1} = I((n - 1)\Delta t)$ and $\xi_{n-1} = \xi((n - 1)\Delta t)$. To introduce the *space* reset after a spike, one have to define $V_n = V_0$ if $V_{n-1} \geq V_{\text{th}}$, instead of using (6). Note that ξ_k is simulated by using $\xi_k = (\Delta t)^{-\frac{1}{2}}\sigma\xi$ where ξ is a standard Gaussian random number. Finally, by means of a suitable R-code based on (6) and the specific reset, we provide simulations of the modeled firing activity.

An Alternative Approach. Following the lines of [2], one can also consider the integral version of the proposed model. Indeed, by using $D_\alpha^* = J_{1-\alpha}D$ on (5) and then the operator $J_{1-\alpha}^{-1}$ on both terms of the same equation we have:

$$DV(t) = J_{1-\alpha}^{-1} \left(-\frac{1}{\theta}(V(t) - V_r) + I(t) + \sigma\xi(t) \right).$$

Furthermore, one can observe that $J_{1-\alpha}^{-1} = DJ_\alpha$ to obtain:

$$V(t) = V_0 + J_\alpha \left(-\frac{1}{\theta}(V(t) - V_r) + I(t) + \sigma\xi(t) \right). \tag{7}$$

Let us consider the integral term in Eq. (7):

$$J_\alpha \left(-\frac{1}{\theta}(V(t) - V_r) + I(t) + \sigma\xi(t) \right)$$

$$= \frac{1}{\Gamma(\alpha)} \int_0^t (t - s)^{\alpha-1} \left(-\frac{1}{\theta}(V(s) - V_r) + I(s) + \sigma\xi(s) \right) ds.$$

Let us proceed as done before, by using a discretization of the integrand:

$$J_\alpha \left(-\frac{1}{\theta}(V(t) - V_r) + I(t) + \sigma\xi(t) \right)$$

$$\simeq \frac{(\Delta t)^\alpha}{\Gamma(\alpha + 1)} \sum_{k=0}^{n-1} \left(-\frac{1}{\theta}(V_k - V_r) + I_k + \sigma\xi_k \right) [(n - k)^\alpha - (n - k - 1)^\alpha].$$

Using this term in Eq. (7), we have for $n > 0$

$$V_n = V_0 + \frac{(\Delta t)^\alpha}{\Gamma(\alpha + 1)} \sum_{k=0}^{n-1} \left(-\frac{1}{\theta}(V_k - V_r) + I_k + \sigma\xi_k \right) [(n-k)^\alpha - (n-k-1)^\alpha]. \tag{8}$$

Now, we have to introduce a specific reset in the simulation procedure based on the Eq. (8). To do that, one can also impose a sort of *space-time reset* (or memory reset), if $\tilde{n} - 1$ is the last spiking time, by setting $V_{\tilde{n}} = V_0$ if $V_{\tilde{n}-1} \geq V_{\text{th}}$, and then for $m > \tilde{n}$:

$$V_m = V_0 + \frac{(\Delta t)^\alpha}{\Gamma(\alpha+1)} \sum_{k=\tilde{n}}^{m-1} \left(-\frac{1}{\theta}(V_k - V_r) + I_k + \sigma\xi_k \right) [(m-k)^\alpha - (m-k-1)^\alpha].$$

(9)

In the next section we provide simulation results by means an R-code based on the Eqs. (8) and (9).

Briefly, we remark that the two simulation procedures based on (6) and (9), respectively, include reset mechanism and memory terms in different ways. The first of them (6) is subject to a simple space reset, then the process continues its evolution *remembering* also what happened before the last spike time. The second (9) does not remember exactly what happened before the spike, but it records the last spike time taking into account the elapsed time. Moreover, these different reset mechanisms produce a difference in complexity and execution time, making (6) much more slower than (9). In order to clarify these differences, we provide some simulation results.

3   Simulation Results

For simulations, we used parameters as in Table 1. In particular, we simulated sample paths, first passage times and firing rates by using firstly a constant current $I(t) = \frac{I_0}{C_m}$ and then a periodic current $I(t) = \frac{I_0}{2C_m}(1 + \cos(\omega t + \varphi))$. Firing rates are calculated using the following formula:

$$F_r(t) = \frac{N_{\text{sp}}(t)}{t}$$

(10)

where $N_{\text{sp}}(t)$ is the number of spikes until the time t. In simulations we also include a refractory time τ_{ref}, such that if $V(t) > V_{\text{th}}$ then $V(t^+) = V_0$ and $V(t + s) = V_0$ for all $s \in [0, \tau_{\text{ref}}]$.

Table 1. Parameters for simulations

$g_L = 25\,\text{nS}$	$\alpha = 0.4$	$V_r = -70\,\text{mV}$	$\omega = \frac{2\pi}{20}\,\text{rad/s}$
$C_m = 0.5\,\text{nF}$	$\sigma = 1\,\text{mV(s)}^{\frac{1}{2}}$	$V_{\text{th}} = -50\,\text{mV}$	$\varphi = 0\,\text{rad}$
$\theta = \frac{C_m}{g_L}\,\text{ms}$	$I_0 = 12\,\text{nA}$	$V_0 = -70\,\text{mV}$	$\tau_{\text{ref}} = 0.5\,\text{ms}$

As suggested in [13], we separately simulated V_1 as:

$$V_1 = V_0 - \frac{1}{\theta}(V_0 - V_r + \frac{\theta I_0}{C_m})\Delta t + \sigma\xi_0\sqrt{\Delta t}.$$

Furthermore, for the histogram of the FPTs, we provide 10000 simulated trajectories for each process.

Let us consider sample paths for (6) and (9), for constant $I(t)$ and then periodic $I(t)$ (Fig. 1). We have a different number of noise generation (gaussian random numbers) in each algorithm, so we can not fix the seed before generating paths. Then, even if very similar, the obtained paths by using the two algorithms are not the same. This is more evident in paths corresponding to the periodic $I(t)$, when the current is sufficiently low to generate a time interval without spikes.

In Fig. 2 we compare the two corresponding firing rates for both kind of currents and equations. For the constant $I(t)$, one can see that firing rates stabilize themselves on different values. In particular, the spiking is more frequent in (9), where the process is subject also to a memory reset. Instead for periodic $I(t)$ we have an evident *adaptation* phenomenon, in which firing rates stabilize on the same value, but with a *different velocity*, as can be seen in Fig. 2.

We also investigated how the firing rates change when we change α (Fig. 3). We can see that when α increases, firing rates stabilize to a lower value.

One can simulate FPTs (Fig. 4). These are not influenced by the reset mechanism; indeed, one can see that the histograms are quite similar in shape. In particular, from Table 2 it is possible to observe that simulations results from (9) tend to provide FPT anticipating that from (6), whereas their coefficient of variation do not change significantly. We can also observe high FPTs for increasing values of α.

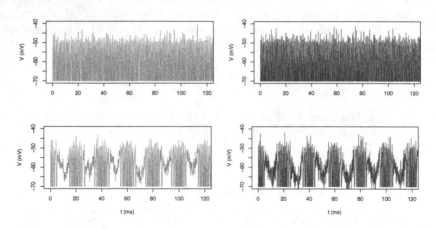

Fig. 1. Sample paths with reset for (6) in red and (9) in blue: on top with constant $I(t)$, on bottom with periodic $I(t)$ and for different values of α. (Color figure online)

Finally, it appears evident that a more detailed analysis of the two considered models and a more extensively investigation of the simulation results is essential and it will be the object of a future work.

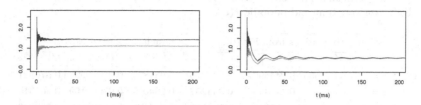

Fig. 2. Overlapping firing rates for (6) in red and (9) in blue: on left with constant $I(t)$, on right with periodic $I(t)$. (Color figure online)

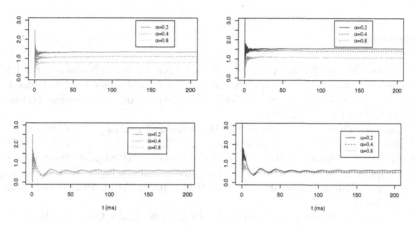

Fig. 3. Firing rates of (6) in red and (9) in blue, on top with constant $I(t)$, on bottom with periodic $I(t)$. (Color figure online)

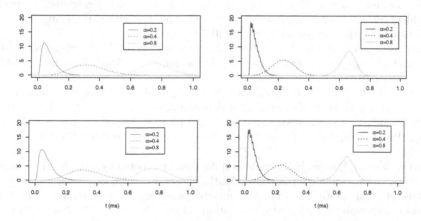

Fig. 4. FPT probability densities from simulated sample paths of (6) in red and (9) in blue, on top with constant $I(t)$, on bottom with periodic $I(t)$. (Color figure online)

Table 2. Sample means, standard deviations and coefficient of variations for the simulated FPTs for constant and periodic $I(t)$.

$I(t)$		Simulations from Eq. (6)			Simulations from Eq. (9)		
		$\overline{M}$	$\overline{SD}$	$\overline{CV}$	$\overline{M}$	$\overline{SD}$	$\overline{CV}$
Const.	$\alpha = 0.2$	0.075555	0.0445907	0.5901754	0.051871	0.02960907	0.5708213
	$\alpha = 0.4$	0.33334	0.1114196	0.3342521	0.234057	0.07057159	0.3015145
	$\alpha = 0.8$	0.760905	0.0917167	0.1205363	0.654228	0.04860864	0.07429924
Per.	$\alpha = 0.2$	0.076838	0.04563277	0.5938829	0.052412	0.02966665	0.5660279
	$\alpha = 0.4$	0.33296	0.1113753	0.3345006	0.233499	0.07055257	0.3021536
	$\alpha = 0.8$	0.767176	0.09418679	0.1227708	0.657651	0.0487794	0.07417217

References

1. Armanyos, M., Radwan, A.G.: Fractional-order Fitzhugh-Nagumo and Izhikevich neuron models. In: 2016 13th International Conference on Electrical Engineering/Electronics, Computer, Telecommunications and Information Technology (ECTI-CON), pp. 1–5 (2016)
2. Bazzani, A., Bassi, G., Turchetti, G.: Diffusion and memory effects for stochastic processes and fractional Langevin equations. Phys. A: Stat. Mech. Appl. **324**(3), 530–550 (2003). https://doi.org/10.1016/S0378-4371(03)00073-6. ISSN 0378-4371
3. Bernido, C.C., Carpio-Bernico, M.V.: On a fractional stochastic path integral approach in modelling interneuronal connectivity. Int. J. Mod. Phys. Conf. Ser. **17**, 23–33 (2012)
4. Buonocore, A., Caputo, L., Pirozzi, E., Ricciardi, L.M.: On a stochastic leaky integrate-and-fire neuronal model. Neural Comput. **22**, 2558–2585 (2010)
5. Buonocore, A., Caputo, L., Pirozzi, E., Ricciardi, L.M.: The first passage time problem for Gauss-Diffusion processes: algorithmic approaches and applications to lif neuronal model. Methodol. Comput. Appl. Prob. **13**, 29–57 (2011)
6. Buonocore, A., Caputo, L., Nobile, A.G., Pirozzi, E.: Restricted Ornstein-Uhlenbeck process and applications in neuronal models with periodic input signals. J. Comput. Appl. Math. **285**, 59–71 (2015)
7. Buonocore, A., Caputo, L., Carfora, M.F., Pirozzi, E.: A leaky integrate-and-fire model with adaptation for the generation of a spike train. Math. Biosci. Eng. **13**(3), 483–493 (2016)
8. D'Onofrio, G., Pirozzi, E., Magnasco, M.O.: Towards stochastic modeling of neuronal interspike intervals including a time-varying input signal. In: Moreno-Díaz, R., Pichler, F., Quesada-Arencibia, A. (eds.) EUROCAST 2015. LNCS, vol. 9520, pp. 166–173. Springer, Cham (2015). https://doi.org/10.1007/978-3-319-27340-2_22
9. D'Onofrio, G., Pirozzi, E.: Successive spike times predicted by a stochastic neuronal model with a variable input signal. Math. Biosci. Eng. **13**(3), 495–507 (2016)
10. Kim, H., Shinomoto, S.: Estimating nonstationary inputs from a single spike train based on a neuron model with adaptation. Math. Bios. Eng. **11**, 49–62 (2014)
11. Kobayashi, R., Tsubo, Y., Shinomoto, S.: Made-to-order spiking neuron model equipped with a multi-timescale adaptive threshold. Front. Comput. Neurosci. **3**, 1–11 (2009). Article 9

12. Kumar, R.U., Mondal, A.: Dynamics of fractional order modified Morris-Lecar neural model. Netw. Biol. **5**(3), 113–136 (2015)
13. Teka, W., Marinov, T.M., Santamaria, F.: Neuronal spike timing adaptation described with a fractional leaky integrate-and-fire model. PLoS Comput. Biol. **10**(3), e1003526 (2014)
14. Teka, W.W., et al.: Fractional-order leaky integrate-and-fire model with long term memory and power law dynamics. Neural Netw. **93**, 110–125 (2017). https://doi. org/10.1016/j.neunet.2017.05.007
15. Touboul, J., Faugeras, O.: Characterization of the first hitting time of a double integral processes to curved boundaries. Adv. Appl. Prob. **40**, 501–528 (2008)

On the Imputation of Missing Values in Univariate PM_{10} Time Series

G. Albano[(✉)], M. La Rocca, and C. Perna

Department of Economics and Statistics, University of Salerno,
via Giovanni Paolo II, 84084 Fisciano, SA, Italy
{pialbano,larocca,perna}@unisa.it

Abstract. Missing data frequently happen in environmental research, usually due to faults in data acquisition, inadequate sampling or measurement error. They make difficult to determine whether the limits set by the European Community on certain indicators of air quality are fulfilled or not. Indeed, due to missing values, the number of exceedances per year of PM_{10}, that is particulate matter $10\,\mu m$ or less in diameter, and other air quality indicators are often heavily underestimated, and no environmental policy is applied to protect citizen health.

In this paper, we propose a non-parametric method to impute missing values in PM_{10} time series. It is primarily based on a local polynomial estimator of the trend-cycle in time series. We also compare the proposed method with other methods usually used in literature for the imputation of missing values in univariate time series and implemented in the R package *imputeTS*.

1 Introduction

Particulate matter is one of the main indicators of air quality, in particular PM_{10} (particles with a diameter less than $10\,\mu m$) is the major air pollutant monitored in Europe since the exposure to particulate matter can contribute to heart and lung diseases, and can lead to premature death [3,6]. Due to this motivations, European Union set some limits to the emissions of PM_{10} concentrations into the air. Precisely, current EU legislation regulating the PM_{10} concentration in ambient air is given in the EU directive 1999/30/EC, establishing for PM_{10} two binding limit values that have to be respected from 1 January 2005:

- a daily limit of $50\,\mu g/m^3$ not to be exceeded on more than 35 days within a calendar year;
- an annual mean value of $40\,\mu g/m^3$.

In this context, missing data, usually due to faults in data acquisition, is a significant problem in environmental research. They are generally due to monitoring station being down, and this can cause a gap in the data a single value to many consecutive values long.

In particular, missing values make difficult to determine whether the limits set by the European Community on certain indicators of air quality are fulfilled

© Springer International Publishing AG 2018
R. Moreno-Díaz et al. (Eds.): EUROCAST 2017, Part II, LNCS 10672, pp. 12–19, 2018.
https://doi.org/10.1007/978-3-319-74727-9_2

or not. Indeed, due to missing values, the number of exceedances per year of PM_{10} and other air quality indicators is often massively underestimated, and no environmental policy is applied to protect citizen health (see, for example, [1,2]).

Many single imputation methods have been proposed to estimate missing data, going from simple mean substitution to regression-based imputation and neural network (see [4,5,8,9]).

In this paper, we discuss a non-parametric method to impute missing values in PM_{10} time series. It is essentially based on a local polynomial estimator of the trend-cycle in time series first proposed by [7]. This method can handle with data in which the residuals has no particular structure, so it seems to work well in the context of environmental data in which, besides the cycle trend, an autocorrelation in the conditional mean and variance is expected. We also compare the proposed method with other methods usually used in literature for the imputation of missing values in univariate time series and implemented in the R package *imputeTS*.

The paper is organized as follows: Sect. 2 introduces the estimation method and briefly describes the imputation methods implemented in the R package *imputeTS* used to evaluate the performance of the method. Section 3 describes the data used for evaluating the procedure. In Sect. 4 a simulation study is performed to simulate additional missing values in the time series and comparison between the different imputation techniques are provided using some performance indicators. Some concluding remarks close the paper.

2   Imputing Missing Values

Let us consider a univariate time series $\{Y_t, t = t_0, t_1, \ldots, t_n = T\}$ in which some missing values, also consecutive, can occur. Following [7], to impute missing values in Y_t, we use a local polynomial estimator in which the missing values are estimated by looking at within a sliding window. That is, the unknown regression function $m(t) = \mathbb{E}[Y_t]$ is estimated for each fitting point τ by $\hat{m}(t)$, obtained by minimizing the function

$$\sum_{t=0}^{T} W\left(\frac{t_i - \tau}{h_T}\right)\left[Y_{t_i} - (a_0 + a_1(t_i - \tau) + \ldots + a_k(t_i - \tau)^k)\right]^2, \qquad k \in \mathbb{N}. \quad (1)$$

The kernel function $W(\cdot)$ is assumed to be symmetric with bounded support and Lipschitz continuous; h_T is a smoothing parameter, generally depending on T and such that $\lim_{T \to \infty} h_T = 0$, $\lim_{T \to \infty} Th_T = \infty$. The local least squares criterion (1) is minimized to procedure estimates $\hat{a}_0, \hat{a}_1, \ldots, \hat{a}_k$. Precisely, let us define the following vectors and matrices:

$$\mathbf{T}_\tau = \begin{pmatrix} 1 & (t_0 - \tau) & \ldots & (t_0 - \tau)^k \\ 1 & (t_1 - \tau) & \ldots & (t_1 - \tau)^k \\ \vdots & \vdots & \ddots & \vdots \\ 1 & (T - \tau) & \ldots & (T - \tau)^k \end{pmatrix},$$

$$\mathbf{Y} = (Y_{t_0}, Y_{t_1}, \ldots, Y_T)^{\mathrm{T}}$$

$$\mathbf{a} = (a_0, a_1, \ldots, a_k)^{\mathrm{T}}$$

$$\mathbf{W}_\tau = \begin{pmatrix} W_1\left(\frac{t_0 - \tau}{h_{1T}}\right) & 0 & \cdots & 0 \\ 0 & W_1\left(\frac{t_1 - \tau}{h_{1T}}\right) & \cdots & 0 \\ \vdots & \vdots & \ddots & \vdots \\ 0 & 0 & \cdots & W_1\left(\frac{T - \tau}{h_{1T}}\right) \end{pmatrix}.$$

The least squares problem is then to minimize the weighted sum-of-squares function

$$(\mathbf{Y} - \mathbf{T}_\tau \mathbf{a})^{\mathrm{T}} \mathbf{W}_\tau (\mathbf{Y} - \mathbf{T}_\tau \mathbf{a})$$

with respect to the parameters $\mathbf{a}$. The solution is

$$\widehat{\mathbf{a}} = (\mathbf{T}_\tau^{\mathrm{T}} \mathbf{W}_\tau \mathbf{T}_\tau)^{-1} (\mathbf{T}_\tau^{\mathrm{T}} \mathbf{W}_\tau \mathbf{Y})$$

The quantity $m(\tau)$ is then estimated by the fitted intercept parameter (i.e. by $\widehat{a}_0$) as this defines the position of the estimated local polynomial curve at the point τ. By varying the value of τ, we can build up an estimate of the function $m(t)$. We have:

$$\widehat{m}(t) = \mathbf{e}_1^{\mathrm{T}} (\mathbf{T}_\tau^{\mathrm{T}} \mathbf{W}_\tau \mathbf{T}_\tau)^{-1} (\mathbf{T}_\tau^{\mathrm{T}} \mathbf{W}_\tau \mathbf{Y})$$

where the vector $\mathbf{e}_1$ has length $k+1$ and has 1 in the first position and 0's elsewhere. The estimate $\widehat{m}(t)$ is then used to recostruct the time series as following:

$$\tilde{Y}_t = \delta_t Y_t + (1 - \delta_t) \widehat{m}(t), \qquad t = t_0, \ldots, T, \tag{2}$$

where the fuction δ_t is the indicator of the missing observations, thus

$$\delta_t = \begin{cases} 0 \text{ if } Y_t \text{ is observed} \\ 1 \text{ if } Y_t \text{ is missing.} \end{cases}$$

We will indicate LP the above method. To compare the performance of the proposed method with other methods proposed in the literature, we use the R package *imputeTS*, in which the following methods are implemented:

Interp linear interpolation in which missing values get replaced by values of an approx interpolation;
Spline cubic (or Hermite) spline interpolation in which the spline used is that of Forsythe, Malcolm and Moler (an exact cubic is fitted through the four points at each end of the data, and this is used to determine the end conditions);
Stine Stineman interpolation function in which missing values are replaced by a piecewise rational interpolation according to the algorithm of Stineman based on an interpolating circle (1980);
locf Missing Value Imputation by Last Observation Carried Forward;
nocb Missing Value Imputation by Next Observation Carried Backwards;

Clearly, a performance comparison can only be done for simulated missing values. To do this, we will consider the time series of PM_{10} concentration in Alessandria province, located in Piemonte region, in the north of Italy. For the considered time series the percentages of missing values are low (between 0 and 10%), and from them, data points are artificially removed. Later on, imputed values and real values can be compared.

For our simulation, we choose a Missing Completely At Random (MCAR) data mechanism. Precisely, we use an exponential missing data distribution, whereas the amount of missing data is assumed geometric distributed.

To compare the performance of the above methods, we will refer to Mean Absolute Error (MAE), Root Mean Square Error (RMSE) and

$$d = 1 - \frac{\sum_{i=1}^{N}(I_i - O_i)^2}{\sum_{i=1}^{N}(|I_i - \bar{O}| + |O_i - \bar{O}|)^2} \tag{3}$$

where N is the number of imputed values, O_i represents the observed value, and I_i is the corresponding imputed value.

3 Alessandria Dataset

We analyse daily PM_{10} data (in $micrograms/meter^3$) measured from 1 January 2015 to 19 October 2016 (658 days) by gravimetric instruments at seven sites in the italian province of Alessandria, in Piemonte region, located in the north of Italy. Data are provided by Agenzia Regionale Protezione Ambientale (ARPA) Piemonte. Figure 1 shows the distribution of PM_{10} concentration from each station. We can see that all the distributions are positively skewed. Table 1 illustrates the percentage of missing values for each location and the number of observations. Only Arquata Scrivia Minzoni and Tortona Carbona

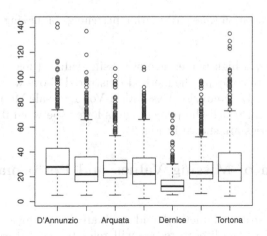

Fig. 1. PM_{10} concentration distribution by station.

Table 1. Percentage of missing values and number of observations in the considered time series.

Station	% of missing	Number of observations
Alessandria - D'Annunzio	4.1	631
Alessandria - Volta	0	658
Arquata Scrivia - Minzoni	26.6	483
Casale Monferrato - Castello	2.7	640
Dernice - Costa	0.3	656
Novi Ligure - Gobetti	9.9	593
Tortona - Carbone	26.4	484

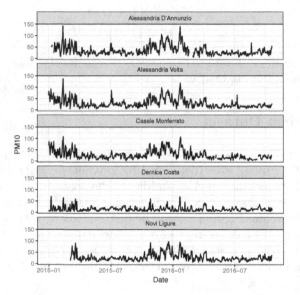

Fig. 2. PM_{10} concentration time series with a percentage of missing values between 0 and 10%.

are characterized by a high percentage of missing values (26.6 and 26.4, respectively), so in order to compare the method illustrated above we will focus on the stations Alessandria-D'Annunzio, Alessandria-Volta, Casale Monferrato-Castello and Dernice-Costa in which the percentage is lower (between 0 and 10%). The considered time series are shown in Fig. 2.

4   Simulation of Missing Values and Performance Indicators

To illustrate the proposed method and compare its performance with other methods proposed in the literature, we will refer to the dataset illustrated in the previous section, consisting of PM_{10} concentration levels in the province of

Alessandria in Piemonte. In particular, we will refer to the stations with a lower percentage of missing values, i.e. Alessandria-D'Annunzio, Alessandria-Volta, Casale Monferrato-Castello and Dernice-Costa.

Table 2. Results of mean of MAE, RMSE and d for three different choices of the simulation parameters ($\lambda = 0.02, 0.05$ and 0.1, $p = 0.2$). MNM denotes the mean number of missing values artificially generated in the series.

Alessandria - D'Annunzio

	$\lambda = 0.02, p = 0.2$ (MNM $= 58.43$)			$\lambda = 0.05, p = 0.2$ ($\bar{x} = 91.35$)			$\lambda = 0.1, p = 0.2$ ($\bar{x} = 144.25$)		
	MAE	RMSE	d	MAE	RMSE	d	MAE	RMSE	d
LP	0.037	0.214	0.973	0.120	0.420	0.940	**0.122**	**0.366**	**0.917**
Interp	**0.028**	**0.161**	0.083	**0.008**	**0.205**	**0.954**	0.156	0.460	0.897
Spline	0.054	0.305	0.927	0.124	0.512	0.822	0.237	0.727	0.661
Stine	0.029	0.166	0.982	0.071	0.279	0.949	0.127	0.383	0.909
locf	0.870	1.122	**0.997**	0.871	1.123	0.942	0.871	1.126	0.793
nocb	0.892	1.154	0.993	0.887	1.149	0.948	0.882	1.145	0.898

Alessandria - Volta

	$\lambda = 0.02, p = 0.2$ ($\bar{x} = 33.24$)			$\lambda = 0.05, p = 0.2$ ($\bar{x} = 66.66$)			$\lambda = 0.1, p = 0.2$ ($\bar{x} = 119.58$)		
	MAE	RMSE	d	MAE	RMSE	d	MAE	RMSE	d
LP	**0.014**	0.100	0.990	0.056	0.231	0.947	**0.080**	**0.251**	0.925
Interp	**0.014**	**0.081**	0.989	**0.044**	**0.182**	**0.966**	0.082	0.272	**0.937**
Spline	0.027	0.152	0.951	0.187	0.781	0.766	0.193	0.651	0.536
Stine	0.015	0.085	0.987	0.046	0.189	0.962	0.085	0.262	0.932
locf	0.017	0.099	**0.996**	0.053	0.215	0.940	0.099	0.303	0.932
nocb	0.017	0.098	0.993	0.058	0.223	0.953	0.097	0.297	0.899

Casale Monferrato - Castello

	$\lambda = 0.02, p = 0.2$ ($\bar{x} = 49.37$)			$\lambda = 0.05, p = 0.2$ ($\bar{x} = 89.98$)			$\lambda = 0.1, p = 0.2$ ($\bar{x} = 133.5$)		
	MAE	RMSE	d	MAE	RMSE	d	MAE	RMSE	d
LP	0.082	0.441	0.954	0.299	0.874	0.809	**0.170**	**0.591**	0.912
Interp	**0.051**	**0.277**	0.980	**0.122**	**0.454**	**0.948**	0.198	0.603	**0.915**
Spline	0.096	0.513	0.912	0.286	1.116	0.608	0.419	1.338	0.531
Stine	0.053	0.287	0.978	0.128	0.478	0.942	0.208	0.621	0.906
locf	0.760	1.191	0.990	0.827	1.266	0.898	0.890	1.314	0.927
nocb	0.750	1.193	**0.996**	0.804	1.248	0.929	0.870	1.311	0.944

Dernice - Costa

	$\lambda = 0.02, p = 0.2$ ($\bar{x} = 32.6$)			$\lambda = 0.05, p = 0.2$ ($\bar{x} = 67.04$)			$\lambda = 0.1, p = 0.2$ ($\bar{x} = 120.12$)		
	MAE	RMSE	d	MAE	RMSE	d	MAE	RMSE	d
LP	**0.009**	**0.050**	**0.992**	**0.024**	**0.098**	**0.970**	**0.042**	**0.120**	**0.951**
Interp	0.011	0.064	0.986	0.025	0.104	0.966	0.046	0.141	0.938
Spline	0.020	0.114	0.952	0.053	0.226	0.833	0.127	0.460	0.192
Stine	0.011	0.066	0.985	0.026	0.108	0.963	0.048	0.147	0.933
locf	0.184	0.287	0.967	0.195	0.297	0.964	0.212	0.313	0.849
nocb	0.182	0.284	0.981	0.193	0.296	0.971	0.207	0.308	0.918

Novi Ligure-Gobetti

	$\lambda = 0.02, p = 0.2$ ($\bar{x} = 90.76$)			$\lambda = 0.05, p = 0.2$ ($\bar{x} = 119.54$)			$\lambda = 0.1, p = 0.2$ ($\bar{x} = 166.32$)		
	MAE	RMSE	d	MAE	RMSE	d	MAE	RMSE	d
LP	**0.022**	0.171	0.983	0.069	0.298	0.948	**0.113**	**0.351**	**0.937**
Interp	0.027	**0.158**	0.983	**0.056**	**0.235**	0.967	0.151	0.449	0.927
Spline	0.044	0.257	0.957	0.097	0.411	0.894	0.219	0.701	0.689
Stine	0.028	0.162	0.981	0.059	0.247	0.963	0.119	0.373	0.917
locf	1.044	1.342	0.987	1.051	1.352	0.973	1.061	1.360	0.936
nocb	1.041	1.339	**0.993**	1.039	1.335	**0.975**	1.041	1.334	0.929

We will consider three different simulated missing data patterns, which differ in the distribution of the gap position and for their length. Missing values have been generated drawing the gap length from two distributions, an exponential of the parameter λ and an hypergeometric distribution with parameter p. For each missing data pattern, 100 simulations are performed, and the means of the RMSE, MAE and d are compared.

Table 2 shows the results of our simulation experiment. In the first column it is considered the case $\lambda = 0.02, p = 0.2$, in the second column it is $\lambda = 0.05, p = 0.2$, whereas in the last column it is $\lambda = 0.1, p = 0.2$. For each case, the mean number of missing values in our 100 simulations is denoted by MNM. The best values of RMSE, MAE and d are indicated in bold.

We can see the worse imputation methods are locf and nocb in which the performances are comparable with the others in the only case in which the original time series has zero percentage of missing values (Alessandria-Volta). Moreover, their performances make worse when the percentage of missing values increases in the original time series. The approach based on spline interpolation seems to perform worse than the interpolation and Stine method in all our simulation experiments. Moreover, interpolation and Stine performs almost equivalently in all the cases. We can see that LP works in line with interpolation and Stine method when the mean number of missing values is low, while the results becomes better and better when the percentage of missing values increases, as we can see in the last column of Table 2, in which the percentage of missing values is between 18% and 26%, so it is comparable with the near stations of Arquata Scrivia - Minzoni and Tortona - Carbone (see Table 1).

5   Conclusions

We discuss a local estimator technique for the imputation of missing values sequences in univariate PM_{10} time series. It is based on [7] in which no particular structure is assumed for the residuals of the detrended time series, so it is particularly able to handle with environmental time series in which some auto-correlation in mean and in variance is expected due to the natural link between "near in time" observations. To compare the suggested method, LP, with other methods already used in literature to impute missing values we provide a simulation experiment in which three different missing data patterns, that differ in the distribution of the gap position and their length. Missing values have been generated drawing the gap length from two distributions, an exponential of the parameter λ and an hypergeometric distribution with parameter p. For each missing data pattern, 100 simulations are performed, and the means of the RMSE, MAE and d defined in (3) are compared. The simulation experiment shows that LP works better than the other methods when the percentage of missing values increases in all the simulated cases.

References

1. Albano, G., Perna, C.: A sequential test for evaluating air quality. In: Moreno-Díaz, R., Pichler, F., Quesada-Arencibia, A. (eds.) EUROCAST 2015. LNCS, vol. 9520, pp. 143–149. Springer, Cham (2015). https://doi.org/10.1007/978-3-319-27340-2_19
2. Albano, G., La Rocca, M., Perna, C.: Non linear time series analysis of air pollutants with missing data. In: Bassis, S., Esposito, A., Morabito, F.C., Pasero, E. (eds.) Advances in Neural Networks. SIST, vol. 54, pp. 371–378. Springer, Cham (2016). https://doi.org/10.1007/978-3-319-33747-0_37
3. Health — Particulate Matter — Air and Radiation — US EPA. Epa.gov. 17 November 2010 (2010)
4. Junger, W.L., Ponce de Leon, A.: Imputation of missing data in time series for air pollutants. Atmos. Environ. **102**, 96–104 (2015)
5. Noor, M.N., Al Bakri, A.M.M., Yahaya, A.S., Ramli, N.A., Fitri, N.F.M.Y.: Estimation of missing values in environmental data set using interpolation technique: fitting on lognormal distribution. Aust. J. Basic Appl. Sci. **7**(5), 336–341 (2013)
6. Raaschou-Nielsen, O., et al.: Air pollution and lung cancer incidence in 17 European cohorts: prospective analyses from the European Study of Cohorts for Air Pollution Effects (ESCAPE). Lancet Oncol. **14**(9), 813–822 (2013)
7. Pèrez-Gonzàlez, A., Vilar-Fernàndez, J.M., Gonzlez-Manteiga, W.: Asymptotic properties of local polynomial regression with missing data and correlated errors. Ann. Inst. Stat. Math. **61**, 85–109 (2009)
8. Junninen, H., Niska, H., Tupprainen, K., Ruuskanen, J., Kolehmainen, M.: Methods for imputation of missing values in air quality data sets. Atmos. Environ. **38**, 2895–2907 (2004)
9. Yahaya, A.S., Ramli, N.A., Ahmad, F., Nor, N.M., Bahrim, M.N.H.: Determination of the best imputation technique for estimating missing values when fitting the weibull distribution. Int. J. Appl. Sci. Technol. **1**(6), 278–285 (2011)

On Sharp Bounds on the Rate of Convergence for Finite Continuous-Time Markovian Queueing Models

Alexander Zeifman[1,2,3]($\boxtimes$), Alexander Sipin[1], Victor Korolev[2,4,5],
Galina Shilova[1], Ksenia Kiseleva[1,6], Anna Korotysheva[1], and Yacov Satin[1]

[1] Vologda State University, Vologda, Russia
a_zeifman@mail.ru
[2] Institute of Informatics Problems FRC CSC RAS, Moscow, Russia
[3] ISEDT RAS, Vologda, Russia
[4] Faculty of Computational Mathematics and Cybernetics,
Lomonosov Moscow State University, Moscow, Russia
[5] Hangzhou Dianzi University, Hangzhou, China
[6] RUDN University, Moscow, Russia

Abstract. Finite inhomogeneous continuous-time Markov chains are studied. For a wide class of such processes an approach is proposed for obtaining sharp bounds on the rate of convergence to the limiting characteristics. Queueing examples are considered.

1 Introduction

We deal with sharp bounds of the rate of convergence for a wide class of finite continuous-time Markov chains which describe the corresponding Markovian queueing models.

As it is known, the problem of finding sharp bounds of the rate of convergence to the limiting characteristics for such processes is very important for a number of reasons:

(i) it is easier to calculate the limit characteristics of a process than to find the exact distribution of state probabilities, (see, for instance [1, 6, 16, 20]);
(ii) the best bounds of perturbations require the corresponding best bounds on the rate of convergence, see [9, 19];
(iii) sharp bounds on the rate of convergence are required to obtain truncation bounds which are uniform in time, see [17].

A general approach is closely connected with the notion of the logarithmic norm and the corresponding bounds for the Cauchy matrix. It was first studied for birth-death processes (with possible catastrophes), see details and references in [3, 7, 16], and for Markovian queueing models with batch arrivals and group service, see [18]. An essential component of this approach consists of special transformations of the reduced intensity matrix. These transformations were proposed in [12] and applied to general inhomogeneous birth-death models in [13]. Here we apply this approach and the same transformation to two new classes of inhomogeneous continuous-time Markov chains describing combined queueing systems with state-dependent arrival intensity and batch service; and vice versa for queueing system with batch arrivals and state-dependent service, see, for instance, [8,10]. The case of a finite state space provides the possibility of, first, to simplify the reasoning and, second, to obtain essentially more precise bounds.

Definition. A Markov chain $X(t)$ is called *weakly ergodic*, if

$$\lim_{t \to \infty} \left\| \mathbf{p}^1(t) - \mathbf{p}^2(t) \right\| = 0$$

for any initial conditions $\mathbf{p}^1(0) = \mathbf{p}^1 \in \Omega$, $\mathbf{p}^2(0) = \mathbf{p}^2 \in \Omega$. In this situation one can consider *any* $\mathbf{p}^1(t)$ as a *quasi-stationary distribution* of the chain $X(t)$.

Definition. A Markov chain $X(t)$ has the limiting mean $\phi(t)$, if

$$|E(t; k) - \phi(t)| \to 0$$

as $t \to \infty$ for any k, where $E(t; k)$ is the mathematical expectation (the mean) of the process at the moment t under the initial condition $X(0) = k$.

2 Basic Approaches to Bounding

There are two approaches to the study of the rate of convergence of continuous-time Markov chains.

The first approach is based on the notion of the logarithmic norm of a linear operator function and the corresponding bounds of the Cauchy operator, see the detailed discussion, for instance, in [3]. Namely, if $B(t)$, $t \geq 0$, is a one-parameter family of bounded linear operators on a Banach space $\mathcal{B}$, then

$$\gamma(B(t))_{\mathcal{B}} = \lim_{h \to +0} \frac{\|I + hB(t)\| - 1}{h} \tag{1}$$

is called the logarithmic norm of the operator $B(t)$. If $\mathcal{B} = l_1$ then the operator $B(t)$ is given by the matrix $B(t) = (b_{ij}(t))_{i,j=0}^{\infty}$, $t \geq 0$, and the logarithmic norm of $B(t)$ can be found explicitly:

$$\gamma(B(t)) = \sup_j \left(b_{jj}(t) + \sum_{i \neq j} |b_{ij}(t)| \right), \quad t \geq 0.$$

Hence the following bound on the rate of convergence holds:

$$\|\mathbf{x}(t)\| \leq e^{\int_0^t \gamma(B(\tau))\, d\tau} \|\mathbf{x}(0)\|,$$

where $\mathbf{x}(t)$ is the solution of the differential equation

$$\frac{d\mathbf{x}}{dt} = B(t)\mathbf{x}(t).$$

Here we apply **the second approach**, the detailed consideration of which for the case of a finite state space can be found in our recent paper [20].

A matrix is called *essentially nonnegative*, if all off-diagonal elements of this matrix are nonnegative.

Let

$$\frac{d\mathbf{x}}{dt} = H(t)\mathbf{x}(t), \tag{2}$$

be a differential equation in the space of sequences l_1 with essentially nonnegative for all $t \geq 0$ countable matrix $H(t) = (h_{ij}(t))$ such that the corresponding operator function on l_1 is bounded for almost all $t \geq 0$ and locally integrable on $[0, \infty)$.

Therefore $\mathbf{x}(s) \geq 0$ implies $\mathbf{x}(t) \geq 0$ for any $t \geq s$.

Put

$$h^*(t) = \sup_j \sum_i h_{ij}(t), \quad h_*(t) = \inf_j \sum_i h_{ij}(t). \tag{3}$$

Let $\mathbf{x}(0) \geq 0$. Then $\mathbf{x}(t) \geq 0$, if $t \geq 0$ and $\|\mathbf{x}(t)\| = \sum_i x_i(t)$. Hence (2) implies the inequality

$$\frac{d\|\mathbf{x}(t)\|}{dt} = \frac{d\sum_i x_i(t)}{dt} \leq h^*(t) \sum_j x_j = h^*(t)\|\mathbf{x}\|.$$

Then $\|\mathbf{x}(t)\| \leq e^{\int_0^t h^*(\tau)d\tau}\|\mathbf{x}(0)\|$, if $\mathbf{x}(0) \geq \mathbf{0}$.

Let now $\mathbf{x}(0)$ be arbitrary vector from l_1. Put $x_i^+(0) = \max(x_i(0), 0)$, $\mathbf{x}^+(0) = \left(x_1^+(0), x_2^+(0), \cdots\right)^T$ and $\mathbf{x}^-(0) = \mathbf{x}^+(0) - \mathbf{x}(0)$. Then $\mathbf{x}^+(0) \geq \mathbf{0}$, $\mathbf{x}^-(0) \geq \mathbf{0}$, $\mathbf{x}(0) = \mathbf{x}^+(0) - \mathbf{x}^-(0)$, hence $\|\mathbf{x}(0)\| = \|\mathbf{x}^+(0)\| + \|\mathbf{x}^-(0)\|$.

Finally we obtain the upper bound

$$\|\mathbf{x}(t)\| = \|\mathbf{x}^+(t) - \mathbf{x}^-(t)\| \leq \|\mathbf{x}^+(t)\| + \|\mathbf{x}^-(t)\| \leq$$
$$\leq e^{\int_0^t h^*(\tau)d\tau}\left(\|\mathbf{x}^+(0)\| + \|\mathbf{x}^-(0)\|\right) = e^{\int_0^t h^*(\tau)d\tau}\|\mathbf{x}(0)\|. \tag{4}$$

for any initial condition.

On the other hand, if $\mathbf{x}(0) \geq 0$, then

$$\frac{d\|\mathbf{x}(t)\|}{dt} = \sum_j \left(\sum_i h_{ij}\right) x_j \geq h_*(t) \sum_j x_j = h_*(t)\|\mathbf{x}\|,$$

and we obtain the lower bound

$$\|\mathbf{x}(t)\| \geq e^{\int_0^t h_*(\tau)d\tau}\|\mathbf{x}(0)\|, \tag{5}$$

for any nonnegative initial condition.

On sharpness of bounds

Let us note that if the matrix of system (2) is essentially nonnegative for any t, then one can see that the logarithmic norm of this matrix is equal to our new characteristic, $\gamma\left(H(t)\right) = h^*(t)$.

Let $\{d_i\}$, $i \geq 0$, be a sequence of positive numbers such that $\inf_i d_i = d > 0$. Let $D = diag\left(d_0, d_1, d_2, \dots\right)$ be the corresponding diagonal matrix and l_{1D} be a space of vectors $l_{1D} = \{\mathbf{x} = (x_0, x_1, x_2, \dots)/\|\mathbf{x}\|_{1D} = \|D\mathbf{x}\|_1 < \infty\}$.

Put $\mathbf{z}(t) = D\mathbf{x}(t)$, then (2) implies the equation

$$\frac{d\mathbf{z}}{dt} = H_D(t)\mathbf{z}(t), \tag{6}$$

where $H_D(t) = DH(t)D^{-1}$ with entries $h_{ijD}(t) = \frac{d_i}{d_j}h_{ij}(t)$ is also essentially nonnegative for any $t \geq 0$.

If there exists a sequence $\{d_i\}$ such that

$$h_D^*(t) = \sup_j \sum_i \frac{d_i}{d_j} h_{ij}(t) = \inf_j \sum_i \frac{d_i}{d_j} h_{ij}(t), \tag{7}$$

then the *equality*

$$\|\mathbf{x}(t)\|_D = e^{\int_0^t h_D^*(\tau)d\tau}\|\mathbf{x}(0)\|_D \tag{8}$$

holds for any nonnegative initial condition. Therefore, the bound

$$\|\mathbf{x}(t)\|_D \leq e^{\int_0^t h_D^*(\tau)d\tau}\|\mathbf{x}(0)\|_D, \tag{9}$$

which is correct for any initial condition, is *sharp*.

Note that the construction of such sequences for homogeneous birth-death processes was studied in many papers, see for instance [2,3,7].

3   Classes of Markov Chains

Let $X(t)$ be a continuous-time finite Markov chain with intensity matrix $Q(t)$. Denote by $A(t) = Q^T(t)$ the corresponding transposed intensity matrix. Thus it has the form

$$A(t) = \begin{pmatrix} a_{00}(t) & a_{01}(t) & \cdots & a_{0r}(t) \\ a_{10}(t) & a_{11}(t) & \cdots & a_{1r}(t) \\ a_{20}(t) & a_{21}(t) & \cdots & a_{2r}(t) \\ & \cdots & & \\ a_{r0}(t) & a_{r1}(t) & \cdots & a_{rr}(t) \end{pmatrix}, \tag{10}$$

where $a_{ii}(t) = -\sum_{k \neq i} a_{ki}(t)$.

Now we can apply this approach to the following classes of finite inhomogeneous continuous-time Markov chains.

(I) inhomogeneous birth-death processes; in this case all $a_{ij}(t) = 0$ for any $t \geq 0$ if $|i - j| > 1$; here $a_{i,i+1}(t) = \mu_{i+1}(t)$ and $a_{i+1,i}(t) = \lambda_i(t)$ are birth and death rates respectively;

(II) inhomogeneous chains with 'batch' births and single deaths; here all $a_{ij}(t) = 0$ for any $t \geq 0$ if $i < j - 1$ and moreover, all 'birth' rates do not depend on the size of a 'population', where $a_{i+k,i}(t) = a_k(t)$ for $k \geq 1$ is the rate of 'birth' of a group of k particles, and $a_{i,i+1}(t) = \mu_{i+1}(t)$ is the death rate;

(III) vice-versa, inhomogeneous chains with 'batch' deaths and single births, here all $a_{ij}(t) = 0$ for any $t \geq 0$ if $i > j + 1$ and moreover, all 'death' rates do not depend on the size of a 'population', where $a_{i,i+k}(t) = b_k(t)$ for $k \geq 1$ is the rate of 'death' of a group of k particles, and $a_{i+1,i}(t) = \lambda_i(t)$ is the birth rate;

(IV) inhomogeneous chains with 'batch' births and deaths, here all rates do not depend on the size of a 'population', where $a_{i+k,i}(t) = a_k(t)$, and $a_{i,i+k}(t) = b_k(t)$ for $k \geq 1$ are the rates of 'birth' and 'death' of a group of k particles respectively.

Here for these four classes a unified approach based on a special transformation of reduced infinitesimal matrix is described. This unified approach has already been successfully applied to system from the Ist and IVth class. Namely, for the inhomogeneous Markovian queueing model with a finite number of waiting rooms such as $M|M|S|S+K$ the bounds were firstly obtained in our papers, see for instance [13].

Markovian queueing systems belonging to the IVth class were been studied in the recent papers [11, 18].

Queueing systems with group arrivals *or* services (types II and III respectively) were also studied (see, for example, the so-called $M^x|M|c$ models [10] and the recent paper [8]).

Here we demonstrate that the approach is also suitable for the systems from the IInd and IIIrd class and thus offers a unified means for the analysis of the ergodicity properties of such Markov chains.

4   Transformation

Let $X(t)$ be a finite continuous-time Markov chain and

$$\frac{d\mathbf{p}(t)}{dt} = A(t)\mathbf{p}(t) \tag{11}$$

be the corresponding forward Kolmogorov system, where $\mathbf{p}(t)$ is the vector of state probabilities. Since $p_0(t) = 1 - \sum_{i=1}^{r} p_i(t)$ due to the normalization condition, one can rewrite the system (11) as

$$\frac{d\mathbf{y}(t)}{dt} = B(t)\mathbf{y}(t) + \mathbf{f}(t), \tag{12}$$

where

$$\mathbf{f}(t) = (a_{10}(t), a_{20}(t), \ldots, a_{r0}(t))^T, \quad \mathbf{y}(t) = (p_1(t), p_2(t), \ldots, p_r(t))^T,$$

and

$$B = (b_{ij}(t))_{i,j=1}^{r} = \begin{pmatrix} a_{11}(t) - a_{10}(t) & a_{12}(t) - a_{10}(t) & \cdots & a_{1r}(t) - a_{10}(t) \\ a_{21}(t) - a_{20}(t) & a_{22}(t) - a_{20}(t) & \cdots & a_{2r}(t) - a_{20}(t) \\ \cdots & \cdots & \cdots & \cdots \\ a_{r1}(t) - a_{r0}(t) & a_{r2}(t) - a_{r0}(t) & \cdots & a_{rr}(t) - a_{r0}(t) \end{pmatrix}.$$

Then the bound of the norm of the solution for the corresponding homogeneous system

$$\frac{d\mathbf{y}(t)}{dt} = B(t)\mathbf{y}(t) \tag{13}$$

yields the rate of convergence for $X(t)$.

Denote by T_r the upper triangular matrix of the form

$$T_r = \begin{pmatrix} 1 & 1 & 1 & \cdots & 1 \\ 0 & 1 & 1 & \cdots & 1 \\ 0 & 0 & 1 & \cdots & 1 \\ \vdots & \vdots & \vdots & \ddots & \vdots \\ 0 & 0 & 0 & \cdots & 1 \end{pmatrix}. \tag{14}$$

Then the corresponding matrix $H(t) = T_r B(t) T_r^{-1} = (h_{ij}(t))_{i,j=1}^{r}$ is essentially nonnegative for all of our classes, hence we can apply the second approach and obtain the sharp bounds on the rate of convergence!

Namely, for the first class we have

$$H(t) = \begin{pmatrix} -(\lambda_0(t) + \mu_1(t)) & \mu_1(t) & 0 & \cdots & 0 \\ \lambda_1(t) & -(\lambda_1(t) + \mu_2(t)) & \mu_2(t) & \cdots & 0 \\ & \ddots & \ddots & \ddots & \ddots & \ddots \\ 0 & & \cdots & & \cdots & \lambda_{r-1}(t) - (\lambda_{r-1}(t) + \mu_r(t)) \end{pmatrix}.$$

For the second class we obtain

$$H(t) = \begin{pmatrix} a_{11}(t) - a_r(t) & \mu_1(t) & 0 & \cdots & 0 \\ a_1(t) - a_r(t) & a_{22}(t) - a_{r-1}(t) & \mu_2(t) & \cdots & 0 \\ & \ddots & \ddots & \ddots & \ddots & \ddots \\ a_{r-1}(t) - a_r(t) & & \cdots & & \cdots & a_1(t) - a_2(t) \; a_{rr}(t) - a_1(t) \end{pmatrix},$$

hence, the off-diagonal elements $h_{ij}(t) \geq 0$, if $a_{k+1}(t) \leq a_k(t)$ for any k, t.

For the third class we have

$$H(t) = \begin{pmatrix} -(\lambda_0(t) + b_1(t)) & b_1(t) - b_2(t) & b_2(t) - b_3(t) & \cdots & b_{r-1}(t) - b_r(t) \\ \lambda_1(t) & -(\lambda_1(t) + \sum_{i \leq 2} b_i(t)) \; b_1(t) - b_3(t) & \cdots & b_{r-2}(t) - b_r(t) \\ & \ddots & \ddots & \ddots & \ddots & \ddots \\ 0 & & \cdots & & \cdots & \lambda_{r-1}(t) - (\lambda_{r-1}(t) + \sum_{i \leq r} b_i(t)) \end{pmatrix},$$

hence, the off-diagonal elements $h_{ij}(t) \geq 0$, if $b_{k+1}(t) \leq b_k(t)$ for any k, t.

Finally, for the fourth class we have

$$H(t) = \begin{pmatrix} a_{11}(t) - a_r(t) & b_1(t) - b_2(t) & b_2(t) - b_3(t) & \cdots & b_{r-1}(t) - b_r(t) \\ a_1(t) - a_r(t) & a_{22}(t) - a_{r-1}(t) & b_1(t) - b_3(t) & \cdots & b_{r-2}(t) - b_r(t) \\ & \ddots & \ddots & \ddots & \ddots & \ddots \\ a_{r-1}(t) - a_r(t) & \cdots & & \cdots & a_1(t) - a_2(t) & a_{rr}(t) - a_1(t) \end{pmatrix},$$

hence, the off-diagonal elements $h_{ij}(t) \geq 0$, if $a_{k+1}(t) \leq a_k(t)$ and $b_{k+1}(t) \leq b_k(t)$ for any k, t.

5   Example

As a queueing example we can consider the rate of convergence for the Erlang loss system $M_t|M_t|S|S$ and its generalizations.

The queue-length process $X(t)$ for $M_t|M_t|S|S$ is a birth-death process with $r = S$, arrival and service intensities $\lambda_k(t) = \lambda(t)$ and $\mu_k(t) = k\mu(t)$ respectively. The rate of convergence for this process in the *homogeneous* situation and its asymptotic as $S \to \infty$ were studied in many papers, see for instance [3,5].

The approach considered in this paper was applied to a general inhomogeneous situation in [4,12,14].

Namely, the queue-length process for the ordinary $M_t|M_t|S|S$ queue is weakly ergodic if and only if

$$\int_0^\infty (\lambda(t) + \mu(t)) \, dt = \infty. \tag{15}$$

Bounds for different intensity functions were presented in [14]. Particularly, bound (4) holds for $h^*(t) = \mu(t)$ and for $h^*(t) = S^{-1}\lambda(t)$ for the transformations $d_k = 1$ and $d_k = 1/k$ respectively.

The simplest analogue of the $M_t|M_t|S|S$ queue for a queueing system with group services was introduced and studied in [15]. For this model we suppose that the intensity of arrival of a customer to the queue is also $\lambda(t)$, and the intensity of departure (servicing) of a group of k customers is $b_k(t) = \mu(t)/k$, $1 \leq k \leq S$. The queue-length process $X(t)$ for such model belongs both to the III-d and IY-th classes. We obtained the same criterion (15) of the weak ergodicity of $X(t)$. Moreover, bound (4) holds for $h^*(t) = \mu(t)$ and for $h^*(t) = S^{-1}\lambda(t)$ with the corresponding transformations $d_k = 1$ and $d_k = 1/k$.

The following model is an essential generalization of the Erlang loss system. Let the length of the queue is $X(t) \leq S$, and assume that a group of $k \leq M \leq S$ customers can arrive to the queue with intensity $a_k = \frac{\lambda(t)}{k}$, and a group of $k \leq N \leq S$ customers can leave the queue after being serviced with intensity $\mu_k = \frac{\mu(t)}{k}$, where M and N are fixed natural numbers. The asymptotics as $S \to \infty$ of the rate of convergence for this model in the homogeneous situation was studied in [21]. A general inhomogeneous case was considered in [20]. Namely, if $M = N = S$, then putting $d_k = 1$ we obtain the sharp bound on the rate of convergence (9) with $h^*(t) = \lambda(t) + \mu(t)$.

Acknowledgments. The work was supported by the Ministry of Education of the Russian Federation (the Agreement number 02.a03.21.0008 of 24 June 2016), by the Russian Foundation for Basic Research, projects no. 15-01-01698, 15-07-05316.

References

1. Di Crescenzo, A., Giorno, V., Nobile, A.G., Ricciardi, L.M.: On the M/M/1 queue with catastrophes and its continuous approximation. Queueing Syst. **43**(4), 329–347 (2003)
2. Van Doorn, E.A.: Conditions for exponential ergodicity and bounds for the decay parameter of a birth-death process. Adv. Appl. Probab. **17**, 514–530 (1985)
3. Van Doorn, E.A., Zeifman, A.I., Panfilova, T.L.: Bounds and asymptotics for the rate of convergence of birth-death processes. Theory Probab. Appl. **54**, 97–113 (2010)
4. Van Doorn, E.A., Zeifman, A.I.: On the speed of convergence to stationarity of the Erlang loss system. Queueing Syst. **63**, 241–252 (2009)
5. Fricker, C., Robert, P., Tibi, D.: On the rate of convergence of Erlang's model. J. Appl. Probab. **36**, 1167–1184 (1999)
6. Giorno, V., Nobile, A.G., Spina, S.: On some time non-homogeneous queueing systems with catastrophes. Appl. Math. Comput. **245**, 220–234 (2014)
7. Granovsky, B.L., Zeifman, A.I.: The N-limit of spectral gap of a class of birthdeath Markov chains. Appl. Stoch. Models Bus. Ind. **16**(4), 235–248 (2000)
8. Li, J., Zhang, L.: Decay property of stopped Markovian bulk-arriving queues with c-servers. Stoch. Models **32**, 674–686 (2016)
9. Mitrophanov, A.Y.: The spectral gap and perturbation bounds for reversible continuous-time Markov chains. J. Appl. Probab. **41**, 1219–1222 (2004)
10. Nelson, R., Towsley, D., Tantawi, A.N.: Performance analysis of parallel processing systems. IEEE Trans. Softw. Eng. **14**(4), 532–540 (1988)
11. Satin, Y.A., Zeifman, A.I., Korotysheva, A.V.: On the rate of convergence and truncations for a class of Markovian queueing systems. Theory Probab. Appl. **57**, 529–539 (2013)
12. Zeifman, A.I.: Some properties of a system with losses in the case of variable rates. Autom. Remote Control **50**(1), 82–87 (1989)
13. Zeifman, A.I.: Upper and lower bounds on the rate of convergence for nonhomogeneous birth and death processes. Stoch. Proc. Appl. **59**, 157–173 (1995)
14. Zeifman, A.I.: On the nonstationary Erlang loss model. Autom. Remote Control **70**(12), 2003–2012 (2009)
15. Zeifman, A.I., Korotysheva, A., Satin, Y., Shilova, G., Pafilova, T.: On a queueing model with group services. Lect. Notes CCIS. **356**, 198–205 (2013)
16. Zeifman, A., Satin, Y., Panfilova, T.: Limiting characteristics for finite birthdeath-catastrophe processes. Math. Biosci. **245**(1), 96–102 (2013)
17. Zeifman, A., Satin, Y., Korolev, V., Shorgin, S.: On truncations for weakly ergodic inhomogeneous birth and death processes. Int. J. Appl. Math. Comput. Sci. **24**, 503–518 (2014)
18. Zeifman, A., Korotysheva, A., Korolev, V., Satin, Y., Bening, V.: Perturbation bounds and truncations for a class of Markovian queues. Queueing Syst. **76**, 205–221 (2014)
19. Zeifman, A.I., Korolev, V.Y.: On perturbation bounds for continuous-time Markov chains. Stat. Probab. Lett. **88**, 66–72 (2014)

20. Zeifman, A.I., Korolev, V.Y.: Two-sided bounds on the rate of convergence for continuous-time finite inhomogeneous Markov chains. Stat. Probab. Lett. **103**, 30–36 (2015)
21. Zeifman, A.I., Shilova, G.N., Korolev, V.Y., Shorgin, S.Y.: On sharp bounds of the rate of convergence for some queueing models. In: Proceedings 29th European Conference on Modeling and Simulation, ECMS 2015, Varna, Bulgaria, pp. 622–625 (2015)

A Random Tandem Network with Queues Modeled as Birth-Death Processes

Virginia Giorno and Amelia G. Nobile$^{(\boxtimes)}$

Dipartimento di Informatica, Università di Salerno,
Via Giovanni Paolo II, n. 132, 84084 Fisciano, SA, Italy
{giorno,nobile}@unisa.it

Abstract. We consider a tandem network consisting of an arbitrary but finite number R_m of queueing systems, where R_m is a discrete random variable with a suitable probability distribution. Each queueing system of the tandem network is modeled via a birth-death process and consists of an infinite buffer space and of a service center with a single server.

1 Introduction

Stochastic queueing networks are successfully used to describe a variety of different real systems, such as telecommunications systems, computer systems, call centers and manufacturing systems (cf., for instance, [3,9,11,12]). In particular, a tandem network consists of some finite number of queueing systems in series. For instance, in manufacturing systems, a serial production line consists of machines and buffers; between any two subsequent machines there is a buffer and between any two subsequent buffers there is a machine.

Figure 1 shows a Jackson tandem network (cf. [8]) that consists of the components $C_1, C_2, \ldots, C_r$ ($r \leq m$). Each component C_j represents a queueing system modeled via a birth-death process with state-dependent service rates.

We assume that the tandem network in Fig. 1 satisfies the following assumptions: (i) there is only one class of customers in the network; (ii) the overall number of customers in the network is unlimited; (iii) the customers arrive to the first queueing system from an external source according to a Poisson process of rate λ; (iv) the service discipline at all nodes is FCFS and there is a single server available to each node; (v) the service times at C_j are exponentially distributed with mean $[\mu_j(n_j)]^{-1}$ when there are n_j customers in C_j just before the departure of a customer.

In [10], Reich extends the Burke's result (cf. [4]) showing that in a queueing system, described via a birth-death process with constant arrival rate $\lambda > 0$, in steady-state regime the departure process is a Poisson process with the same rate as the arrival process.

We denote by N_j the random variable describing the number of customers at the component C_j ($j = 1, 2, \ldots, r$) in steady state regime and by $q_j(n_j) =$

This paper is partially supported by G.N.C.S.- INdAM.

R. Moreno-Díaz et al. (Eds.): EUROCAST 2017, Part II, LNCS 10672, pp. 29–37, 2018.
https://doi.org/10.1007/978-3-319-74727-9_4

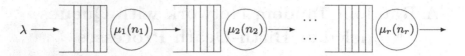

Fig. 1. Tandem queueing network with state dependent service rates.

$P(N_j = n)$ $(n = 0, 1, \ldots)$ the steady-state probability. Furthermore, let $(N_1, N_2, \ldots, N_r)$ be the random vector indicating the number of customers in all the components of the tandem network and let $q(n_1, n_2, \ldots, n_r) = P(N_1 = n_1, N_2 = n_2, \ldots, N_r = n_r)$ be the joint steady-state probability. An extension of Jackson's theorem asserts that in steady-state regime the joint steady-state probability can be still expressed as a product-form solution (cf. [2,3,8]). Indeed, under the assumptions *(i)–(v)*, if

$$S_j = 1 + \sum_{n=1}^{+\infty} \frac{\lambda^n}{\mu_j(1)\mu_j(2)\ldots\mu_j(n)} < +\infty \qquad (j = 1, 2, \ldots, r)$$

then, in steady-state regime, the joint distribution of $(N_1, N_2, \ldots, N_r)$ can be factorized into the product of the marginal distributions of each queueing system:

$$q(n_1, n_2, \ldots, n_r) = q_1(n_1)q_2(n_2)\ldots q_r(n_r), \tag{1}$$

where for $j = 1, 2, \ldots, r$ one has:

$$q_j(0) = S_j^{-1}, \qquad q_j(n) = \frac{\lambda^n}{\mu_j(1)\mu_j(2)\ldots\mu_j(n)} q_j(0) \quad (n = 1, 2, \ldots). \tag{2}$$

Some special birth-death processes are considered in [3,5,6,11].

In Sects. 2 and 3, we calculate the probability distribution of the total number of customers in a tandem network with a fixed number of components and with a random number of components in steady-state regime.

2   Tandem Network with a Fixed Number of Components

We denote by $M_r = N_1 + N_2 + \ldots + N_r$ $(1 \leq r \leq m)$ the total number of customers in the tandem network with r components in steady-state regime. Then, since $N_1, N_2, \ldots, N_r$ are independent random variables, one has

$$P(M_r = n) = \sum_{\mathbf{n} \in \mathcal{D}_r} q_1(n_1)q_2(n_2)\ldots q_r(n_r) \qquad (n = 0, 1, \ldots), \tag{3}$$

with $\mathcal{D}_r = \{\mathbf{n} = (n_1, n_2, \ldots, n_r) : n_i \geq 0 \ (i = 1, 2, \ldots, r), \sum_{i=1}^{r} n_i = n\}$. The probability generating function of M_r is:

$$G_{M_r}(z) = E(z^{N_1+N_2+\ldots+N_r}) = G_1(z)G_2(z)\ldots G_r(z) \quad (r = 1, 2, \ldots), \tag{4}$$

where $G_j(z) = E(z^{N_j})$ is the probability generating function of N_j $(j = 1, 2, \ldots, r)$.

In the sequel we consider two scenarios: *(a)* tandem network with negative binomial components, *(b)* tandem network with logarithmic components.

(a) Tandem network with negative binomial components
In this case the arrival rate is λ and the service rates of the component C_j are:

$$\mu_j(n) = \frac{\mu_j\, n}{\beta_j + n} \qquad (n = 1, 2, \ldots; j = 1, 2, \ldots, r) \tag{5}$$

where $\mu_j > 0$ and $\beta_j \geq 0$. In particular, if $\beta_j = 0$ the birth-death process corresponds to the $M/M/1$ queue. We note that $\mu_j(n) \leq \mu_j$ and $\lim_{n \to +\infty} \mu_j(n) = \mu_j$.

When the traffic intensities $\varrho_j = \lambda/\mu_j < 1$ for $j = 1, 2, \ldots, r$, from (2), one obtains the steady-state distribution of the component C_j:

$$q_j(n) = P(N_j = n) = \frac{\varrho_j^n}{n!}(\beta_j + 1)_n(1 - \varrho_j)^{\beta_j + 1} \qquad (n = 0, 1, 2, \ldots), \tag{6}$$

where $(\gamma)_n$ denotes the Pochhammer symbol, defined as $(\gamma)_0 = 1$ and $(\gamma)_n = \gamma(\gamma + 1) \ldots (\gamma + n - 1)$ if $n = 1, 2, \ldots$. We remark that

- if $\beta_j = 0$, (6) corresponds to the geometric steady-state distribution of $M/M/1$ queueing system: $q_j(n) = (1 - \varrho_j)\, \varrho_j^n$ $(n = 0, 1, 2, \ldots)$;
- if $\beta_j = k - 1$ $(k = 2, 3, \ldots)$, (6) becomes a negative binomial distribution: $q_j(n) = \binom{n+k-1}{k-1}(1 - \varrho_j)^k \varrho_j^n$ $(n = 0, 1, 2, \ldots)$, that corresponds to the sum of $k - 1$ independent geometric random variables.

Since $M_r = N_1 + N_2 + \ldots + N_r$, if $\max(\varrho_1, \varrho_2, \ldots, \varrho_r) < 1$ it follows:

$$E(M_r) = \sum_{j=1}^{r} \frac{\varrho_j}{1 - \varrho_j}(1 + \beta_j), \quad \mathrm{Var}(M_r) = \sum_{j=1}^{r} \frac{\varrho_j}{(1 - \varrho_j)^2}(1 + \beta_j). \tag{7}$$

Assuming that $\mu_1 = \mu_2 = \ldots = \mu_r = \mu$, from (4) for $\varrho = \lambda/\mu < 1$ one has:

$$G_{M_r}(z) = \left(\frac{1 - \varrho}{1 - z\varrho}\right)^{\beta_1 + \beta_2 + \ldots + \beta_r + r} \qquad (r = 1, 2, \ldots). \tag{8}$$

Expanding (8) in a power series, one obtains the probability function of M_r:

$$P(M_r = n) = \frac{(\beta_1 + \ldots + \beta_r + r)_n}{n!}(1 - \varrho)^{\beta_1 + \ldots + \beta_r + r}\, \varrho^n \qquad (n = 0, 1, \ldots). \tag{9}$$

Note that as $\beta_1 = \ldots = \beta_r = 0$ from (9) one obtains the distribution of the total number of customers in a tandem network that consists of r queues $M/M/1$:
$P(M_r = n) = \binom{n+r-1}{r-1}(1 - \varrho)^r \varrho^n$ $(n = 0, 1, \ldots)$.

In Fig. 2, we consider a tandem network with negative binomial components and identical service rates (5) for $\beta_j = \beta$ and $\mu_j = \mu$ $(j = 1, 2, \ldots, r)$.

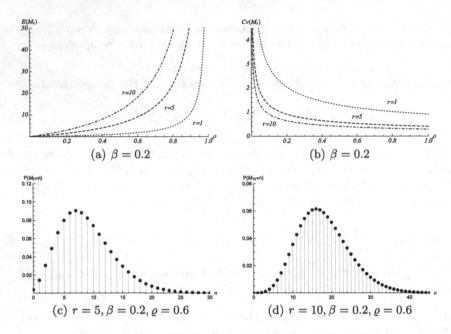

Fig. 2. Tandem network with negative binomial components and identical state-dependent service rates. In (a) and (b), $E(M_r)$ and $\mathrm{Cv}(M_r)$ are plotted. In (c) and (d), the probabilities $P(M_r = n)$ are shown for $r = 5$ and $r = 10$.

(b) Tandem network with logarithmic components

In this case the arrival rate is λ and the service rates of the component C_j are:

$$\mu_j(1) = \mu_1, \qquad \mu_j(n) = \frac{n\,\mu_j}{n-1} \quad (n = 2, 3, \ldots), \tag{10}$$

where $\mu_j > 0$. We note that $\mu_j(n) \geq \mu_j$ and $\lim_{n \to +\infty} \mu_j(n) = \mu_j$. When $\varrho_j = \lambda/\mu_j < 1$ for $j = 1, 2, \ldots, r$, from (2) one has (cf. [7]):

$$q_j(0) = \left[1 - \ln(1 - \varrho_j)\right]^{-1}, \qquad q_j(n) = \frac{\varrho_j^n}{n}\left[1 - \ln(1 - \varrho_j)\right]^{-1} \quad (n = 1, 2, \ldots). \tag{11}$$

Furthermore, since $M_r = N_1 + N_2 + \ldots + N_r$, if $\max(\varrho_1, \varrho_2, \ldots, \varrho_r) < 1$ it follows:

$$E(M_r) = \sum_{j=1}^{r} \frac{\varrho_j}{1 - \varrho_j}\, q_j(0), \qquad \mathrm{Var}(M_r) = \sum_{j=1}^{r} \frac{\varrho_j}{(1 - \varrho_j)^2} q_j(0)\left[1 - \varrho_j q_j(0)\right]. \tag{12}$$

When $\mu_j = \mu$ $(j = 1, 2, \ldots, r)$, from (4) for $\varrho = \lambda/\mu < 1$ one obtains:

$$G_{M_r}(z) = \left[\frac{1 - \log(1 - z\varrho_j)}{1 - \log(1 - \varrho_j)}\right]^r \qquad (r = 1, 2, \ldots). \tag{13}$$

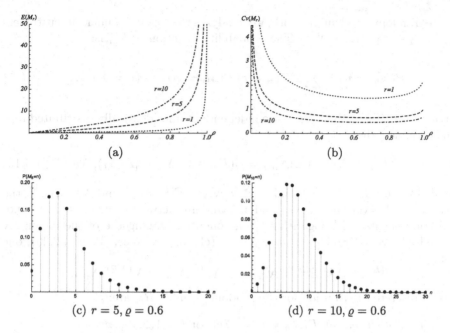

(a)   (b)

(c) $r = 5, \varrho = 0.6$ (d) $r = 10, \varrho = 0.6$

Fig. 3. As in Fig. 2 for a tandem network with logarithmic components and identical state-dependent service rates.

Expanding (13) in a power series and recalling that (cf. [1], p. 824)

$$[\ln(1 + x)]^k = k! \sum_{i=k}^{+\infty} S_i^{(k)} \frac{x^i}{i!} \qquad (|x| < 1),$$

one obtains the probability function of M_r:

$$P(M_r = n) = \frac{(-\varrho)^n}{n! \, [1 - \ln(1 - \varrho)]^r} \sum_{j=0}^{\min(n,r)} \binom{r}{j} (-1)^j \, j! \, S_n^{(j)} \quad (n = 0, 1, \ldots), \quad (14)$$

where $S_n^{(j)}$ denotes the Stirling number of first kind. Figure 3 refers to a tandem network with logarithm components and identical state-dependent service rates.

3   Tandem Network with a Random Number of Components

We assume that the number of components of the tandem network is random and let R_m be a discrete random variable that assumes the values $1, 2, \ldots, m$ with an assigned probability distribution. We denote by $(N_1, N_2, \ldots, N_{R_m})$ the vector that describes the state of the random tandem network in equilibrium regime and we assume the independence between $N_1, N_2, \ldots, N_m$ and R_m.

In the random tandem network we analyze the total number of customers $\mathcal{N}_m = N_1 + N_2 + \ldots + N_{R_m}$. The probability function of $\mathcal{N}_m$ is:

$$P(\mathcal{N}_m = n) = \sum_{r=1}^{m} P(R_m = r)\, P(M_r = n) \qquad (n = 0, 1, \ldots). \tag{15}$$

Furthermore, if $N_1, N_2, \ldots, N_m$ are independent and identically distributed random variables, then

$$E(\mathcal{N}_m) = E(R_m)E(N_1), \quad \mathrm{Var}(\mathcal{N}_m) = E(R_m)\mathrm{Var}(N_1) + [E(N_1)]^2\mathrm{Var}(R_m). \tag{16}$$

We denote by $E^{(NB)}(\mathcal{N}_m)$ and $V^{(NB)}(\mathcal{N}_m)$ [$E^{(L)}(\mathcal{N}_m)$ and $V^{(L)}(\mathcal{N}_m)$] the mean and the variance in a random tandem network with negative binomial components [with logarithmic components]. Making use of the inequality $x < -\ln(1-x) < x/(1-x)$ $(0 < x < 1)$ (cf., for instance, [1], n. 4.2.29), one has:

$$E^{(L)}(\mathcal{N}_m) < E^{(NB)}(\mathcal{N}_m), \qquad V^{(L)}(\mathcal{N}_m) < V^{(NB)}(\mathcal{N}_m).$$

Special discrete distribution for the random variable R_m are:

- *Uniform distribution*: $P(R_m = r) = 1/m$ for $r = 1, 2, \ldots, m$;
- *Truncated geometric distribution*: $P(R_m = r) = (1-\xi)\,\xi^{r-1}/(1-\xi^m)$ for $r = 1, 2, \ldots, m$, with $\xi > 0$ and $\xi \neq 1$. Note that as $\xi \to 1$ one obtains the uniform distribution.

(a) Random tandem network with negative binomial components
We consider a random tandem network with negative binomial components and identical state-dependent service rates $\mu_j(n) = \mu n/(\beta + n)$, with $\mu > 0, \beta \geq 0$. Recalling (7) and (9), from (15) and (16), for $\varrho = \lambda/\mu < 1$ one obtains:

$$P(\mathcal{N}_m = n) = \frac{\varrho^n}{n!} \sum_{r=1}^{m} P(R_m = r)(1 - \varrho)^{r(\beta+1)}\left(r(\beta + 1)\right)_n \qquad (n = 0, 1, \ldots),$$

$$E(\mathcal{N}_m) = \frac{\varrho}{1 - \varrho}\,(\beta + 1)\,E(R_m), \tag{17}$$

$$\mathrm{Var}(\mathcal{N}_m) = \frac{\varrho}{(1 - \varrho)^2}\,(\beta + 1)\Big\{E(R_m) + \varrho(\beta + 1)\mathrm{Var}(R_m)\Big\}.$$

Note that if $\beta = 0$, (17) refers to a random tandem network with $M/M/1$ identical components. In Fig. 4, we take in account a random tandem network with negative binomial components by choosing the uniform ($\xi = 1$) and the truncated geometric distributions ($\xi = 0.5, 1.5$) for R_m.

(b) Random tandem network with logarithm components
We consider a random tandem network with logarithm components and identical state-dependent service rates $\mu_j(1) = \mu, \mu_j(n) = n\mu/(n-1)$, with $\mu > 0$. By

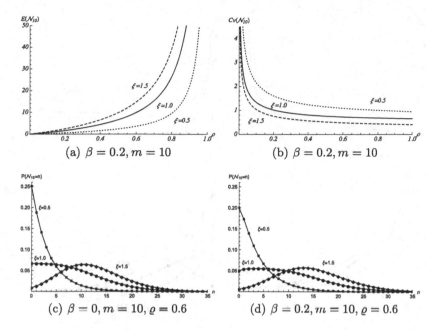

(a) $\beta = 0.2, m = 10$

(b) $\beta = 0.2, m = 10$

(c) $\beta = 0, m = 10, \varrho = 0.6$

(d) $\beta = 0.2, m = 10, \varrho = 0.6$

Fig. 4. Random tandem network with negative binomial components, in which R_{10} has an uniform ($\xi = 1$) or a truncated geometric distribution ($\xi = 0.5, 1.5$). In (a) and (b), $E(\mathcal{N}_{10})$ and $\mathrm{Cv}(\mathcal{N}_{10})$ are plotted. In (c) and (d), $P(\mathcal{N}_{10} = n)$ are shown.

virtue of (12) and (14), from (15) and (16), for $\varrho = \lambda/\mu < 1$ one obtains:

$$P(\mathcal{N}_m = n) = \frac{\varrho^n}{n!} \sum_{r=1}^{m} \frac{r! \, P(R_m = r)}{[1 - \ln(1 - \varrho)]^r} \sum_{j=0}^{\min(n,r)} \frac{(-1)^{n+j}}{(r-j)!} S_n^{(j)} \qquad (n = 0, 1, \ldots),$$

$$E(\mathcal{N}_m) = \frac{\varrho}{1 - \varrho} \frac{1}{1 - \ln(1 - \varrho)} E(R_m), \tag{18}$$

$$\mathrm{Var}(\mathcal{N}_m) = \frac{\varrho}{(1 - \varrho)^2} \frac{[1 - \varrho - \ln(1 - \varrho)] \, E(R_m) + \varrho \, \mathrm{Var}(R_m)}{[1 - \ln(1 - \varrho)]^2}.$$

In Fig. 5 we consider a random tandem network with logarithmic components in which R_m has an uniform ($\xi = 1$) or a truncated geometric distributions ($\xi = 0.5, 1.5$).

The uniform and truncated geometric distributions belong to the class of *truncated discrete distributions*

$$P(R_m = r) = \frac{\varphi(m - r + 1)}{\sum_{k=1}^{m} \varphi(k)} \qquad (r = 1, 2, \ldots, m), \tag{19}$$

where the function $\varphi(x)$ is positive and not depend by m. In particular, the uniform distribution is obtained by choosing $\varphi(x) = 1$ in (19), whereas if

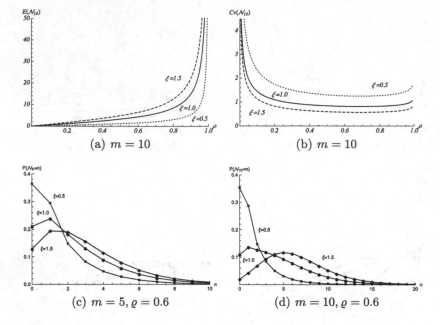

Fig. 5. As in Fig. 4 for a random tandem network with logarithmic components, in which R_{10} has an uniform ($\xi = 1$) or a truncated geometric distribution ($\xi = 0.5, 1.5$).

$\varphi(x) = \xi^{-x}$, with $\xi > 0$ and $\xi \neq 1$, in (19) one has the truncated geometric distribution. A truncated discrete distribution satisfies the iterative relation:

$$P(R_m = r) = [1 - P(R_m = 1)]\, P(R_{m-1} = r - 1) \quad (r = 2, 3, \ldots, m),$$

so that the probabilities (15) and the moments of $\mathcal{N}_m$ can be iteratively computed with respect to m as follows:

$$P(\mathcal{N}_m = n) = P(R_m = 1)q_1(n) + [1 - P(R_m = 1)]\sum_{j=0}^{n} q_1(j)P(\mathcal{N}_{m-1} = n - j)$$

$$(n = 0, 1, \ldots; m = 2, 3, \ldots) \qquad (20)$$

$$E(\mathcal{N}_m^k) = P(R_m = 1)E(N_1^k) + [1 - P(R_m = 1)]\sum_{i=0}^{k} \binom{k}{i} E(\mathcal{N}_{m-1}^i)E(N_1^{k-i})$$

$$(k = 1, 2, \ldots; m = 2, 3, \ldots)$$

with $P(\mathcal{N}_1) = P(N_1 = n) = q_1(n)$ and $E(\mathcal{N}_1^k) = E(N_1^k)$ for $k = 1, 2, \ldots$.

A future study will concern the analysis of a more general random tandem network and the construction of relative suitable continuous approximations.

References

1. Abramowitz, M., Stegun, I.A.: Handbook of Mathematical Functions with Formulas, Graphs, and Mathematical Tables. Dover, New York (1992)
2. Baskett, F., Chandy, K.M., Muntz, R., Palacios, F.G.: Open, closed, and mixed networks of queues with different classes of customers. J. ACM **22**(2), 248–260 (1975)
3. Bose, S.K.: An Introduction to Queueing Systems. Springer Science+Business Media, New York (2002)
4. Burke, P.J.: The output of a queueing system. Oper. Res. **4**(6), 699–704 (1956)
5. Di Crescenzo, A., Giorno, V., Nobile, A.G.: Constructing transient birth-death processes by means of suitable transformations. Appl. Math. Comput. **281**, 152–171 (2016)
6. Giorno, V., Nobile, A.G., Spina, S.: On some time non-homogeneous queueing systems with catastrophes. Appl. Math. Comput. **245**, 220–234 (2014)
7. Giorno, V., Nobile, A.G., Pirozzi, E.: A state-dependent queueing system with asymptotic logarithmic distribution. J. Math. Anal. Appl. **458**, 949–966 (2018)
8. Jackson, J.R.: Networks of waiting lines. Oper. Res. **5**(4), 518–521 (1957)
9. Lemoine, A.J.: State-of-the-art - Networks of queues - a survey of equilibrium analysis. Manag. Sci. **24**(4), 464–481 (1997)
10. Reich, E.: Waiting times when queues are in tandem. Ann. Math. Stat. **28**(3), 768–773 (1957)
11. Serfozo, R.: Introduction to Stochastic Networks. Springer Science+Business Media, New York (1999)
12. Williams, R.J.: Stochastic processing networks. Annu. Rev. Stat. Appl. **3**, 323–345 (2016)

Precise Parameter Synthesis for Generalised Stochastic Petri Nets with Interval Parameters

Milan Češka Jr.[1](✉), Milan Češka[1], and Nicola Paoletti[2]

[1] Faculty of Information Technology, Brno University of Technology,
Brno, Czech Republic
ceskam@fit.vutbr.cz
[2] Department of Computer Science, Stony Brook University, Stony Brook, USA

Abstract. We consider the problem of synthesising parameters affecting transition rates and probabilities in generalised Stochastic Petri Nets (GSPNs). Given a time-bounded property expressed as a probabilisitic temporal logic formula, our method allows computing the parameters values for which the probability of satisfying the property meets a given bound, or is optimised. We develop algorithms based on reducing the parameter synthesis problem for GSPNs to the corresponding problem for continuous-time Markov Chains (CTMCs), for which we can leverage existing synthesis algorithms, while retaining the modelling capabilities and expressive power of GSPNs. We evaluate the usefulness of our approach by synthesising parameters for two case studies.

1 Introduction

Various extensions of Stochastic Petri Nets (SPNs), e.g. generalised SPNs [12] (GSPNs), have been introduced to model complex and concurrent systems in many areas of science. In biochemistry, quantitative models of genetic networks can be expressed as SPNs [8]. In engineering, GSPNs are used to study various reliability and performance aspects of manufacturing processes, computer networks and communication protocols [1,3]. Assuming certain restrictions on their structure, the dynamics of SPNs as well as GSPNs can be described using finite-state continuous-time Markov chains (CTMCs) [12]. This allows modellers and designers to perform quantitative analysis and verification using well-established formal techniques for CTMCs, above all probabilistic model checking [11].

Traditionally, formal analysis of SPNs and GSPNs assumes that transition rates and probabilities are known a priori. This is often not the case and one has to consider ranges of parameter values instead, for example, when the parameters result from imprecise measurements, or when designers are interested in finding parameter values such that the model fulfils a given specification.

This work has been supported by the Czech Grant Agency grant No. GA16-24707Y and the IT4Innovations Excellence in Science project No. LQ1602.

R. Moreno-Díaz et al. (Eds.): EUROCAST 2017, Part II, LNCS 10672, pp. 38–46, 2018.
https://doi.org/10.1007/978-3-319-74727-9_5

In this paper, we tackle the parameter synthesis problem for GSPNs, described as follows:

"Given a time-bounded temporal formula describing the required behaviour and a parametric GSPN (pGSPN) whose transition rates and probabilities are functions of the parameters, automatically find parameter values such that the satisfaction probability of the formula meets a given threshold, is maximised, or minimised".

Importantly, this problem requires effective reasoning about uncountable sets of GSPNs, arising from the presence of continuous parameter ranges. We show that, under restrictions on the structure of pGSPNs (i.e. requiring a finite number of reachable markings or avoiding Zeno behaviour) and on predicates appearing in the temporal formulas, we can describe the dynamics of a pGSPN by a finite-state parametric CTMC (pCTMC). The parameter synthesis problem for pGSPNs can be then reduced to the equivalent problem for pCTMCs and thus, we can employ existing synthesis algorithms that combine computation of probability bounds for pCTMCs with iterative parameter space refinement in order to provide arbitrarily precise answers [5]. We further demonstrate that pGSPNs provide an adequate modelling formalism for designing complex systems where parameters of the environment (e.g., request inter-arrival times) and those inherent to the system (e.g. service rates) can be meaningfully expressed as intervals. We also show that pGSPNs can be used for the *in silico* analysis of stochastic biochemical systems with uncertain kinetic parameters.

The main contribution of the paper can be summarised as follows:

- formulation of the parameter synthesis problem for GSPNs using pGSPNs;
- solution method based on translation of pGSPNs into pCTMCs; and
- evaluation on two case studies from different domains, through which we demonstrate the usefulness and effectiveness of our method.

2   Problem Formulation

In our work we consider the problem of parameter synthesis for Generalised Stochastic Petri Nets (GSPNs). GSPNs naturally combine stochastic (i.e. timed) transitions and immediate (i.e. untimed and probabilisitic) transitions [4] and thus provide an adequate formalism for modelling engineered and biological systems alike. Below we introduce the model of *parametric* GSPNs (*pGSPNs*), which extends GSPNs with parameters that affect transitions rates and probabilities.

Definition 1 (pGSPN). *Let K be a set of parameters. A pGSPN over K is a tuple $(L, T, A, M_{in}, \mathbf{R}, \mathbf{P})$, where:*

- *L is a finite set of places inducing a set of markings M, where for each $m \in M, m = (m_1, m_2, \ldots, m_n) \in \mathbb{R}^n$, with $n = |L|$;*
- *$T = T_{st} \cup T_{im}$ is a finite set of transitions partitioned into stochastic transitions T_{st} and immediate transitions T_{im};*
- *$A \subseteq (P \times T) \cup (T \times P)$ is a finite set of arcs connecting transitions and places;*

- $m_{in} \in M$ *is the initial marking;*
- $\mathbf{R}:T_{st} \rightarrow (M \rightarrow \mathbb{R}[K])$ *is the parametric, marking-dependent rate matrix, where* $\mathbb{R}[K]$ *is the set of polynomials over the reals* $\mathbb{R}$ *with variables* $k \in K;$
- $\mathbf{P}:T_{im} \rightarrow (M \rightarrow \mathbb{R}[K])$ *is the parametric, marking-dependent probability matrix.*

The domain of each parameter $k \in K$ is given by a closed real interval describing the range of possible values, i.e., $[k^{\perp}, k^{\top}] \subseteq \mathbb{R}$. The parameter space $\mathcal{P}$ induced by K is defined as the Cartesian product $\mathcal{P} = \times_{k \in K}[k^{\perp}, k^{\top}]$. Subsets of $\mathcal{P}$ are called *subspaces*. Given a pGSPN and a parameter space $\mathcal{P}$, we denote with $\mathcal{N}_{\mathcal{P}}$ the set $\{\mathcal{N}_p \mid p \in \mathcal{P}\}$ where $\mathcal{N}_p = (L, T, A, M_{in}, \mathbf{R}_p, \mathbf{P}_p)$ is the instantiated GSPN obtained by replacing the parameters in $\mathbf{R}$ and $\mathbf{P}$ with their valuation in p. The rate (probability) matrix for marking m and parameter valuation p is denoted by $\mathbf{R}_{m,p}$ ($\mathbf{P}_{m,p}$).

For all markings $m \in M$ reachable from m_{in}, we require that: (1) For all $t \in T_{st}$, it holds that either $\mathbf{R}_{p,m}(t) > 0$ for all $p \in \mathcal{P}$, or $\mathbf{R}_{p,m}(t) = 0$ for all $p \in \mathcal{P}$. (2) For all $t \in T_{im}$ it holds that either $\mathbf{P}_{p,m}(t) > 0$ for all $p \in \mathcal{P}$ or $\mathbf{P}_{p,m}(t) = 0$ for all $p \in \mathcal{P}$, and $\sum_{t \in T_{im}} \mathbf{P}_{p,m}(t) = 1$. Note that $\mathbf{R}_{p,m}(t) > 0$ ($\mathbf{P}_{p,m}(t) > 0$, respectively) if and only if the transition t is enabled, i.e. there is a sufficient number of tokens in each of its input places. In other words, parameters do not affect the enabledness of transitions. Further, we use the notion of *capacity* C to indicate, for each place $l \in L$, the maximal number of tokens $C(l)$ in l, thus determining when a transition is enabled.

Vanishing and tangible markings. As in GSPNs, a marking $m \in M$ is called *vanishing* if there is an immediate transition $t \in T_{in}$ that is enabled in m, or *tangible* otherwise. In a vanishing marking m, all stochastic transitions are blocked, and the enabled immediate transitions are fired in zero time according the probability distribution $\mathbf{P}_{p,m}$. To avoid Zeno behaviour, we require that there are no cycles over vanishing markings. In a tangible marking m, the sojourn time is exponentially distributed with average time $E_p(m)^{-1}$ where $E_p(m) = \sum_{t \in T_{st}} \mathbf{R}_{p,m}(t)$ is the exit rate. The probability that a transition $t \in T_{st}$ is fired first is given by $\mathbf{R}_{p,m}(t) \cdot E_p(m)^{-1}$.

Specification language. We consider the time-bounded fragment of Continuous Stochastic Logic (CSL) [2] to specify behavioural properties of GSPNs, with the following syntax. A state formula Φ is given by $\Phi ::= \text{true} \mid a \mid \neg\Phi \mid \Phi \wedge \Phi \mid P_{\sim r}[\phi] \mid P_{=?}[\phi]$, where ϕ is a path formula, given by $\phi ::= \mathsf{X}\,\Phi \mid \Phi\,\mathsf{U}^I\,\Phi$, a is an atomic proposition defined over the markings $m \in M$, $\sim \in \{<, \leq, \geq, >\}$, $r \in [0, 1]$ is a probability threshold, and $I \in \mathbb{R}$ is a bounded interval. As explained in [5], our parameter synthesis methods also support time-bounded rewards [11], which we omit in the following for the sake of clarity.

Let ϕ be a CSL path formula and $\mathcal{N}_{\mathcal{P}}$ be a pGSPN over a space $\mathcal{P}$. We denote with $\Lambda_\phi : \mathcal{P} \rightarrow [0, 1]$ the *satisfaction function* such that $\Lambda_\phi(p) = P_{=?}[\phi]$, that is, $\Lambda_\phi(p)$ is the probability of ϕ being satisfied over the GSPNs $\mathcal{N}_p$. Note that the path formula ϕ may contain nested probabilistic operators, and therefore the satisfaction function is, in general, not continuous.

Synthesis problems. We consider two parameter synthesis problems: the *threshold synthesis* problem that, given a threshold $\sim r$ and a CSL path formula ϕ, asks for the subspace where the probability of ϕ meets $\sim r$; and the *max synthesis* problem that determines the subspace within which the probability of the input formula is guaranteed to attain its maximum, together with probability bounds that contain that maximum. Solutions to the threshold synthesis problem admit parameter points left undecided, while, in the max synthesis problem, the actual set of maximising parameters is contained in the synthesised subspace. Importantly, the undecided and maximising subspaces can be made arbitrarily precise through user-defined tolerance values.

For $\mathcal{N}_\mathcal{P}, \phi$, an initial state s_0, a threshold $\sim r$ and a volume tolerance $\varepsilon > 0$, the *threshold synthesis* problem is finding a partition $\{\mathcal{T}, \mathcal{U}, \mathcal{F}\}$ of $\mathcal{P}$, such that: $\forall p \in \mathcal{T} : \Lambda_\phi(p) \sim r; \forall p \in \mathcal{F} : \Lambda_\phi(p) \not\sim r$; and $\mathrm{vol}(\mathcal{U})/\mathrm{vol}(\mathcal{P}) \leq \varepsilon$, where $\mathcal{U}$ is an undecided subspace and $\mathrm{vol}(A) = \int_A 1 d\mu$ is the volume of A.

For $\mathcal{N}_\mathcal{P}, \phi, s_0$, and a probability tolerance $\epsilon > 0$, the *max synthesis* problem is finding a partition $\{\mathcal{T}, \mathcal{F}\}$ of $\mathcal{P}$ and probability bounds $\Lambda_\phi^\perp, \Lambda_\phi^\top$ such that: $\forall p \in \mathcal{T} : \Lambda_\phi^\perp \leq \Lambda_\phi(p \leq \Lambda_\phi^\top; \exists p \in \mathcal{T} : \forall p' \in \mathcal{F} : \Lambda_\phi(p) > \Lambda_\phi(p')$; and $\Lambda_\phi^\top - \Lambda_\phi^\perp \leq \epsilon$. The min synthesis problem is defined in a symmetric way to the max case.

3 Parameter Synthesis for Stochastic Petri Nets

First, we introduce a novel automated translation from pGSPNs into parametric CTMCs (pCTMCs), able to preserve important quantitative temporal properties. This allows us to reduce the pGSPN synthesis problem to the equivalent pCTMC synthesis problem. Second, we provide an overview of our recent results on parameter synthesis for CTMCs [5].

3.1 Translation of pGSPNs into pCTMCs

CTMCs represent purely stochastic processes and thus, in contrast to GSPNs, they do not allow any immediate transitions. The dynamics of a CTMC is given by a transition rate matrix defined directly over its set of states S. *Parametric CTMCs* [5] extend the notion of CTMCs by allowing transitions rates to depend on model parameters. Formally, for a set of parameters K, the parametric rate matrix is defined as $\mathbf{M} : S \times S \rightarrow \mathbb{R}[K]$. Similarly as in the case pGSPNs, for a given parameter space $\mathcal{P}$ the pCTMC $\mathcal{C}_\mathcal{P}$ defines an uncountable set $\{\mathcal{C}_p \mid p \in \mathcal{P}\}$ where $\mathcal{C}_p = (S, \mathbf{M}_p, s_0)$ is the instantiated CTMC obtained by replacing the parameters in $\mathbf{M}$ with their valuation in p and where s_0 denotes the initial state.

We introduce a translation method from pGSPNs to pCTMCs that builds on the translation for non-parametric GSPNs [4] and exploits the fact that parameters do not affect the enabledness of transitions and thus do not alter the set of reachable markings. Therefore we can map the set of markings M in the pGSPN $\mathcal{N}_\mathcal{P}$ to the set of states S in the pCTMC $\mathcal{C}_\mathcal{P}$. This mapping allows us to construct the parametric rate matrix $\mathbf{M}$ of the pCTMC such

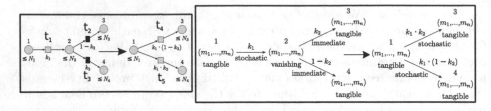

Fig. 1. *Left:* Merging at the level of places. *Right:* Merging at the level of markings.

that $\mathcal{C}_\mathcal{P}$ preserves the dynamic of pGSPN over tangible markings. Formally, for $p \in \mathcal{P}, \mathbf{M}_p(m, m') = \sum_{t \in T(m,m')} \mathbf{R}_{p,m}(t)$, where $T(m,m') = \{t \in T_{st} \mid m'$ is a marking obtained by firing t in marking $m\}$.

The crucial difficulty of this translation lies in handling the vanishing markings. Since any state in a CTMC has a non-zero waiting time, in order to map pGSPN markings into pCTMC states we need to eliminate the vanishing markings. Specifically, for each vanishing marking, we merge the incoming and outgoing transitions and combine the corresponding parameters (if present).

Although we merge at the level of markings, we first explain the intuition of the merging at the level of places, which is illustrated in Fig. 1 (left). The reasoning behind this operation is that the immediate probabilistic transitions t_2 and t_3 take zero time and thus they do not affect the total exit rate in the resulting, merged transitions t_4 and t_5, i.e., $\mathbf{R}(t_1) = \mathbf{R}(t_4) + \mathbf{R}(t_5)$. The transition probabilities k_2 and $1 - k_2$ of t_2 and t_3, respectively, are used to determine the probability that the transition t_4 and t_5, respectively, is fired when the sojourn time in the place 1 is passed.

Since the transition rates and probabilities are marking-dependent, elimination and merging is actually performed at the level of markings. Figure 1 (right) illustrates these operations. The principle is same as in the case of merging at the level of places, but it allows us to reflect the dependencies on the markings and thus to preserve the dynamic of pGSPN over tangible markings. Clearly, we cannot reason about the vanishing markings and thus trajectories in pGSPN that differ only in vanishing markings are indistinguishable in the resulting pCTMC. To preserve the correctness of our approach we have to restrict the set of properties. In particular, we support only properties with atomic propositions defined over the set of tangible markings. This allow us to compute the satisfaction function Λ_ϕ using the satisfaction function Γ_ϕ defined over the pCTMC.

3.2   Parameter Synthesis for pCTMCs

We now give an overview of the parameter synthesis method for pCTMCs [5]. The method builds on the computation of safe bounds for the satisfaction function Γ_ϕ. In particular, for a pCTMC $\mathcal{C}_\mathcal{P}$, the procedure efficiently computes an interval $[\Gamma_\perp^\mathcal{P}, \Gamma_\top^\mathcal{P}]$ that is guaranteed to contain the minimal and maximal probability that pCTMC $\mathcal{C}_\mathcal{P}$ satisfies ϕ, i.e. $\Gamma_\perp^\mathcal{P} \leq \min_{p \in \mathcal{P}} \Gamma_\phi(p)$ and $\Gamma_\top^\mathcal{P} \geq \max_{p \in \mathcal{P}} \Gamma_\phi(p)$.

This safe over-approximation is computed through an algorithm that extends the well-established time-discretisation technique of *uniformization* [11] for the transient analysis of CTMCs, which is the foundation of CSL model checking. Our technique is efficient because the computation of maximal and minimal probabilities is based on solving a series of local and independent optimisation problems at the level of each transition and each discrete uniformisation step. This approach is much more feasible than solving the global optimisation problem, which reduces to the optimisation of a high-degree multivariate polynomial.

The complexity of our approach depends on the degree of the polynomials appearing in the transition matrix $\mathbf{M}$. We restrict to *multi-affine* polynomial rate functions for which we have a complexity of $\mathcal{O}(2^{n+1} \cdot t_{CSL})$, where n is the number of parameters and t_{CSL} is the complexity of the standard non-parametric CSL model-checking algorithm, which mainly depends on the size of the underlying model and the number of discrete time steps (the latter depends on the maximum exit rate of the CTMC and time bound in the CSL property). For linear rate functions, we have an improved complexity of $\mathcal{O}(n \cdot t_{CSL})$.

Crucially, the approximation error of this technique depends linearly on the volume of the parameter space and exponentially on the number of discrete time steps [5], meaning that the error can be controlled by refining the parameter space (i.e. the error reduces with the volume of the parameter region). In particular, the synthesis algorithms are based on the iterative refinement of the parameter space $\mathcal{P}$ and the computation of safe bounds for Γ_ϕ (as per above).

Threshold synthesis. We start with $\mathcal{U}$ containing $\mathcal{P}$ (i.e. the whole parameter space is undecided). For each $\mathcal{R} \in \mathcal{U}$, we compute the safe bounds $\Gamma_\perp^\mathcal{R}$ and $\Gamma_\top^\mathcal{R}$. Then, assuming a threshold $\leq r$, if $\Gamma_\top^\mathcal{R} \leq r$ then for all $p \in \mathcal{R}$ the threshold on the property ϕ is satisfied and $\mathcal{R}$ is added to $\mathcal{T}$. Similarly, if $\Gamma_\perp^\mathcal{R} > r$ then $\mathcal{R}$ is added to $\mathcal{F}$. Otherwise, $\mathcal{R}$ is decomposed into subspaces that are added to $\mathcal{U}$. The algorithm terminates when $\mathcal{U}$ satisfies the required volume tolerance ε. Termination is guaranteed by the shape of approximation error.

Max synthesis. The algorithm starts with $\mathcal{T} = \mathcal{P}$ and iteratively refines $\mathcal{T}$ until the probability tolerance ϵ on the bounds $\Lambda_\phi^\perp = \Gamma_\perp^\mathcal{T}$ and $\Lambda_\phi^\top = \Gamma_\top^\mathcal{T}$ is satisfied. To provide faster convergence, at each iteration it computes an under-approximation M of the actual maximal probability, by randomly sampling a set of probability values and setting M to the maximal sample. In this way, each subspace whose maximal probability is below M can be safely added to $\mathcal{F}$. Otherwise, $\mathcal{R}$ is added to a newly constructed set $\mathcal{T}$ and the bounds $\Lambda_\phi^\perp$ and $\Lambda_\phi^\top$ are updated accordingly. Since that the satisfaction function Λ_ϕ is in general discontinuous, the algorithm might not terminate. This is detected by extending the termination criterion using the volume tolerance ε.

The complexity of the synthesis algorithms depends mainly on the number of subspaces to analyse in order to achieve the desired precision, since this number directly affects how many times the procedure computing the bounds on Γ_ϕ has to be executed. Therefore, complex instances of the synthesis problem can be computationally very demanding. To overcome this problem, we redesigned the sequential algorithms to enable state space and parameter space parallelisation, resulting in data-parallel algorithms providing dramatic speed-up on many-core architectures [6]. Thanks to the translation procedure previously described, we can exploit parallelisation also for the parameter synthesis of GSPNs.

4   Experimental Results

All the experiments were performed using an extended version of the GPU-accelerated tool PRISM-PSY [6]. They run on a Linux workstation with an AMD PhenomTM II X4 940 Processor @ 3 GHz, 8 GB DDR2 @ 1066 MHz RAM and an NVIDIA GeForce GTX 480 GPU with 1.5 GB of GPU memory.

Google File System. We consider a case study of the replicated file system used in the Google search engine known as Google File System (GFS). The model, introduced in [3] as a non-parametric GSPN, reproduces the life-cycle of a single chunk (representing a part of the file) within the file system. The chunk exists in several copies, located in different chunk servers. There is one master server that is responsible for keeping the locations of the chunk copies and replicating the chunks if a failure occurs. In [6] we introduced a parametric version of the model, manually derived the corresponding pCTMC, and performed parameter synthesis using the tool PRISM-PSY.

In this paper we exploit the modelling capabilities of pGSPNs and introduce an extension of the model. In particular, we integrate into the model a message monitoring inspired by the original paper on GFS [7]: the master server periodically send so-called *HeartBeat* messages to check for chunk inconsistency, i.e. when a write request occur between a failure and its acknowledgement. Figure 2 (left) depicts the pGSPN describing the parametric model, where the part in the green-bordered box corresponds to the extension.

We are interested in the probability that the first chunk inconsistency occurs minutes 15 and 45, and how this probability depends on the rate of the HeartBeat messages (c_check) and the rate of the chunk server failure (c_fail). We solve this as a threshold synthesis problem with threshold $\geq 30\%$, path formula $\phi \equiv (C_{inc} = 0) \ U^{[15,45]} \ (C_{inc} > 0)$, parameter intervals c_check $\in [0.01, 10]$ and c_fail $\in [0.01, 0.11]$, and volume tolerance $\varepsilon = 10\%$.

Results are shown in Fig. 2 (right), namely, the decomposition of the parameter space into subspaces satisfying the property (green), not satisfying (red), and uncertain (yellow). The pGSPN has around 139 K states and 740 K transitions, and the synthesis algorithm required around 11 K time steps and produced 460 final subspaces. The data-parallel GPU computation took 25 min, corresponding to more than 7-fold speedup with respect to the sequential algorithm.

Mitogen-Activated-Protein-Kinases cascade. In our second case study we consider the Mitogen-Activated-Protein-Kinases (MAPK) cascade [10], one of the most important signalling pathways that controls molecular growth through activation (i.e. phosphorylation) cascade of kinases. We use a SPN model introduced

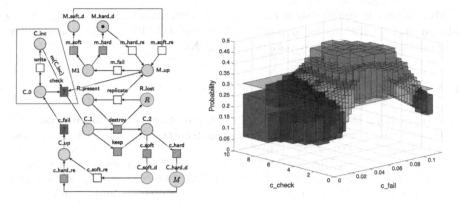

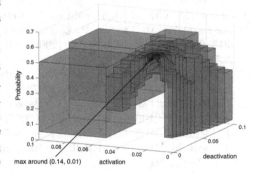

Fig. 2. *Left:* *p*GSPN for a new variant of the GFS model [3]. Red boxes with question marks indicate parametric transitions. *Right:* Results of the threshold synthesis. (Color figure online)

in [9] and study how two key reactions, namely activation by MAPKK-PP and deactivation by Phosphatase, affect the activation of the final kinases. We want to find rates of these reactions that maximise the probability that, within 50 and 55 min, the number of the activated kinases is between 25% and 50%. To this purpose, we formulate a max synthesis problem for property $\phi \equiv G[50, 55]$ $(25\% \leq \gamma \leq 50\%)$, where γ is the percentage of the activated kinases. The interval for both reaction rates is $[0.01, 0.1]$ and the probability tolerance $\epsilon = 5\%$.

Figure 3 illustrates the results of max synthesis, namely, it shows the decomposition of the parameter space into subspaces maximising the property (green) and not maximising (red). The bounds on maximal probability are 57.4% and 62.4%. The *p*GSPN has around 100 K states and 911 K transitions, and the synthesis algorithm required around 121 K time steps and produced 259 final subspaces. The parallel GPU computation took 5 h, corresponding to more than 22-fold speedup with respect to the sequential CPU algorithm.

Fig. 3. Results of the max synthesis for the MAPK cascade. (Color figure online)

5   Conclusion

We have developed efficient algorithms for synthesising parameters in GSPNs, building on the automated translation of parametric GSPNs into parametric

CTMCs. The experiments show that our approach allows us to exploit existing data-parallel algorithms for scalable synthesis of CTMCs, while retaining the modelling power provided by parametric GSPNs.

References

1. Al-Jaar, R.Y., Desrochers, A.A.: Performance evaluation of automated manufacturing systems using generalized stochastic petri nets. IEEE Trans. Robot. Autom. **6**(6), 621–639 (1990)
2. Aziz, A., Sanwal, K., Singhal, V., Brayton, R.: Verifying continuous time Markov chains. In: Alur, R., Henzinger, T.A. (eds.) CAV 1996. LNCS, vol. 1102, pp. 269–276. Springer, Heidelberg (1996). https://doi.org/10.1007/3-540-61474-5_75
3. Baier, C., et al.: Model checking for performability. Math. Struct. Comput. Sci. **23**(4), 751–795 (2013)
4. Balbo, G.: Introduction to generalized stochastic Petri nets. In: Bernardo, M., Hillston, J. (eds.) SFM 2007. LNCS, vol. 4486, pp. 83–131. Springer, Heidelberg (2007). https://doi.org/10.1007/978-3-540-72522-0_3
5. Češka, M., et al.: Precise parameter synthesis for stochastic biochemical systems. Acta Informatica **54**(6), 589–623 (2017)
6. Češka, M., Pilař, P., Paoletti, N., Brim, L., Kwiatkowska, M.: PRISM-PSY: precise GPU-accelerated parameter synthesis for stochastic systems. In: Chechik, M., Raskin, J.-F. (eds.) TACAS 2016. LNCS, vol. 9636, pp. 367–384. Springer, Heidelberg (2016). https://doi.org/10.1007/978-3-662-49674-9_21
7. Ghemawat, S., Gobioff, H., Leung, S.-T.: The Google file system. In: Proceedings of the SOSP 2003, pp. 29–43. ACM (2003)
8. Goss, P.J.E., Peccoud, J.: Quantitative modeling of stochastic systems in molecular biology by using stochastic petri nets. PNAS **95**(12), 6750–6755 (1998)
9. Heiner, M., Gilbert, D., Donaldson, R.: Petri nets for systems and synthetic biology. In: Bernardo, M., Degano, P., Zavattaro, G. (eds.) SFM 2008. LNCS, vol. 5016, pp. 215–264. Springer, Heidelberg (2008). https://doi.org/10.1007/978-3-540-68894-5_7
10. Huang, C., Ferrell, J.: Ultrasensitivity in the mitogen-activated protein kinase cascade. Proc. Natl. Acad. Sci. **93**, 10078–10083 (1996)
11. Kwiatkowska, M., Norman, G., Parker, D.: Stochastic model checking. In: Bernardo, M., Hillston, J. (eds.) SFM 2007. LNCS, vol. 4486, pp. 220–270. Springer, Heidelberg (2007). https://doi.org/10.1007/978-3-540-72522-0_6
12. Marsan, M.A., et al.: Modelling with Generalized Stochastic Petri Nets. Wiley, Hoboken (1994)

On a Non-homogeneous Gompertz-Type Diffusion Process: Inference and First Passage Time

Giuseppina Albano[1](✉), Virginia Giorno[2], Patricia Román-Román[3], and Francisco Torres-Ruiz[3]

[1] Dip. di Scienze Economiche e Statistiche, Università di Salerno, Fisciano, Italy
pialbano@unisa.it
[2] Dip. di Informatica, Università di Salerno, Fisciano, Italy
giorno@unisa.it
[3] Dpto. de Estadística e Investigación Operativa, Universidad de Granada, Granada, Spain
{proman,fdeasis}@ugr.es

Abstract. A stochastic diffusion model based on a generalized Gompertz deterministic growth in which the carrying capacity depends on the initial size of the population is considered. The growth parameter of the process is then modified by introducing a time-dependent exogenous term. The first passage time problem is considered and a two-step procedure to estimate the model is proposed. Simulation study is also provided for suitable choices of the exogenous term.

1 Introduction

The models for the description of growth phenomena, originally associated to the evolution of animals populations, currently play an important role in several fields like that economic, biological, medical, ecological, among others. For this reason many efforts are oriented to the development of always more sophisticated mathematical models for the description of a particular type of behaviour. The most representative curves for modeling growth are of exponential-type as the logistic and Gompertz curves because they are characterized by a finite carrying capacity, that represents, in general terms, the limitation of the natural resources. Specifically, the Gompertz curve is frequently used because in several contexts it seems to fit the experimental data in enough precise way. It is described by the equation:

$$x(t) = \exp\left\{ \frac{m}{\beta}[1 - e^{-\beta(t-t_0)}] + \ln x_0\, e^{-\beta(t-t_0)} \right\}, \tag{1}$$

where m and β are positive constants that represent the growth parameters of the population. Equation (1) is able to describe growth dynamics in a lot of contexts

---

This work was supported in part by the Ministerio de Economia y Competitividad, Spain, under Grant MTM2014-58061-P and partially by INDAM-GNCS.

R. Moreno-Díaz et al. (Eds.): EUROCAST 2017, Part II, LNCS 10672, pp. 47–54, 2018.
https://doi.org/10.1007/978-3-319-74727-9_6

so, for instance animal, vegetable, tumor growth. Equation (1) is solution of the following ordinary differential equation:

$$\frac{dx}{dt} = m\beta - \beta x \ln x, \qquad x(t_0) = x_0.$$

We note that in (1) the carrying capacity is $K = \lim_{t \to \infty} x(t) = \exp\{m/\beta\}$. However, in several contexts, the carrying capacity can depend from the initial size of the population (cf. [4,6,8], for example). In order to take into account this aspect, in [8] the authors modified Eq. (1) as follows

$$x(t) = x_0 \exp\left\{\frac{m}{\beta}\left[e^{-\beta t_0} - e^{-\beta t}\right]\right\}. \tag{2}$$

Equation (2) is the solution of

$$\frac{dx}{dt} = m\,e^{-\beta t}x, \qquad x(t_0) = x_0. \tag{3}$$

We note explicitly that the limit for long time of (2) depends from the initial size of the population being $K = x_0 \exp\{m/\beta\,e^{-\beta t_0}\}$.

In this paper, we consider the stochastic diffusion process associated to (2), denoted by $X(t)$. Then we derive the process $X_C(t)$ by modifying the growth parameter m to introduce the effect of a *therapy* interpreted as a continuous time dependent function $C(t)$. We study both the processes by focusing on the First Passage Time (FPT) problem. Moreover, in experimental studies the effect of a new therapy has to be tested so the term $C(t)$ is usually unknown. The knowledge of such functional form is fundamental since it allows to introduce an external control to the system and to explain how the therapy acts. Further, the study of some problems related to the process $X_C(t)$, i.e. modeling and forecasting, requires the knowledge of $C(t)$. For these reasons, the functional form of $C(t)$ has to be estimated. In this direction we propose a two-step estimation procedure applicable when data from a control group, modeled by means of $X(t)$, and from one or more treated groups, described by $X_C(t)$, are available. In the first step the parameters of the control group, are estimated by maximum likelihood method (see [8,9]). In the second step the function $C(t)$ is estimated making use of relationships between the processes $X(t)$ and $X_C(t)$. Finally, in order to evaluate the goodness of the proposed procedure a simulation study is presented.

The paper is organized as follows. In Sect. 2 the stochastic model $X_C(t)$ is introduced, the transition distribution and the related moments are provided. In Sect. 3 we study the FPT through suitable boundaries of interest in real applications. In Sect. 4 a two-step procedure is proposed to estimate the parameters of $X_C(t)$. A simulation is also provided to validate the procedure. Some conclusions close the paper.

2   The Stochastic Model

In the following we consider the stochastic version of the Eq. (3). Precisely, let $X(t)$ be a stochastic process defined in $\mathbb{R}^+$ described by the following stochastic differential equation (SDE)

$$dX(t) = m\,e^{-\beta t}X(t)dt + \sigma X(t)\,dW(t), \qquad X(t_0) = x_0 \text{ a.s.} \qquad (4)$$

It can be obtained from (3), following a standard procedure (see, for instance [5]). The parameters m and β are positive constants that represent the growth rates of the population $X(t)$, $\sigma > 0$ is the width of random fluctuations and $W(t)$ represents a standard Brownian motion.

In real-life situations, intrinsic growth rates of the population can be modify by means of exogenous terms generally depending on time. Examples of such situations could be suitable food treatments in growth of animals (see [1]) or therapeutic treatments in tumor growth (see for instance [2,3,7]). In order to model such situations, we modify the drift of $X(t)$ by introducing a continuous time dependent function $C(t)$ to model the effect of an exogenous factor, namely therapy, obtaining the stochastic process $X_C(t)$ described by the following SDE

$$dX_C(t) = [m - C(t)]e^{-\beta t}X_C(t)dt + \sigma X_C(t)dW(t), \qquad X_C(t_0) = x_0 \text{ a.s.} \quad (5)$$

The solution of (5) is a diffusion process with sample paths

$$X_C(t) = x_0 \exp\left\{\int_{t_0}^{t}[m - C(s)]\,e^{-\beta s}\,ds - \frac{\sigma^2}{2}(t - t_0) + \sigma\,[W(t) - W(t_0)]\right\}.$$
$$(6)$$

Clearly, the solution of (4) can be obtained by (6) choosing $C(t) = 0$. Equation (5) defines a stochastic diffusion process taking values in $\mathbb{R}^+$, characterized by drift and infinitesimal variance given by

$$A_1^C(x,t) = [m - C(t)]e^{-\beta t}x, \qquad A_2^C(x) = \sigma^2 x^2,$$

respectively. Let $f_C(x,t|y,\tau) = \frac{\partial}{\partial x}P[X_C(t) \leq x|X_C(\tau) = y]$ be the transition probability density function (pdf) of $X_C(t)$. The function $f_C(x,t|y,\tau)$ is solution of the Fokker-Planck equation:

$$\frac{\partial f_C(x,t|y,\tau)}{\partial t} = -[m - C(t)]e^{-\beta t}\frac{\partial}{\partial x}[xf_C(x,t|y,\tau)] + \frac{\sigma^2}{2}\frac{\partial}{\partial x^2}[x^2 f_C(x,t|y,\tau)] \quad (7)$$

and of the Kolmogorov equation:

$$\frac{\partial f_C(x,t|y,\tau)}{\partial \tau} + [m - C(\tau)]e^{-\beta \tau}y\frac{\partial f_C(x,t|y,\tau)}{\partial y} + \frac{\sigma^2 y^2}{2}\frac{\partial f_C(x,t|y,\tau)}{\partial y^2} = 0. \quad (8)$$

Furthermore, $f_C(x,t|y,\tau)$ satisfies the initial delta condition: $\lim_{t \to \tau} f_C(x,t|y,\tau) = \lim_{\tau \to t} f_C(x,t|y,\tau) = \delta(x - y)$. Note that the transformation

$$z = \ln x + d(t), \qquad z_0 = \ln y + d(\tau)$$

with

$$d(t) = \frac{\sigma^2}{2}t + \frac{m}{\beta}e^{-\beta t} + \int^{t}C(\theta)e^{-\beta \theta}d\theta,$$

reduces (7) and (8) to the analogous equations for a Wiener process $Z(t)$ defined in $\mathbb{R}$ with drift and infinitesimal variance $B_1 = 0, B_2 = \sigma^2$, respectively. So one obtains

$$f_C(x,t|y,\tau) = \frac{1}{x\sqrt{2\pi\sigma^2(t-\tau)}} \exp\left\{-\frac{[\ln(x/y)+d(t)-d(\tau)]^2}{2\sigma^2(t-\tau)}\right\}.$$

Moreover one has

$$E[X_C^k(t)|X_C(\tau) = y] = y\,\exp\left\{-k\left[\frac{m}{\beta}\left(e^{-\beta t} - e^{-\beta\tau}\right)\right.\right.$$
$$\left.\left. + \int_\tau^t C(\theta)e^{-\beta\theta}\,d\theta\right] + \frac{k(k-1)}{2}\sigma^2(t-\tau)\right\}.$$

3   First Passage Time Through a Single Boundary

Let

$$T_{x_0,S(t)}^C = \begin{cases} \inf\limits_{t\geq t_0}\{t \,:\, X_C(t) > S(t)|X_C(t_0) = x_0\}, & x_0 < S(t_0) \\ \inf\limits_{t\geq t_0}\{t \,:\, X_C(t) < S(t)|X_C(t_0) = x_0\}, & x_0 > S(t_0) \end{cases}$$

be the FPT of $X_C(t)$ through the boundary $S(t)$ and let $g_C[S(t),t|x_0,t_0]$ be the FPT pdf. If $S(t) \in C^2[t_0, +\infty)$ the FPT pdf $g_C[S(t),t|x_0,t_0]$ is solution of the following second kind Volterra integral equation:

$$g_C[S(t),t|x_0,t_0] = 2\rho\left[\psi_C[S(t),t|x_0,t_0] - \int_{t_0}^t g_C[S(\tau),\tau|x_0,t_0]\,\psi_C[S(t),t|S(\tau),\tau]d\tau\right]$$

with

$$\psi_C[S(t),t|y,\theta] = \frac{f_C[S(t),t|y,\theta]}{2}\left\{S'(t) - [m - C(t)]e^{-\beta t}S(t)\right.$$
$$\left. - S(t)\frac{\ln[S(t)/S(\theta)] - \int_\theta^t[m - C(s)]e^{-\beta s}\,ds}{t-\theta}\right\}$$

Note that if

$$S(t) = A\exp\left\{Bt + \int^t [m - C(s)]\,e^{-\beta s}\,ds\right\}$$

with $A > 0$ and $B \in \mathbb{R}$, then $\psi_C[S(t),t|S(\tau),\tau] = 0, \forall\tau \in [t_0,t]$, so that the FPT pdf can be expressed in the following closed form:

$$g_C[S(t),t|x_0,t_0] = \frac{\left|\ln\frac{S(t_0)}{x_0}\right|}{\sqrt{2\pi\sigma^2(t-t_0)^3}}\exp\left\{-\frac{\left[(\frac{\sigma^2}{2}+B)(t-t_0)+\ln\frac{S(t_0)}{x_0}\right]^2}{2\sigma^2(t-t_0)}\right\}, \quad S(t_0) \neq x_0.$$

Moreover, by choosing in (9) $C(t) = Be^{\beta t}$ and $A = px_0\exp\left\{\frac{m}{\beta}e^{-\beta t_0}\right\}$, one has

$$S(t) = pE[X(t)|X(t_0) = x_0] = px_0\exp\left\{\frac{m}{\beta}\left(e^{-\beta t_0} - e^{-\beta t}\right)\right\} \qquad (9)$$

that, for $0 < p < 1$, represents a percentage of the mean of the process $X(t)$. In other words, for the process $\{X_C(t); t_0 \leq t\}$ characterized by infinitesimal moments

$$A_1^C(x,t) = [m - Be^{\beta t}]e^{-\beta t}x, \quad A_2^C(x) = \sigma^2 x^2, \tag{10}$$

the FPT pdf through boundary (9) is given by

$$g_C[S(t), t|x_0, t_0] = \frac{|\ln p|}{\sqrt{2\pi\sigma^2(t - t_0)^3}} \exp\left\{-\frac{[(\sigma^2/2 + B)(t - t_0) + \ln p]^2}{2\sigma^2(t - t_0)}\right\}. \tag{11}$$

For the process defined in (10) with $m = 0.75, \beta = 0.18, \sigma = 0.07$, in Fig. 1 the FPT pdf (11) is plotted for $p = 0.7$ (on the left) and $p = 0.85$ (on the right) for various choices of B.

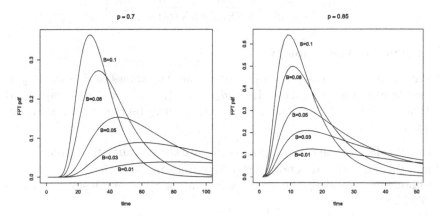

Fig. 1. FPT pdf (11) for $X_C(t)$ in (10) with $m = 0.75, \beta = 0.18, \sigma = 0.07$ through $S(t)$ in (9) for $p = 0.7$ (left) and $p = 0.85$ (right) and some values of B.

4   Inference

In this section we propose a two step estimation procedure that can be used when data from a control group and from one or more treated groups are available. In the first step, from the control group, modeled by means of $X(t)$, the parameters m, β and σ are estimated by maximum likelihood method (see [8,9]). In the second step the function $C(t)$ is estimated making use of some relationships relating the process $X(t)$ describing the control group, i.e. an untreated group, and $X_C(t)$ modeling the treated group. The idea is to take the model $X(t)$ as a starting point and then to use the information provided by the treated group to fit the function $C(t)$. Therefore, after estimating the parameters of $X(t), C(t)$ is estimated by the trajectories of the treated and non treated groups by means of suitable relations between the two models.

In order to relate the trajectories of the processes $X_C(t)$ and $X(t)$, we assume that $X_C(t_0) = X(t_0) = x_0$, i.e. the therapy is applied from time t_0, so that from Eq. (6) one obtains:

$$X_C(t) = \exp\left\{ -\int_{t_0}^{t} C(s)\, e^{-\beta s} ds \right\} X(t). \tag{12}$$

From (12), looking at the conditional mean functions, we find

$$E[X_C(t)|X_C(t_0) = x_0] = \exp\left\{ -\int_{t_0}^{t} C(s)\, e^{-\beta s} ds \right\} E[X(t)|X(t_0) = x_0]$$

from which we have

$$C(t) = -e^{\beta t} \frac{d}{dt} \ln\left(\frac{E[X_C(t)|X_C(t_0) = x_0]}{E[X(t)|X(t_0) = x_0]} \right).$$

4.1   Proposed Methodology

The data required for the proposed strategy are the values of d_1 sample paths of a non-treated group $(x_{ij}, i = 1, \ldots, d_1,\ j = 1, \ldots, n)$ and d_2 sample paths of a treated group $(x_{ij}^C, i = 1, \ldots, d_2,\ j = 1, \ldots, n)$, observed in the same time instants $t_1, \ldots, t_n$.

- From the data of the control group, estimate the parameters of process $X(t)$. From this first step, we obtain ML estimations $\widehat{m}, \widehat{\beta}$ and $\widehat{\sigma}^2$.
- Denoting by x_j and x_j^C the means of x_{ij} and of x_{ij}^C at any instant t_j, respectively, i.e.

$$x_j = \frac{1}{d_1} \sum_{i=1}^{d_1} x_{ij}, \qquad x_j^C = \frac{1}{d_2} \sum_{i=1}^{d_2} x_{ij}^C,$$

  we obtain

$$\gamma_j = -\ln\left[\frac{x_j^C}{x_j} \right].$$

- Interpolate the points γ_j for $j = 1, 2, \ldots n$ (for example by using cubic spline interpolation) obtaining the function $\widehat{\gamma}(t)$. Finally, consider the following function as an approximation of $C(t)$.

$$\widehat{C}(t) = -e^{\widehat{\beta} t}\, \widehat{\gamma}\,'(t).$$

4.2   A Simulation Study

In order to evaluate the goodness of the proposed procedure we present a simulation study. We consider some specific functions for therapies: constant, linear, logarithmic and periodic. 50 sample paths of the control group $X(t)$ with $m = 0.1$, $\beta = 0.01$, $\sigma = 0.01$ and $t_0 = 0$ have been simulated assuming a random initial state x_0 chosen according $\Lambda(1, 0.16)$.

The paths include 300 observations of the process starting from $t_1 = t_0 = 0$ with $t_i - t_{i-1} = 2$.

The first step of the procedure gives the estimation of the control group parameters: $\widehat{m} = 0.09578, \widehat{\beta} = 0.01085$ and $\widehat{\sigma} = 0.0157$. Hence, as control group we consider the stochastic process $X(t)$ with infinitesimal moments

$$A_1(x,t) = 0.09578\,e^{-0.01085\,t}\,x, \qquad A_2(x) = 0.0157^2\,x^2. \tag{13}$$

Then, 50 sample paths of the treated group $X_C(t)$ have been simulated with $X_C(t_0) = X(t_0) = x_0$ and considering the following therapies: $C(t) = -0.005, C(t) = \pm 0.001t, C(t) = 0.005\sin(t), C(t) = 0.02\ln(1 + 0.15t)$. The results obtained by applying the proposed procedure are shown in Fig. 2. The dashed curves represent the functions $C(t)$ whereas the full curves represent the corresponding estimation $\widehat{C}(t)$. We note explicitly that in the considered cases the proposed procedure is able to capture the trend of $C(t)$. To evaluate the performance of the proposed procedure we calculate the mean absolute error (MAE), root mean square error (RMSE) and

$$d = 1 - \frac{\sum_{i=1}^{N}(\widehat{C}(t_i) - C(t_i))^2}{\sum_{i=1}^{N}\left(|\widehat{C}(t_i) - \overline{C(t)}| + |C(t_i) - \overline{C(t)}|\right)^2}$$

where N is the number of estimated values for the considered cases and $\overline{C(t)}$ is the mean of function $C(t)$. The results are shown in Table 1 for the aforementioned cases. For all the choices of the function $C(t)$ the procedure provides very satisfactory estimates of the function $C(t)$.

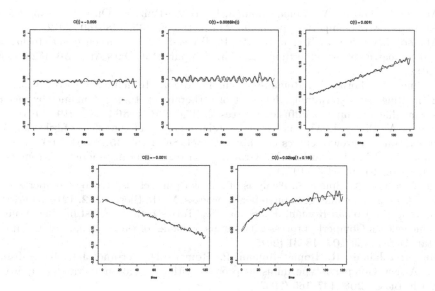

Fig. 2. For the process (13), $C(t)$ and its estimate are shown for different cases.

Table 1. MAE, RMSE and d for the considered therapies.

$C(t)$	MAE	RMSE	d
-0.005	0.095	0.216	0.999
$0.005\sin t$	0.081	0.169	0.995
$0.001t$	0.0918	0.219	0.999
$-0.001t$	0.085	0.175	0.999
$0.02\log(1+0.15t)$	0.086	0.181	0.998

5   Conclusions

We have analyzed a non-homogeneous Gompertz-type stochastic diffusion processes characterized by a carrying capacity depending on the initial state. For such a process we have considered a perturbation of a growth parameter by introducing the effect of a exogenous term $C(t)$ generally depending on time and we have studied the first passage time of the process through a time dependent boundary. Moreover, a two-step procedure has been proposed in order to estimate the model parameters and to fit the function $C(t)$ when data from a control group and one or more treated groups are available. Our simulation study has shown that the proposed procedure is able to capture the trend of $C(t)$ in a very satisfactory way.

References

1. Albano, G., Giorno, V., Román-Román, P., Torres-Ruiz, F.: On the therapy effect for a stochastic Gompertz-type model. Math. Biosci. **235**, 148–160 (2012)
2. Albano, G., Giorno, V., Román-Román, P., Torres-Ruiz, F.: Inference on a stochastic two-compartment model in tumor growth. Comput. Stat. Data Anal. **56**, 1723–1736 (2012)
3. Albano, G., Giorno, V., Román-Román, P., Román-Román, S., Torres-Ruiz, F.: Estimating and determining the effect of a therapy on tumor dynamics by means of a modified Gompertz diffusion process. J. Theor. Biol. **364**, 206–219 (2015)
4. Blasco, A., Piles, M., Varona, L.: A Bayesian analysis of the effect of selection for growth rate on growth curves in rabbits. Genet. Sel. Evol. **35**, 21–41 (2003)
5. Capocelli, R.M., Ricciardi, L.M.: Growth with regulation in random environment. Kibernetik **15**, 147–157 (1974)
6. Di Crescenzo, A., Spina, S.: Analysis of a growth model inspired by Gompertz and Korf laws, and an analogous birth-death process. Math. Biosci. **282**, 121–134 (2016)
7. Giorno, V., Román-Román, P., Spina, S., Torres-Ruiz, F.: Estimating a non-homogeneous Gompertz process with jumps as model of tumor dynamics. Comput. Stat. Data Anal. **107**, 18–31 (2017)
8. Gutiérrez-Jáimez, R., Román-Román, P., Romero, D., Serrano, J.J., Torres-Ruiz, F.: A new Gompertz-type diffusion process with application to random growth. Math. Biosci. **208**, 147–165 (2007)
9. Román-Román, P., Romero, D., Rubio, M.A., Torres-Ruiz, F.: Estimating the parameters of a Gompertz-type diffusion process by means of simulated annealing. Appl. Math. Comput. **218**, 5121–5131 (2012)

Tsallis and Kaniadakis Entropy Measures
for Risk Neutral Densities

Muhammad Sheraz[1]([✉]) [iD], Vasile Preda[2,3] [iD], and Silvia Dedu[4,5] [iD]

[1] Department of Mathematical Sciences, Department of Economics and Finance,
Institute of Business Administration Karachi, KU Circular Road,
Karachi 75270, Pakistan
msheraz@iba.edu.pk
[2] National Institute for Economic Research "Costin C. Kiritescu",
Romanian Academy, 13 Septembrie 13, 050711 Bucharest, Romania
[3] "Gheorghe Mihoc-Caius Iacob" Institute of Mathematical Statistics
and Applied Mathematics of Romanian Academy,
13 Septembrie 13, 050711 Bucharest, Romania
[4] Department of Applied Mathematics,
Bucharest University of Economic Studies,
Calea Dorobanți 15-17, 010552 Bucharest, Romania
silvia.dedu@csie.ase.ro
[5] School of Advanced Studies of the Romanian Academy,
Calea Victoriei 125, 010071 Bucharest, Romania

Abstract. Concepts of Econophysics are usually used to solve problems related to uncertainty and nonlinear dynamics. The risk neutral probabilities play an important role in the theory of option pricing. The application of entropy in finance can be regarded as the extension of both information entropy and probability entropy. It can be an important tool in various financial issues such as risk measures, portfolio selection, option pricing and asset pricing. The classical approach of stock option pricing is based on Black-Scholes model, which relies on some restricted assumptions and contradicts with modern research in financial literature. The Black-Scholes model is governed by Geometric Brownian Motion and is based on stochastic calculus. It depends on two factors: no arbitrage, which implies the universe of risk-neutral probabilities and parameterization of risk-neutral probability by a reasonable stochastic process. Therefore, risk-neutral probabilities are vital in this framework. The Entropy Pricing Theory founded by Gulko represents an alternative approach of constructing risk-neutral probabilities without depending on stochastic calculus. Gulko applied Entropy Pricing Theory for pricing stock options and introduced an alternative framework of Black-Scholes model for pricing European stock options. In this paper we derive solutions of maximum entropy problems based on Tsallis, Weighted-Tsallis, Kaniadakis and Weighted-Kaniadakies entropies, in order to obtain risk-neutral densities.

Keywords: Entropy measures · Risk neutral densities
Entropy pricing theory · Tsallis entropy · Kaniadakis entropy

© Springer International Publishing AG 2018
R. Moreno-Díaz et al. (Eds.): EUROCAST 2017, Part II, LNCS 10672, pp. 55–63, 2018.
https://doi.org/10.1007/978-3-319-74727-9_7

1   Introduction

The concept of entropy plays crucial role to extract universal features of a system from its microscopic details. In statistical mechanics entropy is defined as the logarithm of total number of microstates multiplied by a constant coefficient or alternatively it is written in terms of the probability to occupy the microstates.

The Shannon entropy [14] can be used in particular manners to evaluate the entropy of probability density distribution around some points, which model specific events like deviation from the mean or any sudden news in the case of stock market [16]. At this point additional information is needed. The concept of entropy can be generalized. In 1988 Tsallis introduced a new entropy measure, which successfully describes the statistical features of complex systems [18]. Some other examples of entropy measures that depend on power of the probability introduced by generalizing Shannon entropy include Kaniadakis [9], Rényi [12], Shafee [15] and Ubriaco [19] entropy measures. The application of entropy in Finance can be regarded as the extension of both information entropy and probability entropy. Since last two decades, it has become a very important tool for portfolio selection and asset pricing. In mathematical finance, a risk-neutral measure, also called an equivalent martingale measure, is heavily used in the pricing of financial derivatives. In the theory of option pricing the risk-neutral densities plays very important role and stochastic calculus is vital in this framework. The Entropy Pricing Theory (EPT) was introduced by Gulko as an alternative method for constructing risk-neutral probabilities without relying on stochastic calculus [6, 7]. The famous Black-Scholes model [1] assumes the condition of no arbitrage, which implies the universal risk-neutral probabilities. The uniqueness of these risk-neutral probabilities is crucial. The stock price process is controlled by Geometric Brownian Motion in Black-Scholes model and in this framework stochastic calculus is vital. The EPT provides an alternative approach to construct risk-neutral probabilities without depending on stochastic calculus.

The Principle of Maximum Entropy was used to estimate the distribution of an asset from a set of option prices [8]. Beside this work, the maximum entropy principle was used to retrieve the risk-neutral density of future stock risks or other asset risks [13]. Preda and Sheraz [10] have used recently the Shafee entropy measure for obtaining risk-neutral densities. Recently Preda et al. [11] used Tsallis and Kaniadakis entropy measures for the case of semi-Markov regime switching interest rate models. The maximum entropy approximations were used for the cases of assessing stock market volatility, criterion to select a pricing measure for solving the valuation problem in incomplete markets and ecological bias and there exist many other examples for the use of entropy measures in various fields [4, 18].

In this paper we use Tsallis and Kaniadakis entropy measures to derive the risk-neutral densities using the framework of EPT for stock options. We introduce the weighted cases for both Tsalllis and Kanidakis entropy measures. In Sect. 2 we present the introduction of EPT and Sect. 3 is dedicated to the formulation of our problems, and then we further develop the structure to obtain our new results. In Sect. 4 we present our results using Tsallis and Weighted Tsallis entropies to get risk-neutral density of stock options. In Sect. 5 we present our results by using Kaniadakis and Weighted Kaniadakis entropy measures for the underlying entropy maximization problems. Section 6 concludes our results.

2 Risk Neutral Densities

Risk managers are always interested to use set of tools such that risks that are not profitable can be avoided. One of the most common market risk measures used is volatility, evaluated by standard deviation. The concept of risk involves uncertainty, which is represented by the probability of scenario and its consequence resulting from the happening of the scenario. Financial instruments such as derivatives enables firms and risk managers to manage the risk. These instruments promise payoffs that are derived the underlying which include goods prices, stock prices, exchange rates, bond prices, interest rates, and stock market indices. We consider the European Options (financial derivative) to obtain Risk-Neutral-Densities (RND). European call (put) option are based on a stock provides its buyer with a right to purchase (sell) a predetermined amount of stocks at a contracted price ("strike price or exercise price") on or by a specific date ("maturity").

In finance, particularly pricing derivatives the estimation of correct RND implied by the option prices, remains one of the most important problem. Most of the theoretical and empirical studies, which are aimed to improve the performance of the Black-Scholes model, have focused on recovering the correct RND implied by option prices [13]. Risk-Neutral Densities (RND) are asset price distributions in a risk-neutral world at a future point of time estimated today; also referred to as risk-neutral probability density functions as investors are risk-neutral. If $f(S_T)$ is the RND, then the value of the call option in the Black-Scholes framework is given by:

$$\text{Call} = e^{-r(T-t)} \int_K^\infty f(S_T)(S_T - K)\mathrm{d}S_T \tag{1}$$

In the above equation r, K, T, S_T denote risk free interest rate, strike price, time to maturity and stock price at maturity respectively. In a discrete time setting if there are n possible states for S_T and $\pi_1, \pi_2, \ldots, \pi_n$ are affiliated probabilities to states of S_T where $\sum_{i=1}^n \pi_i = 1$, therefore we can write under the risk neutral framework the price of the call option is given by:

$$\text{Call} = \sum_{i=1}^n \pi_i(\max(S_T - K, 0))/r \tag{2}$$

The expected value of the underlying price S_T will give us the mean of the RND. We can use RND to price other derivatives written on the same asset, hedging derivatives and adjusting interest rates. Usually parametric and non-parametric methods are used to recover the RND. Typically lognormal or mixture of a lognormal functions are used in parametric methods while non-parametric methods use polynomials and approximations such as Edgeworth expansion.

Information theory bases on Shannon entropy is the most efficient way to store an information I out of given set of i possible information is to code every information using $\log_b\left(\frac{1}{f(\mathrm{I})}\right)$, where $f(\mathrm{I})$ is the information relative frequency. The idea from information theory is, that if our assumption shall include the least knowledge possible, the information generated by some random draw should be maximal. Formally we look for some density f for which the expression

$$\mathrm{H}(X) = \sum_{x \in \mathbf{X}} f(x) \ln\left(\frac{1}{f(x)}\right) \tag{3}$$

where $\mathbf{X}$ is the sample space of the random variable X. Equivalently in continuous form is given by

$$\mathrm{H}(X) = \int_{\mathbf{X}} f(x) \ln\left(\frac{1}{f(x)}\right) dx. \tag{4}$$

Since introduced in 1988 [18], the Tsallis entropy attracted a special interest inside the scientific community and for practitioners as well. Now we give the definition of the relative Tsallis entropy.

Definition 1 [17]. The Tsallis entropy of order q corresponding to a probability measure p on a finite set $\mathbf{X}$ is given by:

$$\mathrm{H}_q(p) = \frac{1}{1-q}\left[1 - \sum_{i \in \mathbf{X}} p_i^q\right] \tag{5}$$

For $q \to 0$ this measure reduces to Shannon entropy.

Definition 2 [9]. The Kaniadakis entropy of order k corresponding to a probability measure p on a finite set $\mathbf{X}$ is given by:

$$\mathrm{H}_k(p) = -\sum_{i \in \mathbf{X}} \frac{p_i^{k+1} - p_i^{-k+1}}{2k}. \tag{6}$$

For $k \to 0$ this entropy reduces to Shannon entropy.

3   Problem Formulation

According to Gulko [6], the term market belief is vital in option pricing and current price of any risky asset indicates this belief. The future picture of the market up (down) reflects a state of maximum possible uncertainty therefore market belief for the future performance of an efficient prices can be characterized by maximum uncertainty. Consider a risky asset on time interval $[0, T]$: We denote by Y_T the asset price process at time T, S_T the asset price at future time T, $\mathfrak{p}$ the state space, a subset of real line $\mathbb{R}$, $g(S_T)$ a probability density on $\mathfrak{p}$, $f(S_T)$ the efficient market belief and $\mathrm{H}(g)$ is the index of market uncertainty about Y_T. The index $\mathrm{H}(g)$ is defined on the set of beliefs $g(S_T)$.

Therefore, the efficient market belief $f(S_T)$ maximizes H(g) and given H(g) with some relevant information about current price of S we can determine $f(S_T)$. The index of the market uncertainty about Y_T is H(g) and it will be modeled by using Tsallis, Weighted-Tsallis, Kaniadakis and Weighted-Kaniadakies entropies. The Tsallis and Kaniadakis entropy measures are given by

$$H^T(g) = -\frac{1}{1-q} E^g \left[g^{q-1}(Y_T) - 1 \right] \tag{7}$$

and

$$H^K(g) = -E^g \left[\frac{g^k(Y_T) \quad g^{-k}(Y_i)}{2k} \right] \tag{8}$$

respectively, where $q \neq 1$ and $k \neq 0$. For $q \to 1$ or $k \to 0$ we get the Shannon entropy. The Tsallis logarithm with parameter q is given by:

$$\log_{\{q\}}^T(x) = \frac{x^{q-1} - 1}{q - 1}. \tag{9}$$

The Kaniadakis logarithm with parameter $k \in [-1, 1]$ is given by:

$$\log_{\{k\}}^K(x) = \frac{x^{k-1} - x^{k-1}}{k}. \tag{10}$$

The term H(g) is a functional defined on the set of market beliefs (S_T), where q and k are Tsallis and Kaniadakis parameters respectively. The market belief that maximizes H(g) is called the maximum entropy belief and H(g) is called the entropy of Y_T, which is used to evaluate the degree of uncertainty of $g(S_T)$. The weighted entropy was first introduced by Belis and Guiasu [3], by considering the two basic concepts of objective probability and subjective utility. Guiasu [5] derived the principle of maximum information obtaining the probability distribution maximizing the weighted Shannon entropy. We introduce similarly the weighted Tsallis and weighted Kaniadakis entropies for our problem, respectively

$$H^{WT}(g) = -\frac{1}{1-q} E^g \left[u(Y_T)(g^{q-1}(Y_T) - 1) \right] \tag{11}$$

$$H^{WK}(g) = -E^g \left[u(Y_T) \frac{g^k(Y_T) - g^{-k}(Y_T)}{2k} \right] \tag{12}$$

where $u > 0$ are positive weights. The maximum entropy characterizes the market beliefs regardless of the subjective or aggregate risk preferences and it is useful to determine the risk neutral beliefs in incomplete arbitrage free markets. Consider I the prior information set available. Then the maximum entropy market belief $f(S_T)$ as a solution to the maximum entropy problem can be written as follows:

$$f = \arg\max[H(g), g \in \mathcal{G}]. \tag{13}$$

where $\mathcal{G}$ is the set of density functions and in order to determine the maximum entropy belief $f(S_T)$ derived from the prior information I, let us consider one risky asset S and riskless bond with price P, þ the price space and T is the terminal time, σ^2 the variance of random prices S_T.

4 Risk Neutral Densities Using Tsallis Entropy

We denote by $f(S_T)$ the unknown density function to be determined. More specifically, if $f(S_T)$ is the risk-neutral probability density function for the terminal price of the underlying asset satisfying the given constraints, where $g(S_T) > 0$ on þ: The constraints involved in the next theorem are density normalization, the risk neutral pricing and variance respectively and þ is a continuous subset of the real line for example the stock price þ $= [0, 1)$ and for normalized bond prices þ $= [0, +\infty)$. We consider the analogue of the discrete case of Tsallis entropy. We suppose that all expectations are also well defined and underlying optimization problems admit solutions for some continuous cases. More details can be found in [2]. We consider some constrained entropy optimization problems in order to derive risk neutral densities.

Theorem 1. The unique solution of the Tsallis entropy maximization problem

$$\max_{g} -E^g[\log_{\{q\}} g(Y_T)] \tag{14}$$

subject to the following constraints

$$E^g\left[I_{\{Y_T > 0\}}\right] = 1 \tag{15}$$

$$E^g[(Y_T)] = \frac{S}{P} \tag{16}$$

$$E^g\left[Y_T^2\right] = \sigma^2 + \left(\frac{S}{P}\right)^2 \tag{17}$$

is given by:

$$f(S_T) = \left(\frac{1 + q(\lambda_0 + \lambda_1 S_T + \lambda_2 S_T^2)}{q - 1}\right)^{\frac{1}{q}}, \tag{18}$$

where $\lambda_0, \lambda_1, \lambda_2$ are chosen so that $f(S_T)$ satisfies the price and variance constraints.
Now we consider the case of Weighted Tsallis entropy maximization problem.

Theorem 2. The unique solution of the Tsallis entropy maximization problem with weights $u > 0$

$$\max_{g} -E^g[u(Y_T)\log_{\{q\}}g(Y_T)]. \tag{19}$$

Subject to constraints (15), (16) and (17) is given by:

$$f(S_T) = \left(\frac{u(S_T) + q(\lambda_0 + \lambda_1 S_T + \lambda_2 S_T^2)}{u(S_T)(q+1)}\right)^{\frac{1}{q}}, \tag{20}$$

where $\lambda_0, \lambda_1, \lambda_2$ are chosen so that $f(S_T)$ satisfies the price and variance constraints.

5 Risk Neutral Densities Using Kaniadais Entropy

In this section we present the solution of maximum entropy problems for the cases of Kaniadakis and Weighted Kaniadakis framework. We consider the following constrained Kaniadakis entropy optimization problem in order to derive risk neutral densities.

Theorem 3. The unique solution of the Kaniadakis entropy maximization problem

$$\max_{g} -E^g[\log_{\{k\}}g(Y_T)]. \tag{21}$$

Subject to the constraints (15), (16) and (17) is given by:

$$f(S_T) = \left(\frac{k(\lambda_0 + \lambda_1 S_T + \lambda_2 S_T^2) + \sqrt{k^2(\lambda_0 + \lambda_1 S_T + \lambda_2 S_T^2 - 1) + 1'}}{k+1}\right)^{\frac{1}{k}}, \tag{22}$$

where $\lambda_0, \lambda_1, \lambda_2$ are chosen so that $f(S_T)$ satisfies the price and variance constraints and $k \neq 0$.

Now we consider the case of Weighted Kaniadakis entropy maximization problem and we look for the solution of risk neutral density $f(S_T)$.

Theorem 4. The unique solution of the Kaniadakis entropy maximization problem with weights $u > 0$

$$\max_{g} -E^g[u(Y_T)\log_{\{k\}}g(Y_T)]. \tag{23}$$

Subject to constraints (15), (16) and (17) is given by:

$$f(S_T) = \left[\frac{k(\lambda_0 + \lambda_1 S_T + \lambda_2 S_T^2) + \sqrt{[(k(\lambda_0 + \lambda_1 S_T + \lambda_2 S_T^2) - 1)] + 1}}{(k+1)u(S_T)}\right]^{\frac{1}{k}}, \tag{24}$$

where $\lambda_0, \lambda_1, \lambda_2$ are chosen so that $f(S_T)$ satisfies the price and variance constraints.

6 Conclusions

In this paper we have obtained solutions for maximum entropy problems constructed in some general frameworks based on Tsallis, weighted Tsallis, Kaniadakis and weighted Kanidakis entropy measures, in order to obtain risk-neutral densities using entropy pricing theory of stock options. The problem of extracting implied volatilities from market price of the options has always attained the concentration of researchers in option pricing but this is the single statistic which can be extracted and depends on the option pricing model. The problem of getting risk-neutral density without specifying any model has become crucial and entropy pricing theory represent an alternative approach to solve such problems. The use of Tsallis and Kaniadakis entropy measures is more general comparative to the classical one, based on Shannon entropy.

Acknowledgements. "This work was partially supported by Ningbo Natural Science Foundation (No. 2016A610077) and K.C. Wong Magna Fund in Ningbo University."

References

1. Black, F., Scholes, M.: The pricing of options and corporate liabilities. J. Polit. Econ. **81**, 637–659 (1973)
2. Borwein, J., Choksi, R., Maréchal, P.: Probability distributions of assets inferred from option prices via the principle of maximum entropy. J. Soc. Ind. Appl. Math. **14**, 464–478 (2003)
3. Belis, M., Guiasu, S.: A quantitative-qualitative measure of information in cybernetic systems. IEEE Trans. Inf. Theory **14**(4), 593–594 (1968)
4. Cressie, N., Richardson, S., Jaussent, I.: Ecological bias: use of maximum entropy approximations. ANZ J. Stat. **46**(2), 233–255 (2004)
5. Guiasu, S.: Weighted entropy. Rep. Math. Phys. **2**(3), 165–179 (1971)
6. Gulko, L.: Dart boards and asset prices: introducing the entropy pricing theory. In: Fomby, T.B., Hill, R.C. (eds.) Advances in Econometrics. JAI Press, Greenwich (1997)
7. Gulko, L.: The Entropy Theory of Bond Option Pricing, Working Paper, Yale School of Management, New Haven, CT, October 1995
8. Guo, W.Y.: Maximum entropy in option pricing: a convex-spline smoothing method. J. Futures Markets **21**, 819–832 (2001)
9. Kaniadakis, G.: Non-linear kinetics underlying generalized statistics. Phys. A **296**, 405–425 (2001)
10. Preda, V., Sheraz, M.: Risk-neutral densities in entropy theory of stock options using lambert function and a new approach. Proc. Rom. Acad. **16**(1), 20–27 (2015)
11. Preda, V., Dedu, S., Sheraz, M.: New measure selection for Hunt-Devolder semi-Markov regime switching interest rate models. Phys. A **407**, 350–359 (2014)
12. Rényi, A.: On measures of entropy and information. In: Proceedings of the 4th Berkely Sympodium on Mathematics of Statistics and Probability, vol. 1, pp. 547–561. Berkeley University Press, Berkeley (1961)
13. Rompolis, L.S.: Retrieving risk neutral densities from European option prices based on the principle of maximum entropy. J. Empir. Finan. **17**, 918–937 (2010)
14. Shanon, C.E., Weaver, W.: The Mathematical Theory of Communication. University of Illinois Press, Urbana (1963)

15. Shafee, F.: Lambert function and a new non-extensive form of entropy. J. Appl. Math. **72**, 785–800 (2007)
16. Sheraz, M., Dedu, S., Preda, V.: Entropy measures for assessing volatile markets. Procedia Econ. Finan. **22**, 655–662 (2015)
17. Tsallis, C.: Possible generalization of Boltzmann-Gibbs statistics. J. Stat. Phys. **52**, 479–487 (1988)
18. Trivellato, B.: Deformed exponentials and applications to finance. Entropy **15**(9), 3471–3489 (2013)
19. Ubriaco, M.R.: Entropies based on fractional calculus. Phys. Lett. A **373**, 2516–2519 (2009)

A Note on Diffusion Processes with Jumps

Virginia Giorno[1] and Serena Spina[2(✉)]

[1] Dipartimento di Informatica, Università di Salerno, Fisciano, SA, Italy
giorno@unisa.it
[2] Dipartimento di Matematica, Università di Salerno, Fisciano, SA, Italy
sspina@unisa.it

Abstract. We focus on stochastic diffusion processes with jumps occurring at random times. After each jump the process is reset to a fixed state from which it restarts with a different dynamics. We analyze the transition probability density function, its moments and the first passage time density. The obtained results are used to study the lognormal diffusion process with jumps which is of interest in the applications.

1 Introduction and Description of the Model

Stochastic processes with jumps play a relevant role in many fields of applications. For example, in [3,7,10], diffusion processes with jumps are studied in order to model an intermittent treatment for tumor diseases, in [4] birth-and-death processes with jumps are analyzed as queuing models with catastrophes, in [5] a non-homogeneous Ornstein-Uhlenbeck with jumps is considered in relation to neuronal activity. In these contexts, a jump is random event that changes the state of the process leading it to another random state from which the dynamics restarts with the same or different law.

We consider diffusion processes assuming that the jumps occur at random times chosen with a fixed probability density function (pdf). After each jump the process is reset to a fixed state from which it restarts with a different dynamics. Let $\{\widetilde{X}_k(t), t \geq t_0 \geq 0\}$ $(k = 0, 1, \ldots)$ be a stochastic diffusion process. Following [6], we construct the stochastic process $X(t)$ with random jumps. Starting from the initial state $\rho_0 = X(t_0)$, the process $X(t)$ evolves according to $\widetilde{X}_0(t)$ until a jump occurs that shifts the process to a state ρ_1. From here, $X(t)$ restarts according to $\widetilde{X}_1(t)$ until another jump occurs resetting the process to ρ_2 and so on. The effect of the k-th jump $(k = 1, 2, \ldots)$ is to shift the state of $X(t)$ in ρ_k. Then, the process evolves like $\widetilde{X}_k(t)$, until a new jump occurs.

$X(t)$ consists of cycles, whose durations, I_1, $I_2, \ldots$, representing the time intervals between two consecutive jumps, are independent random variables distributed with pdf $\psi_k(\cdot)$. We denote by $\Theta_1, \Theta_2, \ldots$ the times in which the jumps occur. We set $\Theta_0 = t_0$ that corresponds the initial time and for $k = 1, 2, \ldots$, let $\gamma_k(\tau)$ be the pdf of the random variable Θ_k. The variables I_k and Θ_k are

---

This work was supported by INDAM-GNCS.

R. Moreno-Díaz et al. (Eds.): EUROCAST 2017, Part II, LNCS 10672, pp. 64–71, 2018.
https://doi.org/10.1007/978-3-319-74727-9_8

related, indeed $\Theta_1 = I_1$ and for $k > 1$ it results $\Theta_k = I_1 + I_2 + \ldots I_k$. Hence, the pdf $\gamma_k(\cdot)$ of Θ_k and the pdf $\psi_k(\cdot)$ of I_k are related, indeed $\gamma_1(t) = \psi_1(t)$ and $\gamma_k(t) = \psi_1(t) * \psi_2(t) * \cdots * \psi_k(t)$, where "$*$" denotes the convolution operator.

In the paper we study $X(t)$ by analyzing the transition pdf, its moments and the first passage time of $X(t)$ through a constant boundary. We consider some particular cases when the inter-jumps I_k are deterministic or exponentially distributed. Finally, the lognormal diffusion process with jumps is studied.

2 Some Probabilistic Features of the Process

Let $f(x,t|y,\tau) = \frac{\partial}{\partial x} P[X(t) \leq x | X(\tau) = y]$, $\tilde{f}_k(x, t|y, \tau) = \frac{\partial}{\partial x} P[\tilde{X}(t) \leq x | \tilde{X}(\tau) = y]$ be the transition pdf's of $X(t)$ and $\tilde{X}_k(t)$, respectively. The densities f and $\tilde{f}_k$ are related. Indeed, considering the age of the process with jumps, we have the following expression of the transition pdf of the process $X(t)$

$$f(x,t|\rho_0,t_0) = \left(1 - \int_0^{t-t_0} \psi_1(s)\,ds\right) \tilde{f}_0(x,t|\rho_0,t_0)$$
$$+ \sum_{k=1}^{\infty} \int_{t_0}^{t} \left(1 - \int_0^{t-\tau} \psi_k(s)ds\right) \tilde{f}_k(x,t|\rho_k,\tau)\,\gamma_k(\tau)\,d\tau. \qquad (1)$$

We analyze the right hand side of (1). The first term represents the case in which there are not jumps in the interval (t_0, t), so that $X(t)$ evolves as $\tilde{X}_0(t)$. The factor $1 - \int_0^{t-t_0} \psi_1(s)\,ds$ represents the probability that the first jump occurs after the time t. The sum in (1) concerns the circumstance that one or more jumps occur in (t_0, t). In this case, the last jump, the k-th one, occurs at the time $\tau \in (t_0, t)$ with probability $1 - \int_0^{t-\tau} \psi_k(s)\,ds$; then $X(t)$ evolves according to $\tilde{X}_k(t)$ to reach x at time t, starting from ρ_k.

Denoting by $m^{(n)}(x,t|y,\tau) = E[X^n(t)|X(\tau) = y]$ and $\tilde{m}_k^{(n)}(x,t|y,\tau) = E[\tilde{X}^n(t)|\tilde{X}(\tau) = y]$ the conditional moments of $X(t)$ and $\tilde{X}_k(t)$, respectively, from (1) it follows

$$m^{(n)}(t|\rho_0,t_0) = \left(1 - \int_0^{t-t_0} \psi_1(s)\,ds\right) \tilde{m}_0^{(n)}(t|\rho_0,t_0)$$
$$+ \sum_{k=1}^{\infty} \int_{t_0}^{t} \left(1 - \int_0^{t-\tau} \psi_k(s)ds\right) \tilde{m}_k^{(n)}(t|\rho_k,\tau)\,\gamma_k(\tau)\,d\tau. \qquad (2)$$

To analyze the first passage time (FPT) of $X(t)$, we consider a state $S > \rho_k$ ($k = 0, 1, 2 \ldots$). For $X(t_0) < S$ we denote by $T_{\rho_0}(t_0) = \inf\{t \geq t_0 : X(t) > S\}$ the random variable FPT of $X(t)$ through S and with $g(S, t|\rho_0, t_0) = \frac{\partial}{\partial t} P\{T_{\rho_0}(t_0) < t\}$. Similarly let $\tilde{T}_{\rho_0}(t_0) = \inf\{t \geq t_0 : \tilde{X}(t) > S\}$ with $\tilde{X}(t_0) < S$ be the FPT for $\tilde{X}(t)$ through S and $g(S, t|\rho_0, t_0) = \frac{\partial}{\partial t} P\{T_{\rho_0}(t_0) < t\}$. Recalling that $X(t)$

consists of independent cycles and that $\widetilde{X}_k(t)$ evolves in I_k, the following relation can be obtained

$$g(S, t|\rho_0, t_0) = \left[1 - \int_0^{t-t_0} \psi(s)\, ds\right] \widetilde{g}_0(S, t|\rho_0, t_0) \tag{3}$$

$$+ \sum_{k=1}^{\infty} \int_{t_0}^{t} \left[1 - \int_0^{t-\tau} \psi(s)ds\right] \widetilde{g}_k(S, t|\rho_k, \tau)\gamma_k(\tau)\, d\tau \left\{\prod_{j=0}^{k-1}\left[1 - P(\widetilde{T}_j(\Theta_j) < \Theta_{j+1})\right]\right\},$$

where the product $\prod_{j=0}^{k-1}\left[1 - P(\widetilde{T}_j(\Theta_j) < \Theta_{j+1})\right]$ represents the probability that none of the processes $\widetilde{X}_0(t), \widetilde{X}_1(t), \ldots, \widetilde{X}_{k-1}(t)$ crosses S before τ.

3   Deterministic Inter-jumps

We assume that the jumps occur at fixed times denoted by $\tau_1, \tau_2, \ldots, \tau_N$. Therefore, $X(t)$ consists of a combination of processes $\widetilde{X}_k(t)$ with $\widetilde{X}_k(\tau_k) = X(\tau_k) = \rho_k$. Assuming that $\tau_0 = t_0$, $\tau_{N+1} = \infty$, one has

$$X(t) = \sum_{k=0}^{N} \widetilde{X}_k(t)\mathbf{1}_{(\tau_k, \tau_{k+1})}(t) \quad \text{with} \quad \mathbf{1}_{(\tau_k, \tau_{k+1})}(t) = \begin{cases} 1, & t \in (\tau_k, \tau_{k+1}) \\ 0, & t \notin (\tau_k, \tau_{k+1}). \end{cases}$$

After the time τ_N, the process $X(t)$ evolves as $\widetilde{X}_N(t)$. For $k = 0, 1, \ldots, N$, $\Theta_k = \tau_k$ a.s. and I_k are degenerate random variables; in particular, denoting by δ the delta Dirac function, the pdf's of Θ_k and of I_k are

$$\gamma_k(t) = \delta(t - \tau_k), \qquad \psi_k(t) = \delta\left[t - (\tau_k - \tau_{k-1})\right],$$

respectively. We note that

$$\int_a^b \delta(s - \tau_k)ds = H(b - a - \tau_k), \qquad H(x) = \int_{-\infty}^{x} \delta(u)\, du = \begin{cases} 0, & x < 0 \\ 1, & x > 0, \end{cases} \tag{4}$$

where $H(\cdot)$ is the Heaviside unit step function. Hence, from (1) one has:

$$f(x, t|\rho_0, t_0) = \left[1 - H(t - \tau_1)\right] \widetilde{f}_0(x, t|\rho_0, t_0)$$

$$+ \sum_{k=1}^{\infty} H(t - \tau_k)\left[1 - H(t - \tau_k - (\tau_k - \tau_{k-1}))\right] \widetilde{f}_k(x, t|\rho_k, \tau_k),$$

from which, recalling (4), it follows:

$$f(x, t|\rho_0, t_0) = \sum_{k=0}^{N} \widetilde{f}_k(x, t|\rho_k, \tau_k)\mathbf{1}_{(\tau_k, \tau_{k+1})}(t). \tag{5}$$

Similarly, from (2) the conditional moments of $X(t)$ can be obtained.

Concerning the FPT problem, we note that since $\Theta_k = \tau_k$ *a.s.*, one has

$$1 - P[\widetilde{T}_k(\tau_k) < \tau_{k+1}] = 1 - \int_{\tau_k}^{\tau_{k+1}} \widetilde{g}_k(S, \tau | \rho_k, \tau_k) d\tau;$$

so, following the procedure used to obtain (5), one has:

$$g(S, t | \rho_0, t_0) = \begin{cases} \widetilde{g}_0(S, t | \rho_0, t_0), & t \in \mathcal{I}_1 \\ \prod_{j=0}^{k-1} \left[1 - \int_{\tau_j}^{\tau_{j+1}} \widetilde{g}_j(S, \tau | \rho_i, \tau_j) d\tau \right] \widetilde{g}_k(S, t | \rho_k, \tau_k), & t \in \mathcal{I}_k \\ & (k = 2, 3, \ldots). \end{cases} \quad (6)$$

4 Exponentially Distributed Inter-jumps

We assume that, for $k \geq 1$, $\rho_k = \rho$ and I_k are identically distributed with pdf $\psi_k(s) = \psi(s) = \xi e^{-\xi s}$ for $s > 0$. In this case the pdf of Θ_k is an Erlang distribution with parameters (k, ξ), i.e. $\gamma_k(t) = \xi^k t^{k-1} e^{-\xi t} / (k-1)!$ for $t > 0$. From (1) and (2) the transition pdf and the conditional moments of $X(t)$ result:

$$f(x, t | \rho, t_0) = e^{-\xi(t-t_0)} \widetilde{f}_0(x, t | \rho, t_0) + e^{-\xi t} \sum_{k=1}^{\infty} \int_{t_0}^{t} \frac{\xi^k \tau^{k-1}}{(k-1)!} \widetilde{f}_k(x, t | \rho, \tau) d\tau, \quad (7)$$

$$m^{(n)}(t | \rho, t_0) = e^{-\xi(t-t_0)} \widetilde{m}_0^{(n)}(t | \rho, t_0) + e^{-\xi t} \sum_{k=1}^{\infty} \int_{t_0}^{t} \frac{\xi^k \tau^{k-1}}{(k-1)!} \widetilde{m}_k^{(n)}(t | \rho, \tau) d\tau, \quad (8)$$

respectively. Moreover, concerning the FPT pdf, from (3) one has:

$$g(S, t | \rho, t_0) = e^{-\xi(t-t_0)} \widetilde{g}(S, t | \rho, t_0)$$
$$+ \sum_{k=1}^{\infty} \int_{t_0}^{t} \frac{(\xi \tau)^{k-1} e^{-\xi \tau}}{(k-1)!} \xi e^{-\xi(t-\tau)} \widetilde{g}_k(S, t | \rho, \tau) d\tau \left\{ \prod_{j=0}^{k-1} \left[1 - P(\widetilde{T}_j(\Theta_j) < \Theta_{j+1}) \right] \right\}. \quad (9)$$

We assume that each $\widetilde{X}_k(t)$ evolves as $\widetilde{X}(t)$, from (7) and (8) one obtains:

$$f(x, t | \rho, t_0) = e^{-\xi(t-t_0)} \widetilde{f}(x, t | \rho, t_0) + \xi \int_{t_0}^{t} e^{-\xi(t-\tau)} \widetilde{f}(x, t | \rho, \tau) d\tau; \quad (10)$$

$$m^{(n)}(t | \rho, t_0) = e^{-\xi(t-t_0)} \widetilde{m}^{(n)}(t | \rho, t_0) + \xi \int_{t_0}^{t} e^{-\xi(t-\tau)} \widetilde{m}^{(n)}(t | \rho, \tau) d\tau, \quad (11)$$

in agreement with the analogue results in [1,2]. Moreover, if the involved processes are time homogeneous, one has that $P(\widetilde{T}_j(\Theta_j) < \Theta_{j+1}) = P(\widetilde{T}(0) <$

$I_{j+1}) = P(\widetilde{T}(0) < I)$, where $\widetilde{T}(0)$ is the FPT of $\widetilde{X}_0(t)$ through the threshold S and $I_k \overset{d}{=} I$. Therefore, Eq. (9) becomes:

$$g(S, t - t_0|\rho) = e^{-\xi(t-t_0)}\widetilde{g}(S, t - t_0|\rho) + \sum_{k=1}^{\infty} \int_0^{t-t_0} \frac{\left(\xi\tau\left[1 - P(\widetilde{T}(0) < I)\right]\right)^{k-1}}{(k-1)!}$$

$$\times \; xie^{-\xi t}\widetilde{g}(S, t - \tau|\rho)d\tau \left[1 - P(\widetilde{T}(0) < I)\right]$$

$$= e^{-\xi(t-t_0)}\widetilde{g}(x, t - t_0|\rho)$$

$$+ \xi\left[1 - P(\widetilde{T}(0) < I)\right]e^{-\xi t}\int_0^{t-t_0} e^{\xi\tau\left[1 - P(\widetilde{T}(0) < I)\right]}\widetilde{g}(S, t - \tau|\rho)d\tau. \quad (12)$$

5   The Lognormal Process with Jumps

We construct a new process with jumps on the lognormal process. This is an interesting process to study because it and its transformations are largely used in the applications. For example, in [8] a gamma-type diffusion process is transformed in a lognormal process to model the trend of the total stock of the private car-petrol. So, the study is performed on a lognormal process to provide a statistical methodology by which it can be fitted real data and obtain forecasts that, in statistical term, are quite accurate. In this context, a process with jumps can take into consideration the possibility of stock collapsed and a threshold can represent a control value. Also such process with stock collapses can be studied to give forecasts and, eventually, prevent problems. More recently, in [9], a gamma diffusion process with exogenous factors is transformed also in a lognormal process to describe the electric power consumption during a period of economic crisis. The transformation in the lognormal process allows to infer on parameters to give forecasts and, moreover, an application on the total consumption in Spain is considered. In this context, we can construct a process with jumps to take into consideration the possibility of a breakdown. Regarding this process with breakdowns, a threshold can represent a control value that gives an alarm in some cases which can be of interest for the authority.

Let $\widetilde{X}_k(t)$ be the lognormal time homogeneous diffusion processes with drift $A_1^k(x) = a_k x$ and infinitesimal variance $A_2^k(x) = \sigma_k^2 x^2$. For $\widetilde{X}_k(t)$ one has

$$\widetilde{f}_k(x, t|\rho_k, t_k) = \frac{1}{x\sigma_k\sqrt{2\pi(t - \tau_k)}} \exp\left\{-\frac{[\log(\frac{x}{\rho_k}) - (a_k - \frac{\sigma_k^2}{2})(t - \tau_k)]^2}{2\sigma_k^2(t - \tau_k)}\right\}, \quad (13)$$

$$\widetilde{m}_k^{(n)}(t|\rho_k, \tau_k) = \exp\left\{n\left[\log\rho_k + \left(a_k - \frac{\sigma_k^2}{2}\right)(t - \tau_k)\right] + \frac{n^2}{2}\sigma_k^2(t - \tau_k)\right\}, \quad (14)$$

$$\widetilde{g}_k(S, t|\rho_k, \tau_k) = \frac{|\log(\frac{S}{\rho_k})|}{\sqrt{2\pi\sigma_k^2(t - \tau_k)^3}}exp\left\{-\frac{[\log(\frac{S}{\rho_k}) - (a_k - \frac{\sigma_k^2}{2})(t - \tau_k)]^2}{2\sigma_k^2(t - \tau_k)}\right\}. \quad (15)$$

5.1   Lognormal Process with Deterministic Jumps

Let $\tau_1, \tau_2, \ldots, \tau_N$ be the instants in which jumps occur. From (5), recalling (13) one obtains the transition pdf of $X(t)$:

$$f(x, t|\rho_0, t_0) = \sum_{k=0}^{N} \frac{1_{(\tau_k, \tau_{k+1})}(t)}{x\sqrt{2\pi\sigma_k^2(t - \tau_k)}} \exp\left\{ -\frac{[\log(\frac{x}{\rho_k}) - (a_k - \frac{\sigma_k^2}{2})(t - \tau_k)]^2}{2\sigma_k^2(t - \tau_k)} \right\}$$

and, making use of (14), the conditional moments of $X(t)$ can be obtained from (2). Moreover, the FPT pdf is obtainable by (6) by remarking that from (15) one has:

$$\int_{\tau_j}^{\tau_{j+1}} \tilde{g}_j(S, \tau|\rho_j, \tau_j)\tau = \frac{1}{2}\text{Erfc}\left[\frac{\log(\frac{S}{\rho_j}) + (a_j - \frac{\sigma_j^2}{2})(t_{j+1} - t_j)}{\sqrt{2\sigma_j^2(t_{j+1} - t_j)}} \right]$$

$$+ \frac{1}{2}\exp\left\{ -\frac{2(a_j - \frac{\sigma_j^2}{2})\log(\frac{S}{\rho_j})}{\sigma_j^2} \right\}\text{Erfc}\left[\frac{\log(\frac{S}{\rho_j}) - (a_j - \frac{\sigma_j^2}{2})(t_{j+1} - t_j)}{\sqrt{2\sigma_j^2(t_{j+1} - t_j)}} \right],$$

where $\text{Erfc}(x) = (2/\sqrt{\pi})\int_x^\infty e^{-t^2}dt$ denotes the complementary error function.

5.2   Lognormal Process with Exponentially Distributed Jumps

Let I_k be identically distributed random variables with $\psi_k(s) \equiv \psi(s) = \xi e^{-\xi s}$, so that the expression (7) holds, with $\tilde{f}_k(x, t|\rho, \tau_k)$ defined in (13). Moreover, making use of the moments of the single process $\tilde{X}_k(t)$, also the moments of $X(t)$ can be evaluated via (8). Similarly, recalling (15), from (9) the FPT pdf can be written.

Now we consider the special case in which $\rho_k = \rho$ and $\tilde{X}_k(t) \stackrel{d}{=} \tilde{X}(t)$ with $A_1^{(k)}(x) = ax$ and $A_2^{(k)}(x) = \sigma^2 x^2$. In this case, from (10) and (13) one has:

$$f(x, t|\rho, t_0) = \frac{e^{-\xi(t - t_0)}}{x\sqrt{2\pi\sigma^2(t - t_0)}} \exp\left\{ -\frac{[\log(\frac{x}{\rho}) - (a - \frac{\sigma^2}{2})(t - t_0)]^2}{2\sigma^2(t - t_0)} \right\}$$

$$+ \xi\int_{t_0}^t \frac{e^{-\xi(t - \tau)}}{x\sqrt{2\pi\sigma^2(t - \tau)}} \exp\left\{ -\frac{[\log(\frac{x}{\rho}) - (a - \frac{\sigma^2}{2})(t - \tau)]^2}{2\sigma^2(t - \tau)} \right\}d\tau,$$

where

$$\int_{t_0}^t \frac{e^{-\xi(t - \tau)}}{x\sqrt{2\pi\sigma^2(t - \tau)}} \exp\left\{ -\frac{[\log(\frac{x}{\rho}) - (a - \frac{\sigma^2}{2})(t - \tau)]^2}{2\sigma^2(t - \tau)} \right\}d\tau$$

$$= \frac{e^{\log(\frac{x}{\rho})\left(a - \frac{\sigma^2}{2} - \sqrt{(a - \frac{\sigma^2}{2})^2 + 2\sigma^2\xi}\right)}}{2x\sqrt{\mu^2 + 2\sigma^2\xi}}\left[\text{Erfc}\left(\frac{\log(\frac{x}{\rho}) - (t - t_0)\sqrt{(a - \frac{\sigma^2}{2})^2 + 2\sigma^2\xi}}{\sqrt{2(t - t_0)\sigma^2}} \right) \right.$$

$$\left. - e^{\frac{2\log(\frac{x}{\rho})\sqrt{(a - \frac{\sigma^2}{2})^2 + 2\sigma^2\xi}}{\sigma^2}}\text{Erfc}\left(\frac{\log(\frac{x}{\rho}) + (t - t_0)\sqrt{(a - \frac{\sigma^2}{2})^2 + 2\sigma^2\xi}}{\sqrt{2(t - t_0)\sigma^2}} \right) \right].$$

Moreover, the conditional moments of $X(t)$ can be evaluated from (11) with $\widetilde{m}^{(n)}(t|\rho, t_0)$ given in (14); so it follows:

$$m^{(n)}(t|\rho, t_0) = e^{-\xi(t-t_0)}\widetilde{m}^{(n)}(t|\rho, t_0) + \frac{\xi\rho^n}{n(a - \frac{\sigma^2}{2}) + n^2\frac{\sigma^2}{2} - \xi}$$
$$\times \left[\exp\left\{ [n\left(a - \frac{\sigma^2}{2}\right) + n^2\frac{\sigma^2}{2} - \xi](t - t_0) \right\} - 1 \right]. \qquad (16)$$

Concerning the FPT pdf, recalling that $\widetilde{g}_j(S, \tau|\rho, t_j) = \widetilde{g}(S, \tau|\rho, t_j)$ is defined in (15), from (12) one has:

$$g(S, t|\rho, t_0) = e^{-\xi(t-t_0)}\frac{\log(\frac{S}{\rho})}{\sqrt{2\pi\sigma^2(t - t_0)^3}} \exp\left\{ -\frac{[\log(\frac{S}{\rho}) - (a - \frac{\sigma^2}{2})(t - t_0)]^2}{2\sigma^2(t - t_0)} \right\}$$
$$+ \xi e^{-\xi(t-t_0)}\left[1 - P(\widetilde{T}(0) < I) \right] \int_{t_0}^{t} e^{\xi\tau\left[1 - P(\widetilde{T}(0)<I)\right]} \frac{\log(\frac{S}{\rho})}{\sqrt{2\pi\sigma^2(t - \tau)^3}}$$
$$\times \exp\left\{ -\frac{[\log(\frac{S}{\rho}) - (a - \frac{\sigma^2}{2})(t - \tau)]^2}{2\sigma^2(t - \tau)} \right\} d\tau,$$

with

$$P(\widetilde{T}(0) < I) = -\frac{1}{2}\xi\left\{ L\left[\mathrm{Erfc}\left(\frac{-\log(\frac{S}{\rho}) + (a - \frac{\sigma^2}{2})\theta}{\sqrt{2\sigma^2\theta}} \right) \right]\right.$$
$$\left. + L\left[\mathrm{Erfc}\left(\frac{-\log(\frac{S}{\rho}) - (a - \frac{\sigma^2}{2})\theta}{\sqrt{2\sigma^2\theta}} \right) \right] \right\},$$

where L is the Laplace Transform.

In Fig. 1 the mean of $X(t)$ (full line) and the mean of $\widetilde{X}_0(t)$ (dashed line) are plotted for the deterministic jumps (on the left) and for exponential jumps (on the right).

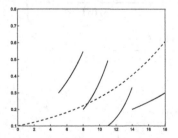 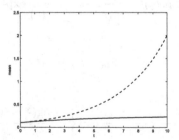

Fig. 1. The mean of $X(t)$ (full line) and the mean of $\widetilde{X}_0(t)$ (dashed line) are plotted for the deterministic jumps (on the left) and for exponential jumps (on the right). For the deterministic case $\rho_k = 0.1, 0.3, 0.2, 0.1, 0.2$, $a_k = 0.1, 0.2, 0.3, 0.4, 0.1$, $\tau_k = 0, 5, 8, 11, 14$ for $k = 0, 1, 2, 3, 4$. For the exponential case the parameters are $\rho = 0.1$, $a = 0.3$ and $\xi = 0.2$. In both cases $\sigma = 1$.

6 Conclusion and Future Developments

Stochastic diffusion processes with jumps occurring at random times have been studied by analyzing the transition pdf and its moments, the FPT density in the presence of constant and exponential distributed jumps. Particular attention has been payed on the lognormal process with jumps.

As future develops one could insert a dead time after a jump representing a delay period after that the process re-starts. This period can be represented by a random variable and the expressions for the transition pdf, the conditional moments and the FPT density can be obtained. Moreover, one can consider other probability distributions for the inter-jump intervals. In general, one can construct other processes with jumps, unknown in literature, on diffusion processes that are of interest in the applications. Finally, a general methodology to infer on parameters could be interesting to fit real data and provide forecasts in application context.

References

1. di Cesare, R., Giorno, V., Nobile, A.G.: Diffusion processes subject to catastrophes. In: Moreno-Díaz, R., Pichler, F., Quesada-Arencibia, A. (eds.) EUROCAST 2009. LNCS, vol. 5717, pp. 129–136. Springer, Heidelberg (2009). https://doi.org/10.1007/978-3-642-04772-5_18
2. Giorno, V., Nobile, A.G., di Cesare, R.: On the reflected Ornstein Uhlenbeck process with catastrophes. Appl. Math. Comput. **218**, 11570–11582 (2012)
3. Giorno, V., Spina, S.: A Stochastic Gompertz model with jumps for an intermittent treatment in cancer growth. In: Moreno-Díaz, R., Pichler, F., Quesada-Arencibia, A. (eds.) EUROCAST 2013. LNCS, vol. 8111, pp. 61–68. Springer, Heidelberg (2013). https://doi.org/10.1007/978-3-642-53856-8_8
4. Giorno, V., Nobile, A.G., Spina, S.: A note on time non-homogeneous adaptive queue with catastrophes. Appl. Math. Comput. **245**, 220–234 (2014)
5. Giorno, V., Spina, S.: On the return process with refractoriness for non-homogeneous Ornstein-Uhlenbeck neuronal model. Math. Biosci. Eng. **11**(2), 285–302 (2014)
6. Giorno, V., Spina, S.: Some remarks on stochastic diffusion processes with jumps. In: Lecture Notes of Seminario Interdisciplinare di Matematica, vol. 12, pp. 161–168 (2015)
7. Giorno, V., Román-Román, P., Spina, S., Torres-Ruiz, F.: Estimating a non-homogeneous Gompertz process with jumps as model of tumor dynamics. Comput. Stat. Data Anal. **107**, 18–31 (2017)
8. Gutierrez, R., Gutierrez-Sanchez, R., Nafidi, A.: The trend of the total stock of the private car-petrol in Spain: stochastic modelling using a new gamma diffusion process. Appl. Energy **86**, 18–24 (2009)
9. Nafidi, A., Gutierrez, R., Gutierrez-Sanchez, R., Ramos-Abalos, E., El Hachimi, S.: Modelling and predicting electricity consumption in Spain using the stochastic Gamma diffusion process with exogenous factors. Energy **113**, 309–318 (2016)
10. Spina, S., Giorno, V., Román-Román, P., Torres-Ruiz, F.: A stochastic model of cancer growth subject to an intermittent treatment with combined effects: reduction of tumor size and rise of growth rate. Bull. Math. Biol. (2014). https://doi.org/10.1007/s11538-014-0026-8

Some Remarks on the Mean of the Running Maximum of Integrated Gauss-Markov Processes and Their First-Passage Times

Marco Abundo and Mario Abundo[(✉)]

Tor Vergata University, Rome, Italy
marco.abundo@gmail.com, abundo@mat.uniroma2.it

Abstract. Explicit formulae for the mean of the running maximum of conditional and unconditional Brownian motion are found; these formulae are used to obtain the mean, $a(t)$, of the running maximum of an integrated Gauss-Markov process $X(t)$. Moreover, the connection between the moments of the first-passage-time of $X(t)$ and $a(t)$ is investigated. Some explicit examples are reported.

Keywords: Running maximum · First-passage time
Gauss-Markov process

1 Introduction

Let $X(t)$ be a continuous stochastic process, with $X(0) = 0$, and let $S(t) := \max_{s \in [0,t]} X(s)$ its running maximum process; in [7] it was found a connection between its mean $a(t) = E[S(t)]$ and that of the first-passage time (FPT) τ_r of X through a threshold $r > 0$. Precisely, assuming that $a(t)$ is strictly increasing (i.e. the inverse a^{-1} exists), it was shown that, for any non-decreasing function $g : [0, +\infty) \longrightarrow [0, +\infty)$, it results $E[g(\tau_r)] \geq \int_0^1 g(a^{-1}(rt))dt$, and that this bound is sharp. In particular, for any integer n, one gets:

$$E\left[(\tau_r)^n\right] \geq \int_0^1 \left(a^{-1}(rt)\right)^n dt. \tag{1}$$

Inequality (1) is very useful when exact calculation of the moments of the FPT of X through the level boundary $r > 0$ is not possible, since it provides lower bounds to them in terms of the mean of the running maximum process.

In this note, motivated by the result of [7], we find the function $a(t)$ for integrated Gauss-Markov (GM) processes; in particular, for integrated Brownian motion and integrated Ornstein-Uhlenbeck process. In the cases when it is possible to calculate exactly $E[(\tau_r)^n]$, we compare it with its lower bound given by (1) (see also [1]). Notice that integrated GM processes have important applications, e.g. in computational neuroscience (see e.g. [10] and references therein).

R. Moreno-Díaz et al. (Eds.): EUROCAST 2017, Part II, LNCS 10672, pp. 72–79, 2018.
https://doi.org/10.1007/978-3-319-74727-9_9

Other applications can be found also in queueing theory, economy, and finance mathematics (see e.g. the discussion in [2,3]).

Now, we recall the definition of GM process, and its integrated process.

Let $m(t)$, $h_1(t)$, $h_2(t)$ be continuous functions of $t \geq 0$, which are C^1 in $(0, +\infty)$, and such that $h_2(t) \neq 0$ and $\rho(t) = h_1(t)/h_2(t)$ is a non-negative, differentiable and increasing function, with $\rho(0) = 0$. If $B(t) = B_t$ denotes standard Brownian motion (BM), then for $t \geq 0$, $Y(t) = m(t) + h_2(t)B(\rho(t))$ is a continuous GM process with mean $m(t)$ and covariance $c(s,t) = h_1(s)h_2(t)$, for $0 \leq s \leq t$.

Besides BM, a noteworthy case of GM process is the Ornstein-Uhlenbeck (OU) process, and in fact any continuous GM process can be represented in terms of a OU process (see e.g. [9]).

For a continuous GM process Y, its integrated process, starting from $X(0)$ is defined by $X(t) = X(0) + \int_0^t Y(s)ds$.

2   The Running Maximum of $B_t, |B_t|$, and the Brownian Bridge

We recall the formulae for the distributions of the running maximum of B_t, in the interval $[0,t]$ i.e. $S_B(t) = \max_{s \in [0,t]} B_s$, and the running maximum of $|B_t|$, i.e. $S_B^*(t) = \max_{s \in [0,t]} |B_s|$; then, we calculate $E[S_B(t)]$ and $E[S_B^*(t)]$. Furthermore, the running maximum of Brownian bridge is studied.

As far as B_t is concerned, it is well known that, for $x \geq 0$, $P(S_B(t) \leq x) = 2\Phi\left(\frac{x}{\sqrt{t}}\right) - 1$, where Φ stands for the standard normal distribution function. Therefore, the probability density of $S_B(t)$ is $f_{S_B}(x) = \frac{2}{\sqrt{2\pi t}} e^{-x^2/2t}$, $x \geq 0$. By straightforward calculation, one gets:

$$E(S_B(t)) = \int_0^{+\infty} \frac{2x}{\sqrt{2\pi t}} e^{-x^2/2t} dx = \sqrt{\frac{2t}{\pi}}. \tag{2}$$

As for the distribution of the maximum absolute value of BM in the interval $[0,t]$, for $x \geq 0$ one has (see e.g. formula (3.6) of [5], or formula (1.3) of [6]):

$$P(S_B^*(t) \leq x) = \sum_{k=-\infty}^{+\infty} (-1)^k \left[\Phi\left(\frac{x(2k+1)}{\sqrt{t}}\right) - \Phi\left(\frac{x(2k-1)}{\sqrt{t}}\right) \right]. \tag{3}$$

Then, the following holds:

Proposition 2.1 (see [1] for the proof). *The mean of the maximum absolute value of BM in the interval $[0,t]$ is*

$$E(S_B^*(t)) = \frac{\pi}{2} E(S_B(t)) = \sqrt{\frac{\pi t}{2}}. \tag{4}$$

$\square$

Remark 2.2. Equation (4) specifies and improves the estimate contained in Example 4 of [8]. In both cases, when $X(t) = B_t$ and $X(t) = |B_t|$, one gets that $a(t) = E[\max_{s\in[0,t]} X(s)]$ is concave, since $a(t) = const \cdot \sqrt{t}$.

Now, we study the running maximum of the Brownian bridge, that is, the diffusion process $X(t)$, starting from zero, and conditioned to take the value β at $t = 1$. Its explicit form is $X(t) = \beta t + (1-t)\widetilde{B}\left(\frac{t}{1-t}\right), 0 \leq t \leq 1$, where $\widetilde{B}$ is BM. So, for $0 \leq t \leq 1, X$ is a GM process with mean $m(t) = \beta t$, and $h_1(t) = t, h_2(t) = 1-t, \rho(t) = \frac{t}{1-t}, c(s,t) = h_1(s)h_2(t)$ ($\rho(t)$ is defined only in $[0,1)$). The following formula holds for the distribution of the maximum of $X(t)$ in the interval $[0,t], t < 1$ (see e.g. Eq. (2.1) of [6]):

$$P\left(\max_{s\in[0,t]} X(s) \leq z\right) = \Phi\left(\frac{z - \beta t}{\sqrt{t(1-t)}}\right) - e^{-2z(z-\beta)}\Phi\left(\frac{(2z - \beta)t - z}{\sqrt{t(1-t)}}\right), \quad z > 0.$$
(5)

The mean of the running maximum of the Brownian bridge in the interval $[0,t]$ is:

$$a(t) = E(\max_{s\in[0,t]} X(s)) = \int_0^{+\infty} P\left(\max_{s\in[0,t]} X(s) > z\right) dz$$

$$= \int_0^{+\infty} \left[1 - \Phi\left(\frac{z - \beta t}{\sqrt{t(1-t)}}\right) + e^{-2z(z-\beta)}\Phi\left(\frac{(2z - \beta)t - z}{\sqrt{t(1-t)}}\right)\right] dz. \quad (6)$$

Since is not possible to analytically calculate the integral in (6), for every $t \in (0,1)$, we have numerically computed it; notice however that, for $t = 1$ and $\beta > 0$, one has $a(1) = \beta + \sqrt{\frac{\pi}{8}}e^{\beta^2/2}[1 - \Phi(\beta/\sqrt{2})]$ (see [1]). In the Fig. 1, we report the graph of $a(t)$, numerically evaluated, as a function of $t \in (0,1)$, for several values of $\beta > 0$. Note that, the shape of $a(t)$ appears to be like $const \cdot \sqrt{t}, t \in (0,1)$, this suggesting that $a(t)$ is concave, as in the cases of B_t and $|B_t|$.

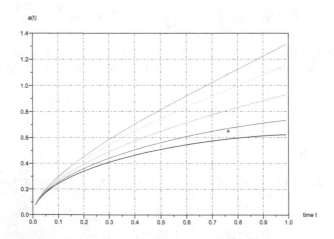

Fig. 1. Graph of the mean $a(t)$ of the maximum of Brownian bridge, as a function of $t \in (0,1)$, for some values of β; from top to bottom, $\beta = 1, 0.8, 0.5, 0.2, 0$.

3 The Maximum of an Integrated Gauss-Markov Process and Lower Bounds to the FPT Moments

For a GM process $Y(t)$, we consider the integrated GM process $X(t) = x + \int_0^t Y(s)ds$. We recall from [2,3] the following:

Theorem 3.1. *Let Y be a GM process, then $X(t) = x + \int_0^t Y(s)ds$ is normally distributed with mean $x + M(t)$ and variance $\gamma(\rho(t))$, where $M(t) = \int_0^t m(s)ds$, $\gamma(t) = \int_0^t (R(t) - R(s))^2 ds$ and $R(t) = \int_0^t h_2(\rho^{-1}(s))/\rho'(\rho^{-1}(s))ds$. Moreover, if $\gamma(+\infty) = +\infty$, then there exists a BM $\widehat{B}$ such that $X(t) = x + M(t) + \widehat{B}(\widehat{\rho}(t))$, where $\widehat{\rho}(t) = \gamma(\rho(t))$. Thus, the integrated process X can be represented as a GM process with respect to a different BM.* □

In the sequel, we assume that the condition $\gamma(+\infty) = +\infty$ holds, and we also suppose that $M(t) = 0$, so, the integrated GM process $X(t) = x + \int_0^t Y(s)ds$ takes the form $X(t) = x + \widehat{B}(\widehat{\rho}(t))$; note that $\widehat{\rho}(t)$ is increasing and $\widehat{\rho}(0) = 0$.

First, we focus on the FPT of $X(t)$ through a single barrier r, that is, $\tau_r(x) = \inf\{t > 0 : X(t) = r | X(0) = x\}$, with $x < r$. Under the above conditions, we obtain that

$$\max_{s \in [0,t]} (X(t) - x) = \max_{s \in [0,t]} \widehat{B}(\widehat{\rho}(t)) = \max_{u \in [0,\widehat{\rho}(t)]} \widehat{B}_u = S_{\widehat{B}}(\widehat{\rho}(t)). \tag{7}$$

From (2), we get

$$a(t) = E[\max_{s \in [0,t]} (X(t) - x)] = \sqrt{\frac{2\widehat{\rho}(t)}{\pi}}, \tag{8}$$

and so $a^{-1}(u) = \widehat{\rho}^{-1}\left(\frac{\pi u^2}{2}\right)$. One has $\tau_r(x) = \inf\{t > 0 : \widehat{B}(\widehat{\rho}(t)) = r - x\}$, and, since the function $\widehat{\rho}$ is increasing, $\widehat{\tau}_r(x) := \widehat{\rho}(\tau_r(x)) = \inf\{s > 0 : \widehat{B}_s = r - x\}$, from which it follows (see also [2]) that the density of $\tau_r(x)$ is:

$$f_{\tau_r}(t) = \frac{(r - x)\widehat{\rho}'(t)}{\sqrt{2\pi}\,\widehat{\rho}(t)^{3/2}}\, e^{-(r-x)^2/2\widehat{\rho}(t)}. \tag{9}$$

From (9) we get that the n-th order moment of $\tau_r(x)$, if it exists finite, is explicitly given by:

$$E[(\tau_r(x))^n] = \int_0^{+\infty} t^n \frac{(r - x)\widehat{\rho}'(t)}{\sqrt{2\pi}\,\widehat{\rho}(t)^{3/2}}\, e^{-(r-x)^2/2\widehat{\rho}(t)} dt. \tag{10}$$

As easily seen, $E[\tau_r^n(x)]$ is finite if and only if the function $t^n \widehat{\rho}'(t)/\widehat{\rho}(t)^{3/2}$ is integrable in $(c, +\infty), c > 0$. Assume that $\alpha > 0$ exists, such that $\widehat{\rho}(t) \sim const \cdot t^\alpha$, as $t \to +\infty$; then, in order that $E[\tau_a^n(x)] < \infty$, it must be $\alpha = 2(n + \delta)$, for some $\delta > 0$ (see [2]).

On the other hand, (1) implies:

$$E[(\tau_r(x))^n] \geq \int_0^1 \left[a^{-1}((r - x)u)\right]^n du = \int_0^1 \left[\widehat{\rho}^{-1}\left(\frac{\pi(r-x)^2 u^2}{2}\right)\right]^n du. \tag{11}$$

Now, we consider the first exit time (FET) of $X(t)$ from the interval (r_1, r_2), that is, $\tau_{r_1,r_2}(x) = \inf\{t > 0 : X(t) \notin (r_1, r_2)|X(0) = x\}, x \in (r_1, r_2)$; for the sake of simplicity, we assume that $r_2 = b > 0, r_1 = -b$, and $X(0) = x = 0$. Thus, we consider $\tau_{-b,b}(0) = \inf\{t > 0 : |\widehat{B}(\widehat{\rho}(t))| = b\}$, or, $\widehat{\rho}(\tau_{-b,b}(0)) = \inf\{s > 0 : |\widehat{B}_s| = b\}$, from which it follows (see [2]) that the density of $\tau_{-b,b}(0)$ turns out to be $\widehat{f}_{-b,b}(\widehat{\rho}(t)|0)\widehat{\rho}'(t)$, where

$$\widehat{f}_{-b,b}(t|x) = \frac{\pi}{b^2} \sum_{k=0}^{\infty} (-1)^k \left(k + \frac{1}{2}\right) \cos\left[\left(k + \frac{1}{2}\right)\frac{\pi x}{\alpha}\right] \exp\left[-\left(k + \frac{1}{2}\right)^2 \frac{x^2 t}{2b^2}\right]$$
(12)

is the density of the FET of $x + \widehat{B}_t$ from the interval $(-b, b)$. Moreover (see [2]):

$$E\left[(\tau_{-b,b}(0))^n\right] = \frac{\pi}{b^2} \sum_{k=0}^{\infty} \left[(-1)^k \left(k + \frac{1}{2}\right) \int_0^{+\infty} e^{-(k+1/2)^2 \pi^2 t/2b^2} (\widehat{\rho}^{-1}(t))^n dt\right].$$
(13)

On the other hand, from (4), we get

$a(t) = E\left(\max_{s\in[0,t]} |\widehat{B}(\widehat{\rho}(s))|\right) = E\left(\max_{s\in[0,\widehat{\rho}(t)]} |\widehat{B}(s)|\right) = \sqrt{\frac{\pi\widehat{\rho}(t)}{2}}$, and so $a^{-1}(u) = \widehat{\rho}^{-1}\left(\frac{2u^2}{\pi}\right)$. Then, (1) implies:

$$E\left[(\tau_{-b,b}(0))^n\right] \geq \int_0^1 \left[a^{-1}(bu)\right]^n du = \int_0^1 \left[\widehat{\rho}^{-1}\left(\frac{2b^2 u^2}{\pi}\right)\right]^n du. \quad (14)$$

3.1   Integrated BM (IBM)

Let be $X(t) = x + \int_0^t B_s ds$; from Theorem 3.1 (see also [2]), it follows that there exists a BM $\widehat{B}$ such that $X(t) = x + \widehat{B}(\widehat{\rho}(t))$, where $\widehat{\rho}(t) = t^3/3$; moreover, for $x < r$, $E[\tau_r(x)] = \left(\frac{3}{2}\right)^{1/3} \Gamma\left(\frac{1}{6}\right) \frac{(r-x)^{2/3}}{\sqrt{\pi}} \approx 3.595 \cdot (r - x)^{2/3}$, while the moments of $\tau_r(x)$ of order $n > 1$ are infinite (see [2]). From (8) we obtain $a(t) = E[\max_{s\in[0,t]}(X(s)-x)] = \sqrt{\frac{2t^3}{3\pi}}$, and so, for $u \geq 0$, $a^{-1}(u) = \left(\frac{3}{2}\pi\right)^{1/3} u^{2/3}$. Thus, from (1) we get:

$$E\left[\tau_r(x)\right] \geq \int_0^1 a^{-1}((r-x)u)du = \left(\frac{3}{2}\pi\right)^{1/3} \cdot \frac{3}{5}(r-x)^{2/3} \approx 1.0059 \cdot (r-x)^{2/3}.$$
(15)

If we compare this lower bound to $E[\tau_r(x)]$ with its exact value given above, we see that the obtained estimate captures the right power of $r - x$, giving an evaluation up to a multiplicative constant.

Explicit formulae for the first two moments of the FET of IBM from the interval $(-b, b)$, when starting from $x \in (-b, b)$, are given in [2]. Taking $b = 1$ and $x = 0$, for the sake of simplicity, one gets

$$E\left[\tau_{-1,1}(0)\right] = \frac{8 \cdot 3^{1/3} \Gamma(\frac{4}{3})}{\pi^{5/3}} \sum_{k=0}^{\infty} (-1)^k \frac{1}{(2k+1)^{5/3}}. \tag{16}$$

$$E\left[(\tau_{-1,1}(0))^2\right] = \frac{192}{\pi^4} \sum_{k=0}^{\infty} (-1)^k \frac{1}{(2k+1)^4}. \tag{17}$$

Formula (4) for the maximum of $|B_t|$ provides $a(t) = E\left(\max_{s \in [0,t]} |\widehat{B}(\widehat{\rho}(s))|\right) = \sqrt{\frac{\pi}{6}} \, t^{3/2}$, and so, for $u \geq 0$, $a^{-1}(u) = \left(\frac{6}{\pi}\right)^{1/3} u^{2/3}$. Thus, from (1) we get $E\left[(\tau_{-1,1}(0))^n\right] \geq \int_0^1 \left[a^{-1}(u)\right]^n du$. For $n = 1$, the last integral is equal to $\left(\frac{6}{\pi}\right)^{1/3} \cdot \frac{3}{5} \approx 0.7444$, while the exact value of $E\left[\tau_{1,1}(0)\right]$, numerically calculated by (16) is 1.3518.

For $n = 2$, the integral above is $\left(\frac{6}{\pi}\right)^{2/3} \cdot \frac{3}{7} \approx 0.6599$, while the exact value of $E((\tau_{-1,1}(0))^2)$, numerically calculated by (17) is 1.95 (see the values reported in the Fig. 2 of [2]). Although for IBM the lower bounds to $E\left[\tau_r(x)\right]$ and $E\left[\tau_{-1,1}(0)\right]$ are somewhat far from the true values, for more general processes, the corresponding lower bounds are useful estimates, when no other information is available.

3.2   Integrated Ornstein-Uhlenbeck (IOU) Process

For $\mu, \sigma > 0$ and $\beta \in \mathbb{R}$, the OU process is the diffusion $Y(t)$ driven by the SDE $dY(t) = -\mu(Y(t) - \beta)dt + \sigma dB_t, Y(0) = y$, whose explicit solution is:

$$Y(t) = \beta + e^{-\mu t}[y - \beta + \widetilde{B}(\rho(t))] \tag{18}$$

where $\widetilde{B}$ is Brownian motion and $\rho(t) = \frac{\sigma^2}{2\mu}\left(e^{2\mu t} - 1\right)$. Y is a GM process with $m(t) = \beta + e^{-\mu t}(y - \beta), h_1(t) = \frac{\sigma^2}{2\mu}\left(e^{\mu t} - e^{-\mu t}\right), h_2(t) = e^{-\mu t}$ and $c(s,t) = h_1(s)h_2(t)$. For the integrated process $X(t) = x + \int_0^t Y(s)ds$, the various functions in Theorem 3.1 have been calculated in [3]. We consider the special case when $y = \beta = 0$, so that $M(t) = 0$. Since $\lim_{t \to +\infty} \gamma(t) = +\infty$, by Theorem 3.1, we conclude that there exists a BM $\widehat{B}$ such that $X(t) = x + \widehat{B}(\widehat{\rho}(t))$, where $\widehat{\rho}(t) = \gamma(\rho(t))$. Since $\widehat{\rho}(t) \sim const \cdot t$, as $t \to +\infty$, we obtain that $E\left[\tau_r(x)\right] = +\infty$ for any x, while $E\left[(\tau_{-b,b}(x)\right]^n$ is finite for any n and $x \in (-b,b)$ (see [2]). It is interesting to compare the first two moments of $\tau_{-b,b}(0)$, numerically calculated in [2], with their lower bounds given by (14); for instance, for $\mu = 2$ and $\sigma = 1$, the calculation of the mean exit time from $(-1, 1)$ starting from $x = 0$ furnishes the value 4.74, while its lower bound turns out to be 1.44; the second order moment of the exit time is 3.319, while its lower bound results to be 2.012.

In the Fig. 2, we report a comparison between the mean exit time from $(-1, 1)$ of IOU process with $\beta = 0, \sigma = 1$ and starting from $x = 0$, and its lower bound given by (14), as a function of $\mu \in [0, 2]$.

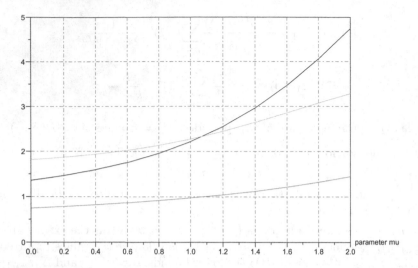

Fig. 2. Comparison between the mean exit time from $(-1, 1)$ of integrated OU process with $\beta = 0, \sigma = 1$ and starting from $x = 0$, and its lower bound, given by (14), as a function of the parameter $\mu \in [0, 2]$. The black line (top) represents $E(\tau_{-1,1}(0))$, the blue line (bottom) its lower bound (LB) and the green line (middle) represents $E(\tau_{-1,1}(0))/LB$; on the horizontal axes $\mu \in [0, 2]$. (Color figure online)

4   Conclusions and Final Remarks

The argument used so far can also be applied to a transformed GM process $X(t)$ driven by the SDE $dX(t) = -\frac{\rho'(t)u''(X(t))}{2(u'(X(t)))^3} dt + \frac{\sqrt{\rho'(t)}}{u'(X(t))} dB_t$, where u is an increasing twice-differentiable function with $u'(x) \neq 0, u(0) = 0$, and $\rho(t)$ is non-negative, increasing and differentiable with $\rho(0) = 0$ and $\rho(+\infty) = +\infty$ (the case $\rho(t) = t$ corresponds to a diffusion conjugated to BM, according to the definition given in [4], and it includes e.g. the Feller, or Cox–Ingersoll–Ross process, and the Wright-Fisher process). We can conclude (see also [1]) that there exists a BM $\widetilde{B}_t$ such that $X(t) = u^{-1}\left(\widetilde{B}(\rho(t)) + u(x)\right)$, i.e. a space transformation of time-changed BM. Then, for $r > x$ we get $\tau_r(x) = \inf\{t > 0 : X(t) \geq r | X(0) = x\} = \inf\{t > 0 : \widetilde{B}(\rho(t)) \geq u(r) - u(x)\}$, that is, $\widetilde{\tau}_r(x) := \rho(\tau_r(x)) = \inf\{s > 0 : \widetilde{B}_s \geq u(r) - u(x)\}$. From this it follows (see [2] and (9)) that the density of $\tau_r(x)$ is $f_{\tau_r}(t) = \frac{(u(r)-u(x))\rho'(t)}{\sqrt{2\pi}\,\rho(t)^{3/2}} e^{-(u(r)-u(x))^2/2\rho(t)}$. In the cases when the FPT moments are finite, but their analytical calculations turn out to be prohibitive, we can use (1) to obtain a lower bound to $E[\tau_r(x)^n]$. In fact, $a(t)$ can be found, by using the density of $\max_{s \in [0,t]} \widetilde{B}_s$. Moreover, since by hypothesis u^{-1} is monotonically increasing, $S(t) = \max_{s \in [0,t]} X(s) = \max_{s \in [0,t]} u^{-1}\left(\widetilde{B}(\rho(t)) + u(x)\right) = u^{-1}\left(u(x) + \max_{s \in [0,\rho(t)]} \widetilde{B}_s\right)$. Analogous considerations allow to obtain lower bounds to the moments of the FET from the interval (r_1, r_2). In this regard, we show two examples of diffusion conjugated to BM, namely $\rho(t) = t$.

Example 1. The Wright-Fisher process, which is driven by the SDE $dX(t) = (1/4 - X(t)/2)dt + \sqrt{X(t)(1 - X(t))}dB_t, X(0) = x \in [0, 1]$, is a diffusion in $[0, 1]$ which is conjugated to BM via the function $u(x) = 2\arcsin\sqrt{x}$, i.e. $X(t) = \sin^2(B_t/2 + \arcsin\sqrt{x})$. Let us consider the first hitting time of $X(t)$ to the boundaries of $[0, 1]$, when starting from $x \in (0, 1)$, that is, $\tau_{0,1}(x) = \inf\{t > 0 : X(t) = 0, \text{ or } X(t) = 1 | X(0) = x\} = \inf\{t > 0 : B_t + u(x) = 0, \text{ or } B_t + u(x) = \pi\}$. If e.g. $x = 1/2$, since $u(1/2) = \pi/2$, one obtains $\tau_{0,1}(1/2) = \inf\{t > 0 : |B_t| = \pi/2\}$, which, as well-known, is equal to $\pi^2/4$. On the other hand, $a(t) = E(\max_{s \subset [0,t]} |B_s|) = \sqrt{\pi t/2}$, then, by (1) $E(\tau_{0,1}(1/2)) \geq \int_0^1 a^{-1}(\pi t/2)dt = \pi/6$, which is a lower bound to the true value $\pi^2/4$.

Example 2. The diffusion $X(t)$ driven by the SDE $dX(t) = \frac{1}{3}X(t)^{1/3}dt + X(t)^{2/3}dB_t, X(0) = x$, is conjugated to BM via the function $u(x) = 3x^{1/3}$, that is, $X(t) = (x^{1/3} + B_t/3)^3$. Then, for $x = 0$, we have $\tau_{-1,1}(0) = \inf\{t > 0 : |X(t)| = 1 | X(0) = 0\} = \inf\{t > 0 : |B_t| = 3\}$, which, as well-known, is equal to 9. On the other hand, $a(t) = E(\max_{s \in [0,t]} |B_s|) = \sqrt{\pi t/2}$, then, by (1) $E(\tau_{-1,1}(0)) \geq \int_0^1 a^{-1}(3t)dt = 6/\pi$, which is a lower bound to the true value 9.

References

1. Abundo, M.: The mean of the running maximum of an integrated Gauss-Markov process and the connection with its first-passage time. Stoch. Anal. Appl. **35**(3), 499–510 (2017). https://doi.org/10.1080/073629942015.1099047
2. Abundo, M.: On the first-passage time of an integrated Gauss-Markov process. Scientiae Mathematicae Japonicae Online e-2015 **28**, 1–14 (2015)
3. Abundo, M.: On the representation of an integrated Gauss-Markov process. Scientiae Mathematicae Japonicae Online e-2013 **26**, 719–723 (2013)
4. Abundo, M.: Limit at zero of the first-passage time density and the inverse problem for one-dimensional diffusions. Stoch. Anal. Appl. **24**(1), 1119–1145 (2006)
5. Abundo, M.: Some conditional crossing results of Brownian motion over a piecewise-linear boundary. Stat. Probab. Lett. **58**(2), 131–145 (2002)
6. Beghin, L., Orsingher, E.: On the maximum of the generalized Brownian bridge. Lith. Math. J. **39**(2), 157–167 (1999)
7. Brown, M., de la Pena, V.H., Klass, M., Sit, T.: On an approach to boundary crossing by stochastic processes. Stoch. Process. Appl. (2016, in press). https://doi.org/10.1016/j.spa.2016.04.027
8. Brown, M., de la Pena, V.H., Klass, M., Sit, T.: From boundary crossing of non-random functions to boundary crossing of stochastic processes. Probab. Eng. Inf. Sci. **29**, 345–359 (2015). https://doi.org/10.1017/S0269964815000030
9. Nobile, A.G., Pirozzi, E., Ricciardi, L.M.: Asymptotics and evaluations of FPT densities through varying boundaries for Gauss-Markov processes. Sci. Math. Jpn. **67**(2), 241–266 (2008)
10. Touboul, J., Faugeras, O.: Characterization of the first hitting time of a double integral processes to curved boundaries. Adv. Appl. Probab. **40**, 501–528 (2008)

Applications of the Quantile-Based Probabilistic Mean Value Theorem to Distorted Distributions

Antonio Di Crescenzo[1(✉)], Barbara Martinucci[1], and Julio Mulero[2]

[1] Dipartimento di Matematica, Università di Salerno, Fisciano, Italy
{adicrescenzo,bmartinucci}@unisa.it
[2] Departamento de Matemáticas, Universidad de Alicante, Alicante, Spain
julio.mulero@ua.es

Abstract. Distorted distributions were introduced in the context of actuarial science for several variety of insurance problems. In this paper we consider the quantile-based probabilistic mean value theorem given in Di Crescenzo *et al.* [4] and provide some applications based on distorted random variables. Specifically, we consider the cases when the underlying random variables satisfy the proportional hazard rate model and the proportional reversed hazard rate model. A setting based on random variables having the 'new better than used' property is also analyzed.

Keywords: Quantile function · Distorted distribution
Mean value theorem

1 Introduction and Background

A probabilistic generalization of the Taylor's theorem was proposed and studied by Massey and Whitt [8] and Lin [7], showing that for a nonnegative random variable X and a suitable function f one has

$$\mathbb{E}[f(t+X)] = \sum_{k=0}^{n-1} \mathbb{E}[X^k] \frac{f^{(k)}(t)}{k!} + \mathbb{E}[f^{(n)}(t+X_e)] \frac{f^{(n)}(t)}{n!}, \qquad t > 0,$$

where X_e is a random variable possessing the equilibrium distribution of X. This result was employed by Di Crescenzo [3] in order to obtain the following probabilistic version of the well-known mean value theorem:

$$\mathbb{E}[g(Y)] - \mathbb{E}[g(X)] = \mathbb{E}[g'(Z)][\mathbb{E}(Y) - \mathbb{E}(X)],$$

where X and Y are nonnegative random variables such that $X \leq_{st} Y$, i.e., $\mathbb{P}(X > x) \leq \mathbb{P}(Y > x)$, for all $x \geq 0$, and $\mathbb{E}(X) < \mathbb{E}(Y)$. Moreover, the random variable Z is a generalization of X_e and has density function

$$\frac{\mathbb{P}(Z \in dx)}{dx} = \frac{\mathbb{P}(Y > x) - \mathbb{P}(X > x)}{\mathbb{E}(Y) - \mathbb{E}(X)}.$$

© Springer International Publishing AG 2018
R. Moreno-Díaz et al. (Eds.): EUROCAST 2017, Part II, LNCS 10672, pp. 80–87, 2018.
https://doi.org/10.1007/978-3-319-74727-9_10

Various related results have been exploited recently, such as the fractional probabilistic Taylor's and mean value theorems (see Di Crescenzo and Meoli [5]), and a quantile-based version of the probabilistic mean value theorem (see Di Crescenzo et al. [4]). The latter involves a distribution that generalizes the Lorenz curve, and allows the construction of new distributions with support $(0, 1)$. Specifically, for any random variable X, let $F(x) = \mathbb{P}(X \le x)$, $x \in \mathbb{R}$, denote the distribution function, and let

$$Q(u) = \inf\{x \in \mathbb{R} : F(x) \ge u\}, \qquad 0 < u < 1 \tag{1}$$

be the quantile function, with $Q(0) := \lim_{u \downarrow 0} Q(u)$ and $Q(1) := \lim_{u \uparrow 1} Q(u)$.

Definition 1. *Let $\mathcal{D}$ be the family of all absolutely continuous random variables having finite nonzero mean, and such that the quantile function (1) satisfies $Q(0) = 0$ and is differentiable, in order that the following quantile density function exists:*

$$q(u) = Q'(u), \qquad 0 < u < 1.$$

We remark that if $X \in \mathcal{D}$ then $F[Q(u)] = u$, $0 < u < 1$. According to Di Crescenzo et al. [4], if $X \in \mathcal{D}$, let X^L denote an absolutely continuous random variable taking values in $(0, 1)$ with distribution function

$$L(p) = \frac{1}{\mathbb{E}[X]} \int_0^p Q(u)\, du, \qquad 0 \le p \le 1, \tag{2}$$

and density

$$f_{X^L}(u) = \frac{Q(u)}{\mathbb{E}[X]}, \qquad 0 < u < 1.$$

Note that the function given in (2) is also known as the Lorenz curve of X, and deserves large interest in mathematical finance for the representation of the distribution of income or of wealth. Then, for $X \in \mathcal{D}$, if $g : (0, 1) \to \mathbb{R}$ is n-times differentiable and $g^{(n)} \cdot Q$ is integrable on $(0, 1)$ for any $n \ge 1$, then

$$\mathbb{E}[\{g(1) - g(U)\} q(U)] = \sum_{k=1}^{n-1} \frac{1}{k!} \mathbb{E}\left[g^{(k)}(U)(1 - U)^k q(U)\right]$$

$$+ \frac{1}{(n-1)!} \mathbb{E}\left[g^{(n)}(X^L)(1 - X^L)^{n-1}\right] \mathbb{E}[X],$$

where U is uniformly distributed in $(0, 1)$. Furthermore, the following result can be viewed as a quantile-based analogue of the probabilistic mean value theorem (cf. Theorem 3 in Di Crescenzo et al. [4]). In particular, given $X, Y \in \mathcal{D}$ such that $X \le_{st} Y$ and a differentiable function $g : (0, 1) \to \mathbb{R}$ with $g' \cdot Q_X$ and $g' \cdot Q_Y$ integrable on $(0, 1)$, then

$$E[\{g(1) - g(U)\} \{q_Y(U) - q_X(U)\}] = E\left[g'(Z^L)\right] \{E[Y] - E[X]\}, \tag{3}$$

where U is uniformly distributed in $[0,1]$ and q_X and q_Y are the quantile densities of X and Y, respectively. Moreover, Z^L denotes a random variable having distribution function

$$L_{X,Y}(p) = \frac{1}{E[Y] - E[X]} \int_0^p [Q_Y(u) - Q_X(u)]du, \qquad 0 \le p \le 1,$$

which is a suitable extension of (2). Note also that, under the previous assumptions, $E\left[g'(Z^L)\right]$ in (3) is finite.

Stimulated by the above mentioned results, in this paper we construct new relationships involving distorted random variables that deserve interest in utility theory and can be applied for assessing stochastic dominance among risks (we refer, for instance, to the recent papers by Balbás *et al.* [1] and Sordo *et al.* [12,14]).

Let Γ be the set of continuous, nondecreasing and piecewise differentiable functions $h : [0,1] \to [0,1]$ such that $h(0) = 0$ and $h(1) = 1$. These functions are called distortion functions. See Sordo and Suárez-Llorens [13] for applications of distortion functions to classes of variability measures, and Gupta *et al.* [6] for the use of distortion functions for the analysis of random lifetimes of coherent systems. We denote by $\overline{F}(x) = 1 - F(x)$, $x \in \mathbb{R}$, the survival function of X.

Definition 2. *For each distortion function $h \in \Gamma$, and any survival function $\overline{F}(x)$, the position*

$$\overline{F}_h(x) = h(\overline{F}(x)), \qquad x \in \mathbb{R},$$

defines a survival function associated to a new random variable X_h, which is called the distorted random variable induced by h.

Distorted distributions were introduced by Denneberg [2] and Wang [15,16] in the context of actuarial science for several variety of insurance problems. One of the most important applications is in the rank dependent expected utility model (see Quiggin [10], Yaari [17], Schmeidler [11]).

It is easy to see that given a random variable X and a distortion function $h \in \Gamma$, the distorted random variable induced by h, say X_h, has quantile function given by

$$Q_h(u) := Q(1 - h^{-1}(1 - u)), \qquad 0 < u < 1,$$

where Q is the quantile function of X.

Proposition 1. *Let X be a nonnegative random variable, and $h, l \in \Gamma$ two distortion functions. Then,*

$$X_h \le_{st} X_l \quad \Leftrightarrow \quad h(x) \le l(x), \qquad 0 < x < 1.$$

Proof. The proof immediately follows noting that $X_h \le_{st} X_l$ holds if, and only if, $h(\overline{F}(x)) \le l(\overline{F}(x))$ or, equivalently, $h(x) \le l(x)$, for all $0 < x < 1$.

2   Results Based on Distorted Random Variables

In this section we provide some applications of (3) based on the comparisons of distorted random variables.

Theorem 1. *Let $X \in \mathcal{D}$ be a nonnegative random variable with quantile function Q and quantile density q. Let h and l be two distortion functions such that $h(x) \leq l(x)$, for all $0 < x < 1$, one has $\mathbb{E}[X_l] < \mathbb{E}[X_h] < +\infty$. Then, for a random variable U uniformly distributed in $(0,1)$ we have that*

$$\mathbb{E}\left[\left(\frac{q(1 - l^{-1}(1 - U))}{l'(l^{-1}(1 - U))} - \frac{q(1 - h^{-1}(1 - U))}{h'(h^{-1}(1 - U))}\right)(g(1) - g(U))\right]$$
$$= \mathbb{E}[g'(Z^L)](\mathbb{E}[X_l] - \mathbb{E}[X_h]),$$

where $\mathbb{E}[X_h] = \int_0^\infty h(\overline{F}(t))\mathrm{d}t$ and Z^L is the random variable with density function

$$f_{Z^L}(x) = \frac{Q(1 - l^{-1}(1 - x)) - Q(1 - h^{-1}(1 - x))}{\mathbb{E}[X_l] - \mathbb{E}[X_h]}, \qquad 0 < x < 1. \qquad (4)$$

Proof. Denoting as q_h and q_l the quantile densities corresponding to Q_h and Q_l, respectively. Then, for $0 < u < 1$, one has

$$q_h(u) = \frac{q(1 - h^{-1}(1 - u))}{h'(h^{-1}(1 - u))} \text{ and } q_l(u) = \frac{q(1 - l^{-1}(1 - u))}{l'(l^{-1}(1 - u))}.$$

Hence, the thesis follows from (3).

There exist several types of distortion functions that leads to special cases of interest. For instance, if $h(t) = t^\alpha$, then X_h and X_l correspond to the proportional hazard rate model (see, for instance, Balbás *et al.* [1] and Navarro *et al.* [9]). This suggests our first application.

Application 1. Let us consider the distortion functions $h(t) = t^\alpha$ and $l(t) = t^\beta$ for $0 < \beta < \alpha$ and $0 < t < 1$, so that $h(t) \leq l(t)$ for $0 < t < 1$. We can consider the distorted random variables induced by h and l, say X_h and X_l, respectively. Hence, due to Proposition 1, we have $X_h \leq_{st} X_l$. It is easy to see that the survival and quantile functions of X_h and X_l are

$$\overline{F}_h(x) = (\overline{F}(x))^\alpha \text{ and } \overline{F}_l(x) = (\overline{F}(x))^\beta, \qquad x > 0,$$

respectively. With straightforward calculations, we can obtain the corresponding quantile functions

$$Q_h(u) = Q(1 - (1 - u)^{1/\alpha}) \text{ and } Q_l(u) = Q(1 - (1 - u)^{1/\beta}),$$

and the quantile densities

$$q_h(u) = \frac{1}{\alpha}q(1 - (1 - u)^{1/\alpha})(1 - u)^{\frac{1}{\alpha} - 1} \text{ and } q_l(u) = \frac{1}{\beta}q(1 - (1 - u)^{1/\beta})(1 - u)^{\frac{1}{\beta} - 1},$$

for $0 < u < 1$, where Q and q are respectively the quantile and quantile density function of X. Then, from Theorem 1, we have

$$\mathbb{E}\left[\left(\frac{q(1 - (1 - U)^{1/\beta})}{\beta(1 - U)^{1-\frac{1}{\beta}}} - \frac{q(1 - (1 - U)^{1/\alpha})}{\alpha(1 - U)^{1-\frac{1}{\alpha}}}\right)(g(1) - g(U))\right]$$
$$= \mathbb{E}[g'(Z^L)](\mathbb{E}[X_l] - \mathbb{E}[X_h]),$$

where the density of Z^L, given in (4), becomes

$$f_{Z^L}(x) = \frac{Q(1 - (1 - x)^{1/\alpha}) - Q(1 - (1 - x)^{1/\beta})}{\mathbb{E}[X_l] - \mathbb{E}[X_h]}, \qquad 0 < x < 1,$$

with $\mathbb{E}[X_l] = \int_0^\infty (\overline{F}(x))^\alpha dx$ and $\mathbb{E}[X_h]$ similar. For instance, if X is uniformly distributed in $(0, 1)$, then $Q(u) = u$, and $q(u) = 1$, $0 < u < 1$, so that $\mathbb{E}[X_l] = (\alpha + 1)^{-1}$ and $\mathbb{E}[X_h] = (\beta + 1)^{-1}$. Therefore,

$$\mathbb{E}\left[\left(\frac{(1 - U)^{\frac{1}{\beta}-1}}{\beta} - \frac{(1 - U)^{\frac{1}{\alpha}-1}}{\alpha}\right)(g(1) - g(U))\right] = \mathbb{E}[g'(Z^L)]\frac{\alpha - \beta}{(\alpha + 1)(\beta + 1)},$$

where, for $0 < \beta < \alpha$,

$$f_{Z^L}(x) = \frac{(\alpha + 1)(\beta + 1)}{\alpha - \beta}((1 - x)^{1/\alpha} - (1 - x)^{1/\beta}), \qquad 0 < x < 1.$$

Next we consider $h(t) = 1 - (1 - t)^n$, $0 < t < 1$, for some positive integer n. Note that in this case $\mathbb{E}[X_h] = \mathbb{E}[\max\{X_1, \ldots, X_n\}]$, where $X_1, \ldots, X_n$ are i.i.d. random variables.

Application 2. Let us consider the distortion functions $h(t) = 1 - (1 - t)^n$ and $l(t) = 1 - (1 - t)^m$, $0 < t < 1$, for integers $1 \le n \le m$. It is not hard to see that these distortion functions refer to the proportional reversed hazard rate model. Hence, since $h(t) \le l(t)$, $0 < t < 1$, the corresponding distorted random variables satisfy $X_h \le_{st} X_l$ due to Proposition 1. The survival and quantile functions of X_h and X_l are respectively

$$\overline{F}_h(x) = 1 - F^n(x) \text{ and } \overline{F}_l(x) = 1 - F^m(x), \qquad x > 0,$$

$$Q_h(u) = Q(1 - u^{1/n}) \text{ and } Q_l(u) = Q(1 - u^{1/m}), \qquad 0 < u < 1,$$

where $F(x)$ and $Q(u)$ are the distribution and the quantile function of the i.i.d. random variables $X_1, X_2, \ldots$, respectively. If $M_k := \max\{X_1, \ldots, X_k\}$, for $k \ge 1$, then from Theorem 1 we have

$$\mathbb{E}\left[\left(\frac{q(U^{\frac{1}{m}})}{mU^{1-\frac{1}{m}}} - \frac{q(U^{\frac{1}{n}})}{nU^{1-\frac{1}{n}}}\right)(g(1) - g(U))\right] = \mathbb{E}[g'(Z^L)](\mathbb{E}[M_m] - \mathbb{E}[M_n]), \quad (5)$$

where

$$f_{Z^L}(x) = \frac{Q(x^{\frac{1}{m}}) - Q(x^{\frac{1}{n}})}{\mathbb{E}[M_m] - \mathbb{E}[M_n]}, \qquad 0 < x < 1.$$

For instance, if X is exponentially distributed with parameter $\lambda > 0$, it is known that, for each integer $n \geq 1$,

$$\mathbb{E}[M_n] = \frac{1}{\lambda} \int_0^1 \frac{1 - t^n}{1 - t} dt = \frac{1}{\lambda} H_n,$$

where $H_n := \sum_{k=1}^n \frac{1}{k}$ is the nth harmonic number. Hence, for $1 \leq n \leq m$, from (5) and given that $q(u) = [\lambda(1 - u)]^{-1}$, for all $0 < u < 1$, we have

$$\mathbb{E}\left[\frac{1}{U}\left(\frac{1}{m}\frac{1}{U^{-1/m} - 1} - \frac{1}{n}\frac{1}{U^{-1/n} - 1}\right)(g(1) - g(U))\right] = \mathbb{E}[g'(Z^L)](H_m - H_n),$$

where

$$f_{Z^L}(x) = \frac{1}{H_m - H_n} \log \frac{1 - x^{1/n}}{1 - x^{1/m}}, \qquad 0 < x < 1.$$

The following theorem involves a random variable having the NBU property. We recall that a nonnegative random variable X is said NBU if its survival function $\overline{F}$ satisfies $\overline{F}(s)\overline{F}(t) \geq \overline{F}(s + t)$, for all $s \geq 0$ and $t \geq 0$.

Theorem 2. *Let X be a nonnegative random variable with quantile density q, and having the NBU property, and let h be a distortion function. Then, for a random variable U uniformly distributed in $(0,1)$, and for all $t > 0$, one has*

$$\mathbb{E}\left[\left(\frac{q(1 - h^{-1}(1 - U))}{h'(h^{-1}(1 - U))} - \overline{F}(t)\frac{q(1 - h^{-1}(1 - U)\overline{F}(t))}{h'(h^{-1}(1 - U))}\right)(g(1) - g(U))\right]$$
$$= \mathbb{E}[g'(Z^L)](\mathbb{E}[X_h] \quad \mathbb{E}[(X_t)_h]),$$

where $\mathbb{E}[X_h] = \int_0^\infty h(\overline{F}(t))dt$ and, given $t > 0$, $(X_t)_h$ and Z^L are respectively the random variables having survival and density functions given by

$$\overline{F}_{(X_t)_h}(x) = h\left(\frac{\overline{F}(x + t)}{\overline{F}(t)}\right), \qquad x \geq 0,$$

$$f_{Z^L}(x) = \frac{Q(1 - h^{-1}(1 - x)) - Q(1 - h^{-1}(1 - x)\overline{F}(t)) + t}{\mathbb{E}[X_h] - \mathbb{E}[(X_t)_h]}, \qquad 0 < x < 1.$$

Proof. Denoting as q_h and $q_{t,h}$ the quantile densities of X_h and $(X_t)_h$, respectively, for $0 < u < 1$ we have

$$q_h(u) = \frac{q(1 - h^{-1}(1 - u))}{h'(h^{-1}(1 - u))} \text{ and } q_{t,h}(u) = \overline{F}(t)\frac{q(1 - h^{-1}(1 - u)\overline{F}(t))}{h'(h^{-1}(1 - u))}.$$

The thesis then follows from Theorem 1. $\qquad\square$

Application 3. Let X be an uniformly distributed random variable in $(0, 1)$. It is well known that X is NBU. The survival and quantile function of X_h are given respectively by

$$\overline{F}_h(x) = h(1 - x), \quad 0 < x < 1, \text{ and } Q_h(u) = 1 - h^{-1}(1 - u), \quad 0 < u < 1.$$

Similarly, for the distorted random variable $(X_t)_h$ given $t \in (0, 1)$, we have

$$\overline{F}_{(X_t)_h}(x) = h\left(\frac{1 - (t + x)}{1 - t}\right), \qquad 0 < x < 1 - t,$$

$$Q_{t,h}(u) = 1 - (1 - t)h^{-1}(1 - u), \qquad 0 < u < 1.$$

The corresponding quantile densities of X_h and $(X_t)_h$ are

$$q_h(u) = \frac{1}{h'(h^{-1}(1 - u))} \text{ and } q_{t,h}(u) = \frac{1 - t}{h'(h^{-1}(1 - u))}, \quad 0 < u < 1.$$

From Theorem 2, we have

$$\mathbb{E}\left[\left(\frac{1}{h'(h^{-1}(1 - U))} - \frac{1 - t}{h'(h^{-1}(1 - U))}\right)(g(1) - g(U))\right]$$
$$= \mathbb{E}[g'(Z^L)](\mathbb{E}[X_h] - \mathbb{E}[(X_t)_h]), \tag{6}$$

where Z^L is the random variable having density

$$f_{Z^L}(x) = \frac{(1 - t)h^{-1}(1 - x) - h^{-1}(1 - x)}{\mathbb{E}[X_h] - \mathbb{E}[(X_t)_h]}, \qquad 0 < x < 1.$$

Let us now consider $h(t) = \min\left\{\frac{t}{1-p}, 1\right\}$, for a fixed $p \in (0, 1)$. It can be seen that h is a proper distortion function, and that $\mathbb{E}[X_h] = \mathbb{E}[X|X > Q(p)]$ for any random variable X having quantile function $Q(p)$. Specifically, if X is uniformly distributed in $(0, 1)$, we have

$$\mathbb{E}[X|X > Q(p)] = \frac{1 + p}{2} \text{ and } \mathbb{E}[X_t|X_t > Q_t(p)] = \frac{(1 - t)(1 + p)}{2}.$$

Hence, since $h^{-1}(u) = (1 - p)u$ and $h'(u) = \frac{1}{1-p}$, $0 < u < 1$, (6) gives

$$(1 - p)\mathbb{E}[g(1) - g(U)] = \mathbb{E}[g'(Z^L)]\frac{1 + p}{2}, \qquad 0 < p < 1,$$

where the density of Z^L is

$$f_{Z^L}(x) = 2\frac{1 - (1 - p)(1 - x)}{1 + p}, \qquad 0 < x < 1.$$

3   Concluding Remarks

The quantile-based probabilistic mean value theorem proposed in [4] has been shown to be useful (i) to construct new probability densities with support $(0, 1)$ starting from suitable pairs of stochastically ordered random variables, and (ii) to obtain equalities involving uniform-$(0, 1)$ distributions and quantile functions. On this ground, further applications have been provided in the present paper based on distorted distributions, with special care to the cases when the underlying random variables satisfy the proportional hazard rate model, the proportional reversed hazard rate model, and the 'new better than used' property.

Acknowledgements. The research of A. Di Crescenzo and B. Martinucci has been performed under partial support by the Group GNCS of INdAM. J. Mulero acknowledges support received from the Ministerio de Economía, Industria y Competitividad under grant MTM2016-79943-P (AEI/FEDER, UE).

References

1. Balbás, A., Garrido, J., Mayoral, S.: Properties of distortion risk measures. Methodol. Comput. Appl. Probab. **11**, 385–399 (2009)
2. Denneberg, D.: Premium calculation: why standard deviation should be replaced by absolute deviation. ASTIN Bull. **20**, 181–190 (1990)
3. Di Crescenzo, A.: A probabilistic analogue of the mean value theorem and its applications to reliability theory. J. Appl. Probab. **36**, 706–719 (1999)
4. Di Crescenzo, A., Martinucci, B., Mulero, J.: A quantile-based probabilistic mean value theorem. Probab. Eng. Inf. Sci. **30**, 261–280 (2016)
5. Di Crescenzo, A., Meoli, A.: On the fractional probabilistic Taylor's and mean value theorems. Fract. Calc. Appl. Anal. **19**, 921–939 (2016)
6. Gupta, N., Misra, N., Kumar, S.: Stochastic comparisons of residual lifetimes and inactivity times of coherent systems with dependent identically distributed components. Eur. J. Oper. Res. **240**, 425–430 (2015)
7. Lin, G.D.: On a probabilistic generalization of Taylor's theorem. Stat. Prob. Lett. **19**, 239–243 (1994)
8. Massey, W.A., Whitt, W.: A probabilistic generalization of Taylor's theorem. Stat. Prob. Lett. **16**, 51–54 (1993)
9. Navarro, J., del Águila, Y., Sordo, M.A., Suárez-Llorens, A.: Preservation of stochastic orders under the formation of generalized distorted distributions. Applications to coherent systems. Methodol. Comput. Appl. Prob. **18**, 529–545 (2016)
10. Quiggin, J.: A theory of anticipated utility. J. Econ. Behav. Organ. **3**, 323–343 (1982)
11. Schmeidler, D.: Subjective probability and expected utility without additivity. Econometrica **57**, 571–587 (1989)
12. Sordo, M.A., Navarro, J., Sarabia, J.M.: Distorted Lorenz curves: models and comparisons. Soc. Choice Welf. **42**, 761–780 (2014)
13. Sordo, M.A., Suárez-Llorens, A.: Stochastic comparisons of distorted variability measures. Insur. Math. Econ. **49**, 11–17 (2011)
14. Sordo, M.A., Suárez-Llorens, A., Bello, A.J.: Comparison of conditional distributions in portfolios of dependent risks. Insur. Math. Econ. **61**, 62–69 (2015)
15. Wang, S.: Insurance pricing and increased limits ratemaking by proportional Hazards transforms. Insur. Math. Econ. **17**, 43–54 (1995)
16. Wang, S.: Premium calculation by transforming the layer premium density. ASTIN Bull. **26**, 71–92 (1996)
17. Yaari, M.E.: The dual theory of choice under risk. Econometrica **55**, 95–115 (1987)

Model-Based System Design, Verification and Simulation

One Degree of Freedom Copter

Peter Ťapák$^{(\boxtimes)}$ and Mikuláš Huba

Slovak University of Technology in Bratislava,
Ilkovičova 3, 812 19 Bratislava, Slovakia
{peter.tapak,mikulas.huba}@stuba.sk

Abstract. The paper presents laboratory model of unmanned aerial vehicle. The plant design, construction and building was a part of a project in the course on autonomous mechatronic systems. The goal was to make an exercise in mechatronics, resulting in plants which could be used in other project or classes, using the same hardware as common UAVs.

Keywords: Control education using laboratory equipment
Balance issues of theoretical-versus-practical training
Teaching curricula developments for control and other engineers

1 Introduction

The unmanned aerial vehicles (UAV) have been very popular for several years already. It is very common practice to use PD controllers for attitude stabilization and altitude control. In this paper, very simplified laboratory model of an UAV (Fig. 1) consisting of a platform with only one propeller is presented. The plant construction was inspired by the papers [1,5]. The altitude of the platform is measured by an ultrasonic sensor. Since the most popular UAVs, multicopters, do spend a lot of time in hover mode while performing their tasks, the altitude control of this vehicle ranks among the tasks requiring special attention. The altitude control algorithm presented in this paper uses ultra local model of the plant for tuning. The least squares method was employed in identification process. The controller performance has been evaluated taking into account the total variance of control signal, in other words the sum of all control signal changes which gives the reader nice overview on power consumption. The paper is organized as follows. Section 2 describes the plant construction, Sect. 3 presents the control loop programming. The Sect. 4 discusses the mathematical model of the plant, in Sect. 5 the real experiments results considering various filter used in the closed loop are presented, the Sect. 6 presents results of closed loop control by various controllers and their tunings. Contributions of the paper are summarized in Conclusions.

M. Huba—This work has been supported by the grant APVV-0343-12 Computer aided robust nonlinear control design.

© Springer International Publishing AG 2018
R. Moreno-Díaz et al. (Eds.): EUROCAST 2017, Part II, LNCS 10672, pp. 91–98, 2018.
https://doi.org/10.1007/978-3-319-74727-9_11

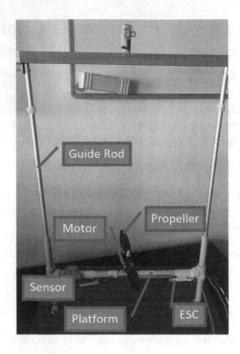

Fig. 1. One degree of freedom copter model

2   Plant Parts

The parts which the model consists of were chosen to be the same as can be used on a real multicopter. However, the cheap ones were used, because there was no need to lift a lot of payload. The model is equipped with BLDC motor A2212 for the copter propulsion. The motors came with the electronic speed controller (ESC) and a pair of propellers. To measure the platform altitude the SR04 ultrasonic sensor was bought. The Arduino UNO R3 microcontroller is used for control and communication with PC.

3   Real Time Control and Measurement

As the plant uses embedded controller, it is a crucial part of the design to be able to program the measurement and control algorithm in real time. It is necessary to choose the sampling time properly. The actuator/powertrain is brushless direct current motor, its outer body contains permanent magnets. The ESC is used to switch the phase of the windings to make the rotor run properly. The input of most of the basic ESCs is the square signal at the frequency of 50 Hz, where the pulse width corresponds to a desired speed of the rotor. The width of 1000 μs usually corresponds to the lowest speed, the 2000 μs corresponds to the full speed. Since the ESC input has usually the frequency of 50 Hz it would be enough to use the same sampling frequency for the controller. Nevertheless there is necessary to check if the measurement can be made within one period. We use the ultrasonic sensor, the maximum altitude is not more than 80 cm. Considering the speed of sound $v_s = 340.29$ [m/s], to measure the distance of 80 cm the sound has to travel twice the distance. It is trivial to calculate the time needed to travel the distance of 1.6 m, $t_s = 1.6/v_s = 0.0047$ [s], which is below 20 ms which is the sampling time at 50 Hz. It would be possible to measure the altitude more times within one sampling period, however it is measured only once for simplicity. To keep the timing as accurate as possible, timer interrupt was used. Arduino's CPU provides three timers: Timer0, Timer1 and Timer2. The 16bit Timer1 is used by servo library to generate the signal to control the ESC connected to digital pin 9 of the microcontroller board. 8bit Timer2 was chosen to generate interrupts for the control loop timing. The AVR CPU of Arduino board runs at 16 MHz. The maximal prescaler value for Timer2 is 128. To work at 50 Hz the prescaler was set to its maximum value 128, and it was set to CTC mode making interrupt and reseting the counter at the timer value 249, this corresponds to sampling time of 2 ms. To get 20 ms sampling, another counter of these interrupts was used and at every 10th interrupt when this counter is reset, the control action is applied the measurment is made and new control signal value is calculated. The standard deviation of the sampling time is 4.3865 µs, maximum deviation was 24 µs. These results are quite satisfactory for this low cost hardware and sufficient when considering the sampling time of 20000 µs.

4 Mathematical Model

In many papers, books and articles e.g. [2,4] the mathematical model of quadcopter starts with the rigid body dynamics.

Therefore, in this paper, the double integrator plus dead time model

$$G(s) = \frac{K_1}{s^2} e^{-T_D s} \tag{1}$$

is considered.

Since the body can not make any significant rotations, only one degree of freedom, movement up and down, was taken into account in the paper. The dynamics of the propeller is neglected for simplicity.

The following Fig. 2 shows the model (1) with parameters

$$K_1 = 0.97, T_D = 0.12 \tag{2}$$

obtained using least squares method matching the unfiltered measured data shifted to zero. The measurement was performed by making a step change of the system input at an altitude stabilized by PD controller.

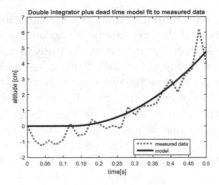

Fig. 2. Double integrator plus dead time model matching the measured data

5   Real Experiments - Filters

In this section the real time control using PD and PID control employing various filters is presented.

The PID controller's algorithm corresponds to the following equation

$$u = u_f + K_P e(t) + K_I \int_0^t e(\tau)d\tau - K_D \frac{dy(t)}{dt} \qquad (3)$$

where

- u_f is the bias compensating the gravitational force,
- K_P is proportional action gain,
- K_I is integral action gain,
- K_D is derivative action gain,
- $e = w - y$ is the control error,
- w is the setpoint,
- y is the altitude,
- u is the control signal.

The PD controller algorithm is the same considering the integral action gain to be $K_I = 0$. The first experiments were performed using no additional filter in the control loop. The system output derivative was realized as a difference. The second experiments were performed using the finite impulse response (FIR) filter in the form of the simple moving average using unweighted last five samples to calculate the filtered system output and its derivative. The third experiments were performed using 3rd order binomial filter

$$H_y(s) = \frac{1}{(T_f + 1)^3} \qquad (4)$$

with the time constant $T_f = 0.04$ for the system output filtration.

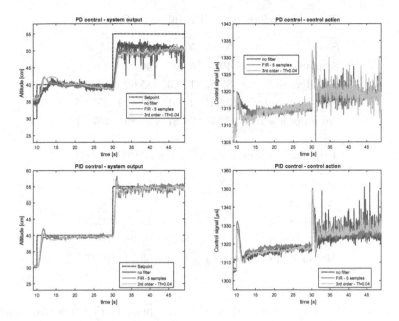

Fig. 3. Real experiment using PD controller with various filters - altitude (upper left) and control signal (upper right) - PID controller with various filters - altitude (lower left) and control signal (lower right)

The system output derivative was obtained from the measured output by the following filter

$$H_d(s) = \frac{s}{(T_f + 1)^3} \tag{5}$$

The discretized realizations of the transfer functions (4, 5) were used in the real time experiments. In Fig. 3 the PD controller with bias was used. All experiments start at steady state at 30 cm altitude. Then the setpoint step is made to 40 cm. Then at time 30 s the setpoint step to 55 cm was made. In the case of PD control, all the transients show steady state error, which is ok since they do not use any kind of integral action. The transients have the most noise in the cases when the FIR filter was used. In Fig. 3 one can see the experiment results where the PID controller with bias was used. In the case of PID control, lower P and D action gains are used due the presence of integral action. In these experiments the most noise can be seen using no filter. As it was mentioned in Sect. 1, the UAVs spend a lot of times in hover. Despite the fact they look barely moving in this mode, there is a lot of movement going on, keeping the vehicle in the spot due to the external disturbances and the system natural instability. It can be observed in the transients that the altitude and control signal are changing permanently during the flight. To evaluate the performance by the means of the energy efficiency total variance (TV) integral criterion can be used. It sums all

Table 1. Performance - TV values - PD control

Signal	Y	Y	U	U
Setpoint	40	55	40	55
No filter	85.62	178.35	85.62	178.35
FIR - 5 samples	22.31	40.31	140.79	253.24
3rd order low pass	7.54	24.87	95.89	287.37

Table 2. Performance - TV values - PID control

Signal	Y	Y	U	U
Setpoint	40	55	40	55
No filter	6.67	40.52	196.5	922.55
FIR - 5 samples	19.86	80.26	87.28	403.46
3rd order low pass	7.51	32.2	50.32	203.93

the changes of the signal over a specified time period. The following formula represents its discrete realization

$$TV(y) = \sum_{i=1}^{n} |y(n) - y(n-1)| \qquad (6)$$

where

- y represents the samples over the time period,
- n is the number of samples.

The performance is summarized in Tables 1 and 2, where two left columns represent the system output (Y) TV values for setpoints at 40 and 55 cm, the two right columns show the TV values of control signal (U) for these two setpoints. In Table 1 one can see that employing the 3rd order low pass filter yielded the lowest TV values altogether with PD controller, except the first setpoint when the loop without any filtration outperforms the loop with this filter. Nevertheless, the TV values for the loop with FIR filter are considerably higher than both the other options. In Table 2 one can see the PID without any filter outperforms the loop with FIR filter when considering only the output of the system, however when considering the TV of the control signal, the loop without filter is obviously the worst.

6   Real Experiments - Controller Design

In this chapter several control algorithms will be used to control the plant.

Since the binomial filter (4) yielded good results in Sect. 5 and can be used in the system output first and second derivative calculation as well, the same filter

Table 3. Controller parameters

Signal	K_P	K_I	K_D
PID_1	0.3	0.02	1
PID_2	1.2	0.2	2
PID_3	0.4	0.01	1.7

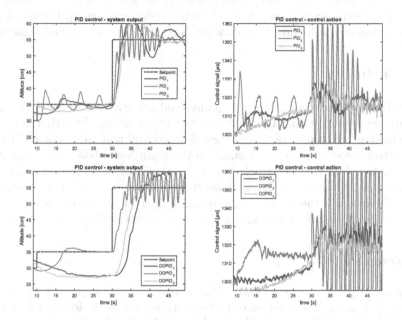

Fig. 4. Real experiment - PID tuning - altitude (upper left) and control signal (upper right) - DOPID tuning - altitude (lower left) and control signal (lower right)

but with the time constant $T_f = 0.1$ is going to be used in all the experiments in this chapter. The Matlab's pidtool was used to tune the controller based on the plant model (1). Three controller tunings summarized in Table 3 are presented in the paper. Figure 4 represents results obtained using the controller (3) and the PD controller expanded by the input disturbance observer based on the inverse plant model was used (DOPID). The input disturbance is obtained according to the following formula (see e.g. [3])

$$\hat{d}_i = 1/K_1 \frac{d^2 y(t)}{dt^2} - u(t) \tag{7}$$

where

- d_i is the input disturbance,
- $u(t)$ is system input,
- K_1 is process gain from (1),
- $y(t)$ is the altitude system output.

From the presented results there is obvious the tuning with the highest gains yields oscillatory behavior. The PID controller with the lowest integral action gain shows the least oscillations. In the case of DOPID controller, the tuning with the low proportional gain is not able to achieve the setpoint. The system output of the control loop with the DOPID with higer proportional gain oscillates around the setpoint. However the problem is the controller uses the second derivative of the system output which is hard to obtain using the ultrasonic sensor. So these experiments show that there is necessary to equip the plant with the accelerometer in the future to allow users to apply more complex algorithms.

7   Conclusion

The developing and controlling the plant enabled students to perform many tasks in the area of mechatronics. Presented experiments show that the plant can be used as a starting point before going to control real multicopter. The real time control based on the inverse model of the plant have shown the necessity of equipping the plant with the accelerometer to obtain the higher order derivatives of the altitude. Nevertheless, it will make the plant control even more close to the real UAVs since it is the same way the common inertial measurement units work and it will bring broader spectrum of the experiments available to the students.

References

1. Bermúdez-Ortega, J., Besada-Portas, E., López-Orozco, J., Bonache-Seco, J., de la Cruz, J.: Remote web-based control laboratory for mobile devices based on EJsS, Raspberry Pi and Node.js. IFAC-PapersOnLine 48(29), 158–163 (2015)
2. Garcia Carrillo, L.R., Dzul, A., Lozano, R., Pegard, C.: Quad Rotorcraft Control. Vision-Based Hovering and Navigation. Springer, Heidelberg (2012). http://hal.archives-ouvertes.fr/hal-00937561
3. Huba, M.: Designing robust controller tuning for dead time systems. In: International Conference on System Structure and Control. IFAC, Ancona (2010)
4. Nonami, K., Kendoul, F., Suzuki, S., Wang, W., Nakazawa, D.: Autonomous Flying Robots: Unmanned Aerial Vehicles and Micro Aerial Vehicles, 1st edn. Springer Publishing Company, Incorporated, New York (2010). https://doi.org/10.1007/978-4-431-53856-1
5. Sanchez, D.: Vertical rotor for the implementation of control laws. In: 9th IFAC Symposium Advances in Control Education. Nizni Novgorod, Russia (2012)

Prediction of Coverage of Expensive Concurrency Metrics Using Cheaper Metrics

Bohuslav Křena, Hana Pluháčková[✉], Shmuel Ur, and Tomáš Vojnar

IT4Innovations Centre of Excellence, FIT, Brno University of Technology,
Brno, Czech Republic
{krena,ipluhackova,vojnar}@fit.vutbr.cz, shmuel.ur@gmail.com

Abstract. Testing of concurrent programs is difficult since the scheduling non-determinism requires one to test a huge number of different thread interleavings. Moreover, a simple repetition of test executions will typically examine similar interleavings only. One popular way how to deal with this problem is to use the noise injection approach, which is, however, parametrized with many parameters whose suitable values are difficult to find. To find such values, one needs to run many experiments and use some metric to evaluate them. Measuring the achieved coverage can, however, slow down the experiments. To minimize this problem, we show that there are correlations between metrics of different cost and that one can find a suitable test and noise setting to maximize coverage under a costly metrics by experiments with a cheaper metrics.

1 Introduction

With the current massive use of multicore processors, concurrent programming has become widespread. Such programming is, however, far more challenging since apart from errors that one can cause in sequential code, there is a number of synchronization-related errors specific for concurrent code. What is worse, such errors are easy to cause but difficult to find since they often manifest under some very specific conditions only. Therefore, advanced approaches for finding such errors are highly needed.

A traditional, yet still dominating approach to finding errors is *testing*. However, for testing to be effective in the context of concurrent code, a special care must be taken to cope well with the nondeterminism of thread scheduling.

First, to steer the testing process, various *coverage metrics* are often used. When testing concurrent code, traditional coverage metrics (such as statement coverage) are not sufficient as they do not reflect how well the concurrent behaviour of the program under test has been exercised. Instead, one needs to use specialised metrics, such as coverage of concurrently executing instructions [4], synchronisation coverage [21], or coverage of internal states of dynamic analysers while chasing for various concurrency-related bugs [15]. Sometimes, maximizing coverage under several metrics at the same time is even used since they characterize different aspects of concurrent behaviour.

© Springer International Publishing AG 2018
R. Moreno-Díaz et al. (Eds.): EUROCAST 2017, Part II, LNCS 10672, pp. 99–108, 2018.
https://doi.org/10.1007/978-3-319-74727-9_12

To maximize coverage under a chosen concurrency coverage metric (or a combinations of such metrics), the space of possible thread schedules has to be properly examined. For that, simple repetitive execution of the same tests in the same environment does not help much [9]. Indeed, despite the scheduling is non-deterministic, some schedules may prevail. To deal with this problem, one can use the approach of *noise injection* [7,9] which influences the thread scheduling by injecting different kinds of noise (e.g., context switches or delays) into a program execution. However, there are many different heuristics for *generating noise* and for deciding *where to place it*, which are, moreover, heavily parametrized. This, in turn, leads to a need of solving the so-called *test and noise configuration setting* (TNCS) problem, which consists in finding the right parameters of noise generation together with the right test cases and suitable values of their possible parameters [11].

If the TNCS problem is not solved properly, the usage of noise can even decrease the obtained coverage [9]. However, solving the TNCS problem is not an easy task. Sometimes, its solution is not even attempted, and purely random noise generation is used. Alternatively, one can use genetic algorithms or data mining [1,11,12]. These approaches can outperform the purely random approach, but finding suitable test and noise settings this way can be quite costly. The aim of this paper is to make the cost of this process cheaper.

The approach which we propose builds on the facts that (1) maximizing coverage under different metrics may have different *costs*, and that (2) one can find *correlations* between test and noise settings that are suitable for maximizing coverage under different metrics. Moreover, such correlations may link even metrics for which the process of maximizing coverage is expensive but which are highly informative for steering the testing process and metrics for which the process of maximizing coverage is cheaper but which are less efficient when used for steering the testing process. We confirm all these facts through a set of our experiments. In particular, we identify the correlations by building a *predictive model* between several expensive metrics (under which one may want to simultaneously maximize coverage) and several cheap metrics.

Using the above facts, we suggest to *optimize the testing process* in the following way. Given some expensive but informative metrics, one may find suitable values of test and noise parameters for maximizing coverage under these metrics by experimenting with coverage under some cheap metric (or a combination of such metrics) and then use this setting for testing with the expensive metrics. We show on a set of experiments that this approach can indeed increase the efficiency of noise-based testing.

Our contribution is thus threefold: (1) An experimental categorisation of various concurrency-related metrics to cheap and expensive ones according to the price of maximizing coverage under these metrics. (2) The observation and experimental confirmation of correlations between test and noise settings suitable for testing under metrics of different cost. (3) The idea of exploiting the above facts for more efficient noise-based testing of concurrent programs and its experimental evaluation.

Related Work. There exist many ways how to verify concurrent programs such as systematic testing [13], coverage-driven testing [22], or various kinds of static analysis and model checking [16]. Compared with these techniques, noise-based testing, considered in this paper, is probably the most light-weight. In our previous works [1,11,12], we focused on solving the TNCS problem via genetic algorithms and data mining. Here, we propose an orthogonal optimisation based on solving the TNCS problem for expensive concurrency metrics by using cheaper ones, which is justified by existence of a predictive model between the expensive and cheap metrics. Prediction is used in various other areas of software testing, e.g., to predict bug severity [17] or to link concurrency-related code revisions with the corresponding issues and characterize bugs [5]. None of these works, however, builds on prediction in a similar way as this work.

2 Preliminaries

In this section, we briefly introduce noise injection, concurrency coverage metrics, as well as the benchmark programs and experimental setting used in the rest of the paper.

2.1 Noise Injection

The main principle of noise injection when testing concurrent programs is to influence the scheduling of concurrently executing threads by inserting various delays, context switches, temporary blocking of some threads, and/or additional synchronization into the execution. Two main questions to be resolved when applying noise injection are *how* to generate noise and *when* to generate it, which are referred to as the so-called *noise seeding* and *noise placement* problems. For solving these problems, a number of heuristics has been proposed [9]. In this work, we, in particular, use noise seeding and noise placement heuristics implemented within IBM ConTest [6] or on top of it.

2.2 Concurrency Coverage Metrics

A number of concurrency coverage metrics has been proposed in the literature. We build primarily on a selection of those discussed in [15], including both metrics concentrating on various generic aspects of concurrency behaviour as well as on metrics focusing on behavioural aspects relevant when chasing for various synchronisation defects. The former are represented by the *ConcurPairs*, *Synchro*, *WSynchro*, and *HBPair* metrics while the latter by *Avio*, *Avio**, *GoodLock* [3], *WEraser**, *GoldiLockSC**, and *Datarace*. Below, we briefly characterize those of these metrics that we use the most.

Coverage tasks of the *ConcurPairs* metric [4] consist of pairs of program locations that are checked to be reached in the given order and a Boolean flag indicating whether a context switch happened in between. Coverage tasks of the *Avio* metric are based on the Avio atomicity violation detector [18] and track which pairs of locations of one thread were interleaved with which locations reached in another thread. The *Avio** metric is the same as *Avio* up its tasks are enriched with an abstract identification of the involved threads (reflecting, e.g., their type).

The *WEraser** and *GoldiLockSC** metrics are based on coverage of internal states of the Eraser [19] and GoldiLocks [8] data race detectors (the latter with the so-called short-circuit checks), both of them extended with an abstract identification of the involved threads. The *Datarace* metric measures the number of data race notifications raised by the GoldiLock detector.

2.3   Benchmarks and Experimental Setting

The experimental results presented below are based on the following 10 multi-threaded benchmark programs written in Java: *Airlines* (0.3 kLOC), *Cache4j* (1.7 kLOC), *Animator* (1.5 kLOC), *Crawler* (1.2 kLOC), *Elevator* (0.5 kLOC), *HEDC* (12.7 kLOC), *Montecarlo* (1.4 kLOC), *Rover* (5.4 kLOC), *Sor* (7.2 kLOC) and *TSP* (0.4 kLOC). More details about these benchmarks can be found in [1,2]. All these benchmarks are available on the Internet[1]. All our experiments were performed using the IBM ConTest tool [6] on a machine with Intel Xeon E3-1240 v3 processors at 3.40 GHz, 32 GiB RAM, running Linux Debian 3.16.36, and using OpenJDK version 1.8.0_111.

3   Distinguishing Cheap and Expensive Metrics

We now explain our way of distinguishing cheap and expensive metrics, i.e., metrics for which collecting coverage is cheaper or more expensive, respectively.

For the classification of the cost of the metrics, we first ran a series of 1000 test runs of each of our benchmark programs without collecting any coverage. These tests were, however, run already in the ConTest environment, using its random noise setting, which already slows the programs down. This way, we obtained the so-called *bottom case*. The running time of the tests in the bottom case was around 93 s for one execution when averaging over all our case studies.

Second, for each metric, we performed 100 test runs while collecting coverage under the given metric, again using ConTest with random noise injection. We then compared the time needed for the bottom case with the times of the experiments with each single metric. We classify metrics into three groups: cheap metrics, expensive metrics, and others (i.e., metrics with medium slowdown). In particular, we mark metrics with the slowdown between 10% and 30% as cheap metrics and those with the slowdown 50% and more as expensive metrics.

Experimental Results. The obtained classification is shown in Table 1 and used in the further experiments.

---

[1] http://www.fit.vutbr.cz/research/groups/verifit/benchmarks/.

Table 1. Cheap and expensive metrics.

	Slowdown in %	Metrics
Cheap metrics	$10\% \leq x < 30\%$	Avio, Avio*, Concurpairs, HBPair, GoodLock, ShvarPair*, Synchro, WSynchro
Expensive metrics	$50\% \leq x$	Datarace, WEraser*, GoldiLockSC*

4 Discovering Correlations Between Cheap and Expensive Metrics

Next, we aim at automatically finding correlations between metrics that will allow us to find suitable test and noise settings for testing under expensive metrics by experimenting with cheaper ones. Due to multiple metrics are often used in testing of concurrent programs (each of them stressing somewhat different aspects of the behaviour), we, in fact, aim at correlations between sets of expensive metrics and sets of cheap metrics.

For the above, one can use *multi-variable regression* on the cumulative coverage of the different metrics obtained from multiple test runs (i.e., coverage based on a union of the sets of coverage tasks covered in the different runs). However, we, instead, decided to use the so-called *lasso* (least absolute shrinkage and selection operator) algorithm [10,14,20] to build a *predictive model* between cheap and expensive metrics. The algorithm selects suitable cheap metrics and constructs their linear combination capable of predicting a given expensive metric, hence showing correlation among the metrics. In our experience, this approach gives more stable results than normal correlation.

In more detail, we use the lasso algorithm to search for a combination of cheap metrics which has a high partial correlation coefficient with a chosen expensive metric. The algorithm iteratively increases the partial correlation and selects a subset of cheap metrics with the highest partial correlation. The obtained predictive model then looks as follows: $expMetric = \beta_0 + \beta_1 * cheapMetric^1 + \cdots + \beta_n * cheapMetric^n$.

Note that the above model predicts a single expensive metric based on several cheap ones. However, we said that, in general, we aim at maximizing coverage under *several* expensive metrics based on settings suitable for several cheap metrics. To cope with this, we propose to replace the role of the single expensive metric in the above model by using a *fitness function* representing a weighted combination of the chosen expensive metrics (as often done in genetic algorithms).

Such a combination can have the following form:

$$fitness = \frac{expMetric^1}{expMetric^1_{max}} + \cdots + \frac{expMetric^n}{expMetric^n_{max}}.$$

Here, $expMetric^i$ is the cumulative coverage under the i-th metric obtained in the given series of test runs with the same test and noise setting while

$expMetric^i_{max}$ is the maximum of all cumulative coverage values under the given metric in all so far performed experiments even with different test and noise settings (this way of approximating the maximum is used since there is no exact way of computing it).

Experimental Results. We have decided to experiment with finding suitable test and noise settings maximizing simultaneously coverage under all the three identified expensive metrics, i.e., $GoldiLockSC^*$, $WEraser^*$, and $Datarace$. The first step is to construct the fitness function combining these three metrics for the lasso algorithm. For that, we generated 100 random test and noise settings, ran 5 tests with each of the configurations, cumulating the coverage obtained in these runs. Finally, we took the maximum values of the cumulated coverage from the 100 experiments. This way, we obtained the following fitness function:

$$fitness = \frac{GoldiLockSC^*}{1443} + \frac{WEraser^*}{3862} + \frac{Datarace}{267}.$$

Second, we used the lasso algorithm with forward regression as implemented in the $glmnet()$ function from $glmnet$ package [14] of the R-project tool to obtain the predictive model. We created the predictive model from a cumulation of results from five runs on all the considered case studies.

In the forward lasso algorithm, one can choose how many cheap metrics one wants for the prediction. It is because the algorithm starts with an empty model, and, in each step, it adds one cheap metric to the previously built prediction model. Thus, one can see which cheap metrics form the model in each iteration. For our case, we choose to predict three expensive metrics by only two cheap metrics. In the future, it could be interesting to compare prediction using two, three, or four cheap metrics. We suppose that using more cheap metrics for the prediction could be more precise but also more time consuming.

Using the above approach, we obtained the following predictive model:

$$fitness = 2.9e - 01 + 2.2e - 06 * ConcurPairs + 1.8e - 03 * Avio^*.$$

This predictive model and also the above mentioned fitness function are used in all further experiments described below.

5 Using Correlations of Metrics to Optimize Noise-Based Testing

Once the predictive model is created and we know which set of cheaper metrics can be used to predict coverage under a given (set of) expensive metrics, this knowledge can be used to optimize the noise based testing process. In particular, we can try to find suitable test and noise settings for the given expensive metrics by experimenting with the cheap ones. The experiments can be controlled using a genetic algorithm [11,12], or data mining on the test results can be used [1], all the time evaluating the performed experiments via the chosen cheap metrics, or, more precisely, through the predictive model built. In the simplest case, one can perform just a number of random experiments with different test and noise settings and choose the settings that performed the best in these experiments wrt the predictive model. This is the approach we follow below to show that our approach can indeed improve the noise-based testing process.

Experimental Results. We randomly generated 100 test and noise configurations and executed 5 test runs with each of them for each of our case studies while collecting coverage under the selected cheap metrics (leading to 500 executions for each case study). We cumulated results within the 5 executions of one configuration and then worked with the obtained cumulative value. We chose 20 configurations with the best results wrt the derived predictive model. These 20 configurations were used for further test runs under the three considered expensive metrics. Each of the chosen 20 configurations was executed 200 times, leading to 4000 test executions under the three expensive metrics for each case study. Finally, to compare the efficiency of this approach with the purely random one, we also performed 4500 test runs with random test and noise settings while directly collecting coverage under the expensive metrics for each of the case studies. Hence, both of the approaches were given the same number of test runs.

In Table 2, we compare the random approach with our prediction-based approach. In particular, we aim at checking whether the proposed approach can help to increase the obtained coverage of the expensive metrics when weighted by the consumed testing time. From the table, we can see that this is indeed the case: the coverage over time increased in most of the cases. The average improvement of the obtained cumulative coverage over the testing time across all our case studies ranges from 46% to 62%.

Table 2. A comparison of random and prediction-optimized noise-based testing.

Case studies	GoldiLockSC*		WEraser*		Datarace	
	Random	Predict	Random	Predict	Random	Predict
Airlines	9.46	**22.42**	74.92	**182.59**	0.28	**0.72**
Animator	817.82	**1451.35**	233.20	**291.42**	0.35	**0.46**
Cache4j	0.93	**2.62**	4.14	**10.98**	0.03	**0.10**
Crawler	54.93	**88.69**	351.85	**547.41**	1.90	**2.86**
Elevator	**297.09**	286.30	**756.72**	733.91	**2.31**	2.23
HEDC	**27.50**	19.93	**67.37**	48.73	**0.50**	0.36
Montecarlo	4.24	**5.19**	9.03	**11.35**	0.02	**0.03**
Rover	37.62	**62.89**	174.14	**292.18**	0.08	**0.08**
Sor	3.19	**7.16**	4.93	**12.69**	0.00	**0.00**
TSP	**1.86**	1.40	**15.36**	11.74	**1.14**	0.86
Average impr.		**1.62**		**1.59**		**1.46**

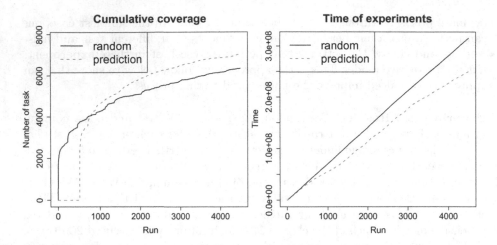

Fig. 1. Cumulative coverage (left) and testing time (right) for an increasing number of test runs.

Finally, Fig. 1(left) compares how the obtained cumulative coverage, averaged over all of our case studies, increases when increasing the number of performed test runs under the purely random noise-based approach and under our optimized approach. Our approach wins despite it has some initial penalty due to using some number of test runs to find suitable test and noise parameters via cheap metrics. The right part of the figure then compares the average time needed by the two approaches over all the case studies. Again, the optimized approach is winning.

6   Conclusion and Future Work

We have proposed an approach that uses correlations between cheap and expensive concurrency metrics to optimize noise-based testing under the expensive metrics by finding suitable values of test and noise parameters for such testing through experiments with the cheap metrics. Our experiments showed that such an approach can improve noise-based testing. In the future, we would like to combine the proposed optimization with previously proposed optimizations of noise-based testing via genetic optimizations and/or data mining. Moreover, it is interesting to generalize the idea of finding suitable noise settings maximizing coverage under an expensive metric via experiments with a cheap one to a context of dealing with other kinds of cheap and expensive analyses some of whose parameters may also be correlated.

Acknowledgement. The work was supported by the Czech Science Foundation (project 17-12465S), the internal BUT project FIT-S-17-4014, and the IT4IXS: IT4Innovations Excellence in Science project (LQ1602).

References

1. Avros, R., Hrubá, V., Křena, B., Letko, Z., Pluháčková, H., Ur, S., Vojnar, T., Volkovich, Z.: Boosted decision trees for behaviour mining of concurrent programs. In: Proceeding of MEMICS 2014. NOVPRESS (2014)
2. Avros, R., Hrubá, V., Křena, B., Letko, Z., Pluháčková, H., Ur, S., Vojnar, T., Volkovich, Z.: Boosted Decision Trees for Behaviour Mining of Concurrent Programs. Extended version of [1], under submission, 2017
3. Bensalem, S., Havelund, K.: Dynamic deadlock analysis of multi-threaded programs. In: Ur, S., Bin, E., Wolfsthal, Y. (eds.) HVC 2005. LNCS, vol. 3875, pp. 208–223. Springer, Heidelberg (2006). https://doi.org/10.1007/11678779_15
4. Bron, A., Farchi, E., Magid, Y., Nir, Y., Ur, S.: Applications of synchronization coverage. In: Proceeding of PPoPP 2005. ACM Press (2005)
5. Ciancarini, P., Poggi, F., Rossi, D., Sillitti, A.: Mining concurrency bugs. Embed. Multi-Core Syst. Mixed Criticality Summit, CPS Week (2016). http://www.artemis-emc2.eu/fileadmin/user_upload/Publications/2016_EMC2_Summit_Wien/15RCiancariniPoggiRossiSillittiConcurrencyBugs.pdf
6. Edelstein, O., Farchi, E., Nir, Y., Ratsaby, G., Ur, S.: Multithreaded Java Program Test Generation. IBM Syst. J. **41**, 111–125 (2002)
7. Edelstein, O., Farchi, E., Goldin, E., Nir, Y., Ratsaby, G., Ur, S.: Framework for testing multi-threaded Java programs. Concurrency Comput.: Pract. Experience **15**(35), 485–499 (2003)
8. Elmas, T., Qadeer, S., Tasiran, S.: Goldilocks: a race and transaction-aware Java runtime. In: Proceeding of PLDI 2007. ACM Press (2007)
9. Fiedor, J., Hrubá, V., Křena, B., Letko, Z., Ur, S., Vojnar, T.: Advances in noise-based testing of concurrent software. Softw. Test. Verification Reliab. **25**(3), 272–309 (2015)
10. Hastie, T., Tibshirani, R., Friedman, J.: The Elements of Statistical Learning. SSS. Springer, New York (2009). https://doi.org/10.1007/978-0-387-84858-7
11. Hrubá, V., Křena, B., Letko, Z., Ur, S., Vojnar, T.: Testing of concurrent programs using genetic algorithms. In: Fraser, G., Teixeira de Souza, J. (eds.) SSBSE 2012. LNCS, vol. 7515, pp. 152–167. Springer, Heidelberg (2012). https://doi.org/10.1007/978-3-642-33119-0_12
12. Hrubá, V., Křena, B., Letko, Z., Pluháčková, H., Vojnar, T.: Multi-objective genetic optimization for noise-based testing of concurrent software. In: Le Goues, C., Yoo, S. (eds.) SSBSE 2014. LNCS, vol. 8636, pp. 107–122. Springer, Cham (2014). https://doi.org/10.1007/978-3-319-09940-8_8
13. Hwang, G.-H., Lin, H.-Y., Lin, S.-Y., Lin, C.-S.: Statement-coverage testing for concurrent programs in reachability testing. J. Inf. Sci. Eng. **30**(4), 1095–1113 (2014)
14. James, G., Witten, D., Hastie, T., Tibshirani, R.: An Introduction to Statistical Learning. STS, vol. 103. Springer, New York (2013). https://doi.org/10.1007/978-1-4614-7138-7
15. Křena, B., Letko, Z., Vojnar, T.: Coverage metrics for saturation-based and search-based testing of concurrent software. In: Khurshid, S., Sen, K. (eds.) RV 2011. LNCS, vol. 7186, pp. 177–192. Springer, Heidelberg (2012). https://doi.org/10.1007/978-3-642-29860-8_14
16. Křena, B., Vojnar, T.: Automated formal analysis and verification: an overview. Int. J. Gen. Syst. **42**(4), 335–365 (2013). Taylor and Francis

17. Kwanghue, J., Amarmend, D., Geunseok, Y., Jung-Won, L., Byungjeong, L.: Bug severity prediction by classifying normal bugs with text and meta-field information. Adv. Sci. Technol. Lett. **129** (2016). Mechanical Engineering
18. Lu, S., Tucek, J., Qin, F., Zhou, Y.: AVIO: detecting atomicity violations via access interleaving invariants. In: Proceeding of ASPLOS 2006. ACM press (2006)
19. Savage, S., Burrows, M., Nelson, G., Sobalvarro, P., Anderson, T.: Eraser: a dynamic data race detector for multi-threaded programs. In: Proceeding of SOSP 1997. ACM press (1997)
20. Tibshirani, R.: Regression shrinkage and selection via the lasso. J. Roy. Stat. Soc. Ser. B **58**(1), 267–288 (1996)
21. Trainin, E., Nir-Buchbinder, Y., Tzoref-Brill, R., Zlotnick, A., Ur, S., Farchi, E.: Forcing small models of conditions on program interleaving for detection of concurrent bugs. In: Proceeding of PADTAD 2009. ACM Press (2009)
22. Yu, J., Narayanasamy, S., Pereira, C., Pokam, G.: Maple: a coverage-driven testing tool for multithreaded programs. In: Proceeding of OOPSLA 2012. ACM Press (2012)

Towards Smaller Invariants for Proving Coverability

Lenka Turoňová and Lukáš Holík[✉]

Faculty of Information Technology, Brno University of Technology,
Brno, Czech Republic
{turonova,holik}@fit.vutbr.cz

Abstract. In this paper, we explore a possibility of improving existing methods for verification of parallel systems. We particularly concentrate on safety properties of *well-structured transition systems*. Our work has relevance mainly to recent methods that are based on finding an *inductive invariant* by a sequence of refinements learned from counterexamples. Our goal is to improve the overall efficiency of this approach by concentrating on choosing refinements that lead to a more succinct invariants. For this, we propose to analyze so called *minimal counterexample runs*. They are digests of counterexamples concise enough to allow for a more detailed analysis. We experimented with a simple refinement algorithm based on analysing minimal runs and succeeded in generating significantly more succinct invariants than the state-of-the-art methods.

1 Introduction

Verification of parallel programs is challenging since they can generate a huge number of interleavings. This is called a *state space explosion* (the size of the state space grows exponentially to the number of processes). On top of that, the number of parallel processes may be unbounded or processes might be dynamically created which render the state space infinite. However, a large class of the parallel programs can be understood as *well structured transition systems* (WSTS) where many properties can be effectively verified. The class of WSTS include for instance Petri nets, lossy channel systems, or various broadcast and mutual exclusion protocols. We are interested especially in safety properties of WSTS, particularly in those that can be formulated as *coverability* of a set of incorrect configurations.

Our work is in the direction originating from or conceptually similar to backward reachability analysis [1], especially close to recent works [5,7,8] that are based on learning a safe inductive invariant of the similar form. A safe inductive invariant is a set of configurations of the system with the three following properties: (a) it contains its initial configurations; (b) it does not intersect with

This work was supported by the Czech Science Foundation project 16-24707Y, the BUT FIT project FIT-S-17-4014, and the IT4IXS: IT4Innovations Excellence in Science project (LQ1602).

© Springer International Publishing AG 2018
R. Moreno-Díaz et al. (Eds.): EUROCAST 2017, Part II, LNCS 10672, pp. 109–116, 2018.
https://doi.org/10.1007/978-3-319-74727-9_13

the target configurations; and (c) it is closed under the transition relation. The properties together are an inductive proof of safety of the system.

The methods [5,7,8] build-up the invariant by iterative steps, refining the current invariant approximation in order to satisfy inductiveness. A crucial component of this process is always a use of abstraction used to accelerate the process and regulated by a variation on counterexample-guided refinement (CEGAR).

The basic variant of CEGAR [4] runs the program within the abstract domain and in the case of reaching the target, the path to the target, so called *counterexample*, is analyzed. If the path is feasible in the real system then the target is reachable. Otherwise, the counterexample is spurious due to a too coarse abstraction and the path is used to refine the abstraction.

The counterexample analysis of [5,7,8] is based on forms of backward exploration of the state space starting in the target, using the operation *pre*. Although the methods differ in many aspects, it can be said that all of them perform essentially an eager backward exploration of the state space—they do not take into account any counterexample path in particular. Almost all found configurations (modulo some local and indeed powerful optimizations such as the "generalization" of algorithms [7,8]) are being used to refine the representation of the invariant approximation. Due to the eagerness, much of the added precision is unnecessary and makes the inferred invariants much more verbose than needed.

We conjecture that heuristics for finding more succinct invariants would improve efficiency of the discussed methods [5,7,8] significantly, for the following reasons: the invariant approximations are the most costly data that the methods work with, and so the price of the analysis depends directly on the size of the invariant representation. Moreover, more succinct invariants can be usually found by less invariant building steps.

Our approach. Our work comes out from the work [5], referred to as GBR. Its backward counterexample analysis is based on an exhaustive backward search through the state space implemented using the precise precondition operator.

We propose a modification of GBR which instead of the exhaustive depth-bounded backward search concentrates on a single entire counterexample path called *minimal counterexample run*. Since it is usually very concise we can analyze it thoroughly. From each such a run, we try to derive a "reason" for why the run could not be executed in the real system. Intuitively, it consists of parts of preconditions of transitions in the run that cannot occur (similarly to interpolants [9]). The minimal reason has a potential to be a part of the minimal inductive invariant since (1) it is necessary to refute the examined spurious counterexample run and (2) it is a "minimal" such "reason". In contrast to the eager strategy, many useless candidates for parts of the invariant and unsafe overapproximations can be avoided in this way.

2   Well-quasi-ordered Transition Systems

The well-quasi-ordered transition systems, *WSTS* for short, are systems with infinitely many configurations with a well-quasi-ordering (*wqo*), and whose transitions satisfy the monotonicity property. It has been shown that the coverability problem (reachability of an UCS) is decidable for $WSTS$ [2].

Formally, according to [8], $WSTS$ is a tuple $(\Sigma, I, \rightarrow, \succeq)$ where Σ stands for a set of configurations, a finite set $I \subset \Sigma$ is a set of initial configurations, $\rightarrow \subset \Sigma \times \Sigma$ is a transition on Σ, and a $\succeq \subset \Sigma \times \Sigma$ is a wqo.

The transition relation $\rightarrow$ between configurations is $monotonic$ wrt. to the relation $\succeq$ if for each two configurations s_0 and s_1 such that $s_0 \succeq s_1$ and a relation $s_0 \rightarrow s_0'$ there is a configuration s_1' such that $s_1 \rightarrow s_1'$.

The pre-order $\succeq$ is defined over a set S such that for any infinite sequence $s_0, s_1, s_1, \ldots$, there are i, j with $i < j$ and $s_i \succeq s_j$. If $\succeq$ is an equivalence relation, then the condition of $\succeq$ being a wqo amounts to the equivalence relation having a finite index [1].

Given a pre-order $\succeq$ defined over a set S set of configurations S, the upward-closure $T\uparrow$ of a subset $T \subseteq S$ and the downward closure $T\downarrow$ are defined as

$$T\uparrow \stackrel{\text{def}}{=} \{s \in S \mid \exists t \in T : t \succeq s\} \quad \text{and} \quad T\downarrow \stackrel{\text{def}}{=} \{s \in S \mid \exists t \in T : s \succeq t\}.$$

We define a set T to be an UCS, respectively a downward closed set (DCS), iff $T\downarrow = T$, respectively $T\uparrow = T$. If T is an UCS, its complement $S\backslash T$ is a DCS, and, conversely, if T is a DCS, its complement is an UCS [3]. Based on the monotocity of $\rightarrow$, for any UCS, the set of its predecessors is an UCS.

Let $x, y \in \Sigma$. If $x \rightarrow y$ we call x a $predecessor$ of y and y a $successor$ of x, and define

$$pre(x) := \{y \mid y \rightarrow x\} \quad \text{and} \quad post(x) := \{y \mid x \rightarrow y\}.$$

For $X \subset \Sigma$, $pre(X)$ and $post(X)$ are defined as usual, i.e.

$$pre(X) = \bigcup_{x \in X} pre(x) \quad \text{and} \quad post(X) = \bigcup_{x \in X} post(x).$$

For sets X and Y of configurations, we use $X \rightarrow Y$ to denote that there are $x \in X$ and $y \in Y$ such that $x \rightarrow y$. If there are configurations $x_0, \ldots, x_k \in \Sigma$ such that $x_0 = x, x_k = y$ and $x_i \rightarrow x_{k+1}$ for $0 \leq i < k$, then we write $x \stackrel{k}{\rightarrow} y$. Furthermore, $\stackrel{*}{\rightarrow}$ represents the reflexive transitive closure of $\rightarrow$. A set X of configurations is said to be reachable if $X_{init} \stackrel{*}{\rightarrow} X$. The set of k-reachable configurations, reachable in at most k steps, is defined as:

$$Reach_k := \{y \in \Sigma \mid \exists k' \leq k, \exists x \in I, x \stackrel{k'}{\rightarrow} y\}.$$

The set of all reachable configurations is formally defined as:

$$Reach := \bigcup_{k \geq 0} Reach_k = \{y \in \Sigma \mid \exists x \in I, x \stackrel{k}{\rightarrow} y\}.$$

Given a $WSTS$ $S = (\Sigma, I, \rightarrow, \succeq)$, we denote by $Cover$ the covering set of S, consisting of all configurations covered by some of the reachable configurations:

$$Cover(S) \stackrel{\text{def}}{=} post^*(I)\downarrow.$$

3   Testing Coverability Using CEGAR

Let us have given a WSTS $S_0 = (\Sigma, I, \rightarrow, \succeq)$ and a target configuration **bad**. The coverability problem means to decide whether a configuration from the set **bad**$\uparrow$ is reachable from the set of initial configurations I or not. A system where **bad**$\uparrow$ is not reachable is called safe. Formally, we say that the set **bad** is coverable if **bad** $\in Cover(S_0)$.

The method of [5] on which we build uses abstract interpretation with the abstraction defined by a set of configurations D, called domain. Every configuration $x \in D$ gives the abstraction means to distinguish configurations in $x\uparrow$ from the rest. We say that D is "expressive enough" when it can express a safe and inductive invariant, that is, if it has a subset V such that the complement of $V\uparrow$ is a safe inductive invariant. All runs of the system then stay within the complement of $V\uparrow$.

The domain is refined using CEGAR until unreachability or reachability of **bad**$\uparrow$ is proven. One iteration of the CEGAR loop takes the current value of D and generates an abstract forward run until it reaches a fixpoint, or until **bad** is reached. If the fixpoint is reached without reaching **bad**, then **bad** is not coverable. If it generates **bad**, then it is a counterexample run starting from the complement of $D\uparrow$ and ending to the target **bad**.

A counterexample run either signifies that **bad** is indeed reachable, or it is a spurious counterexample. A spurious counterexample is generated due to the abstraction not being able to distinguish certain dangerous configurations that can reach **bad** from some configurations that are reachable. The abstraction has then to be refined. This is done by adding the configurations from the counterexample run to D in order to give the abstraction the means of distinguishing the dangerous configurations from the reachable ones. CEGAR then continues by the next iteration with the refined abstraction. Otherwise, if the counterexample was not spurious, then it gives a proof of coverability of **bad**.

In the following sections, we present our modification of forward and backward analysis of the method [5].

3.1   Forward Abstract Interpretation

In this section, we recall the forward abstract interpretation of WSTS of [5] that uses an abstract domain parameterised by a finite set D of configurations. Initially, the input of the algorithm is given as a $WSTS \ S = (\Sigma, I, \rightarrow, \succeq)$, a parameter D and a target **bad**. For simplicity, we assume that D contains **bad** and the set I contains a single initial configuration x_0. The abstract interpretation algorithm runs the system in the abstract domain and decides whether the target **bad** is reachable from the initial set I or not.

Intuitively, a set of configurations E is abstracted into the set of elements form D that are covered by it (in a sense of $\preceq$). Formally, the abstraction is defined as

$$\forall E \in 2^X : \alpha[D](E) \stackrel{\text{def}}{=} E\downarrow \cap D,$$

while the concretisation function is defined as:

$$\forall P \in 2^D : \gamma[D](P) \stackrel{\text{def}}{=} \{x \in X \mid x{\downarrow} \cap D \subseteq P\}.$$

The abstract post operator in the domain defined by D is defined in a standard way as:

$$post^{\sharp}[D] \stackrel{\text{def}}{=} \alpha[D] \circ post \circ \gamma[D].$$

To find all configurations reachable in the abstract system from a set P, the forward steps are carried out till no new elements within the abstract domain defined by D are found, computing the image under the transitive closure of the abstract post:

$$post^{\sharp}[D]^*(P) \stackrel{\text{def}}{=} \bigcup_{i \geq 0} post^{\sharp}[D]^i(P).$$

The concretisation function γ may return an infinitely large set of configurations. Therefore, it is necessary to compute $post^{\sharp}$ in a symbolic manner. In particular, the post image consists of those elements $x \in D$ which have their predecessors in the concretization of the elements from the set P. Formally defined:

$$x \in post^{\sharp}(P) \Leftrightarrow (x \in D \land \neg(pre(x{\uparrow}) \subseteq (D \backslash P){\uparrow})).$$

After the set of reachable configurations is computed, it is necessary to check whether it contains the target **bad**. The target **bad** is unreachable if

$$bad \notin post^{\sharp}[D]^*.$$

Otherwise, if $bad \in post^{\sharp}[D]^*$, then the system can reach the target **bad** or the precision of the abstract domain defined by D is not sufficient and has to be refined.

Algorithm 1. Forward analysis

    input : a $WSTS\ S = (\Sigma, I, \rightarrow, \succeq)$, a parameter D and a target **bad**
    output: Is **bad** reachable?
1 $P_0 = \alpha[D](x_0)$
2 $i = 0$
3 path $= \epsilon$
4 **do**
5   **for** $t \in T$ **do**
6   **if** $t.\texttt{isEnabled}\ (P_i)$ **then**
7   path $=$ path $.\ t$
8           $P_{i+1} = P_i \cup post^{t\sharp}[D](P_i)$
9           $i = i + 1$
10 **while** $P_i \neq P_{i-1}$
11 **if** $bad \in P_i$ **then**
12   **return** UNREACHABLE
13 **else if** $\exists x_0, \ldots, x_k \in D : x_0 \rightarrow \cdots \rightarrow x_k$ and $x_k = bad$ **then**
14   **return** REACHABLE

Algorithm 1 implements the fixpoint computation of $post^{\sharp}[D]^*(P)$ and decides whether **bad** was or was not reached. The function t.isEnabled() returns **true** if there is a configuration in $\gamma[D](P_i)$ from where the transition t can be fired, and **false** otherwise. The function $post^{t\sharp}[D](P_i)$ represents a set of abstract successors of P_i under the transition t. It is a restriction of the post operator $post^{\sharp}$ to a single transition t, namely:

$$x \in post^{t\sharp}(P) \Leftrightarrow (x \in D \wedge \neg(pre^t(x\uparrow) \subseteq (D\backslash P)\uparrow)),$$

where $pre^t(X) = \bigcup_{x\in X} pre^t(x)$ and $pre^t(x) = \{y \mid y \xrightarrow{t} x\}$. Algorithm 1 is analogous to the forward search presented in [5] up that it also records the sequence $P_0, \ldots, P_n$ of the fixpoint approximations that and the sequence $t_1, \ldots, t_n$ of transitions that were taken to compute them. They will be used in the counterexample analysis.

3.2   Counterexample Analysis and Abstraction Refinement

The analysis of the counterexample run is based on the construction of the so called *minimal (abstract) counterexample run*. A minimal counterexample run is considered to be a run within an abstract domain, leading from an abstraction of initial set I to the target **bad** which is executed using a shortest sequence of transitions. The minimal counterexample run records how exactly were the elements of P_i, $i < n$ that were necessary for reaching **bad** generated by the forward analysis.

Algorithm 2. Construction of the graph

---

    input : $P_0, \ldots, P_n$; $t_1, \ldots, t_n$; a parameter D; a target **bad**;
             $G = (V, E)$ where $V = $ **bad** and $E = \emptyset$
    output: a minimal counterexample run represented by a DAG (V, E)

1   $n = $ length(path)
2   $W = \emptyset$
3   $W' = \{\textbf{bad}\}$
4   **for** $i \leftarrow n$ **to** 1 **do**
5      $W = W'$
6      $W' = \emptyset$
7   **for** node $\in W$ **do**
8   Choose t_i from path
9   **if** node $\in post^{t_i\sharp}(P_{i-1})$ **then**
10            $V = V \cup \{pre^{t_i}(\text{node})\} \cup \alpha[D](pre^{t_i}(\text{node}))$
11            $E = E \cup \{(\text{node}, pre^{t_i}(\text{node}))\} \cup \{(pre^{t_i}(\text{node}), x) \mid x \in$
              $\alpha[D](pre^{t_i}(\text{node}))\}$
12            $W = W\backslash\{\text{node}\}$
13            $W' = W' \cup \alpha[D](pre^{t_i}(\text{node}))$
14   break

Algorithm 2 constructs the minimal counterexample run in the form of a graph G based on the records of intermediate configurations and transitions taken during the forward analysis. Its nodes are the elements of $P_i's$ needed for reaching **bad** and also the concrete preconditions of these elements wrt. the transitions which generated them within the forward analysis.

We so far propose only a naive method for the analysis of minimal runs. It is sufficient to generate more succinct invariants, but it is not yet optimized for overall efficiency: From each minimal run, we randomly select a configuration from the DCS of preconditions and use it to extend D.

4 Experiments

We have implemented our method in a prototype tool *MINA* in Python and compared the size of invariants generated by our method with the invariants generated by the state-of-the-art methods BFC [7], IIC [8], and our implementation of the algorithm GBR on several verification tasks from the benchmark of MIST2 [6]. In BFC we have deactivated the coverability oracle which uses simple forward exploration to search for reachable configurations and excludes their downward closure from the candidates for invariant refinement.[1]

Table 1. A comparison of the size of the invariants generated by our method (Random Search), our implementation of the algorithm GBR [5], BFC [7] and IIC [8]. The size of the smallest invariant are typeset bold.

Benchmark name	MINA	GBR	BFC	IIC
basicME.spec	**6**	22	23	7
read-write.spec	**3**	Timeout	456	67
pingpong.spec	**14**	80	31	**14**
newrtp.spec	**45**	45	54	**45**
mesh2 × 2.spec	**10**	Timeout	16454	**10**
mesh3 × 2.spec	**10**	Timeout	Timeout	**10**
lamport.spec	**17**	68	70	33
newdekker.spec	**21**	Timeout	234	45
peterson.spec	**32**	135	191	67
multiME.spec	**7**	Timeout	64	**7**
manufacturing.spec	**6**	Timeout	39	**6**

Since our method chooses invariant refinements randomly, the size of the final invariant is variable. We therefore report the size of the most the most succinct invariant generated in 30 executions. The overall runtime was therefore much higher than that of the other tools. Most importantly, despite the naivety of choosing refinements from minimal counterexample runs, we have succeeded in

---

[1] Optimizations like this are rather orthogonal to the choice of the main algorithm.

generating significantly more succinct invariants than the other tools, as reported in Table 1. This supports our hypothesis that with a more advanced analysis of the minimal counterexample runs, our approach has a potential to be more efficient than the existing methods. Our future research will therefore focus on efficient analysis of minimal counterexample runs.

References

1. Abdulla, P.A.: Well (and better) quasi-ordered transition systems. Bull. Symb. Log. **16**(4), 457–515 (2010)
2. Abdulla, P.A., Cerans, K., Jonsson, B., Tsay, Y.-K.: General decidability theorems for infinite-state systems. In: IEEE Computer Society LICS 1996, pp. 313–321 (1996)
3. Bradley, A.R.: SAT-based model checking without unrolling. In: Jhala, R., Schmidt, D. (eds.) VMCAI 2011. LNCS, vol. 6538, pp. 70–87. Springer, Heidelberg (2011). https://doi.org/10.1007/978-3-642-18275-4_7
4. Clarke, E.M.: SAT-based counterexample guided abstraction refinement. In: Bošnački, D., Leue, S. (eds.) SPIN 2002. LNCS, vol. 2318, pp. 1–1. Springer, Heidelberg (2002). https://doi.org/10.1007/3-540-46017-9_1
5. Ganty, P., Raskin, J.-F., Van Begin, L.: A complete abstract interpretation framework for coverability properties of WSTS. In: Emerson, E.A., Namjoshi, K.S. (eds.) VMCAI 2006. LNCS, vol. 3855, pp. 49–64. Springer, Heidelberg (2005). https://doi.org/10.1007/11609773_4
6. Kaiser, A., Kroening, D., Wahl, T.: Bfc - a widening approach to multi-threaded program verification. http://www.cprover.org/bfc/
7. Kaiser, A., Kroening, D., Wahl, T.: Efficient coverability analysis by proof minimization. In: Koutny, M., Ulidowski, I. (eds.) CONCUR 2012. LNCS, vol. 7454, pp. 500–515. Springer, Heidelberg (2012). https://doi.org/10.1007/978-3-642-32940-1_35
8. Kloos, J., Majumdar, R., Niksic, F., Piskac, R.: Incremental, inductive coverability. In: Sharygina, N., Veith, H. (eds.) CAV 2013. LNCS, vol. 8044, pp. 158–173. Springer, Heidelberg (2013). https://doi.org/10.1007/978-3-642-39799-8_10
9. McMillan, K.L.: Interpolation and SAT-Based model checking. In: Hunt, W.A., Somenzi, F. (eds.) CAV 2003. LNCS, vol. 2725, pp. 1–13. Springer, Heidelberg (2003). https://doi.org/10.1007/978-3-540-45069-6_1

Simplifying Some Characteristics of Manipulators Based on Features of Their Models

Ignacy Duleba[✉] and Iwona Karcz-Duleba

Electronics Faculty, Wroclaw University of Technology,
Janiszewski St. 11/17, 50-372 Wroclaw, Poland
{ignacy.duleba,iwona.duleba}@pwr.edu.pl

Abstract. In this paper a method to simply computations of singular configurations of redundant and nonredundant manipulators was presented. Theoretical and numerical aspects of the method were given. Illustrating examples were carried out on models of planar pendula. The method can be also applied to simplify computations of manipulators' dynamics.

1 Introduction

Basic topics in robotics include forward and inverse kinematics, a Jacobian matrix and singularities of the latter. The first one describes a relationship between a configuration and an end-effector position and orientation, the second one maps velocities at joints into those of the end-effector, the third one checks a rank of the Jacobian matrix aimed at determining configurations where inverse kinematics, based on the Newton algorithm [1], is ill-conditioned. In textbooks on robotics [1,2] two ways to compute singular configurations are advised: either directly calculate the rank of the matrix or to check determinant of a manipulability matrix.

It appears that some characteristics of robots (e.g. singularities) do not depend on the first coordinate of the configuration vector. This useful observation is not pointed out in textbooks [1–3] although Kircansky [4], searching for isotropic configurations of a planar pendulum, noticed that Jacobian matrices are particularly simple at some coordinate frames. To popularize this fact, we will show its theoretical background and original practical implications significantly simplifying computations of the characteristics.

In the paper [5], a usefulness of exploiting a structure of robots' models was shown to gain some cognitive and practical benefits. The same paradigm is present in this paper, being an extended version of [6], which is organized as follows: in Sect. 2 a basic terminology is introduced, a theoretical background presented and a main result given. In Sect. 3 the result was illustrated on models of non-redundant and redundant pendula. In Sect. 4 it was extended to some characteristics of manipulators' dynamics. Section 5 concludes the paper.

© Springer International Publishing AG 2018
R. Moreno-Díaz et al. (Eds.): EUROCAST 2017, Part II, LNCS 10672, pp. 117–125, 2018.
https://doi.org/10.1007/978-3-319-74727-9_14

2   Theory and the Main Result

Forward kinematics is a mapping

$$\mathbb{Q} \ni \boldsymbol{q} \to \boldsymbol{x} = \boldsymbol{K}(\boldsymbol{q}) = \boldsymbol{A}_0^n(q) = \begin{bmatrix} \boldsymbol{R}_0^n & \boldsymbol{T}_0^n \\ \boldsymbol{0}_{1,3} & 1 \end{bmatrix} = \prod_{i=1}^{n} \boldsymbol{A}_{i-1}^i(q_i) \in \mathbb{X} \subset \mathbb{SE}(3), \qquad (1)$$

where $\mathbb{Q}$ is a configuration space (dim $\mathbb{Q} = n$), $\boldsymbol{q}$ is a configuration, $\mathbb{X}$ is a taskspace being a subspace of the special Euclidean group $\mathbb{SE}(3)$, $\boldsymbol{x}$ – a point corresponding to the configuration $\boldsymbol{q}$, $\boldsymbol{0}_{3,1}$ is a (1×3) matrix composed of zeroes, $\boldsymbol{A}_{i-1}^i \in \mathbb{SE}(3)$ is a transformation between the $(i-1)$st and the i-th coordinate frames (0-th frame is the base (global) one, the n-th frame is fixed at the end-effector). Usually coordinates ϕ (e.g. Cartesian for position and RPY/Euler angles for orientation) are introduced into $\mathbb{SE}(3)$ to get kinematics in coordinates $\boldsymbol{k}(\boldsymbol{q}) = \phi(\boldsymbol{K}(\boldsymbol{q}))$.

There exist a few types of Jacobian matrices [7] (in space, in body, geometric, analytic). Each of them transforms velocities from a configuration space appropriately defined velocities of the end-effector and depend on $\dot{\boldsymbol{R}}_0^n$ and $\dot{\boldsymbol{T}}_0^n$ of $\boldsymbol{A}_0^n$ (dot stands for the time derivative).

Based on an analytic Jacobian $\boldsymbol{J}(\boldsymbol{q}) = \partial \boldsymbol{k}(\boldsymbol{q})/\partial \boldsymbol{q}$, a manipulability index is defined

$$m(\boldsymbol{q}) = \sqrt{det(\boldsymbol{M}(\boldsymbol{q}))} = \sqrt{det(\boldsymbol{J}(\boldsymbol{q}) \cdot \boldsymbol{J}^T(\boldsymbol{q}))},$$

where $\boldsymbol{M}(\boldsymbol{q})$ is a manipulability matrix.

Fact 1: Singular configurations do not depend on q_1 coordinate.

A proof on independence of $m(\boldsymbol{q})$ on q_1 coordinate (or, equivalently, the rank of $\boldsymbol{J}(\boldsymbol{q})$ on q_1) can be carried out either geometrically or algebraically. In a geometric approach, it is noticed that at singular configurations columns of the Jacobian matrix (alternatively a manipulability matrix) become dependent in a global frame. An absolute value of determinant of the matrix describes a volume of parallelepiped spanned by the columns of the matrix. The volume is insensitive to rotations and translations of a global coordinate frame in which the Jacobian matrix was calculated. Thus, the global frame (0) can be shifted to the frame (let us call it $0b$) where the motion of the first joint is performed (transformation between the two frames is given by $\boldsymbol{A}_0^{0b} \in \mathbb{SE}(3)$, cf. (2)). If the 1st joint is revolute, so $0b$ frame can be rotated by any angle (q_1 is preferable in our case) without changing the manipulability index. The same argument is valid also for translational joint q_1. Thus, the manipulability index (and consequently singular configurations) can not depend on q_1.   □

The same result is obtained using algebraic arguments. This approach is more powerful because it is extendable to dynamic characteristics of manipulators. Kinematics (1) can be rewritten as

$$\boldsymbol{K}(\boldsymbol{q}) = \boldsymbol{A}_0^n(\boldsymbol{q}) = \boldsymbol{A}_0^{0a} \cdot \boldsymbol{A}_{0a}^{0b}(q_1) \cdot \boldsymbol{A}_{0b}^1 \cdot \boldsymbol{A}_1^2(q_2) \ldots \boldsymbol{A}_{n-1}^n(q_n). \qquad (2)$$

The transformation $A_0^1(q_1)$ was expressed as a composition of three matrices: constant A_0^{0a}, A_{0b}^1 (possible identity ones) and a q_1-dependent matrix $A_{0a}^{0b}(q_1)$ corresponding to a pure rotation or translation. Let us compute $\partial A_0^n(q)/\partial q_i$ for $i = 1, \ldots, n$, to get a structure of the resulting Jacobian matrix. For $i = 2, \ldots, n$, the derivative is of the form

$$\frac{\partial A_0^n(q)}{\partial q_i} = A_0^{0a} \cdot A_{0a}^{0b}(q_1) \cdot B_i(q_2, \ldots, q_n), \tag{3}$$

where B_i is a (4×4) matrix ($\notin \mathbb{SE}(3)$). It will be shown that $\partial A_0^n(q)/\partial q_1$ is also in the form (3). Taking a derivative of matrices from $\mathbb{SE}(3)$ is equivalent to multiplying them by appropriately defined matrix operators [8]. Rotation around canonical versors e_i, $i = 1, 2, 3$ (e.g. $e_2 = (0, 1, 0)^T$)

$$A_{0a}^{0b}(q_1) = Rot(e_i, q_1) \quad \Rightarrow \quad \frac{A_{0a}^{0b}(q_1)}{\partial q_1} = A_{0a}^{0b}(q_1) \begin{bmatrix} [e_i] & 0_{3,1} \\ 0_{1,3} & 0 \end{bmatrix} = A_{0a}^{0b}(q_1) E_R \tag{4}$$

where E_R is the operator for a rotational joint, and a skew-symmetric matrix $[e_i]$ corresponds to e_i (e.g. $[e_3] = \begin{bmatrix} 0 & -1 & 0 \\ 1 & 0 & 0 \\ 0 & 0 & 0 \end{bmatrix}$). For a translational joint

$$A_{0a}^{0b}(q_1) = Trans(e_i, q_1) \quad \Rightarrow \quad \frac{A_{0a}^{0b}(q_1)}{\partial q_1} = A_{0a}^{0b}(q_1) \cdot \begin{bmatrix} 0_{3,3} & e_i \\ 0_{1,3} & 0 \end{bmatrix} = A_{0a}^{0b}(q_1) \cdot E_T, \tag{5}$$

where E_T is an operator for a translational joint. Substituting (4) or (5) to the derivative $\partial A_0^n(q)/\partial q_1$, the form (3) is obtained

$$\frac{\partial A_0^n(q)}{\partial q_1} = A_0^{0a} \cdot \frac{\partial A_{0a}^{0b}(q_1)}{\partial q_1} A_{0b}^1 \cdot A_1^n(q_2, \ldots, n) = A_0^{0a} A_{0a}^{0b}(q_1) B_1(q_2, \ldots, n).$$

Now, we observe that the product $A_0^{0a} A_{0a}^{0b}(q_1) \in \mathbb{SE}(3)$, so all derivations (and also the Jacobian) can be expressed in $0b$ frame by multiplying all the items by the matrix $(A_0^{0a} \cdot A_{0a}^{0b}(q_1))^{-1}$

$$(A_0^{0a} A_{0a}^{0b}(q_1))^{-1} \frac{\partial A_0^n(q)}{\partial q_i} = (A_0^{0a} A_{0a}^{0b}(q_1))^{-1} A_0^{0a} A_{0a}^{0b}(q_1) B_i(q_2, \ldots, q_n) = B_i(q_2, \ldots, n). \tag{6}$$

Consequently, the Jacobian matrix, expressed in the $0b$ frame, does not depend on q_1 coordinate. $\qquad \square$

Fact 2: The $0b$ frame is not the only one in which Fact 1 holds. It is easy to check that at any frame $1, \ldots, n$, following the $0b$ frame, the property also holds and derivatives are equal to, cf. (6)

$$(A_1^i(q_2, \ldots, q_i))^{-1}(A_{0b}^1)^{-1} B_i(q_2, \ldots, n), \quad i = 2, \ldots, n.$$

The aforementioned method is well motivated theoretically but its practical applications are difficult as lead to complicated computations. From an engineering point of view, there is no need to shift and/or rotate the global frame to get all items in $0b$ frame or any other following it. As the manipulability index does not depend on q_1, thus, still remaining in the global frame, we can state the following useful fact.

Fact 3: While computing singularities, any value (denoted as $*$) can be assigned to the coordinate q_1

$$J(q) = J((*, q_2, \ldots, q_n)), \quad \text{or equivalently} \quad m(q) = m((*, q_2, \ldots, q_n)).$$

The selected value of q_1 should simplify computations of the Jacobian and singularities as well.

3   Illustrating Examples

Exemplary calculations were performed on models of $n = 2, 3$ dofs planar pendula, depicted in Fig. 1a, with kinematics in $\mathbb{SE}(3)$ for $n = 2$ given as

$$K(q) = Rot(e_3, q_1)Tran(e_1, a_1)Rot(e_3, q_2)Tran(e_1, a_2) \tag{7}$$

$$= \begin{bmatrix} c_{12} & -s_{12} & 0 & a_1c_1 + a_2c_{12} \\ s_{12} & c_{12} & 0 & a_1s_1 + a_2s_{12} \\ 0_{2,2} & & & I_2 \end{bmatrix},$$

where I_2 denotes the (2×2) identity matrix, $Rot(\text{axis, angle}), Tran(\text{axis, shift})$ denote standard robotic transformations [1], $q = (q_1, \ldots, q_n)^T$, and lengths of links a_i, $i = 1, \ldots, n$. c_{12} abbreviates $\cos(q_1 + q_2)$ and $s_1 = \sin(q_1)$.

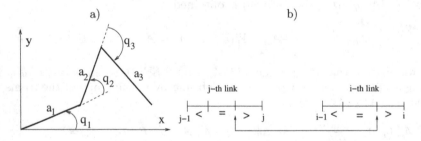

Fig. 1. (a) 2 and 3 dof pendulum, (b) frames, links and transformations

Further on, instead of calculating derivatives $\partial A/\partial q_i$, $i = 1, \ldots, n$, we compute one time derivative including all partial derivatives according to the formula $\dot{A} = \sum_{i=1}^{n} \partial A/\partial q_i \cdot \dot{q}_i$. Derivative of kinematics (7) in the base, global 0-th frame

$$\dot{A}_0^2 = \begin{bmatrix} -s_{12}(\dot{q}_1 + \dot{q}_2) & -c_{12}(\dot{q}_1 + \dot{q}_2) & 0 & -(a_1s_1 + a_2s_{12})\dot{q}_1 - a_2s_{12}\dot{q}_2 \\ c_{12}(\dot{q}_1 + \dot{q}_2) & -s_{12}(\dot{q}_1 + \dot{q}_2) & 0 & (a_1c_1 + a_2c_{12})\dot{q}_1 + a_2c_{12}\dot{q}_2 \\ 0_{2,2} & & & 0_{2,2} \end{bmatrix},$$

leads to the Jacobian in positional coordinates $(x, y)^T$

$$J^0 = \begin{bmatrix} -(a_1s_1 + a_2s_{12}) & -a_2s_{12} \\ a_1c_1 + a_2c_{12} & a_2c_{12} \end{bmatrix}. \tag{8}$$

Derivative of kinematics in the $0b$ frame, cf. (6)

$$(A_0^{0b})^{-1}\dot{A}_0^2 = \begin{bmatrix} -s_2(\dot{q}_1 + \dot{q}_2) & -c_2(\dot{q}_1 + \dot{q}_2) & 0 & -a_2s_2(\dot{q}_1 + \dot{q}_2) \\ c_2(\dot{q}_1 + \dot{q}_2) & -s_2(\dot{q}_1 + \dot{q}_2) & 0 & a_1\dot{q}_1 + a_2c_2(\dot{q}_1 + \dot{q}_2) \\ & \mathbf{0}_{2,2} & & \mathbf{0}_{2,2} \end{bmatrix},$$

in the 1-st frame

$$(A_0^1)^{-1}\dot{A}_0^2 = (A_{0b}^1)^{-1}(A_0^{0b})^{-1}\dot{A}_0^2 = (A_0^{0b})^{-1}\dot{A}_0^2, \tag{9}$$

(the last equality in (9) holds as $(A_{0b}^1)^{-1} = Tran(e_1, -a_1)$ does not change the matrix post-multiplying it) and in the 2-nd frame

$$(A_0^2)^{-1}\dot{A}_0^2 = \begin{bmatrix} 0 & -(\dot{q}_1 + \dot{q}_2) & 0 & a_1s_2\dot{q}_1 \\ \dot{q}_1 + \dot{q}_2 & 0 & 0 & a_1c_2\dot{q}_1 + a_2(\dot{q}_1 + \dot{q}_2) \\ & \mathbf{0}_{2,2} & & \mathbf{0}_{2,2} \end{bmatrix},$$

result in the Jacobians in these frames equal to, respectively

$$J^{0b} = \begin{bmatrix} -a_2s_2 & -a_2s_2 \\ a_1 + a_2c_2 & a_2c_2 \end{bmatrix} = J^1, \quad J^2 = \begin{bmatrix} a_1s_2 & 0 \\ (a_1c_2 + a_2) & a_2 \end{bmatrix},$$

which are independent on q_1 and Fact 1 is confirmed.

In order to show simplifications in computing singularities, let us apply Fact 3 to the Jacobian (8) in the global frame J^0 setting $q_1 = 0$.

$$\text{rank}(J(\mathbf{q}|q_1 = 0)) = \text{rank} \begin{bmatrix} -a_2s_2 & -a_2s_2 \\ a_1 + a_2c_2 & a_2c_2 \end{bmatrix} = \text{rank} \begin{bmatrix} -a_2s_2 & 0 \\ a_1 + a_2c_2 & -a_1 \end{bmatrix}. \tag{10}$$

From (10), it is easy to get the condition $a_1a_2s_2 = 0$ to determine singular configurations. It is met for $q_2 = k\pi/2$, $k \in Z$ and any value of q_1.

Another example illustrates computing singular configurations for a redundant manipulator. Positional kinematics of the 3 dof pendulum, Fig. 1a, with configuration $\mathbf{q} = (q_1, q_2, q_3)^T$

$$(x, y)^T = \begin{bmatrix} a_1c_1 + a_2c_{12} + a_3c_{123} \\ a_1s_1 + a_2s_{12} + a_3s_{123} \end{bmatrix}$$

generates the Jacobian matrix

$$J(\mathbf{q}) = \begin{bmatrix} -(a_1s_1 + a_2s_{12} + a_3s_{123}) & -(a_2s_{12} + a_3s_{123}) & -a_3s_{123} \\ a_1c_1 + a_2c_{12} + a_3c_{123} & a_2c_{12} + a_3c_{123} & a_3c_{123} \end{bmatrix}.$$

The first step of simplifications relies on manipulating with columns (subtracting them) of the Jacobian matrix and excluding constant terms that depend on lengths of links a_1, a_2, a_3 by multiplying columns by factors of $1/a_i$ (all the transformation do not change the rank of a matrix). Thus,

$$\text{rank}(J(\mathbf{q})) = \text{rank} \begin{bmatrix} s_1 & s_{12} & s_{123} \\ c_1 & c_{12} & c_{123} \end{bmatrix}.$$

Then, it follows the step presented in this paper. The value of q_1 is set to 0

$$\text{rank}(J(\boldsymbol{q})) = \text{rank} \begin{bmatrix} 0 & s_2 & s_{23} \\ 1 & c_2 & c_{23} \end{bmatrix} = 1 + \text{rank} \begin{bmatrix} s_2 & s_{23} \end{bmatrix}.$$

At singular configurations, the rank is smaller than 2. So, all singular configurations are characterized by $q_2 = k\pi/2$, $q_3 = l\pi/2$, where k, l are integers and q_1 takes any value.

4   Simplifying Model of Dynamics

The same method can be utilized to simplify computing some terms of a model of dynamics for manipulators. The dynamics is given as [1–3]

$$\boldsymbol{Q}(\boldsymbol{q})\ddot{\boldsymbol{q}} + \boldsymbol{C}(\boldsymbol{q}, \dot{\boldsymbol{q}})\dot{\boldsymbol{q}} + \boldsymbol{D}(\boldsymbol{q}) = \boldsymbol{u}.$$

where is $\boldsymbol{Q}$ a symmetric inertia matrix, $\boldsymbol{C}$ is a matrix of centripetal and Coriolis forces, $\boldsymbol{D}$ is a gravity vector, $\boldsymbol{u}$ are controls at joints. Due to shortage of space, the simplification will be illustrated on computing the $Q_{ij}(\boldsymbol{q})$ term only which is given in a standard Lagrange-based form as follows [1,8]

$$Q_{ij}(\boldsymbol{q}) = \text{tr}(\sum_{s=i}^{n} \frac{\partial \boldsymbol{A}_0^s}{\partial q_i} \boldsymbol{J}_s (\frac{\partial \boldsymbol{A}_0^s}{\partial q_j})^T) = \text{tr}(\sum_{s=i}^{n} (\frac{\partial \boldsymbol{A}_0^s}{\partial q_j})^T \frac{\partial \boldsymbol{A}_0^s}{\partial q_i} \boldsymbol{J}_s), \quad i \geq j. \tag{11}$$

Here $\boldsymbol{J}_s$ stands for a pseudo-inertia matrix of the s-th link and tr is a trace operator. At first a useful notation is introduced to denote splitting any transformation matrix $\boldsymbol{A}_{r-1}^r$ (illustrated in Fig. 1b)

$$\boldsymbol{A}_{r-1}^r(q_r) = \overset{<}{\boldsymbol{A}_{r-1}^r} \overset{=}{\boldsymbol{A}_{r-1}^r}(q_r) \overset{>}{\boldsymbol{A}_{r-1}^r} \tag{12}$$

into fixed (pre- and post-) multiplying matrices, and a matrix (with symbol $=$) corresponding to variable of motion in the r-th joint (example: for the 2D pendulum, cf. (7), the matrix $\boldsymbol{A}_0^1(q_1)$ is split into $\boldsymbol{I}_4$, $Rot(e_3, q_1)$, $Tran(e_1, a_1)$, respectively).

Now, we will concentrate on the product of first two matrices in (4). Using derivative operators, cf. (4), (5) and the form (12), one can get

$$\frac{\partial \boldsymbol{A}_0^s}{\partial q_j} = \boldsymbol{A}_0^{j-1} \overset{<}{\boldsymbol{A}_{j-1}^j} \frac{\partial \overset{=}{\boldsymbol{A}_{j-1}^j}}{\partial q_j} \overset{>}{\boldsymbol{A}_{j-1}^j} \boldsymbol{A}_j^s = \boldsymbol{A}_0^{j-1} \overset{<}{\boldsymbol{A}_{j-1}^j} \overset{=}{\boldsymbol{A}_{j-1}^j} \boldsymbol{E}_j \overset{>}{\boldsymbol{A}_{j-1}^j} \boldsymbol{A}_j^s = \overset{<=}{\boldsymbol{A}_0^j} \boldsymbol{E}_j \overset{>}{\tilde{\boldsymbol{A}}_j^s}. \tag{13}$$

Then, using properties of the special Euclidean group $\mathbb{SE}(3)$, the transposition and respecting $i \geq j$, one gets

$$(\frac{\partial \boldsymbol{A}_0^s}{\partial q_j})^T \frac{\partial \boldsymbol{A}_0^s}{\partial q_i} = \underbrace{(\overset{>}{\tilde{\boldsymbol{A}}_j^s})^T \boldsymbol{E}_j^T (\overset{<=}{\boldsymbol{A}_0^j})^T}_{(1)} \underbrace{\overset{<=}{\boldsymbol{A}_0^i} \boldsymbol{E}_i}_{(2)} \underbrace{\overset{>}{\tilde{\boldsymbol{A}}_i^s}}_{(3)}. \tag{14}$$

where the first term does not depend on variables $1, \ldots, j$, the second may depend only on variables $1, \ldots, i$ and the third does not depend on variables $1, \ldots, i$. As we want to prove that $\boldsymbol{Q}_{ij}$ does not depend on q_j and any q_r, $r < j$, thus only the second term in (14) will be processed.

$$\boldsymbol{E}_j^T (\overset{<=}{\boldsymbol{A}_0^j})^T \overset{<=}{\boldsymbol{A}_0^i} \boldsymbol{E}_i \tag{15}$$

The product of two middle matrices in (15), cf. (1) (we omit the symbol $<=$ to abbreviate notations)

$$\begin{bmatrix} (\boldsymbol{R}_0^j)^T & \boldsymbol{0}_{3,1} \\ (\boldsymbol{T}_0^j)^T & 1 \end{bmatrix} \begin{bmatrix} \boldsymbol{R}_0^i & \boldsymbol{T}_0^i \\ \boldsymbol{0}_{3,1} & 1 \end{bmatrix} = \begin{bmatrix} \tilde{\boldsymbol{R}}_j^i & (\boldsymbol{R}_0^j)^T \boldsymbol{T}_0^i \\ (\boldsymbol{T}_0^j)^T \boldsymbol{R}_0^i & (\boldsymbol{T}_0^j)^T \boldsymbol{T}_0^i + 1 \end{bmatrix}. \tag{16}$$

It is worth noticing, cf. Fig. 1b, that $\tilde{\boldsymbol{R}}_j^i = \overset{>}{\boldsymbol{R}_{j-1}^j} \boldsymbol{R}_j^{i-1} \overset{<=}{\boldsymbol{R}_{i-1}^i}$ does not depend on coordinate q_j. Now a detailed analysis should be performed for any pair (j, i) of rotational/translational joints.

Case 1: rotational (j) – rotational (i) pair. In this case, cf. (15), (16), we get

$$\begin{bmatrix} [e_j]^T & \boldsymbol{0}_{3,1} \\ \boldsymbol{0}_{1,3} & 0 \end{bmatrix} \begin{bmatrix} \tilde{\boldsymbol{R}}_j^i & * \\ * & * \end{bmatrix} \begin{bmatrix} [e_i] & \boldsymbol{0}_{3,1} \\ \boldsymbol{0}_{1,3} & 0 \end{bmatrix} = \begin{bmatrix} [e_j]^T \tilde{\boldsymbol{R}}_j^i [e_i] & \boldsymbol{0}_{3,1} \\ \boldsymbol{0}_{1,3} & 0 \end{bmatrix} \tag{17}$$

where $*$ denotes an unimportant value. As operators $[e_i], [e_j]$ are constant and the matrix $\tilde{\boldsymbol{R}}_j^i$ do not depend on q_j (and on all $q_r, r \leq j$) we conclude, collecting (11)–(17) that $\boldsymbol{Q}_{ij}$ do not depend on any variable q_r with $r \leq j$.

Case 2: rotational (j) – translational (i) pair

$$\begin{bmatrix} [e_j]^T & \boldsymbol{0}_{3,1} \\ \boldsymbol{0}_{1,3} & 0 \end{bmatrix} \begin{bmatrix} \tilde{\boldsymbol{R}}_j^i & * \\ * & * \end{bmatrix} \begin{bmatrix} \boldsymbol{0}_{3,3} & e_i \\ \boldsymbol{0}_{1,3} & 0 \end{bmatrix} = \begin{bmatrix} \boldsymbol{0}_{3,3} & [e_j^T] \tilde{\boldsymbol{R}}_j^i e_i \\ \boldsymbol{0}_{1,3} & 0 \end{bmatrix}.$$

Case 3: translational (j) – rotational (i) pair

$$\begin{bmatrix} \boldsymbol{0}_{3,3} & \boldsymbol{0}_{3,1} \\ e_j^T & 0 \end{bmatrix} \begin{bmatrix} \tilde{\boldsymbol{R}}_j^i & * \\ * & * \end{bmatrix} \begin{bmatrix} [e_i] & \boldsymbol{0}_{3,1} \\ \boldsymbol{0}_{1,3} & 0 \end{bmatrix} = \begin{bmatrix} \boldsymbol{0}_{3,3} & \boldsymbol{0}_{3,1} \\ e_j^T \tilde{\boldsymbol{R}}_j^i [e_i] & 0 \end{bmatrix}.$$

Case 4: translational (j) – translational (i) pair

$$\begin{bmatrix} \boldsymbol{0}_{3,3} & \boldsymbol{0}_{3,1} \\ e_j^T & 0 \end{bmatrix} \begin{bmatrix} \tilde{\boldsymbol{R}}_j^i & * \\ * & * \end{bmatrix} \begin{bmatrix} \boldsymbol{0}_{3,3} & e_i \\ \boldsymbol{0}_{1,3} & 0 \end{bmatrix} = \begin{bmatrix} \boldsymbol{0}_{3,3} & \boldsymbol{0}_{3,1} \\ e_j^T \tilde{\boldsymbol{R}}_j^i [e_i] & 0 \end{bmatrix}.$$

In Cases 2–4 conclusions are the same as in Case 1. Summing up all the cases:

1. $\boldsymbol{Q}_{ij}(q)$ with $i \geq j$ does not depend on all variables q_r, $1 \leq r \leq j$
2. while computing $\boldsymbol{Q}_{ij}(q)$, any values can be set to all variables q_r, $r \leq j$ in auxiliary computations.

The following example illustrate this issue. While computing equations of dynamics, for an exemplary transformation matrix

$$\boldsymbol{A}_0^3 = Rot(\boldsymbol{e}_3, q_1) Tran(\boldsymbol{e}_1, a_1) Rot(\boldsymbol{e}_2, q_2) Tran(\boldsymbol{e}_1, a_2) Rot(\boldsymbol{e}_3, q_3) Tran(\boldsymbol{e}_3, a_3)$$

one needs to evaluate $\partial \boldsymbol{A}_0^3 / \partial q_3$ having known that it does not depend on q_1 and q_2 variables. In this case, instead of processing the expression in a standard way

$$\frac{\partial \boldsymbol{A}_0^3}{\partial q_3} = Rot(\boldsymbol{e}_3, q_1) Tran(\boldsymbol{e}_1, a_1) Rot(\boldsymbol{e}_3, q_2) Tran(\boldsymbol{e}_1, a_2) Rot(\boldsymbol{e}_2, q_3) \, \boldsymbol{E}_3 \, Tran(\boldsymbol{e}_3, a_3),$$

we assume that $q_1 = q_2 = 0$, thus $Rot(\boldsymbol{e}_3, q_1)$ and $Rot(\boldsymbol{e}_3, q_2)$ becomes identity matrices and can be neglected in a chain of multiplications. Finally, computations are simplified significantly $\frac{\partial \boldsymbol{A}_0^3}{\partial q_3} = Tran(\boldsymbol{e}_1, a_1 + a_2) Rot(\boldsymbol{e}_2, q_3) \cdot \boldsymbol{E}_3 \cdot Tran(\boldsymbol{e}_3, a_3)$.

5   Conclusions

In this paper simplifications in computing some kinematic and dynamic characteristics of manipulators were presented. Possible practical applications are based on speeding-up on-line computations of the characteristics and can be exploited either within a computer software (like Mathematica) or by lecturers of robotics at classes. The approach can also extend simplifications presented in [5]. The reasoning presented in this paper is heavily based on a geometric structure of $\mathbb{SE}(3)$, therefore it can not be applied to other concepts of kinematics (like endogenous kinematics for nonholonomic manipulators [7]).

Acknowledgments. The research was supported by WUST within grant no. 0401/0209/16.

References

1. Spong, M., Vidyasagar, M.: Robot Dynamics And Control. Wiley, London (1989)
2. McKerrow, P.J.: Introduction to Robotics. Electronic Systems Engineering Series. Addison-Wesley Publishing, Boston (1991)
3. Murray, R.M., Li, Z., Sastry, S.: A Mathematical Introduction to Robotic Manipulation. CRC Press, Boca Raton (1994)
4. Kircanski, M.: Robotic isotropy and optimal robot design of planar manipulators. In: Proceedings IEEE International Conferene on Robotics and Automation, vol. 2, pp. 1100–1105 (1998)
5. Duleba, I.: Structural properties of inertia matrix and gravity vector of dynamics of rigid manipulators. J. Field Robot. **19**(11), 555–567 (2002)
6. Duleba, I., Karcz-Duleba, I.: Simplifying computations of singular configurations using features of manipulators' models. In: Moreno-Diaz, R., Pichler, F.R., Quesada-Arencibia, A. (eds.) EUROCAST, Las Palmas, pp. 148–149 (2017). (extended abstract)

7. Tchon, K., Jakubiak, J.: Endogenous configuration space approach to mobile manipulators: a derivation and performance assessement of Jacobian inverse kinematics algorithms. Int. J. Control **26**(14), 1387–1419 (2003)
8. Brady, M., Hollerbach, J.M., Johnson, T.L., Lozano-Perez, T., Mason, M.T. (eds.): Robot Motion. Planning and Control. Artificial Intelligence. MIT Press, Cambridge (1982)

A Study of Design Model for IoT System

Atsushi Ito[1(✉)], Yuko Hiramatsu[2], Hiroyuki Hatano[1], Mie Sato[1],
Akira Sasaki[3], Fumihiro Sato[2], and Yu Watanabe[1]

[1] Information System Science Department, Graduate School of Engineering,
Utsunomiya University, Utsunomiya, Tochigi 321-8505, Japan
{at.ito,hatano,mie,yu}@is.utsunomiya-u.ac.jp
[2] Faculty of Economics, Chuo University, Hachioji, Tokyo 192-039, Japan
{susana_y,fsato}@tamacc.chuo-u.ac.jp
[3] GClue Inc., Aizu-wakamatsu, Fukushima 965-8580, Japan
akira@gclue.jp

Abstract. A large amount of information is not always appealing to
tourists. We developed a new application using Bluetooth Low Energy
(BLE) beacons that not only provides a guide to a specific location but
also provides the traditional customs and history of the area as an easter
egg. In this paper, we discuss a system development process for IoT system of development of our sightseeing support system for IoT system. We
extract the development model that is suitable for developing a service
that uses devices embedded in environment and maintaining it.

Keywords: Location-based service · Embedded system
BLE beacon · Sightseeing support application
System development model · World cultural heritage · Zeigarnik effect

1 Introduction

A large amount of information is not always appealing to tourists. We have been
developing a new sightseeing support application using Bluetooth Low Energy
(BLE) beacons that not only provides a guide to a specific location but also
provides the traditional customs and history of the area [1]. Information and
communication technology (ICT) is widely used for travel and tourism and has
now made considerable information available. Tourists get information about
maps, shops, accommodations, museums, events etc. However, the plethora of
information available on the Web is not always appealing to tourists. We, so that,
have to consider what information is appealing to tourists, when they should
receive it and who the target audience for this information should be. In 2014, we
investigated the information needs of tourists in Nikko and tested the provision
of information using a Bluetooth Low Energy (BLE) beacon system. Before
using BLE to provide information for tourists, we re-inspected and analyzed the
information contents and designed behaviour and UX of our system. Based on
this research, we finally decided to use BLE to embed the tourist information in

R. Moreno-Díaz et al. (Eds.): EUROCAST 2017, Part II, LNCS 10672, pp. 126–133, 2018.
https://doi.org/10.1007/978-3-319-74727-9_15

the environment. In this paper, we mainly discuss on the development process of our sightseeing support system based on the result of our experiments. The one of key factors of this system is "beacon". The balance between devices (beacon) and software (smartphone application) is very important to develop this system. And we think that this idea is important for designing IoT system since IoT system consists of devices and software (application). There are several design processes such as Waterfall model, Spiral model and Prototyping model, however, there is no design model for IoT system. There are some unique characteristics in a development process of IoT.

1. In a usual software development process, we consider only interfaces between software module, however, in a development process of IoT, we have to consider both devices and software.
2. Devices are key of the service of IoT, so that it is required to keep valance between development of devices and that of software.
3. It is required to clarify the interaction between devices and software.

So that, we would like to propose our design model based on our experiments. In this paper, we review the design process of the development of sightseeing support application using BLE beacon and propose a design model for IoT system.

2 Design Process

In this section, we analyze the design process of this sightseeing support system using BLE beacons in detail and extract an outline of design process for IoT system.

2.1 Analyze Requirements of Users (Step 1)

We administered questionnaires to the visitors to Nikko [2], in order to know the focus points for our new system, on September 2014. A total of 606 questionnaires (534 in Japanese and 72 in English) were completed. A key finding was that young people reported that they would like to have enjoyed Nikko more completely and that most of them were smartphone users. Cluster analysis was used to confirm this pattern. Young tourists tend to plan active trips to Nikko. Their characteristic tendencies are as follows:

- They come to Nikko by train and navigate Nikko by bus or on foot
- They are smartphone users
- They know little about traditional culture
- They like to experience new things

Responses from foreign tourists showed the same profile.

2.2 Define Service Model (Step 2)

Tourists who know little about the area and the history currently exchange comments using SNS. We therefore addressed the use of beacons to allow residents of the tourism area to recover information from such tourists. In addition, we devised several quizzes using the Zeigarnik effect [3], aimed at young students on school trips. This is a psychological effect. We, human beings remember better an unfinished event or an incomplete one. We defined our service model as shown in Fig. 1.

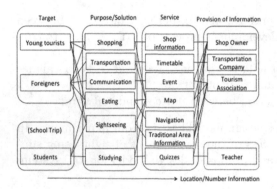

Fig. 1. Service model.   **Fig. 2.** Use case in UML.

2.3 Define Environments (Step 3)

In this section, we mention the environment where this system will be used.

Usage of Features of Smartphone. Current smartphones incorporate a range of sensors and communication devices. Some of the devices can be used to collect information and identify the location of smartphone users. However, foreign tourists rarely use roaming data communication services because of their high cost, so we also designed a service that did not require the use of 3G/LTE. As both GPS and the camera quickly deplete the battery, we designed our service to work without them.

Street, Shops and Users. We assumed the following design constraints in the environment:

- Information should be displayed using the BLE beacon.
- If 3G/LTE is not available, it must be possible to download applications and information using Wi-Fi.

The use case model of our application was as shown in Fig. 2. A beacon sends advertising message with UUID, major, minor to mobile phone application. Such information is key to retrieve data to be displayed to a user.

2.4   Define Architecture (Step 4)

As described in Fig. 3, we defined system architecture. This figure shows relation of each devices and information. The application should provide a full range of information including location, shopping information and bus timetables since foreigners usually do not have wireless connection during walking in Nikko. Each beacon contains information relating to location of a beacon. This architecture is the basis of development of this system.

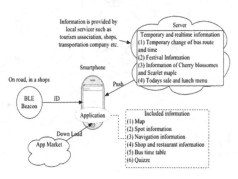

Fig. 3. Software components of the application.

2.5   Develop Devices (Step 5)

In this section, we mention the development process of devices that works with application software.

Design Concept of Devices. The traditional Japanese method of attending services of worship, in which the journey to and from the service are not the same. Before entering a temple or shrine, tourists purify themselves. After leaving the shrine, they eat or go shopping. This is an established Japanese cultural custom, which has long been taken for granted. The installation of new signboards is seldom permitted in Nikko, following the Convention Concerning the Protection of the World Cultural and Natural Heritage (UNESCO, 1972). Using beacons, we were able to show the information on a smartphone. We created a traditional road, 'SANDOU' (it means a road approaching the main temple or shrine in Japanese), for the Nikko cultural heritage site.

Define Requirements. To improve the visitor experience, we imposed the following requirements on the system:

- Reduce power consumption by avoiding the use of GPS
- Provide sightseeing information related to the BLE beacon location
- Have navigation operate in both foreground and background while displaying the distance from the station to the Shinkyo Bridge entrance to Toshogu (the main shrine)
- Display a timetable of main bus routes.

Define Environment. Many trials using BLE beacons have been reported in which location-specific and shopping information was provided in shopping malls and train stations. However, the BLE beacon has rarely been used for outdoor sightseeing, we have to define the physical location of beacon and distance between two consecutive beacons. We finally decided the distance of beacon as follows.

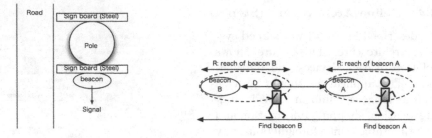

Fig. 4. How to use devices in the environment.

Fig. 5. Distance between devices.

- To receive a signal from beacon in background, it is required to receive a beacon every one minute or longer to avoid a restriction of iPhone.
- Let assume the distance of beacons is D and the range of a beacon is R. We set a beacon to send a message to one direction. (Fig. 4) If D > R, beacons does not make interference each other. Walking speed is about 84 m/min [4] and range of beacon is about 80 m as described in Table 1.
- Finally, we decided D = 100 m and R = 80 m. (Fig. 5).

Fig. 6. Signboard and space for beacon device.

Fig. 7. Screenshots of the software.

Select Parts. We tested several BLE devices. We firstly tested BLE113. The power consumption is high (about 20 mA), so that the battery last in one month. Next we tested MDBT40. The power consumption is about 5 mA and is lower than BLE113. We decided to use MDBT40 for this system.

Table 1. Reach of advertisement of BLE

Location	Reach of advertisement
Between the Nikko station and Shinkyo-bridge	About 80 m
Near the Toshogu-shrine: on the main approach	About 80 m
Near the Toshogu-shrine in the forest	About 40 m
Bus stops near the station	About 80 m
Shops near the station: in front of a shop	About 10 m
Forest of Senjyogahara (summer)	About 20 m
Forest of Senjyogahara (winter)	About 40 m

Develop Device. To install the device on the signboards (see Fig. 6), the space is very limited (5 cm x 5 cm). In addition, we set the parameter of beacon. We set transmission interval as 500 ms to reduce power consumption. This is 5 times longer than usual transmission timing (100 ms). Usually, advertising uses three channels. To reduce power consumption, we modified farm ware to use only one channel for advertising.

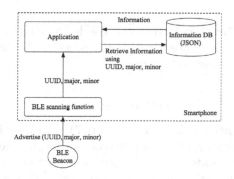

Fig. 8. Software components of the application.

2.6 Develop Software (Step 6)

In this section, we mention the development process of application software.

Design Concept of Software. The target user of this application is foreigner and young people. Functions that are useful for our target users should be installed. Sustainability is important. The research period is three years, however the software should be used longer. It is important to use flexible framework is important. One decision was to use open street map instead of propriety map.

Define Requirements. As described in Fig. 7, Map, Spot information, Navigation information, Shop and restaurant information, Bus timetable are important.

Navigation information, Shop and restaurant information should be displayed using Advertising signal from beacons. Especially foreigner, they do not want to use 3G/LTE since the roaming fee is expansive. It is important to provide all required information in the software and not to access the 3G/LTE network to get information.

Collect Contents and Data. We collected information. It is important to collect high quality information. We asked Tobu bus company, Nikko Tourism Association, Hatsuishi-kai: an association of Nikko shopping streets to provide data.

Design Software. Figure 8 explains the software components of this application. In the operating system (OS), BLE access function always scans advertising message. If the OS catches an advertising message, the information of the advertising message is forwarded to the application. For example, Core Location framework of iOS (7 or later) provides three properties such as proximity UUID, major and minor. Android 5.0 or later also provides similar function. If the information such as UUID, major and minor is received from a beacon, the application retrieves information that matches triples (UUID, major and minor).

Develop Software. We developed software concerning requirements both for iPhone and Android.

2.7 Test and Deployment

Integration Test (Step 7). We tested application using beacon device in the laboratory and checked the every screen transition according to change of beacon devices.

System Test (Step 8). We performed system test after installing beacon devices on the road and shops. We walked from Nikko station to Toshogu-shrine.

Deployment (Step 9). This application was checked by Apple Inc. and opened in the iTune store.

Finally, we obtained a development model for IoT system as described in Fig. 9.

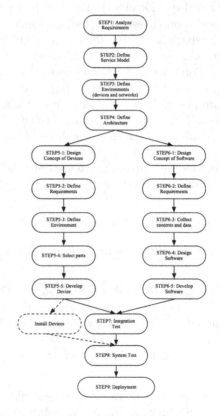

Fig. 9. Development model for IoT system.

3 Discussion

In this section, we mention what we learned from this study, what was the bottleneck of development. We had a lot of unexpected trouble during the development. We learned the following issues from the experience.

- For the IoT service, stability of devices is crucial. In our case, the beacon device sometimes broken by rain (intrusion of water).
- It is also important to estimate battery life under low temperature.
- We have to design devices first. Understanding of performance of devices and limitation of devices are important. We have to design software within the limitation and performance of devices. Devices are not flexible to modify. Also the location where to install devices is restricted by several rules and law.
- The devices should be maintenance free. In our case, the most time consuming work was battery check of devices (beacons).
- Performance and stability of smartphone is different. For example, BLE in Android smartphone was not stable when we started this project.
- Application check by App market takes long time. I our case, it took 4 months. We have to develop simple and flexible architecture to maintain application easily.

4   Conclusion

In this paper, we explained the design and development process of an IoT system that contains both devices and software based on our experience to develop the sightseeing support system using BLE beacons as described in Fig. 9. The development using this model and reuse of knowledge is useful especially for developing devices.

Our main goal to develop this sightseeing support system is to inform tourists about traditional cultures. Our system allows knowledge of the culture of a location to be transmitted to the next generation and to foreigners. Such travel information will inspire tourists and encourage them to treat the culture respectfully. We are planning to extend our collaboration to other world heritage sites as well.

Acknowledgments. We would like to express special thanks to Mr. Funakoshi (Nikko Tourism Association), Mr. Takamura and Mr. Yoshida (Hatsuishi-kai: An association of Nikko shopping streets), Mr. Nakagawa (Kounritsuin Temple). This research was performed as a SCOPE (Strategic Information and Communications R&D Promotion Programme) [5] project funded by the Ministry of Internal Affairs and Communications (MIC) in Japan.

References

1. Hiramatsu, Y., Sato, F., Ito, A., Hatano, H., Sato, M., Watanabe, Y., Sasaki, A.: Designing mobile application to motivate young people to visit cultural heritage sites. Int. J. Soc. Behav. Educ. Econ. Bus. Ind. Eng. 11(1), 121–128 (2017)
2. Nikko City Official Homepage: Tourist Information. http://www.city.nikko.lg.jp.e. tj.hp.transer.com. Accessed 31 May 2017
3. Zeigarnik, B.V.: On finished and unfinished tasks. In: Ellis, W.D. (ed.) A Sourcebook of Gestalt Psychology. Humanities Press, New York (1967)
4. Hori, M., Miyajima, H., Inukai, Y., Oguni, K.: Agent simulation for prediction of post-earthquake mass evacuation. J. Jpn. Soc. Civ. Eng. A 64(4), 1017–1036 (2008)
5. The Ministry of Internal Affairs and Communications: Strategic Information and Communications R&D Promotion Programme. http://www.soumu.go.jp/main_ sosiki/joho_tsusin/scope/. Accessed 31 May 2017

Data-Driven Maritime Processes Management Using Executable Models

Tomáš Richta[1]([⊠]), Hao Wang[2], Ottar Osen[2], Arne Styve[2],
and Vladimír Janoušek[1]

[1] Faculty of Information Technology, IT4Innovations Centre of Excellence,
Brno University of Technology, 612 66 Brno, Czech Republic
{irichta,janousek}@fit.vutbr.cz
[2] Big Data Lab, Department of ICT and Natural Sciences,
Norwegian University of Science and Technology, 6009 Aalesund, Norway
{hawa,ottar.osen,asty}@ntnu.no

Abstract. In this paper we describe a decision support system for maritime traffic and operations, based on formal models and driven by data from the environment. To handle the complexity of system description, we work with a decomposition of the system to set of abstraction levels. At each level, there are specific tools for system functionality specification, respecting particular domain point of view. From the business level point of view, the system consists of processes and vehicles and facilities over those the processes are performed. From the engineering point of view, each process consists of a set of devices, that should be controlled and maintained. Software engineering point of view operates on reading and converting bytes of data, storing them into variables, arrays, collections, databases, etc. For complex trading processes management purposes we need to cover all levels of abstraction by specific description, suitable to model and automate the operations on each particular level. As a case study we use salmon farming in Norway. The system implementation is based on *Reference Petri nets* and interpreted by the *Petri Nets Operating System* (PNOS) engine. This approach brings formal foundations to the system definition as well as dynamic reconfigurability to its runtime and operation.

1 Introduction and Motivation

In this paper we focus on describing the system for maritime traffic and operations support, based on formal methods and driven by the data from environment. Some of the work has already been done in this area. For example Ray et al. base their Decision Support System (DSS) on the idea that it needs to include mechanisms from which operators can define some contextual situations he wants to be detected as suspicious, dangerous or abnormal. They build this mechanism on a rule-based engine approach allowing to formalise rules ensuring the link between the conceptual specification of a situation and its implementation. The main aim of their DSS is a design, where the business logic might be re-configured by a surveillance operator [6].

© Springer International Publishing AG 2018
R. Moreno-Díaz et al. (Eds.): EUROCAST 2017, Part II, LNCS 10672, pp. 134–141, 2018.
https://doi.org/10.1007/978-3-319-74727-9_16

Production rules are defined as fragments of knowledge, that can be expressed in the format: "WHEN conditions are verified THEN perform some actions", where the WHEN part is referred to as the "left-hand side", and the THEN part as the "right-hand side". This format allows experts to express their knowledge in a straightforward way, without using any specific programming language and therefore removing the need for a computer programmer to assist the expert in encoding his knowledge [6]. Some of these ideas were already addressed e.g. by Ludwig Ostermayer and his colleagues [5].

We suggest a decomposition of the problem to a set of abstraction levels to reduce the complexity of a whole problem definition. This approach also allows for separating the concerns of different domains specialists as well as languages and tools they use for particular level specification. Similar ideas could be also found in some literature about expert systems like e.g. [4].

2 Maritime Logistics and Operations

We use salmon farming in Norway as a case study. Salmon production starts with hatching of eggs in freshwater tanks on land. After 1–1.5 years the juvenile salmon goes through a physical transformation process that is called smoltification that prepares the fish for life in seawater. The salmon is now called smolt and is ready to be transferred to the sea cages.

In the sea the salmon is fed pelleted feed for 1 to 2 years. Due to the high concentration of salmon it is common to add oxygen to the water and to remove CO_2. The salmon is harvested when it has reached optimum size. This is usually done by pumping the salmon into a well-boat and shipping the live salmon to the salmon processing plant.

Aquaculture is a profitable business dominated by big companies. In order to maximise the profit there are continuous efforts put on optimising the process. Optimisation of: time at sea (fast growth), fodder, produced biomass vs fodder volume, harvesting time, medicine, O_2 usage and fish quality. In later years sea-lice has been a problem for aquaculture companies, in addition to other pathogens such as toxic algae. In order to succeed a close control of biomass production at every step in the process is vital.

3 Rule-Based Modelling

Each action in the system produces some data that are sent to the particular rule engine that decides, what action should be taken. Rules apply to much more higher number of situations, and they also must be applied first, before the action caused by the task occurrence within the process could take the place. Rules within the system trigger the task fulfilment, and therefore a start of following task.

The important problem is the language used for the rules definition. The main rule clause structure is when-eval-then. But the definition of all these three parts is not constrained at all, or the constraints depend on the environment used for rules execution, like Java in Drools or RPNs in our example.

4 Data-Driven System

Technologies are being adopted for acquiring monitoring data about how the vehicle and different components are behaving. Recently, with the intention of remote ship monitoring for better services for shipping customers, vessel builders started to adopt new sensor technology by installing different sensors for different components on board a vehicle and transmit data using satellite communications to land-based service centres, e.g., *HEalth MOnitoring System* (HEMOS) by Rolls-Royce Marine AS.

These systems provide more accurate and timely operational data, but they also introduce new danger to the operations: *information overload problem* (IOP) [3,9] – the crew members receive a large volume of monitoring information and alert messages that s/he can easily overlook important/vital ones. Therefore, it is urgently needed to develop and implement a new framework to integrate and visualise the monitoring data in an informative way. In this way, the crew members can examine the massive, multi-dimensional, multi-source, time-varying information streams to make effective decisions in time-critical situations.

Our system bases on data flows and their processing according to predefined rules similarly as Ray et al. defined in their system [6], where the AIS (Automatic Identification System) data are processed by the rules engine producing the specific information and warnings about vessels movement and behaviour.

5 Levels of Abstraction

To be able to define the whole system functionality while reducing the complexity of the problem, it is better to separate it by a set of levels of abstraction [4]. Each level could be seen as a sole system, consisting of nodes, communication means and dependencies checking. Each system operates on nodes specified in more detail within the level below it. From the level 3 to 5 the nodes of the system could be taken as actors (ref. Actor model), in levels below, they behave less independently.

5.1 Level 5 - Aquaculture Facilitation

This level represents a set of processes forming the maritime trade. When performing each task, the facilitation system uses services from Level 4. This level of abstraction is intended to be used by the maritime trading management people. The most appropriate way of modelling processes at this level seems to be the sequence of tasks with dependencies among them as well as participant involved. For example using BPMN notation. At this level, basic processes of the system are defined.

5.2 Level 4 - Vessels Chartering, Berthing Process, Etc.

This level defines a set of nodes and communication means involved within trading processes that takes the place by serving as a platform for the Level 5 processes organisation. I.e. this level is an decomposition of participants from the level above. This level is intended for modelling vessels, ports, etc. relationships together with relevant communication channels.

5.3   Level 3 - Vessels, Ports, Etc.

This level describes the functional nodes with independent behaviour that use services of modules from the Level 2 and serve as services for the level 4. This level is defined as Workflow System Specification and could be directly transformed to the interpretable Reference Nets structure for further process management purposes [7]. Typical example of system parts at this level of abstraction are independent units usable for the Level 4 purposes, like vessels, fish farms, or fish factories.

5.4   Level 2 - Modules

This level describes assembled components providing specific set of services within Level 3 models. Modules consist (physically or logically) of components from the Level 1 and are usually controlled by staff, or also using any kind of programming interface, or both. Components communicate among others using defined protocol. Modules could be represented by e.g. navigation module, dynamic positioning module, wellboat pumping and cleaning module.

5.5   Level 1 - Components

This level covers mountable devices with well-defined and encapsulated behaviour defined as a set of primitive operations defining the protocol of the component. The example of a device at this level is pumping component operating over one pipe within pumping facilities. Components operate on parts from the Level 0 and are accessible via programmable interface or some specific of bus. Here the appropriate examples of components belong to thrusters, engines, pumps, etc.

5.6   Level 0 - Sensors and Actuators

At this level, simple parts mounted within the environment take place. Sensors are able to read data from the environment and serve it as raw values, or digitised and calibrated. Actuators have direct effect on the environment, it means these are e.g. multiple types of switches, servos, motors, etc. In PNOS, sensors and actuators are triggered by invoking primitive operations bounded with Reference Petri Nets transitions. These operations produce or consume values in specific strings-based format, which are propagated through the system to particular node they are dedicated. The important part of each node is its ability to store rules for data filtering, before they are directly sent to upper levels of the system. Data could be also modified or combined by these rules.

6   Running Example

System construction process will be described on real-life scenario of wellboat operations and technology. Wellboats carry fish from fish farms to fish factories. Fish are pumped from the farm into the boat and then transported to the factory, where they are pumped back again. The water with fish is treated following some predefined rules to keep the fish in good conditions. While pumping the fish out of the boat, it is possible to separate them according to their size. An example of described process definition could be found in Fig. 1.

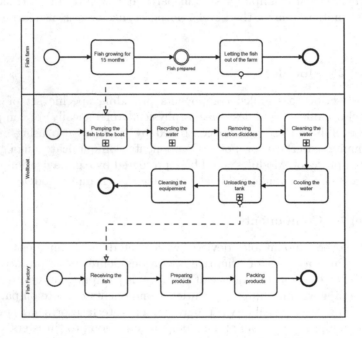

Fig. 1. Wellboat process description (Level 4)

From the point of view of control system structure, there are three control sub-systems of fish farm, the wellboat itself, and the fish factory.

7   System Construction Process

Management of distributed trading processes must take into account many involved nodes and regarding the maritime processes, there is also necessary to take into account the conditions coming from the fact, that processes are undertaken on the sea. One of the main influencing condition is that ships and their crew could in some situations remain without the connection with the land. Therefore it is necessary to count on with adequate control system installation and communication ways.

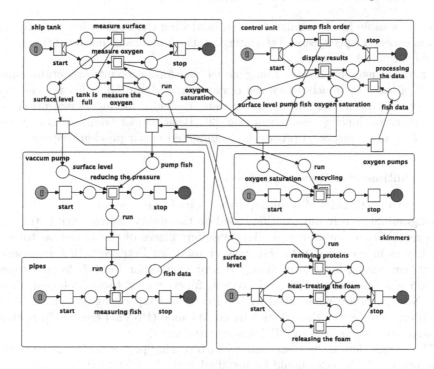

Fig. 2. Maritime system example (Level 3)

Particularly it means that the system must be distributed and all the nodes must be able to behave independently on the connection to other nodes, as well as some particular sets of nodes that operate together should be able to act independently on the rest of the system. This leads to the isolation of particular sub-ecosystems, like the vessel control system, port control system, etc. that together form the process management platform. These ecosystems are defined at each level of abstraction and represented as a subset of PNOS installations.

7.1 Data Propagation and Analysis

There are two ways of data propagation - (A) from the top to the bottom and (B) from the bottom to the top. At each layer of the system decomposition there are PNOS nodes that allow to retrieve commands and Petri nets specifications from the above layers as well as the data produced by layers below. Each PNOS node hosts a set of Petri nets that perform commands. Other Petri nets are responsible for filtering data coming from lower layers.

Model-Driven Engineering moves the software engineering paradigm to the level, where the code itself does not play the central role of the application design and implementation, but more abstract model of the application logic takes the place as a first-class artifact within the development process [8]. This approach makes it possible to distinguish between modelling the application logic by the domain expert or specialist and the interpretation or transformation of the model into its executable form.

There are many papers describing model transformation into executable code, but all of these approaches lack the dynamic reconfigurability features as well as preserving the model during the runtime, therefore the model execution got more attention among researchers now [2]. Basic model transformations and target system construction process is documented in our previous papers [7].

7.2   Runtime and Reconfiguration

The core characteristics of resulting system - its formal basements and dynamic reconfigurability - is in our solution based on the ability of Reference Petri Nets interpretable representations to migrate among places of the system as tokens, similarly as in reference Nets. The new or modified Petri Net, that represents the system partial behaviour change is sent over other Petri Nets to its destination place to change the whole system functionality. In our solution, these Petri Nets parts are maintained by the Petri Nets Operating System (PNOS) and interpreted by the Petri Nets Virtual Machine (PNVM) engine [7]. System decomposition is inspired by MULAN architecture [1].

The installation of the system starts with placing proper nodes to the target environment. Each node should be installed with the PNOS, PNVM and basic platform layer. In our running example the scenario should start with installing the processes for each Workflow Specification and then sending particular sub-processes nets to relevant nodes. All processes of the node could be changed and then passed to its platform to change the behaviour of the node. Finally all the sub-processes nets could be modified and sent to particular nodes processes that reinstall them within the nets place.

8   Conclusion

In this paper we discussed the problem of data driven distributed control and decision support system. The system itself is decomposed into five levels of abstraction to reduce the complexity of its construction. At each level a specific notation or formalism for system functionality description and operation is used. The main idea of system design is to enable target users with the possibility to maintain the system during its runtime by introducing new functionality as well as new rules defining the expert knowledge. The implementation of the system is based on Petri Nets Operating System (PNOS) that is able to interpret textual representation of Reference Petri nets called Petri Nets Byte-Code (PNBC). System itself is constructed as a set of Reference Petri Nets installed within nodes of the system. Parts of the system specification at each node represents rules defining the data propagation from each node to the above layers of the system.

Acknowledgment. This work was supported by The Ministry of Education, Youth and Sports of the Czech Republic from the National Programme of Sustainability (NPU II); project IT4Innovations excellence in science - LQ1602 and partially by the Norwegian Funds under the academic staff mobility programme (NF-CZ07-INP-5-337-2016).

References

1. Cabac, L., Duvigneau, M., Moldt, D., Rölke, H.: Modeling dynamic architectures using nets-within-nets. In: Ciardo, G., Darondeau, P. (eds.) ICATPN 2005. LNCS, vol. 3536, pp. 148–167. Springer, Heidelberg (2005). https://doi.org/10.1007/11494744_10
2. Girault, C., Valk, R.: Petri Nets for System Engineering: A Guide to Modeling, Verification, and Applications. Springer-Verlag, Secaucus (2001)
3. Keim, D., Andrienko, G., Fekete, J.-D., Görg, C., Kohlhammer, J., Melançon, G.: Visual analytics: definition, process, and challenges. In: Kerren, A., Stasko, J.T., Fekete, J.-D., North, C. (eds.) Information Visualization. LNCS, vol. 4950, pp. 154–175. Springer, Heidelberg (2008). https://doi.org/10.1007/978-3-540-70956-5_7
4. Nikolopoulos, C.: Expert Systems: Introduction to First and Second Generation and Hybrid Knowledge Based Systems, 1st edn. Marcel Dekker Inc., New York (1997)
5. Ostermayer, L., Seipel, D.: A prolog framework for integrating business rules into java applications. In: Nalepa, G.J., Baumeister, J. (eds.) Proceedings of 9th Workshop on Knowledge Engineering and Software Engineering (KESE9) co-located with the 36th German Conference on Artificial Intelligence (KI2013), Koblenz, Germany, 17 September 2013, CEUR Workshop Proceedings, vol. 1070. CEUR-WS.org (2013)
6. Ray, C., Grancher, A., Thibaud, R., Etienne, L.: Spatio-temporal rule-based analysis of maritime traffic. In: Third Conference on Ocean & Coastal Observation: Sensors and Observing Systems, Numerical Models and Information (OCOSS), pp. 171–178 (2013)
7. Richta, T., Janousek, V., Kočí, R.: Dynamic software architecture for distributed embedded control systems. In: Moldt, D., Rölke, H., Störrle, H. (eds.) Proceedings of the International Workshop on Petri Nets and Software Engineering (PNSE 2015), Including the International Workshop on Petri Nets for Adaptive Discrete Event Control Systems (ADECS 2015) A Satellite Event of the Conferences: 36th International Conference on Application and Theory of Petri Nets and Concurrency Petri Nets 2015 and 15th International Conference on Application of Concurrency to System Design ACSD 2015, Brussels, Belgium, 22–23 June 2015, CEUR Workshop Proceedings, vol. 1372, pp. 133–150. CEUR-WS.org (2015)
8. Rutle, A., MacCaull, W., Wang, H., Lamo, Y.: A metamodelling approach to behavioural modelling. In: Proceedings of the Fourth Workshop on Behaviour Modelling - Foundations and Applications, pp. 5:1–5:10. BM-FA 2012. ACM, New York, NY, USA (2012). https://doi.org/10.1145/2325276.2325281
9. Wang, H., Zhuge, X., Strazdins, G., Wei, Z., Li, G., Zhang, H.: Data integration and visualisation for demanding marine operations. In: Oceans 2016: MTS/IEEE Oceans Conference (2016)

The Neural System of Monitoring and Evaluating the Parameters of the Elements of an Intelligent Building

Andrzej Stachno[✉]

Department of Control Systems and Mechatronics,
Wroclaw University of Science and Technology, Wroclaw, Poland
andrzej.stachno@pwr.edu.pl

Abstract. In the intelligent buildings more and more features are dependent on the proper operation of equipment powered by electricity. Possible damage to these devices is very detrimental to the functioning of the building and must be repaired or replaced immediately. In many cases a failure prevents the building from being used. Possible equipment failures, however, can be predicted from the analysis of how they operate. Monitoring and analysis of power parameters of individual electrical devices allow to distinguish characteristic parameters of each electrical receiver. Any departure from stable working conditions may be recorded by a neuronal monitoring and evaluation system. The mechanism for such an analysis is the implementation of an adaptive prediction algorithm using artificial neural networks. This method allows adaptation of the decision mechanism to the current working conditions of controlled devices. The measured parameters are the measurements of physical quantities that illustrate the operation of the device. For example, for air handling and ventilation units, this is the electricity consumed. The advantage of power analysis is the identification of common faults and corresponding deformations of power supply parameters. Based on the previously prepared pattern, neural networks identify component damage or predict the predicted critical failure time of the component or control subsystem using MWF. Early forecasting of the failure situation contributes significantly to the comfort and security of intelligent building users.

Keywords: Artificial neuron networks · FFT · Forecasting
Successive values of a time series
Environmental measurements in an intelligent building

1 The Need to Analyze Measurement Data in an Intelligent Building

In an intelligent buildings we need measurement many of parameter: for example: sun light, wind speed, outdoor humidity or outdoor temperature. These parameters are independent and we can only measure, while those parameters are examples of the effects of control by the automation system.

© Springer International Publishing AG 2018
R. Moreno-Díaz et al. (Eds.): EUROCAST 2017, Part II, LNCS 10672, pp. 142–150, 2018.
https://doi.org/10.1007/978-3-319-74727-9_17

For measurements [1] of independent parameters, only measurements and their current analysis or forecasting of future values are possible, e.g. for modifications of control algorithms.

The second type of measurement is the measurement of parameters which are the effect of the regulations developed by the automation system. For example, temperature indoors, energy consumption or gas concentration in the building. Analysis of these parameters, with stable operation of the automation system, gives additional information on the efficiency of all mechanical and electrical equipment included in the building automation system.

2   Methodology of Measurements

The results of measurements of the aforementioned parameters, for their analysis [2], were presented by time series [3]. An example of a time series representing the current consumed by the electric kettle is shown in Fig. 1.

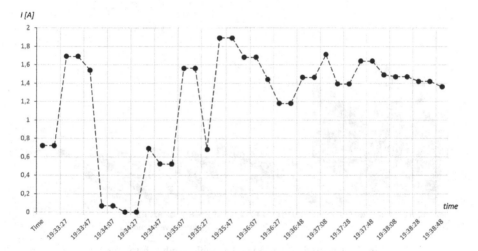

Fig. 1. The current consumed by the electric kettle – time series.

In order to increase the amount of information, the measurement time series was extended with additional information contained in current harmonics. An example of a measuring point - vector of harmonic currents, taking into account the amplitudes of individual harmonic currents from the first to the twenty-first one, is shown in Fig. 2.

Measurement time series, composed of successively occurring, within specified time distances, vector harmonic currents. An example of a series of harmonic current amplitude vectors - harmonic time series (HTS) illustrating the current consumption of a sandwich toaster is shown in Fig. 3.

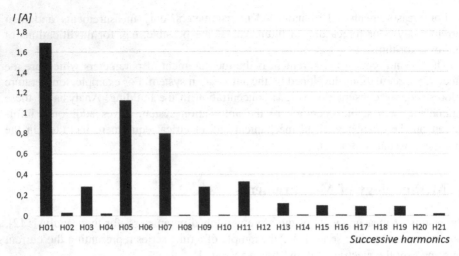

Fig. 2. The components of the amplitudes of harmonic current - example of electric kettle.

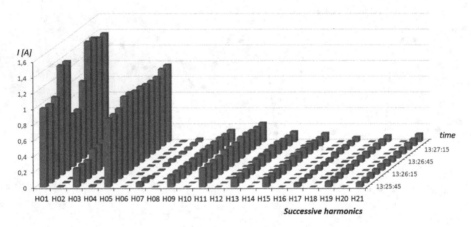

Fig. 3. Harmonic time series. Example sandwich toaster.

3 Analysis of Measurement Results - Profiling of Devices

Figures 4a and b show harmonic time series of two electrical devices: sandwich toaster and coffee maker.

Based on the analysis [4] of the differences in their harmonic time series, it is possible to isolate the individual profiles for each of the devices. In the example presented, the tested devices are a toaster and a coffee maker. Devices with very similar electrical functionality - heating. Despite a very similar electrical construction, there is a significant difference in the seventh harmonic.

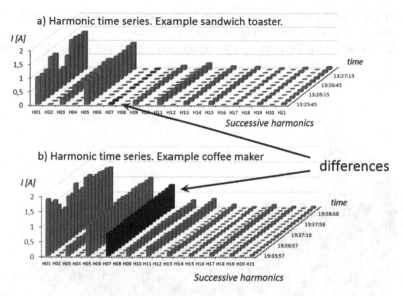

Fig. 4. Harmonic time series illustrating the differences in the seventh harmonic.

For the purpose of the experiments showing differences in harmonic time series, the following devices were measured:

– LED bulb,
– laptop,
– electric motor.

The harmonic time series corresponding to these devices are illustrated in Figs. 5a, b and c.

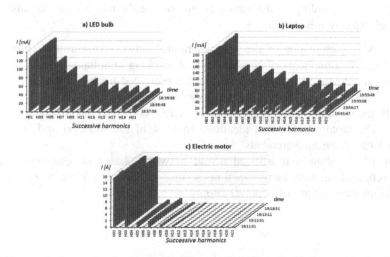

Fig. 5. Harmonic time series for the test equipment: (a) LED bulb, (b) laptop, (c) electric motor.

The purpose of the analysis was to identify the elementary profiles of devices from the harmonic time series consisting of sets of the above devices. An example of harmonic time series for the working simultaneously laptop and LED bulb is shown in Fig. 6.

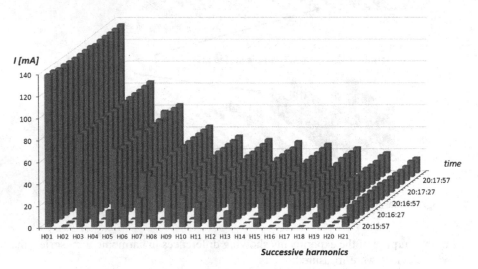

Fig. 6. Harmonic time series. The LED bulb and laptop.

An artificial neural network [5] was used to identify elementary devices. The methodology of analysis and identification was:

Teaching ANN: Presentation of the artificial neural network of individual harmonic profiles of each device working independently (at the ANN input presents a harmonic time series corresponding to the measurement of the device, on the outputs of the network identifying this device).

Testing ANN: Presentation of the artificial neural network of harmonic time series derived from measurements of several devices working simultaneously. ANN at its outputs set the device identifiers detected in the presented harmonic time series.

The results of device identification are shown in Fig. 7.

In all tested harmonics time series, the LED bulb was identified with 100% accuracy. The laptop profile was identified in 98% of the situation and the electric motor in 84% of the measurements.

These results show that artificial neural networks, based on elementary device profiles reflected in their harmonic time series, can identify elementary electrical devices connected to the mains power supply.

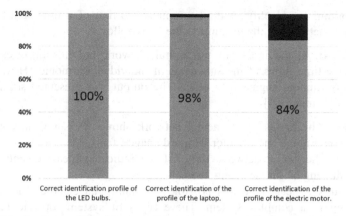

Fig. 7. Correct identification of the elementary profile electrical device.

4 Identification of Defects of Components Attached to the Building Automation System

The extension of the ability to identify elementary electrical devices connected to the mains power supply is the identification of anomalies of power caused by an improperly functioning electrical appliance. With the knowledge of the harmonic profile of the device, and on the basis of its identification among many other devices, it is possible continuous observation of the impact of his work on the harmonic time series of the entire system. Any deviation from proper operation, e.g. due to mechanical damage to the diagnosed device reflected in the measurements. For example, Fig. 8a shows the harmonic time series for a properly functioning fan and fan with a damaged blade (Fig. 8b).

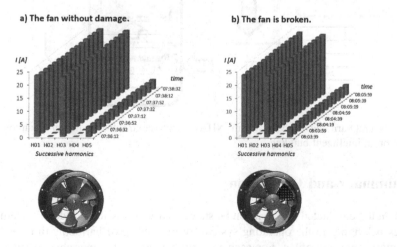

Fig. 8. Comparison of harmonic time series for the fan working properly (a) and damaged (b).

The algorithm for identifying the harmonic distortion of a time series using an artificial neural network in the examined case is as follows:

Teaching ANN: At the inputs of the neural network presents historical data as compiled in the time series of measurements of individual harmonic {H01, H02, ..., H05} correctly running engine with a fan. The output was presented specifying the correct operation index {0}.

Testing ANN: The inputs of the neural network shows measurement data from a properly running engine and a motor with a damaged fan blade.

As a result the neural network 100% of the indicated measurements from the damaged of the engine-fan detecting anomalies power.

These results indicate that it is possible to identify an electrical device that is malfunctioning in a complex system. These types of systems operate in building automation. The tested devices were equivalent to the components of the air conditioning/ventilation unit of the automated building.

An example of the method of identifying a defective component in a building automation system is shown in Fig. 9. This is a typical air conditioning and ventilation control system, enhanced with modules: higher harmonics measurement and neural analysis of higher current harmonics.

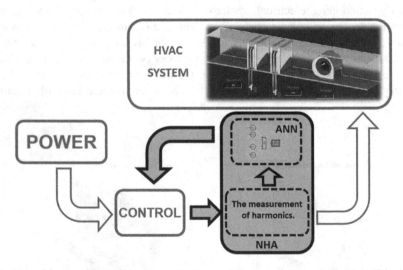

Fig. 9. Neural harmonic analysis system [NHA] used to control the central air conditioning and ventilation in intelligent building.

5 Summary and Conclusion

Based on the conducted research can be stated that it is possible to identify elementary devices belonging to the operating system by measuring of harmonic time series and elementary device profiles. For each identified device, it is possible to carry out a current analysis of the correctness of its operation and to identify the anomaly based on the harmonic time series and the corresponding pattern of the functioning of the device.

Early detection of elementary component damage in complex systems (e.g. HVAC system in intelligent building) allows for faster restoration of the entire system to full efficiency, thus minimizing the cost of replacing more defective elements (damage to one component can lead to failure) and minimizing Downtime of the entire system. Figure 10 shows the sequence of operations in the standard failures of the system without Neural Harmonic Analysis [NHA] and with NHA.

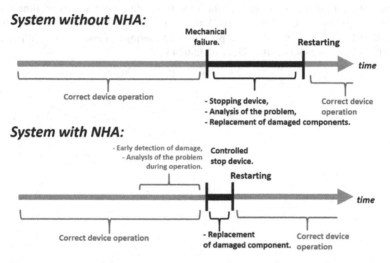

Fig. 10. Sequence of events in the event of a device failure in the system without NHA and NHA.

The above conclusions lead to the presentation of the Algorithm of Neural Harmonic Analysis [NHA] of fault identification in complex building automation system based on neural model:

- Use of a device capable of measuring higher harmonics in the system.
- Determination profile harmonic elementary electrical equipment included in the system.
- Identification of the elementary electric devices based on their harmonic profile using artificial neural network.
- Presentation of measurement of harmonics for a device in the form of time series on the input of artificial neural network.
- Analysis of the measurement of harmonics for each extracted elementary profile electrical device.
- Automatic detection of changes of elementary.

References

1. Suproniuk, M., Kamiński, P., Pawłowski, M., Kozłowski, R., Pawłowski, M.: An intelligent measurement system for the characterisation of defect centres in semi-insulating materials. Electr. Rev. **86**(12), 247–252 (2010)
2. Peitgen, H.O., Jurgens, H., Saupe, D.: Introduction to Fractals and Chaos. PWN, Warsaw (2002)
3. Technical Analysis of the Financial Markets. WIG Press (1999)
4. Jabłoński, A.: Intelligent buildings as distributed information systems. CASYS: Int. J. Comput. Anticip. Syst. **21**, 385–394 (2008)
5. Stachno, A., Jablonski, A.: Hybrid method for forecasting next values of time series for intelligent building control. In: Moreno-Díaz, R., Pichler, F., Quesada-Arencibia, A. (eds.) EUROCAST 2015. LNCS, vol. 9520, pp. 822–829. Springer, Cham (2015). https://doi.org/10.1007/978-3-319-27340-2_101. ISSN 0302-9743

A Study of Precedent Retrieval System
for Civil Trial

Yuya Kiryu[1]([✉]), Atsushi Ito[1], Takehiko Kasahara[2], Hiroyuki Hatano[1],
and Masahiro Fujii[1]

[1] Information System Science Department, Graduate School of Engineering,
Utsunomiya University, 7-1-2 Yoto, Tochigi, Utsunomiya 321-8505, Japan
u8164@gmail.com, {at.ito,hatano,fujii}@is.utsunomiya-u.ac.jp
[2] Toin University of Yokohama, 1614 Kurogane-cyo, Aoba-ku,
Yokohama-shi, Kanagawa 225-8503, Japan
kasahara@toin.ac.jp

Abstract. Recently, ICT has been adopted for use by the judiciaries in
many countries such as Singapore and India, in response to the global
changes in the economy and politics. It is urgently necessary for Japanese
judiciary to adopt ICT to catch up with the global trend. For example,
since almost all judicial records are stored as printed material, it is difficult to access them. Recently, limited number of record are opened on
web site. UI for retrieving is not well designed for ordinary people. For
example, they cannot search precedents with thing in action and what
to prove to win.

In this paper, to solve this problem, we propose an intelligent precedent retrieval system that uses ontology. It allows users to find precedents
with thing in action and similarity of documents. Also, it suggests appropriate laws and what to prove. In other words, our system can support
user's prosecution.

Keywords: Cyber court · High tech court · Legal document · XML
Ontology · Semantic web · Civil trial · Precedent retrieval system

1 Introduction

Recently, multilateral trade has increased and multinationalization of companies
has developed by progress of globalization. In addition, the market of e-commerce
is in a trend of expansion in the world [11].

With this, the disputes that persons, companies of the foreign nationality are
concerned and the disputes to cross between many countries increase. Therefore,
jurisdictions of many countries aim at following economic growth and politi-
cal change to allow anyone to access the judiciary easily by introducing ICT
(Information and Communication Technology). Particularly, we call such justice
system enpowered by ICT as "Cyber Coat".

The pioneer study of a Cyber Court System is Courtroom 21 [1]. It started
in 1993 at William & Mary University as a joint project of that university and

© Springer International Publishing AG 2018
R. Moreno-Díaz et al. (Eds.): EUROCAST 2017, Part II, LNCS 10672, pp. 151–158, 2018.
https://doi.org/10.1007/978-3-319-74727-9_18

the National Center for State Courts in the U.S.A. In addition, the Singaporean Supreme Court introduced ICT in 1998 [2], and High Court of Delhi and District Court of Delhi introduced paperless E-Court. From the standpoint of technological research, there are many studies about models combining law and knowledge structures of law [3], however it is very rare to find a model to create an experimental system for ICT based judiciaries. We have decided that it is very important to develop a prototype of a Cyber Court to evaluate the effectiveness of such an idea.

In Japan, "Judicial Reform Promotion Plan" was decided in 2002 [12]. The prototype for the first civil trial in Japan was developed at Toin University of Yokohama in 2004 [4,5], and the effectiveness, particularly the usefulness in Saibanin system (Japanese jury system) which in late years was introduced, was proved [6]. The experiment of the remote trial was conducted, too [10]. However, judicial records are not yet fully opened and only some decisions are opened. Also, the retrieval tool is not well designed for ordinary people.

In this paper, we mainly discuss the result that we examined how to support to describe legal documents by a citizen ignorant about a law and the possibility to improve the flexibility of civil trial.

We firstly discuss requirements for the records of a lawsuit in the civil trial and extract problems in Sect. 2. In Sect. 3, we suggest a system to solve the problems and explain the structure of ontology that is used in the system in Sect. 4. Then we explain detail of the search system it in Sect. 5. Finally, we mention conclusion in Sect. 6.

2   Requirements of Creating the Case Record

2.1   Presupposed Ultimate Fact

In a civil action, in order to obtain a profitable legal effect, it is necessary for parties to find sentences in law, as a weapon to win, and provide presupposed ultimate facts to satisfy each law. When we take an action, we must describe a cause of the demand in accordance with these presupposed ultimate facts. For example, when a self-possession, such as a house or ground, is occupied illegally, the presupposed ultimate facts to return the possession based on the ownership is as follows [8].

[plaintiff]
- Requirement to return the ground
   a1 The fact that the plaintiff owns the thing concerned
   a2 The fact that defendant occupy the thing concerned
- Requirement for a damage claim as an Attendant claim
   * Injuria
      a3 The fact that the defendant has occupied the thing concerned for a certain period of time
      a4 The fact that the plaintiff has owned the thing concerned in the beginning period mentioned above
      a5 Amount of loss

[Defendant]

d1 Completion of a lease agreement between the plaintiff and the defendant for the thing concerned

d2 The extradition of the thing concerned from the plaintiff to the defendant based on the above

It is necessary for the plaintiff to prove the above their requirements. The plaintiff can get legal effects by proving these presupposed ultimate facts, which accept a demand. In contrast, the defendant proves the above their requirements, so that their demands that is the rejection to the plaintiff's statement (plea). Furthermore, there are the requirements that a plaintiff performs a claim (surre-buttal) to reject the effect of this law. The situation such as "own" and "occupy" should satisfy requirements described in law.

2.2   Decision of Laws

When plaintiffs prosecutes, they need to check whether they can prove the presupposed ultimate facts that they must satisfy to get a law effect, and whether they are more likely to lose the case by a plea beforehand, and then it is necessary to choose an appropriate law effect. However, this is very difficult for people with a little legal knowledge.

As an example, we focus a situation that the presupposed ultimate facts of demand for return based on proprietary rights when a self-possession is occupied illegally (showed in Sect. 2.1). The example in Sect. 2.1 is the return request based on proprietary rights, but plaintiffs can demand it as excessive profit return request when there is a gain of the equivalent amount of rent to get an equal effect [8].

In practice, damages request based on tort (illegal occupation) are often used since a proof is easy [8]. In this case, when plaintiffs prosecute, they must consider possibility that illegality of occupation is rejected by a plea. In other words, they must choose the law that is easy to satisfy presupposed ultimate facts from the situation of the dispute. To pass demands, the people with a little legal knowledge must check the most suitable law effects, and compare situation of litigant parties and the situation of the precedent, and check whether presupposed ultimate facts are satisfied. However, these procedures are difficult for people with a little legal knowledge because there are few clue to search it.

2.3   Access Precedent

Some Japanese case records are now open to access. We can access them online and search them [4, 10]. The search screen is shown in Figs. 1 and 2. We think that the legibility and flexibility of the search function are not so well developed since the search system requires knowledge of the legal system and the court system, such as the court name, case number, trial date and time. So, it is difficult for an ordinary person to use it effectively.

Because of the difficulty about decision of which showed in Sect. 2.1, it is the best for the people with a little legal knowledge to allow them to search precedents whose content is similar to what they intend. About the precision improvement of the similar legal documents search, the trial with the similarity between laws is carried out, too, but the effectiveness is not shown [9].

Fig. 1. Japanese case records searching system

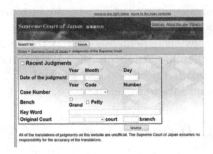

Fig. 2. Japanese case records searching system in English

3 Proposed System

We propose a system described in Fig. 3 to solve the problems in Sect. 2. Proposed system is based on a system based on ontology that we suggested before [7]. As a premise, the records of a lawsuit shall be described according to ontology to suggest (civil trial ontology). Moreover, we use OWL and SPARQL as ontology language and query language. In addition, we make an assumption that is the reference relations between a precedent (records of a lawsuit) and a law (this time the set of presupposed ultimate facts and the law effect), but such meta data are not added to an exhibited precedent at the present. The study of the reference method identification to parse this from plain sentence is conducted now, but the precision is not so high [9]. It is necessary to prepare for reference relations manually when we start use this system. Once this system is started, reference relations can be added automatically, because if a law suggested on proposed system is used, then records of a lawsuit is generated automatically. It is clear that high quality data can be generated in this sequence, once this flow is invoked. The reference relations shall be expressed by using a property defined as owl:refer from a node to expressed as complaint according to civil trial.

The rough structure of this system is as follows: (1) throwing "claim" and "situation" as a query, (2) searching law effects that can demand claim matching with the claim in the query, (3) taking precedents that have high similarity between itself and "situation" in the query as natural language expression, (4) outputting information gained in (1)–(3) and reasoned form this.

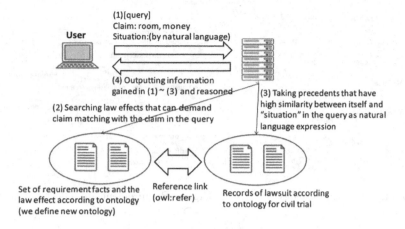

Fig. 3. Structure of proposed system

4   Ontology for Interaction Between Fact and Law

We explain "ontology for interaction between fact and law" to enable the search showed in Sect. 3. We will show effects and usefulness of this ontology in Sect. 5. This ontology is composed of two ontologies that are "Law term ontology" and "Requirement and Effect ontology".

"Requirement and Effect ontology" gathers up presupposed ultimate facts and the law effects showed as an example that self-possession is occupied illegally in Sect. 2.1. This knowledge are searched by procedure (2) in Fig. 3.

"Law term ontology" expresses terms, which are, that are the "state", "action" and "relations" decided by law referred in "ontology for Requirement and Effect". In the case of Sect. 2.1, it is "own" in "The fact that the plaintiff owns the thing concerned" and "occupy" in "The fact that defendant occupy an applicable thing". Figure 4 shows an example of "Law term ontology".

A structure about "own" is drawn in detail, but "occupy", "tort" are also described by a similar method. In the case of "own", it is separated by two patterns (P1, P2). P1 is "opposite party does not go to law about current ownership" and P2 is "opposite party goes to law about current ownership". In the case of P1, it is proved only in "to insist current ownership" (we assume this R1). In addition, P2 has two patterns (P3, P4). In the case of P2–P3, we must insist R2 and R3. Also, if it is a pattern of P2–P4, we must insist R4, R5 and so on. Therefore, leaves of the hierarchy structure are the requirements.

Figure 5 shows "ontology for Requirement and Effect" in the case of demand for return based on proprietary rights when a self-possession is occupied illegally in Sect. 2.1. Each sign was defined in Sect. 2.1. Purpose class has what we can demand by the law effects. Therefore, we refer to this in a search from a claim. We define the reference relations by owl:refer property to express reference about the term referred by each requirement. We explain the detail of the search using these ontologies in the next Section.

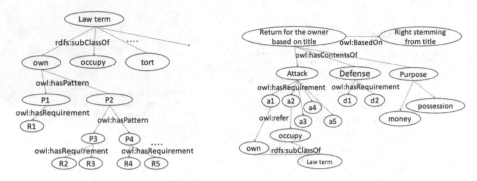

Fig. 4. Law term ontology   **Fig. 5.** Requirement and effect ontology

5   Search by Using Ontology

In Sect. 3, we explain procedure (1)–(3) in Fig. 3. In this Section, we explain the procedure (4) in proposed system. The rule-based is described in accordance with Apache Jena's form. Apache Jena is Java framework to treat semantic web.

In procedure (4), proposed system performs a reasoning to support users. At first, it finds the major difference in requirements of each law. The following rules are given

$$[defferenceRule :$$
$$(?Lowl : hasContentsOf?A), (?Aowl : hasRequirement?B),$$
$$(?Bowl : refer?W) \rightarrow (?Lowl : hasImportant?B)$$

This rule defines requirements referring to law terms as important requirements. The property to point at important requirements is defined as "owl:hasImportance" newly. "Outbreak of the damage and amount of damage" and "the causation" are one of presupposed ultimate facts, but they are not main factors expressing the difference between laws. When we choose laws, what we should check is what it is based on. Therefore, if the system search information that owl:hasImportance property point at and display them, users can choose suitable information easily without reading every proposed presupposed ultimate facts.

Then, we give proposed system possibility of rights of conflict to claim. Rights of conflict to claim mean that there are plural rights to claim in same situation. In the case of the presupposed ultimate facts of demand for return based on proprietary rights in Sect. 2, a relationship between that right and vindication of unjust enrichment is right of conflict to claim. If it is difficult to prove presupposed ultimate facts of demand for return based on proprietary rights, vindication of unjust enrichment is often used.

We give some examples of famous rights of conflict to claim are as follows:

: $DebtDefault$ owl : $CompeteWith$: $Tort$.

: $RightStemmingFromTitle$ owl : $CompeteWith$: $RightToClaimInTheContract$.

: $Tort$ owl : $CompeteWith$: $UnjustEnrichmentClaim$.

: $RightStemmingFromTitle$ owl : $CompeteWith$: $UnjustEnrichmentClaim$.

"owl:CompeteWith" is a property that shows possibility of rights of conflict to claim and is given "owl:SymmetricProperty", which is a property restriction. If users browse "return for the owner based on title", the query with SPARQL as follows shows possibility of rights of conflict to claim.

$$select\ distinct?owhere\{$$
$$"returnfortheownerbasedontitle"$$
$$owl:BasedOn?r$$
$$?rowl:CompeteWith?o.\}$$

As described above, the proposed system supports users with a little knowledge of law. Figure 6 shows dialogue between the system and users. Because it is easy to add relationships of referring right to claim to records of lawsuit made by using proposed system, the precedent database becomes bigger automatically.

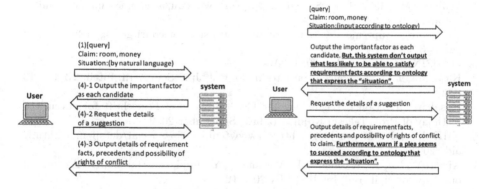

Fig. 6. Dialogue between the system and users

Fig. 7. Expansion of proposed system

6   Conclusion

In current precedents searching system, there is a problem that it targeting judicial officer. It cannot allow ordinary people to search precedent that they want. Therefore, in this paper, we suggested the system that urges people with a little knowledge of law to use justice. Specifically, we suggested the system that support precedent search that is required when users take an action. To implement this system, we defined new ontology that explain presupposed ultimate fact. New ontology allowed proposal system to consider conflict of law and thing in action.

However, it is difficult for the system to recognize the situation of trial enough because the proposed system simply searches similar precedents by natural language processing. Therefore, we are going to design new ontology which can express the situation of trial. Then, an problem arise. It is difficult for ordinary people to apply their situation to new ontology. As the system that users can describe ontology easily, we are going to examine a way which is "To allow users describe ontology graphically described as Figs. 4 and 5" in future. In this way, a good tool with user friendly UI is demanded. Figure 7 shows procedures in a system that uses ontology for situation. It uses this ontology to search similar precedents. Also, we would like to implement function to support users' argument by visualization by using flow chart like Toulmin Model [13] to explain structure of discussion.

Acknowledgments. We express special thanks to Mr. Furumoto, attorney of law, who provides a lot of useful information and suggestions to proceed this research. We also express special thanks to all people who joined the Cyber Court Project.

References

1. William & Mary Law School. http://law.wm.edu/academics/intellectuallife/researchcenters/clct/
2. Singapore Supreme Court. http://www.supremecourt.gov.sg/default.aspx?pgID=361#10
3. JURISIN, Japan, October 2013, 2014
4. Kasahara, T.: Cybercourt - court technology. J. Jpn. Soc. Artif. Intell. **23**(4), 513 (2008)
5. Kasahara, T.: Court Technology for Civil Procedure, Festschrift for seventieth birthday of Prof. Takeshi Kojima, p. 961, September 2009
6. NHK Special Can you judge? http://www6.nhk.or.jp/special/detail/index.html?aid=20050213
7. Kiryu, Y., Ito, A., Kasahara, T., Watanabe, Y., Fujii, M., Hatano, H.: A study of ontology for civil trial. In: JURISIN 2015 (2015)
8. Okaguti, K.: Requirement Facts Manual. Gyohsei (2005)
9. Legal documents analysis using the documents-laws graph: The Association for Natural Language Processing, pp. 921–924 (2016)
10. A study of applying ICT for legal service: Report of a research funded by the Ministry of Ministry of Internal Affairs and Communications in Japan in 2009, March 2010
11. Ministry of Economy, Trade and Industry. http://www.meti.go.jp/policy/it_policy/ec/cbec/cbec_images/crossborderec_houkokusho.pdf#search=%27%E6%B5%B7%E5%A4%96+%E5%95%86%E5%8F%96%E5%BC%95%27
12. Prime Minister of Japan and His Cabinet. http://www.kantei.go.jp/jp/singi/sihou/keikaku/020319keikaku.html
13. Toulmin, S.E.: The Uses of Argument. Cambridge University Press, Cambridge (2003). Updated version, ISBN-13: 978-0521534833

Applications of Signal Processing Technology

Investigations on Sparse System Identification with l_0-LMS, Zero-Attracting LMS and Linearized Bregman Iterations

Andreas Gebhard$^{(\boxtimes)}$ (iD), Michael Lunglmayr (iD), and Mario Huemer

Christian Doppler Laboratory for Digitally Assisted RF Transceivers for Future
Mobile Communications, Institute of Signal Processing,
Johannes Kepler University Linz, Linz, Austria
{andreas.gebhard,michael.lunglmayr,mario.huemer}@jku.at

Abstract. Identifying a sparse system impulse response is often performed with the l_0-least-mean-squares (LMS)-, or the zero-attracting LMS algorithm. Recently, a linearized Bregman (LB) iteration based sparse LMS algorithm has been proposed for this task. In this contribution, the mentioned algorithms are compared with respect to their parameter dependency, convergence speed, mean-square-error (MSE), and sparsity of the estimate. The performance of the LB iteration based sparse LMS algorithm only slightly depends on its parameters. In our opinion it is the favorable choice in terms of achieving sparse impulse response estimates and low MSE. Especially when using an extension called micro-kicking the LB based algorithms converge much faster than the l_0-LMS.

Keywords: Sparse · System identification · Adaptive filter · LMS

1 Introduction

In many applications such as interference cancellation or system identification, the underlying system impulse response is of sparse nature, i.e. it only has a small number of non-zero elements. In the presence of noise, the zero-elements cannot exactly be identified by l_2-norm based estimators such as the least-mean-squares (LMS) algorithm. Extensions to the LMS algorithm such as the zero-attracting least-mean-squares (ZA-LMS) [1], l_0-least-mean-squares (l_0-LMS) [2], and the recently proposed linearized Bregman-based sparse LMS (LB-SLMS) algorithm [3] show significantly better performance in terms of sparsity of the estimate and steady state mean-square-error (MSE). In this contribution, we compare the mentioned algorithms in terms of transient and steady-state performance for different signal-to-noise ratio (SNR) scenarios. We demonstrate the influence on the l_2-norm cost function surface by adding a sparsity promoting l_1- or l_0-norm. The parameter dependency of the algorithms, and the tradeoff between a low MSE and the sparsity in the estimated impulse response is discussed.

© Springer International Publishing AG 2018
R. Moreno-Díaz et al. (Eds.): EUROCAST 2017, Part II, LNCS 10672, pp. 161–169, 2018.
https://doi.org/10.1007/978-3-319-74727-9_19

Furthermore, we present performance plots and improvement strategies for the LB-SLMS based algorithms providing a faster convergence than the original LB-SLMS algorithm as well as sparse estimates.

2　Sparse System Identification

Figure 1 depicts a block diagram to identify the finite impulse response of an unknown system with adaptive filtering. The input sequence $x[n]$ convoluted with the true system impulse response $h_T[n]$ gives the output $y[n] = x[n] * h_T[n]$. Additive white Gaussian noise (AWGN) is added to the output to model the measurement noise in a practical application. The signal $d[n] = y[n] + w[n]$ is used as a desired signal for the adaptive algorithm to estimate the N coefficients of the impulse response vector $\mathbf{h}_T = [h_T[0], \ldots, h_T[N-1]]^T$ using the error signal $e[n] = d[n] - \hat{y}[n]$, where $\hat{y}[n] = \mathbf{x}[n]^T \hat{\mathbf{h}}[n]$ is the estimated output. The traditional LMS uses the instantaneous cost function $J[n] = e[n]^2$ to calculate the gradient, which results in the update equation

$$\hat{\mathbf{h}}[n+1] = \hat{\mathbf{h}}[n] + \mu e[n]\mathbf{x}[n] \tag{1}$$

for the estimated impulse response vector using the step-size μ. The idea of the ZA-LMS and l_0-LMS algorithm is to extend the gradient in the update equation with an algorithm dependent sparsity promoting term $\mathbf{p}(\hat{\mathbf{h}}[n])$ which forces the small values toward zero.

$$\hat{\mathbf{h}}[n+1] = \hat{\mathbf{h}}[n] + \mu e[n]\mathbf{x}[n] - \mu\sigma\,\mathbf{p}(\hat{\mathbf{h}}[n]) \tag{2}$$

This term is a function of the estimated impulse response and is scaled by a factor σ, which accounts for the l_1- or l_0-norm part. As shown in [1,4], this zero-point attracting function causes a biased estimate of the ZA-LMS and l_0-LMS algorithm. The LB-SLMS algorithm enforces sparsity by using a shrinking function to force the coefficient entries which are below a threshold level λ to zero.

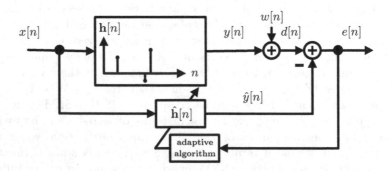

Fig. 1. System identification of a sparse system.

3   Investigated Algorithms

In this section, the ZA-LMS, l_0-LMS and LB-SLMS algorithms are discussed, and the influence of adding an l_1- or l_0-norm part to the l_2-norm cost function is visualized by using the least-squares cost function. Figure 2 shows the least-squares cost function surface with the minimum at the optimal coefficients $\mathbf{h}_{\text{opt}} = [1, 0.5]^T$. Adding an l_1-norm part with the scaling factor σ to the least-squares cost function results in

$$J = \sum_{i=0}^{N-1} e[i]^2 + \sigma \, \|\mathbf{h}[n]\|_1 . \tag{3}$$

Using the same approximation for the l_1-norm as used in the derivation of the ZA-LMS in Sect. 3.1, the cost function surface looks like depicted in Fig. 3. For a better visibility of the l_1-norm's influence, the factor σ is set to a value that is much larger than typically used in practice. It can be observed that the minimum is moved towards a sparse solution. A similar effect is observed for adding an l_0-norm part.

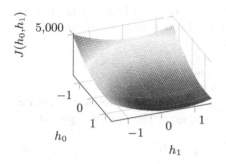

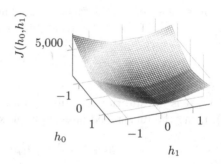

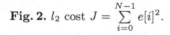

Fig. 2. l_2 cost $J = \sum_{i=0}^{N-1} e[i]^2$.

Fig. 3. Combined l_1/l_2 cost function.

3.1   Zero-Attracting LMS Algorithm

The ZA-LMS algorithm [1,5] aims to minimize the combination of the l_1/l_2-norm

$$J_{\text{ZA}}[n] = \frac{1}{2} e^2[n] + \gamma \, \|\mathbf{h}[n]\|_1 . \tag{4}$$

This results in the estimated impulse response vector update equation

$$\hat{\mathbf{h}}[n+1] = \hat{\mathbf{h}}[n] + \mu e[n]\mathbf{x}[n] - \mu\gamma \, \mathbf{p}(\hat{\mathbf{h}}[n]) \tag{5}$$

where γ is the l_1-norm weighting factor, and $p_i(\hat{h}_i[n]) = \text{sign}(\hat{h}_i[n])$. To prevent excessive forcing of coefficients with small magnitude to zero, the so called reweighted zero-attracting least-mean-squares (RZA-LMS) algorithm was introduced [1]. It uses the cost function

$$J_{\text{RZA}}[n] = \frac{1}{2}e^2[n] + \gamma \sum_i \log\left(1 + \epsilon\,|h_i[n]|\right) \tag{6}$$

with the additional parameter ϵ. This changes the sparsity-promoting term in (5) to

$$p_i(\hat{h}_i[n]) = \frac{\epsilon\,\text{sign}(\hat{h}_i[n])}{1 + \epsilon\left|\hat{h}_i[n]\right|}. \tag{7}$$

Extensions of the ZA-LMS and RZA-LMS with a variable step-size are presented in [6]. Due to the superior performance of the RZA-LMS over the ZA-LMS, below we will only present results for the RZA-LMS.

3.2 l_0-LMS Algorithm

The cost function of the l_0-LMS contains an l_0-norm part of the estimated vector

$$J_{l0}[n] = \frac{1}{2}e^2[n] + \kappa\,\|\mathbf{h}[n]\|_0\,, \tag{8}$$

which is weighted by the factor κ [2]. The nonlinear l_0-norm is approximated by $\|\mathbf{h}[n]\|_0 \approx \sum_i \left(1 - e^{-\alpha|h_i[n]|}\right)$, and the exponential function is linearized with

$$e^{-\alpha|h_i|} = \begin{cases} 1 - \alpha\,|h_i| & \text{for } |h_i| \leq \frac{1}{\alpha} \\ 0 & \text{else} \end{cases} \tag{9}$$

where α is an additional scaling parameter. The gradient of the cost function (8), using the mentioned l_0-norm approximations, gives the update equation

$$\hat{\mathbf{h}}[n+1] = \hat{\mathbf{h}}[n] + \mu e[n]\mathbf{x}[n] + \mu\kappa\mathbf{p}(\hat{\mathbf{h}}[n]) \tag{10}$$

with the sparsity promoting term

$$p_i(\hat{h}_i[n]) = \begin{cases} \alpha^2\hat{h}_i[n] - \alpha\text{sign}(\hat{h}_i[n]) & \text{for } |h_i[n]| \leq \frac{1}{\alpha} \\ 0 & \text{else} \end{cases}. \tag{11}$$

A detailed performance analysis of the l_0-LMS for a Gaussian input signal is presented in [7]. In [8] a variant with an adaptive zero-attractor for applications with time-varying measurement noise is derived.

3.3 Linearized-Bregman Based Sparse LMS Algorithm

The linearized Bregman iteration based sparse LMS algorithm [3,4] can be seen as a simplified variant of the linearized Bregman iteration [9–11]. The coefficient update

$$e[n] = d[n] - \mathbf{x}[n]^T\hat{\mathbf{h}}[n] \tag{12}$$

$$\hat{\mathbf{v}}[n+1] = \hat{\mathbf{v}}[n] + \mu e[n]\mathbf{x}[n] \tag{13}$$

is similar to the standard LMS update, but with an important difference: For the calculation of the instantaneous error $e[n]$, the vector $\hat{\mathbf{h}}[n]$ is used, which is calculated from $\hat{\mathbf{v}}[n]$ via a shrinking function. This shrinking function shrinks each element $\hat{h}_i[n]$ of $\hat{\mathbf{h}}[n]$ by $\hat{h}_i[n] = \mathrm{shrink}\,(\hat{v}_i, \lambda) = \max\,(|\hat{v}_i| - \lambda, 0)\,\mathrm{sign}\,(\hat{v}_i)$, thereby promoting sparsity in the estimate. The threshold parameter λ is a trade-off between achieving sparse solutions and fast convergence of the adaptive filter. Choosing a small value employs a fast convergence but more non-zero elements in the solution and vice versa. We also used a variant of the Linearized-Bregman based Sparse LMS with an improvement called micro-kicking [12]. Its idea is to reuse the gradient vector $\mu e[n]\mathbf{x}[n]$ in the next iteration if an iteration does not significantly change $\hat{\mathbf{h}}[n]$. This saves computations and can lead to a much faster convergence of Linearized-Bregman based sparse LMS algorithms [12].

4   Simulation Results

The simulation results are structured as follows. First, the parameter dependency of each algorithm is investigated. Second, we compare the different algorithms. The step-size μ of each algorithm is adjusted to give the same normalized mean-square-error (NMSE) of the estimated impulse response vector at a signal-to-noise ratio (SNR) of 20 dB and for the particular parameter choices $\epsilon = 5, \gamma = 0.01, \alpha = 0.01, \kappa = 0.01$ and $\lambda = 10$ (Fig. 7a). The system impulse response to identify has 10 coefficients with only three non-zero elements. The location of these non-zero elements is selected randomly for each realization. All figures show the averaged results over 1000 simulated random systems with random uncorrelated input samples from a standard Gaussian distribution.

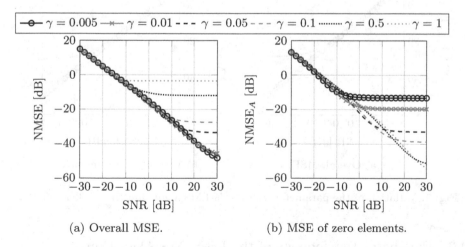

(a) Overall MSE.   (b) MSE of zero elements.

Fig. 4. Variation of the parameter γ in the RZA-LMS algorithm ($\mu = 0.6, \epsilon = 5$).

4.1   Parameter Dependency of the RZA-LMS Algorithm

The RZA-LMS algorithm uses the parameter γ to account for the l_1-norm in the cost function. The overall coefficient $\text{NMSE} = \text{mean} \sum_i (h_T[i] - \hat{h}[i])^2 / \|\mathbf{h}_T\|_2^2$ for $i = 0, 1, \ldots, 9$ at different SNR values and different values for γ is depicted in Fig. 4a. It can be observed that choosing a small γ results in a better NMSE at high SNR values. On the other hand, Fig. 4b shows that with lowering γ, the zero-element $\text{NMSE}_A = \text{mean} \sum_{i \in A} (h_T[i] - \hat{h}[i])^2 / \|\mathbf{h}_T\|_2^2$ of the zero positions $A = \{i \mid h_T[i] = 0\}$ in $\hat{\mathbf{h}}$ increases. This means a tradeoff between a low MSE of the non-zero coefficients and a low MSE of the zero coefficients has to be found.

4.2   Parameter Dependency of the l_0-LMS Algorithm

For the l_0-LMS, the overall coefficient NMSE and the NMSE of the zero-elements depicted in Fig. 5 show a very similar behavior. The NMSE curves in Fig. 5a are slightly higher than the curves in Fig. 5b because the error in the non-zero elements is added. Increasing the l_0-norm weighting factor κ increases the error of the estimation and also the residual power in the zero-elements. Thereby a small value for κ seems optimal, but as can be seen in Fig. 7a, lowering κ also slows down the convergence speed. An influence of the parameter α is omitted due to the limited page space and due to the fact that this parameter has only a small influence on the performance of the l_0-LMS.

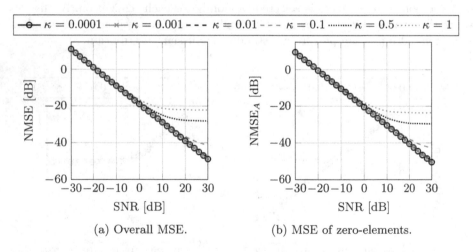

(a) Overall MSE.   (b) MSE of zero-elements.

Fig. 5. Variation of the parameter κ in the l_0-LMS algorithm ($\mu = 0.3, \alpha = 0.01$).

4.3   Parameter Dependency of the LB-SLMS Algorithm

The coefficient NMSE in Fig. 6a shows only a slight dependency of the parameter λ. Increasing the parameter λ decreases the NMSE and also produces more sparse

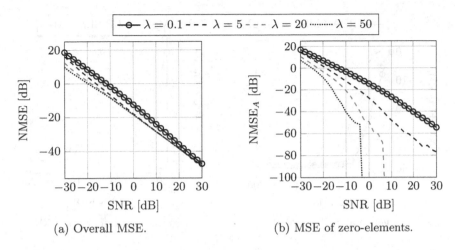

(a) Overall MSE.   (b) MSE of zero-elements.

Fig. 6. Variation of the parameter λ in the LB-SLMS algorithm ($\mu = 1$).

solutions, as depicted in Fig. 6b. Setting the parameter e.g. to $\lambda = 20$, produced a perfectly sparse solution for the simulated test cases at an SNR greater than 7 dB. Consequently, choosing a high value for λ produces sparse estimates with low NMSE. Unfortunately, increasing the value for λ has a negative influence on the initial convergence speed of the adaptive filter as can be seen in Fig. 7a. This drawback can be resolved by extending the LB-SLMS algorithm with the micro-kicking approach. With micro-kicking, the initial convergence is nearly as fast as with the RZA-LMS.

4.4 Algorithm Comparison

In this section the investigated algorithms are compared in terms of initial convergence speed and sparsity of the solution. Figure 7a shows the convergence of the algorithms at an SNR of 20 dB. The step-size of the algorithms is adjusted to have the same steady-state MSE. The RZA-LMS shows the fastest, and the l_0-LMS the slowest reduction of the NMSE to the steady-state value. The LB-SLMS algorithm has a slow initial reduction of the NMSE, which can be improved by the micro-kicking approach. An exemplary result, evaluating the sparsity in terms of the number of (exactly) non-zero elements is shown in Fig. 7b. It demonstrates the drawback of the l_0-LMS and RZA-LMS algorithms, that the zero-elements of the estimated coefficient vectors never exactly become zero. In contrast to that, the sparsity promoting shrinking function of the LB-SLMS based algorithms sets the elements exactly to zero. It can be observed, that the LB-SLMS algorithm with $\lambda = 10$ produces truly sparse estimates for SNR values above 12 dB.

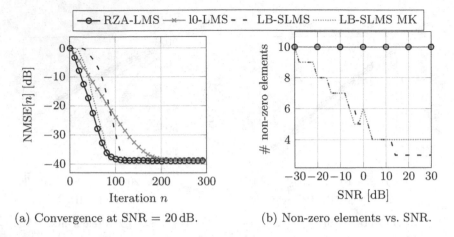

(a) Convergence at SNR = 20 dB.   (b) Non-zero elements vs. SNR.

Fig. 7. Performance comparison ($\epsilon = 5, \gamma = 0.01, \alpha = 0.01, \kappa = 0.01, \lambda = 10$).

5   Conclusion

This paper addresses the issue of sparse system identification and compares the RZA-LMS, l_0-LMS and LB-SLMS algorithms in terms of parameter dependency, NMSE, convergence speed and sparsity of the solution. By augmenting the l2-norm cost function by an l_1- or l_0-norm part, sparsity of the solution can be enforced. The LB-SLMS algorithms shows a low parameter dependency compared to the RZA-LMS and l_0-LMS algorithms, and the drawback of a slow initial convergence speed can be counteracted by using the so called micro-kicking approach. The biggest advantage of the LB-SLMS algorithm is its ability to produce truly sparse estimates where the zero-elements of the coefficient vector are identified as exactly zero. Furthermore, only the single parameter λ needs to be selected compared to two parameters for the other algorithms which simplifies its applicability.

The financial support by the Austrian Federal Ministry of Science, Research and Economy and the National Foundation for Research, Technology and Development is gratefully acknowledged.

References

1. Chen, Y., Gu, Y., Hero, A.O.: Sparse LMS for system identification. In: 2009 IEEE International Conference on Acoustics, Speech and Signal Processing, pp. 3125–3128, April 2009
2. Gu, Y., Jin, J., Mei, S.: l_0 norm constraint LMS algorithm for sparse system identification. IEEE Sig. Process. Lett. **16**(9), 774–777 (2009)
3. Lunglmayr, M., Huemer, M.: Efficient linearized Bregman iteration for sparse adaptive filters and Kaczmarz solvers. In: 2016 IEEE Sensor Array and Multichannel Signal Processing Workshop (SAM), pp. 1–5, July 2016

4. Hu, T., Chklovskii, D.B.: Sparse LMS via online linearized Bregman iteration. In: 2014 IEEE International Conference on Acoustics, Speech and Signal Processing (ICASSP), pp. 7213–7217, May 2014
5. Chen, J., Richard, C., Song, Y., Brie, D.: Transient performance analysis of zero-attracting LMS. IEEE Sig. Process. Lett. **23**(12), 1786–1790 (2016)
6. Salman, M.S., Jahromi, M.N.S., Hocanin, A., Kukrer, O.: A zero-attracting variable step-size LMS algorithm for sparse system identification. In: 2012 IX International Symposium on Telecommunications (BIHTEL), pp. 1–4, October 2012
7. Su, G., Jin, J., Gu, Y.: Performance analysis of L_0-LMS with Gaussian input signal. In: IEEE 10th International Conference on Signal Processing, pp. 235–238, October 2010
8. Wang, C., Zhang, Y., Wei, Y., Li, N.: A new l_0-LMS algorithm with adaptive zero attractor. IEEE Commun. Lett. **19**(12), 2150–2153 (2015)
9. Osher, S., Burger, M., Goldfarb, D., Xu, J., Yin, W.: An iterative regularization method for total variation-based image restoration. Multiscale Model. Simul. **4**(2), 460–489 (2005)
10. Cai, J.F., Osher, S., Shen, Z.: Convergence of the linearized Bregman iteration for ℓ_1-norm minimization. AMS Math. Comput. **78**(268), 2127–2136 (2009)
11. Yin, W.: SIAM J. Imaging Sci. **3**(4), 856–877 (2010)
12. Lunglmayr, M., Huemer, M.: Microkicking for fast convergence of sparse Kaczmarz and sparse LMS. In: 25th European Signal Processing Conference (EUSIPCO) 2017 (2017)

Influence of MEMS Microphone Imperfections on the Performance of First-Order Adaptive Differential Microphone Arrays

Andreas Gaich$^{(\boxtimes)}$ (iD) and Mario Huemer

Institute of Signal Processing, Johannes Kepler University Linz, Linz, Austria
andreas.gaich@jku.at

Abstract. In many speech applications the desired speaker is in the far field, i.e. in teleconferencing, hearing aids, hands-free communication in cars, home voice control, just to name a few. To still capture a clean speech signal in a noisy surrounding an acoustic beamformer can be used. Differential microphone arrays (DMAs) allow for compact microphone arrangements and show a reasonable speech enhancement performance. For an optimal performance the microphones used in the array have to be perfectly matched. In this paper, we investigate the effect of the microphone mismatch on the performance of first-order adaptive DMAs, given model data from state-of-the-art micro-electro-mechanical systems (MEMS) microphones. As an important outcome, our simulations show that the performance becomes independent of the mismatch with an increasing number of microphones used.

Keywords: Beamforming · Differential microphone array
MEMS microphones

1 Introduction

The performance of an acoustic beamformer is dependent upon a number of factors. Major issues are the number of microphones used, the target localization error, the microphone position error in the array and the microphone's amplitude and phase mismatch (denoted as microphone mismatch in the following) to all other microphones in the array. With regard to the last point, for an optimal performance the microphones used in the array have to be perfectly matched.

Micro-electro-mechanical systems (MEMS) microphones offer the advantage to be fabricated simultaneously as a single die providing a much better frequency response matching than conventional electret condenser microphones. Still, there is potential for further improvement for matching the MEMS microphones. The influence of the microphone mismatch is highly dependent on the used beamforming algorithm. A differential microphone array (DMA) enables a theoretically frequency independent beam pattern for small microphone array configurations with the drawback of an increased white noise gain (WNG) [1].

© Springer International Publishing AG 2018
R. Moreno-Díaz et al. (Eds.): EUROCAST 2017, Part II, LNCS 10672, pp. 170–178, 2018.
https://doi.org/10.1007/978-3-319-74727-9_20

The WNG is a measure of the amplification of uncorrelated noise, i.e. sensor noise, thus indicating the robustness of a beamformer to microphone imperfections. To mitigate the WNG in DMAs the minimum norm solution (MNS) was proposed in [1].

In this contribution, the performance of a first-order adaptive differential microphone array (FOADMA) is analyzed, when used in a speech enhancement application. Specifically, we investigate the impact of the number of used microphones, the microphone spacing, the microphone signal-to-noise ratio (SNR) and the microphone mismatch.

2 Adaptive DMAs

An adaptive DMA (ADMA) consists of a fixed and an adaptive part as described in Sect. 2.1. Realizations of ADMAs are presented in [2,3]. Every DMA is realized in endfire direction, meaning that the source of interest lies in line of the uniform line array (ULA). A feature of DMAs is the suppression of interfering sources by nullforming towards the corresponding directions in the back half of the microphone array.

2.1 First-Order ADMA

An efficient implementation of the FOADMA first introduced in [2] is shown in Fig. 1. The conventional first-order implementation needs two microphones.

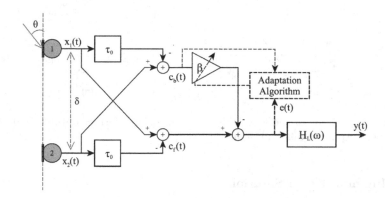

Fig. 1. Schematic implementation of the first-order ADMA [4, p. 15].

The fixed part of the beamformer is realized as two individual DMAs forming the so-called forward-facing cardioid $C_f(\omega, \theta)$ and the backward-facing cardioid $C_b(\omega, \theta)$, given in the frequency domain as

$$C_f(\omega, \theta) = \begin{bmatrix} 1 \ e^{-j\omega\tau_0 \cos\theta} \end{bmatrix} \begin{bmatrix} 1 \\ -e^{-j\omega\tau_0} \end{bmatrix} S(\omega), \tag{1}$$

$$C_{\mathrm{b}}(\omega, \theta) = \left[1 \ e^{-j\omega\tau_0\cos\theta}\right] \begin{bmatrix} -e^{-j\omega\tau_0} \\ 1 \end{bmatrix} S(\omega), \qquad (2)$$

where $S(\omega)$ is the spectrum of the signal source, ω is the angular frequency, θ is the azimuthal angle and $\tau_0 = \delta/c$ is the delay with the speed of sound c and the microphone distance δ (cf. Fig. 2(a)). These two outputs are adaptively combined to obtain the overall output $Y(\omega, \theta)$ calculated as

$$Y(\omega, \theta) = [C_{\mathrm{f}}(\omega, \theta) - \beta C_{\mathrm{b}}(\omega, \theta)] \, H_{\mathrm{L}}(\omega), \qquad (3)$$

where β is a real valued parameter and $H_{\mathrm{L}}(\omega)$ is the compensation filter. The parameter β, ranging between $0 \le \beta \le 1$, controls the resulting beam pattern and is updated by the normalized least mean squares (NLMS) algorithm. The update equation is

$$\beta_{t+1} = \beta_t + \mu \frac{y(t)c_{\mathrm{b}}(t)}{\|c_{\mathrm{b}}(t)^2\| + \Delta}, \qquad (4)$$

with the step-size μ and the regularization parameter Δ. Figure 2(b) illustrates the beam patterns of the overall beamformer output for different values of β.

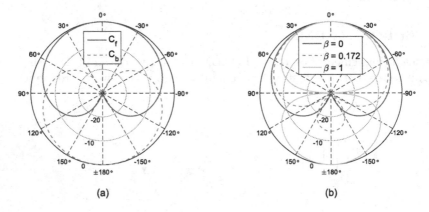

(a)   (b)

Fig. 2. Beam patterns of the FOADMA: (a) Fixed beamformer outputs; (b) Overall beamformer output for different values of β.

2.2   Minimum Norm Solution

As mentioned in the introduction, the drawback of DMAs is the white noise gain (WNG) at low frequencies. This is due to the compensation of the high-pass characteristic of DMAs (6 dB/octave) as shown in Fig. 3. To mitigate the WNG in DMAs the minimum norm solution was proposed in [1]. For that, more than two microphones are used. The resulting filters of the fixed beamformers are then no longer constructed as simple delay elements (cf. Fig. 1) and can be described in closed form as

$$\mathbf{h}(\omega, \boldsymbol{\alpha}, \boldsymbol{\beta}) = \mathbf{D}^T(\omega, \boldsymbol{\alpha}) \left[\mathbf{D}(\omega, \boldsymbol{\alpha})\mathbf{D}^T(\omega, \boldsymbol{\alpha})\right]^{-1} \boldsymbol{\beta}, \qquad (5)$$

where $\mathbf{D}(\omega, \boldsymbol{\alpha})$ is a constraint matrix and $\boldsymbol{\alpha}$ and $\boldsymbol{\beta}$ are the design vectors given as

$$\boldsymbol{\alpha} = \begin{bmatrix} 1 & -1 \end{bmatrix}^T , \tag{6}$$

$$\boldsymbol{\beta} = \begin{bmatrix} 1 & 0 \end{bmatrix}^T , \tag{7}$$

for realization of a cardioid pattern. As an example for $M = 6$ microphones the constraint matrix is given by (cf. [1])

$$\mathbf{D}(\omega, \boldsymbol{\alpha}) = \begin{bmatrix} 1 & e^{-j\omega\tau_0} & e^{-j2\omega\tau_0} & e^{-j3\omega\tau_0} & e^{-j4\omega\tau_0} & e^{-j5\omega\tau_0} \\ 1 & e^{j\omega\tau_0} & e^{j2\omega\tau_0} & e^{j3\omega\tau_0} & e^{j4\omega\tau_0} & e^{j5\omega\tau_0} \end{bmatrix}^T . \tag{8}$$

The authors in [5] extended this concept to ADMAs.

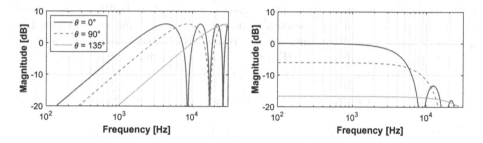

Fig. 3. Directional response of a first-order cardioid for selected angles: (left) Without and (right) with compensation.

3   Model of the MEMS Microphone Frequency Response and Mismatch

The microphones under investigation are state-of-the-art digital MEMS microphones [6]. They utilize an omnidirectional pattern at a small package size of $4 \times 5 \times 1\,\text{mm}$ and feature an SNR up to $66\,\text{dB}$. In general a microphone is an electro-acoustical transducer that operates in three domains: the electrical, mechanical and acoustical domain. The frequency response of the MEMS microphone can be modeled as a linear combination of the corresponding electro-mechanic-acoustic parameters. To obtain the overall frequency response all parameters are usually transformed into the electrical domain and described as lumped elements [7]. Due to confidential contents of the MEMS microphone model of our cooperation partner we are not allowed to show details about the model, but simulations show that the model almost perfectly matches the real MEMS microphone transfer function [6] as depicted in Fig. 4. Given the model for the MEMS microphone, the beamformer output given in (3) is extended to

$$Y_{\text{MEMS}}(\omega, \theta) = [C_{\text{f}}(\omega, \theta) - \beta C_{\text{b}}(\omega, \theta)] H_{\text{L}}(\omega) U(\omega, \theta, \gamma), \tag{9}$$

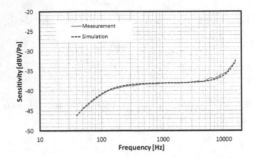

Fig. 4. Comparison of the measured and simulated frequency response [6].

where $U(\omega, \theta, \gamma)$ is the MEMS microphone frequency response. The parameter γ models the frequency response deviation due to manufacturing tolerances in the production process. Measurements reveal that differences in the frequency responses are determined by a variable lower corner frequency between $16\,\mathrm{Hz} \leq f_{c\mathrm{MEMS}} \leq 26\,\mathrm{Hz}$ of the MEMS microphone transfer function. The design of γ is chosen to map $f_{c\mathrm{MEMS}}$ according to

$$-1 \leq \gamma \leq 1 \qquad \propto \qquad 16\,\mathrm{Hz} \geq f_{c\mathrm{MEMS}} \geq 26\,\mathrm{Hz}. \tag{10}$$

4   Simulation Results

To study the influence of the MEMS microphones imperfections on the FOADMA output we will consider two noise scenarios: A low SNR diffuse noise field in Sect. 4.2 and a high SNR diffuse noise field with additional reverberation in Sect. 4.3. Details of the simulation setup are presented in Sect. 4.1.

4.1   Simulation Setup

We investigate FOADMAs with $M = 2, 4, 6$ microphones and microphone spacings of $\delta = 1, 5, 10, 15\,\mathrm{mm}$. The parameters γ_m, $m = 1, \ldots, M$, are equally spaced between $[-1, 1]$ depending on the number of microphones to ensure the maximum distance between the microphones lower corner frequencies.

Implementations are based on block processing with the overlap-add method and 50% overlapping. The used window-type is Hanning and the sample frequency is $f_s = 16\,\mathrm{kHz}$ considering a frame size for the block processing of 2^9 samples. The step-size of the NLMS is set to $\mu = 0.01$ and the regularization constant is $\Delta = 10^{-4}$.

All results are averaged over 50 speakers (female and male) randomly selected from the GRID corpus [8] and are evaluated in terms of perceptual evaluation of speech quality (PESQ) [9], an ITU recommendation for speech quality assessment. The noise files are selected from the NOISEX-92 database [10].

4.2 Scenario 1: Diffuse Babble Noise

In this scenario a target speaker with a sound pressure level (SPL) of 70 dB is located 0.5 m in front of the microphone array and is interfered by diffuse babble noise with a signal-to-interference ratio (SIR) of 6 dB. For the generation of the diffuse noise field we refer to [11].

Figure 5 shows the results of the speech enhancement performance in terms of PESQ in dependence of the microphone SNR. Results are illustrated for perfect microphone matching on the left side, for microphone mismatch on the right side, and are sorted by the number of microphones from top to bottom. The unprocessed (UP) signal is the recording of just one microphone in the array and determines the baseline of the enhancement performance. The other curves show the results of the FOADMA for different microphone spacings. It can be seen that the microphone mismatch significantly degrades the speech enhancement performance for a two-element FOADMA. This effect is more distinct for a smaller microphone spacing and disappears if more microphones are used,

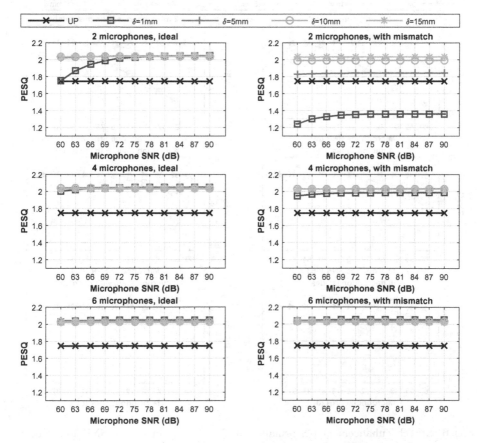

Fig. 5. Speech enhancement performance of FOADMAs in a low SNR babble noise scenario for perfect microphone matching (left) and microphone mismatch (right).

meaning that the performance is independent of the matching for FOADMAs that utilize more than four microphones. In general a performance gain in comparison to the UP signal is achieved, except for the two-element FOADMA with microphone mismatch.

4.3   Scenario 2: Diffuse Living Room Noise with Room Reverberations

The second scenario examines the target speaker with an SPL of 60 dB in a distance of 1 m. The interferer is diffuse ambient living room noise with an SIR of 26 dB. Additionally, we introduce room reverberations with a reverberation time of $T_{60} = 0.3$ s using the image method in [12].

Results are shown in Fig. 6 and are structured in the same way as in the previous section. A similar behavior regarding the microphone mismatch can be

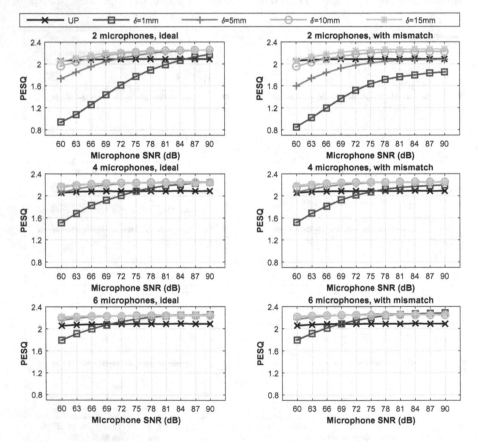

Fig. 6. Speech enhancement performance of FOADMAs in a high SNR living room noise scenario and additional room reverberation of $T_{60} = 0.3$ ms for perfect microphone matching (left) and microphone mismatch (right).

observed, although the speech enhancement performance loss due to the mismatch is less pronounced. However, a performance gain over the entire investigated SNR range is only achieved for FOADMAs with at least four microphones and microphone spacings greater than or equal to $\delta = 5\,\mathrm{mm}$ (with and without the modeled microphone mismatches). For a spacing of $\delta = 1\,\mathrm{mm}$ at least four microphones with an SNR of $78\,\mathrm{dB}$ have to be utilized to improve the speech quality. This can be explained by the strong amplification of the WNG at small microphone spacings leading to a microphone noise floor that exhibits the noise power level of the weak surrounding diffuse living room noise field.

5   Conclusion

This paper addresses the impact of manufacturing tolerances of state-of-the-art MEMS microphones on the performance of FOADMAs. Given the model of the microphone mismatch we investigated two simulation setups covering a low and a high SNR diffuse noise field case. The simulations confirm that a high microphone SNR is important in low ambient noise situations. Furthermore, to obtain speech enhancement for small microphone spacings of $\delta = 1\,\mathrm{mm}$ using a two-element FOADMA, perfect matching of the microphones is necessary. On the contrary, increasing the number of microphones in the microphone array also increases the robustness of the FOADMA to microphone mismatch. In the case of at least six microphones the speech enhancement is independent of the microphone matching for the considered mismatch model.

References

1. Benesty, J., Chen, J.: Study and Design of Differential Microphone Arrays. Springer, Heidelberg (2013). https://doi.org/10.1007/978-3-642-33753-6_3
2. Elko, G.W., Pong, A.T.N.: A simple adaptive first-order differential microphone. In: Proceedings of 1995 Workshop on Applications of Signal Processing to Audio and Accoustics, pp. 169–172 (1995)
3. Elko, G.W., Meyer, J.: Second-order differential adaptive microphone array. In: 2009 IEEE International Conference on Acoustics, Speech and Signal Processing, pp. 73–76 (2009)
4. Messner, E.: Differential Microphone Arrays. Master's thesis, Graz University of Technology (2013)
5. Messner, E., Pessentheiner, H., Morales-Cordovilla, J.A., Hagmüller, M.: Adaptive differential microphone arrays used as a front-end for an automatic speech recognition system. In: 2015 IEEE International Conference on Acoustics, Speech and Signal Processing, pp. 2689–2693 (2015)
6. Dehe, A., Wurzer, M., Füldner, M., Krumbein, U.: The infineon silicon mems microphone. In: AMA Conferences 2013 - SENSORS 2013, pp. 95–99 (2013)
7. Zollner, M., Zwicker, E.: Elektroakustik. Springer, Heidelberg (2013)
8. Cooke, M., Barker, J., Cunningham, S., Shao, X.: An audio-visual corpus for speech perception and automatic speech recognition. J. Acoust. Soc. Am. **120**(5), 2421–2424 (2006)

9. Rix, A.W., Beerends, J.G., Hollier, M.P., Hekstra, A.P.: Perceptual evaluation of speech quality (PESQ) - a new method for speech quality assessment of telephone networks and codecs. In: 2001 IEEE International Conference on Acoustics, Speech and Signal Processing, vol. 2, pp. 749–752 (2001)
10. Varga, A., Steeneken, H.J.M., Tomlinson, M., Jones, D.: The NOISEX-92 study on the effect of additive noise on automatic speech recognition. Technical report, DRA Speech Research Unit (1992)
11. Habets, E.A.P., Cohen, I., Gannot, S.: Generating nonstationary multisensor signals under a spatial coherence constraint. J Acoust. Soc. Am. **124**(5), 2911–2917 (2008)
12. Allen, J.B., Berkley, D.A.: Image method for efficiently simulating smallroom acoustics. J. Acoust. Soc. Am. **65**(4), 943–950 (1979)

Global Decision Making for Wavelet Based ECG Segmentation

Carl Böck[1]([✉])[ID], Michael Lunglmayr[1][ID], Christoph Mahringer[1],
Christoph Mörtl[2], Jens Meier[2], and Mario Huemer[1]

[1] Institute of Signal Processing, Johannes Kepler University Linz, Linz, Austria
carl.boeck@jku.at
[2] Department of Anesthesiology and Critical Care Medicine,
Kepler University Hospital Linz, Linz, Austria

Abstract. In this work, we propose an improvement of an established single lead electrocardiogram (ECG) beat segmentation algorithm based on the wavelet transform. First, for a particular recording a reference beat is determined by averaging over a certain amount of beats. Subsequently, this beat is used to obtain recording specific thresholds and search windows needed for the segmentation of the whole recording. Since noise and artifacts significantly influence the segmentation process, we show that using the information provided by the reference beat positively impacts the results. Specifically, using this global information of the reference beat, the algorithm becomes more robust against transient noise and signal abnormalities. Consequently, the proposed approach leads to an ECG beat segmentation algorithm specifically suited for detecting subtle relative changes of characteristic time intervals and amplitude levels.

Keywords: ECG beat delineation · ECG beat segmentation
ECG characteristic points · Wavelet transform

1 Introduction

The electrocardiogram (ECG) is a well-established and easy to obtain physiological signal of remarkable diagnostic power. It is composed of the concatenation of single ECG beats, which themselves can be split into single waves (P, Q, R, S, T wave). Most of its clinically useful information is given by the amplitudes and durations of the single waves as well as the time intervals between them. Thus, automated ECG beat detection and the subsequent segmentation into the beats' waves has been an important subject of research during the past decades. Algorithms have to be capable of dealing with inter-individual morphology as well as with common artifacts, characteristic for ECG recordings. Consequently, there exist many approaches for ECG beat segmentation. One of the most promising and most extensively evaluated algorithms uses the wavelet transform (WT) and was suggested in [1]. An improved version of the algorithm can be found in [2]. Although this method generally performs very well, Fig. 1a–c

© Springer International Publishing AG 2018
R. Moreno-Díaz et al. (Eds.): EUROCAST 2017, Part II, LNCS 10672, pp. 179–186, 2018.
https://doi.org/10.1007/978-3-319-74727-9_21

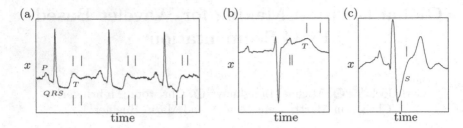

Fig. 1. (a)–(c) Expert and algorithm annotations (top vs. bottom) for different ECG characteristic points. (a) Algorithm detects wrong T_{peak} and T_{end} for third beat due to slightly varying morphology. (b) Wrong detection of T_{peak} and T_{end} due to an artifact and (c) wrong detection of QRS_{end} because of an untypically long S wave.

show some cases where the algorithm fails. This is mainly due to the fact that the delineation, i.e. the determination of the single waves' limits and peaks, is done using local information of the current beat only. This leads to problems in case that wave shapes slightly differ from one beat to another (Fig. 1a) or in case of artifacts (Fig. 1b). Additionally specific thresholds and search windows used for identifying the different waves are set to constant values which could lead to misdetections of single waves given that the wave shapes are untypical (Fig. 1c). In this work we suggest a method capable of dealing with the above mentioned limitations and consequently improving the original approach. Particularly, the new approach decreases beat-to-beat fluctuations of estimated ECG characteristic points, given that the true morphology of the ECG does not change appruptly from one beat to another. Figure 1a, e.g., illustrates that the original approach performs a wrong T wave delineation for the third beat, probably due to transient noise at the end of the T wave. Assuming that we want to track changes of the QT interval ($QT = T_{\text{end}} - QRS_{\text{on}}$) over time, such beat-to-beat fluctuations introduced by the algorithm are very problematic, and are thus decreased by the approach suggested. In order to assess the robustness of the new approach with regard to the inter-variety of different patient's ECGs, the work is evaluated on the QT database (QTBD) [3], a standard database for testing segmentation algorithms, by comparing our results to those of the algorithm originally provided in [2].

2   ECG Beat Delineation

2.1   Wavelet Transform of the ECG

The wavelet transform of the ECG signal $x(t)$ is determined by calculating the inner product of $x(t)$ with dilated as well as shifted versions of a single mother wavelet $\psi(t) \in L_2(\mathbb{R})^1$, defined as

$$W(b, a) = |a|^{-\frac{1}{2}} \int_{-\infty}^{\infty} x(t) \, \psi\left(\frac{t - b}{a}\right) dt. \tag{1}$$

---

[1] $L_2(\mathbb{R})$ is the set of complex valued functions which satisfy $\int_{-\infty}^{\infty} |f(t)|^2 \, dt < \infty$.

Dilating and translating the mother wavelet $\psi(t)$ enables to decompose the signal into a combination of a set of basis functions. As suggested by [1], using the so called biorthogonal spline wavelet, (1) can be rewritten as

$$W(b, a) = -a\frac{d}{db} \int_{-\infty}^{\infty} x(t)\,\phi_a\,(t - b)\,dt, \tag{2}$$

i.e., the WT is proportional to the derivative (with respect to the translation parameter b) of a smoothed version of the ECG [4]. Smoothing is performed by the corresponding scaling filter ϕ at scale a. Hence, for a particular scale a, the zero crossing at a specific translation b corresponds to a local maximum or minimum of the filtered ECG (Fig. 2a). Additionally, maxima and minima of the WT correspond to maximum positive and maximum negative slopes. Since the ECG is composed of rising and falling edges (at different frequencies/scales) as well as local extremes, the WT using the specific mother wavelet is well suited for our application.

As suggested by [1] the WT is only evaluated for a subset of dyadic scales $a = 2^k$, $k = \{1, 2, 3, 4\}$, leading to the so called stationary discrete wavelet transform (SDWT), which results in the scales $d_1 - d_4$ and can be efficiently implemented using a filter bank [1]. Based on these scales $d_1 - d_4$, different valid wave morphologies can be distinguished, and subsequently onset, peak and end of the specific waves can be detected (Fig. 2a). Depending on which wave one is looking for, the information of the according scales $d_1 - d_4$ is exploited, since the P and T waves have significant energy components in scales 3 and 4 (lower frequencies) while the Q and the S wave are mainly associated with scale 1 and 2 (higher frequencies). The R wave usually influences all scales.

2.2 Delineation Algorithm

The QRS complex is the most striking feature in an ECG and is therefore detected at first. For that reason we search for local positive maxima (rising edges) and negative minima (falling edges) in scales $d_1 - d_4$ (Fig. 2a), which will be called modulus maxima (MM) from now on. Chaining the MM of the different scales, whereas the MM lying closest to each other are connected, lead to so called MM lines [5]. These initial MM lines are illustrated as dashed lines in Fig. 2a and are used to identify potential QRS complexes by comparing them to scale dependent thresholds[2]. Applying different exclusion criteria to eliminate non-valid or redundant MM lines, and determining the according zero crossings of scale d_1 leads to a vector of R peak time instants $\mathbf{t}_R$ (Fig. 2b). These positions are used to create an average reference beat by calculating the mean of the first N_b beats of a recording, whereas the beats are sliced and temporally aligned using $\mathbf{t}_R$.

QRS Delineation of the Reference Beat: The algorithm continuous by detecting the peak positions of the reference beat's Q and S waves (t_Q, t_S) as

---

[2] All upcoming relevant thresholds and search windows are listed in Table 1.

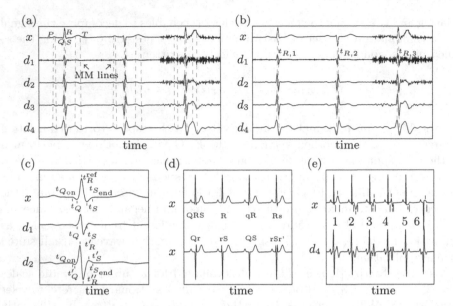

Fig. 2. (a) SDWT for scales $d_1 - d_4$ of different ECG beat morphologies, one covered in noise, (b) resulting vector $\mathbf{t}_R$ of R peak time instants, where $t_{R,n}$ is the n-th element of $\mathbf{t}_R$, (c) QRS delineation, (d) valid QRS morphologies and (e) valid T wave morphologies (normal, inverted, positive biphasic, negative biphasic, ascending, descending).

well as the onset and end of the QRS complex ($t_{Q_{\mathrm{on}}}$, $t_{S_{\mathrm{end}}}$) for the mean beat. As stated in [2] Q and S waves have their main frequency content within scale two. Hence, the search for significant MM is performed for this scale. Figure 2c shows the delineation process of the standard QRS complex, whereas the algorithm starts at the reference beat's R peak location t_R^{ref}, which is bordered by a pair of MM (t_R' and t_R''). This pair is related to the maximal slopes of the R wave's rising and falling edges. Within a predefined search window SW_{QRS}, we then search for significant MM before t_R' and after t_R'' in order to detect the slopes of the Q and S waves (t_Q', t_S'). These significant MM are identified by selecting MM higher than a threshold

$$th_Q = th_q \max(|d_2[n]|), \quad n \in \mathrm{SW}_{QRS}, \text{ and} \tag{3}$$

$$th_S = th_s \max(|d_2[n]|), \quad n \in \mathrm{SW}_{QRS}, \tag{4}$$

whereas th_q, th_s, and SW_{QRS} are determined using an optimization process explained further below. If there exist more than one significant MM, the one best suited is chosen via a criterion based on the distances to t_R' or t_R'' and the absolute values of the competing MM. Similarly to the determination of the R peak, the according zero crossings at scale one are then used to identify the peaks of the two waves (t_Q, t_S).

In order to determine the onset of the QRS complex, the algorithm departs from the earliest significant MM and searches to the left within SW_{QRS}.

Afterwards three potential candidates for the onset of the QRS complex are identified, namely the time instant where the absolute value of d_2 falls below a specific threshold for the first time, the time instant of the first local maximum or minimum found in d_2, and the beginning of the search window. Subsequently, the candidate lying closest to the R peak (the most right one) is defined to be the onset of the QRS complex. In case that no significant MM are detected (no Q wave), the search for the onset simply starts at t'_R. Finally, the end of the QRS complex is determined analogously to the onset. Based on the presence or absence of significant MM and on their signs (positive or negative), we are able to distinguish between the most common types of QRS complex illustrated in Fig. 2d.

P and T Wave Delineation of the Reference Beat: In contrast to the QRS complex, the P and the T waves are slowly varying signals, hence the search for significant MM is performed at scale four. These MM are identified by selecting MM higher than a threshold

$$th_P = th_p \, \text{RMS}(d_4[n]), \text{ and} \tag{5}$$

$$th_T = th_t \, \text{RMS}(d_4[n]), \tag{6}$$

whereas RMS stands for root mean square. The specific thresholds th_p and th_t are determined using an optimization process for the mean beat (explained below). The simplest case would be to find one positive and one negative significant MM leading to the identification of a normal or inverted P/T wave (Fig. 2e, Beat 1–2). However, there might be invalid MM within the identified set of MM, e.g., two or more consecutive positive MM. These invalid MM are eliminated via the following strategies:

- MM lines with a significant MM in scale one (highest frequent scale) are eliminated since this most probably is some kind of artifact.
- MM smaller than one eighth of the most significant positive and negative MM are considered to be irrelevant small deflections and are eliminated.
- In case of a biphasic T wave (Fig. 2e - Beat 3–4), the two surrounding MM must be approximately equal. If this is not the case, one of them is eliminated based on the distance to the center MM and the absolute values of the MM.

The zero-crossing(s) between the remaining MM correspond to the peak(s) of the wave and are identified in scale three due to the better time resolution. Finally, the appropriate type of morphology is determined and onset as well as end of the P/T wave are defined via a similar strategy as used for the QRS complex (Fig. 2e).

Optimizing Thresholds and Search Windows: Depending on the patient and the lead, the single ECG waves vary in height, length, and shape. Hence, e.g., waves can be very small for a specific patient or lead while they are still present. Figure 3a shows an example for a small S wave, which is not detected in case that the threshold th_s is too large. On the other hand, the selection of a too small threshold makes the algorithm more susceptible to noise. Consequently it is important to find an appropriate set of thresholds for a specific recording, which is illustrated in Fig. 3a for selecting th_s. The threshold th_s is optimized such that the beats' shape is approximated best, using only the information given by the detected ECG characteristic points. For that reason we perform a shape-preserving piecewise cubic spline interpolation between the ECG characterisitic points to receive an approximation of the beat. Starting, e.g., with $th_s = 0.09$, as suggested by [2], th_s is decreased in order to minimize the mean square error between the mean beat and the approximation. Simultaneously, the search window of the S wave is varied within a specific interval. The same strategy is used for all other thresholds and search windows needed for detecting significant MM. Table 1 shows all variable and constant thresholds as well as search windows used for the delineation (based on [2]).

Table 1. Defined thresholds and search windows, ($\overline{RR}$ is the mean time distance in s between consecutive R peaks).

th^i_{QRS}, $i = 1,2,3$	th^4_{QRS}	th_q	th_s	th_t	th_p
RMS($d_i[n]$)	0.5 RMS($d_4[n]$)	[0.03, 0.06]	[0.05, 0.09]	[0.15, 0.35]	[0.01, 0.03]
SW$_{QRS,start}$	SW$_{QRS,end}$	SW$_{P,start}$	SW$_{P,end}$	SW$_{T,start}$	SW$_{T,end}$
$[t^{ref}_R - 0.12s$, $t^{ref}_R - 0.06s]$	$[t^{ref}_R + 0.06s$, $t^{ref}_R + 0.15s]$	$t^{ref}_R - \overline{RR}/2$	QRS_{on}	QRS_{end}	$t^{ref}_R + 2\overline{RR}/3$

Beat Delineation of the Recording: After setting the recording-specific thresholds and search windows, ECG beat delineation is carried out for the whole recording. A crucial step for the detection of the waves' correct onset, peak, and end, is the selection of the right MM of the according scales out of several potential candidates. As shown for the P wave in Fig. 3b, there usually exist several MM candidates within a specific search window due to noise or other influences. Assuming that the currently considered beat only deviates slightly from the mean beat, we select those significant minima and maxima which are most similar to those of the mean beat, comparing their amplitudes, their distances to the R peak, and the distances between the MM. If the current beat

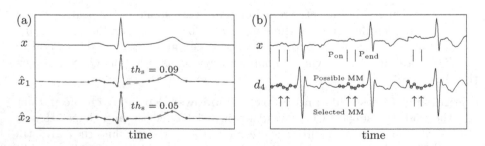

Fig. 3. (a) Optimizing the threshold for the S wave, whereby $\hat{x}_1$ leads to a larger error than $\hat{x}_2$. (b) Selecting the correct MM for the P wave.

strongly deviates from the mean beat, as it would be the case for an extrasytole, the algorithm selects the MM using local information only without considering the mean beat.

3 Evaluation

The QT database (QTDB) provides more than 3000 manually annotated ECG beats of 105 different records designed for the evaluation of segmentation algorithms and includes a broad variety of QRS and $ST-T$ morphologies [3]. Hence, this database is suited to assess the robustness of our algorithm with regard to the inter-variety of different patient's ECGs. All beats, which had been manually annotated by one expert, were extracted and compared to the results provided by the algorithm. For every ECG characteristic point labeled by the expert, the time difference e between the expert and algorithm label was calculated. Subsequently we determined the bias $\bar{e}$ over all beats. Additionally the standard deviation of e was calculated for every recording and averaged over the number of recordings leading to $\bar{\sigma}$. Using these two parameters our results can be compared to those provided in [2].

Table 2 compares the average time difference $\bar{e}$ as well as the average intra-recording standard deviation $\bar{\sigma}$ for single ECG characteristic points. The average time differences $\bar{e}$ between the expert and the algorithm annotations are slightly larger than those of the original approach. However, considering the fact that there is no golden rule for the true positions of the ECG characteristic points, a specific bias to the expert annotations is not considered problematic since we want to focus on tracking relative changes of ECG intervals and amplitudes. For that reason the average of the intra-recording standard deviations $\bar{\sigma}$ was improved for the most of the ECG characteristic points, indicating that the undesired beat-to-beat fluctuations, introduced by the algorithm, were decreased compared to the original approach. An example is shown for the P wave delineation in Fig. 4. In the work of [2] the decision which of the MM to select is based on local beat criteria only, i.e., the selection of the MM only depends on the currently segmented beat. As shown in Fig. 4a this leads to a wrong detection of ECG characteristic points for the second and third beat, which results in high beat-to-beat fluctuations regarding the position of the P wave. This is prevented by the new approach (Fig. 4b). Basically, using a mean reference beat for deciding which MM to select, imitates the behavior of physicians, who would also exploit the information of preceding beats for the delineation of the current one. Consequently this approach is well suited to track subtle changes of ECG wave amplitudes and durations as well as the intervals between them, even if some of the beats are disturbed by transient noise. However, for recordings containing a lot of irregularly shaped beats, e.g. extrasystoles, this approach is clearly not suited since averaging the first N_b beats of the recording would already lead to a non-meaningful reference beat, not appropriate for decision support for any of the subsequent beats.

Table 2. Performance comparison between this work and [2].

	#Ann	This work		Orig. app.	
		$\bar{e}$	$\bar{\sigma}$	$\bar{e}$	$\bar{\sigma}$
-		ms	ms	ms	ms
P_{on}	3194	−12.7	11.9	2.0	14.8
P_{peak}	3194	−4.7	8.9	3.6	13.2
P_{end}	3194	2.7	9.9	1.9	12.8
QRS_{on}	3623	−2.4	7.6	4.6	7.7
QRS_{end}	3623	−4.6	8.7	0.8	8.7
T_{peak}	3542	−2.0	12.6	0.2	13.9
T_{end}	3542	−1.8	16.8	−1.5	18.1

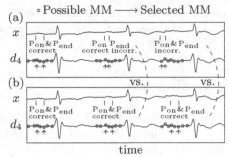

Fig. 4. P wave segmentation (a) of [2] and (b) of this work.

4   Conclusion

Within this work we have implemented and further improved an ECG beat delineation algorithm based on the wavelet transform. The individual determination of the thresholds and search windows as well as the exploitation of the information given by the reference beat lead to a robust algorithm, which is specifically suited for detecting minimal changes in a patient's ECG. The algorithm has been validated using the QTBD and compared to the original approach provided by [2]. The proposed approach is robust against the inter-variety of different patient's ECG and reduces the intra-recording standard deviation of the time differences between the expert and the algorithm annotations.

References

1. Li, C., Zheng, C., Tai, C.: Detection of ECG characteristic points using wavelet transforms. IEEE Trans. Biomed. Eng. **42**, 21–28 (1995)
2. Martinez, J.P., Almeida, R., Rocha, A.P., Laguna, P.: A wavelet-based ECG delineator evaluation on standard databases. IEEE Trans. Biomed. Eng. **51**, 570–581 (2004)
3. Laguna, P., Mark, R.G., Goldberg, A., Moody, G.B.: A database for evaluation of algorithms for measurement of QT and other waveform intervals in the ECG. In: Proceedings of Computers in Cardiology, Lund, pp. 673–676 (1997)
4. Sahambi, J.S., Tandon, S.N., Bhatt, R.K.P.: Wavelet based ST-segment analysis. Med. Biol. Eng. Comput. **36**, 568–572 (1998)
5. Mallat, S.G., Peyre, G.: A Wavelet Tour of Signal Processing: The Sparse Way. Academic Press, Boston (2008)

Heartbeat Classification of ECG Signals Using Rational Function Systems

Gergő Bognár$^{(\boxtimes)}$ and Sándor Fridli

Department of Numerical Analysis, Faculty of Informatics,
ELTE Eötvös Loránd University, Budapest, Hungary
bognargergo@caesar.elte.hu, fridli@inf.elte.hu

Abstract. The main idea of this paper is to show that rational orthogonal function systems, called Malmquist-Takenaka (MT) systems can effectively be used for ECG heartbeat classification. The idea behind using these systems is the adaptive nature of them. Then the constructed feature vector consists of two main parts, called dynamic and morphological parameters. The latter ones contain the coefficients of the orthogonal projection with respect to the MT systems. Then Support Vector Machine algorithm was used for classifying the heartbeats into the usual 16 arrhythmia classes. The comparison test were performed on the MIT-BIH arrhythmia database. The results show that our algorithm outperforms the previous ones in many respects.

Keywords: ECG signals · Heartbeat classification
Rational functions · Malmquist-Takenaka systems
Support vector machine

1 Introduction

The importance of automatic electrocardiogram (ECG) analysis is reflected in the intense research activity in this area. Reliable computer-assisted cardiac disorder detection can efficiently contribute the management of clinical situations. In this paper we concentrate on the classification problem when each heartbeat is associated with one of the predefined 16 arrhythmia [19] types in PhysioNet [8]. There have several methods been published so far using various kinds of feature extraction methods and classification algorithms [2,10,12,17,18,20,22]. A common approach is to combine dynamic and morphological features, where the morphological features are generated by means of a proper transformation of each heartbeat. Then a machine learning classification algorithm is applied. Review of the relevant literature can be found in [14,21]. The methods are usually validated through the standard MIT-BIH arrhythmia database (distributed

G. Bognár—Supported by the New National Excellence Program of the Ministry of Human Capacities of Hungary.
S. Fridli—This research was supported by the Hungarian Scientific Research Funds (OTKA) No. K115804.

© Springer International Publishing AG 2018
R. Moreno-Díaz et al. (Eds.): EUROCAST 2017, Part II, LNCS 10672, pp. 187–195, 2018.
https://doi.org/10.1007/978-3-319-74727-9_22

by the Massachusetts Institute of Technology and the Boston's Beth Israel Hospital) [15] that consists of more than 100 000 heartbeats of 48 signals providing reference annotations for each heartbeat, including the location of the QRS complex and the manually specified class of the heartbeat. The performed evaluation schemes are either 'class-oriented' or 'subject-oriented'.

The novelty in our paper is the use of rational function systems [4,5] to extract morphological features. We note that the use of the well known orthogonal transformations, like the classical trigonometric, wavelet etc. have a long history in this area. The reason why we prefer rational function systems for heartbeat classification is the adaptivity. Namely, the so-called poles that determine the system can be adjusted to the individual signal. Another reason why these system perform well in ECG processing is that the elementary rational functions imitate the shapes of the natural segments, i.e. the P and T waves and the QRS complex of heartbeats. Therefore there is a direct connection between the shape of the ECG curve and the poles and coefficients of the rational system. We note that in recent years, several papers [4–7,11,13] have appeared on using rational transforms for heartbeat detection, and for compressing, approximating ECG signals.

We tested the heartbeat classification method developed in this paper on the MIT-BIH arrhythmia database with 16 classes using the 'class-oriented' evaluation scheme, in which followed the common methodology described in [2,14]. The classification was made for a combination of morphological as dynamic features with using support vector machine (SVM) classifier, and two-lead fusion. Finally we performed comparison test on our and on the other known methods.

2   Rational Orthogonal Systems

Our method is based on modeling the signal by rational functions. To this order let us take the basic rational functions of the form

$$r_{a,k}(z) = \frac{1}{(1 - \overline{a}z)^k} \qquad (a \in \mathbb{D},\ k \in \mathbb{N},\ k \geq 1),$$

where $\mathbb{D}$ is the unit disc of complex numbers. a is called the inverse pole and k is the degree of $r_{a,k}$. Suppose we have $n \in \mathbb{N}$ distinct inverse poles $\{a_k \in \mathbb{D} : k = 0, \ldots, n-1\}$, and each of them is associated with multiplicity $m_k \in \mathbb{N}$ ($k = 0, \ldots, n-1$). They define the set of basic rational functions $\{r_{a_k,j_k} : k = 0, \ldots, n-1,\ j_k = 1, \ldots, m_k\}$. We consider these functions on the torus $\mathbb{T} = \{z \in \mathbb{C} : |z| = 1\}$, which can be naturally associated with the interval $[-\pi, \pi)$ (Fig. 1). Then we take the usual scalar product $\langle f, g \rangle = \frac{1}{2\pi} \int_{-\pi}^{\pi} f(e^{it})\overline{g(e^{it})}\, dt$, where f, and g be complex valued square integrable functions on $\mathbb{T}$. Using the Gram–Schmidt orthogonalization process with respect to the above collection of basic functions and scalar product we obtain an $\sum_{k=0}^{n-1} m_k$ dimensional Malmquist–Takenaka type orthonormal system. We note that it is not necessary to perform the orthogonalization process since there is an explicit form for the Malmquist-Takenaka functions expressed by the so called Blaschke functions. For general

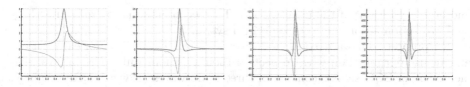

Fig. 1. Real (blue) and imaginary (red) part of rational basic functions (inverse pole magnitude 0.8, order 1 to 4). (Color figure online)

reference on rational systems we refer the reader to [9]. In summary, if the poles and multiplicities are given then a rational orthogonal system on $[-\pi, \pi]$ is determined.

3   Methodology

In this section we provide a detailed description about the proposed method, that consists of six stages: preprocessing, heartbeat segmentation, pole identification for the rational transform, feature extraction, classification and two-lead fusion.

3.1   Preprocessing and Heartbeat Segmentation

Raw ECG signals suffer from several type of distortions, like noise, power-line interference, baseline wander, electrode displacement and muscle movement artifacts. To suppress these effects, we applied baseline wander correction using wavelet-transform [23], and low-pass filtering with cutoff frequency of 35 Hz to remove high-frequency noise [2].

ECG signals are quasi-periodic, containing subsequent heartbeats, where one beat typically consists of three wave components (P wave, QRS complex, and T wave). Although the durations of heartbeats vary with time we will use a fixed window size corresponding to the typical length as it was done in [22]. Namely, we take 99 samples (0.28 s) preceding and 200 samples (0.56 s) succeeding the annotated R peak, i.e. a total of 300 samples. This may raise problems, e.g. when the heart rate is high, the next heartbeat can overlap. On the other side the window is normalized and contains the relevant information about the heartbeat. We also note that in real application, a reliable automatic R peak detector is required.

3.2   Rational Transform

In this study, heartbeat segments are represented by rational functions using real MT-systems, defined in Sect. 2. This process is called adaptive transformation technique consisting of two steps. Namely, it includes a non-linear optimization problem, where the task is to find the optimal inverse pole combination that determines the system itself. Here the questions are the number, multiplicities and actual positions of the inverse poles. The second step is to perform orthogonal projection according to the specified system.

We propose to use a fixed (2, 4, 2) inverse pole configuration for each patient and to use optimization with respect to their positions only. The natural segmentation of ECG signals consists of three main waves (P wave, QRS complex, T wave) therefore it is reasonable to work with three poles. The arguments of the poles hold information about the locations, and the magnitudes carry information about the shapes of the waves. Choosing a pole order 4 for the QRS complex is justified by its higher complexity. With these fixed pole combination, we suggest to find one pole-triplet assigned to individual patients. There are several options available to find proper patient-specific pole combinations. The optimization depends on the optimization method, the objective function, and the initial guess. Our choice for the optimization method was the Hyperbolic Nelder-Mead algorithm [6,16]. It is a fast and effective derivate-free method that has already been utilized to pole identification problems. We modified the original concept in order to apply the method to a set of heartbeats at the same time. The objective function is the summarized L_2 errors of the individual heartbeats from the first 5 min. For a good initial guess we took the virtual heartbeat, which was the average of the heartbeats from the first 5 min. After this smoothing, the P and T peaks are determined as the locations of the most prominent local maxima before and after the QRS complex. Finally, the initial values of the complex arguments of the inverse poles are defined according to these locations, and fixed values are used for magnitudes. This idea is based on the time-localization property of the basic rational functions, and our tests show that it is a good initial guess. The optimization method was used to each of the records independently, then the rational transform with the patient-adapted pole triplet is applied to each heartbeats of the record. Figure 2 demonstrates the result of pole identification and heartbeat representation applied to the first record (#100) of the MIT-BIH database.

This results in an adaptive and stable pole selection. Moreover, the poles hold information about the average behaviour of the patient's heart. Furthermore, each heartbeat of a patient's record is represented in the same, patient-specific rational basis, thus the coefficients of the projection have a comparable meaning.

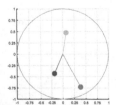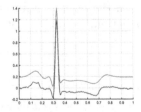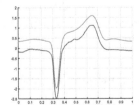

Fig. 2. Optimal inverse poles of signal #100; a normal and a ventricular ectopic (PVC) heartbeat (blue) and their rational approximations (red, shifted). (Color figure online)

3.3 Feature Vector Extraction

We propose to generate morphological features based on the adaptive rational transform. As we described in the previous sections, the rational transform holds frequency and time information about the signal. Furthermore, direct morphological information can be extracted from the projection, due to the suggested patient-adapted pole identification method and the connection between ECG waveforms and the basic rational functions. In this study, after the patient-specific optimal pole combination is identified, the rational orthogonal projection is calculated using the three poles with (2, 4, 2) multiplicities. The 16 coefficients of the real MT system are used as morphological features.

Beside the morphological descriptors, the dynamic RR interval features is added to the feature vector, to describe the dynamic behaviour of the heart. According to the literature, 4 RR interval features are considered, the length of the previous and next RR interval (pre-RR and post-RR), the mean of the RR intervals in the previous 10 s and 5 min (local and average RR).

3.4 Classification

The performance evaluation of the proposed method was performed following the 'class-oriented' scheme on the MIT-BIH arrhythmia database. The heartbeats of the whole database are clustered according to the arrhythmia classes and split into training and test dataset. From each cluster a certain fraction of heartbeats are selected randomly to the training dataset, the rest of the heartbeats are put to the test dataset. The selected fraction is 13% for normal beats, 40% for bigger classes, and 50% for smaller classes, a total of 22% of the whole database. The details are given in Table 1.

In this study, support vector machine (SVM) classifier [3] is utilized to classify the heartbeats into one of the 16 classes. We applied the multi-class extension of SVM following the one-against-one (OAO) approach, with radial basis function (RBF) kernel, where the model parameters (regularization parameter and kernel width) are automatically selected using tenfold cross-validation on the training dataset. Then an SVM classifier model is trained using the training dataset. Finally, the heartbeats of the test dataset are classified based on the trained model, providing a class prediction and probability estimates. We used the LIBSVM package [1] to implement the SVM-related algorithms.

3.5 Lead Configuration and Fusion

The MIT-BIH arrhythmia database provides two-lead signals, the proposed method could be applied to both of the leads independently. The final classification decision can be improved if we combine the results derived from the two leads. Following the literature, we considered different approaches. The *Bayesian fusion approach* combine the probability estimates of the two classifier to fused probability estimates, the final class prediction is decided according to this fusion. The *rejection approach* rejects heartbeats where the two classifier gave different class predictions. A small amount of heartbeats will be excluded from classification (i.e. no class prediction is made), but for the remaining beats the predictions are more reliable. Later, the rejected heartbeats can be reviewed by physicians.

We note that instead of using individual pole combination for the two leads, it is also reasonable to use the same pole combination for both of the leads. This modification is not discussed in details, since the results are almost identical, yet the concept can be interesting in a future research.

4 Results

In this section we evaluate the proposed classification method and discuss the results. Our method yields accuracy of 99.05% based on the first lead, and 98.65% based on the second. The combined accuracy is 99.38% with Bayesian fusion approach, and 99.66% with rejection approach (1.56%, 1377 beats were rejected). A detailed evaluation is provided in Table 1. Here two performance metric is calculated for each heartbeat classes, based on the number of true positives (TP), false positives (FP), and false negatives (FN): the sensitivity (Se = TP/(TP + FN)) and the positive predictivity (+P = TP/(TP + FP)).

Table 1. Summary of classes, training and test datasets, and performance evaluation

Heartbeat		Total	Train		Test	Bayesian fusion			Rejection			
Class	a.	Num.	%	Num.	Num.	TP	Se%	+P%	TP	R	Se%	+P%
NOR	N	75016	13	9753	65263	65010	99.61	99.75	64233	898	99.79	99.87
LBBB	L	8072	40	3229	4843	4839	99.92	99.94	4816	26	99.98	100.00
RBBB	R	7255	40	2902	4353	4347	99.86	99.72	4342	7	99.91	99.84
APC	A	2546	40	1019	1527	1436	94.04	94.97	1388	85	96.26	96.86
PVC	V	7129	40	2852	4277	4225	98.78	96.53	4084	167	99.37	98.08
PACED	/	7024	40	2810	4214	4210	99.91	99.88	4198	15	99.98	99.95
AP	a	150	50	75	75	59	78.67	92.19	51	14	83.61	100.00
VF	!	472	50	236	236	229	97.03	100.00	221	13	99.10	100.00
VFN	F	802	50	401	401	347	86.53	89.20	310	66	92.54	94.51
BAP	x	193	50	97	96	93	96.88	100.00	89	6	98.89	100.00
NE	j	229	50	115	114	104	91.23	64.60	96	12	94.12	73.85
FPN	f	982	50	491	491	485	98.78	96.61	474	15	99.58	97.93
VE	E	106	50	53	53	48	90.57	100.00	48	2	94.12	100.00
NP	J	83	50	42	41	39	95.12	88.64	33	7	97.06	91.67
AE	e	16	50	8	8	2	25.00	66.67	1	3	20.00	100.00
UN	Q	33	50	17	16	0	–	–	0	4	–	–
Total	16	110108	22	24100	86008	85473	99.38	99.38	84384	1340	99.66	99.66

Information in this table (columns from left to right): heartbeat class and annotation; total number of processed beats; fraction and number of training beats; number of test beats; number of true positives (TP), sensitivity (Se), positive predictivity (+P) of Bayesian fusion approach; number of rejected beats (R), TP, Se, +P of rejection approach.

Table 2. Comparison of the proposed method and the reference works

Method	Feature vector	Classifier	Accuracy
Lagerholm et al. [12]	Hermite	SOM	98.49%
Prasad and Sahambi [18]	Wavelet + RR	ANN	96.77%
Osowski et al. [17]	HOS + Hermite	SVM	98.18%
Rodriguez et al. [20]	Waveform	DT	96.13%
Jiang et al. [10]	Wavelet + ICA	SVM	98.86%
Ye et al. [22]	Wavelet + ICA + RR	SVM	99.32%
Proposed method	Rational + RR	SVM	99.38%

We compared our method to earlier results published [10,12,17,18,20,22] which also adopt the 'class-oriented' scheme with 16 classes. The comparison is provided in Table 2. We can conclude that the proposed method improves the performance of the previous results, reaching higher overall accuracy level.

Finally, we conducted a robustness test. Following the literature in [14,22], we experimented the robustness of the feature extraction method by displacing the QRS locations with an artificial jitter (Gaussian-distributed random variable with zero mean and standard deviation of five samples). This test simulates the uncertainty of the automatic R-peak detectors. An overall accuracy of 99.06% (99.54% with 2.44% rejection) is reached, that confirms the desired robustness of our method.

5   Conclusion

In this paper, we proposed an automatic heartbeat classification approach for arrhythmia types. The novelty of our concept is to apply an adaptive rational transform to the heartbeats for feature extraction. We represented the beats with a rational orthogonal projection using patient-specific poles, then we extracted morphological features as the coefficients of this projection. The performance evaluation with SVM classifier and two-lead fusion on the benchmark MIT-BIH arrhythmia database yields an accuracy of 99.38% (99.66% with 1.56% rejection).

Our future research interests include to improve the classification accuracy (e.g. incorporate adaptive segmentation techniques), to develop patient-specific methods, and to build a real application out of our concept.

References

1. Chang, C.C., Lin, C.J.: LIBSVM: a library for support vector machines. ACM Trans. Intell. Syst. Techol. **2**, 27: 1–27: 27 (2011). https://www.csie.ntu.edu.tw/~cjl in/libsvm/
2. de Chazal, P., O'Dwyer, M., Reilly, R.B.: Automatic classification of heartbeats using ECG morphology and heartbeat interval features. IEEE Trans. Biomed. Eng. **51**(7), 1196–1206 (2004). https://doi.org/10.1109/tbme.2004.827359

3. Cortes, C., Vapnik, V.N.: Support-vector networks. J. Mach. Learn. **20**(3), 1–25 (1995). https://doi.org/10.1023/A:1022627411411
4. Fridli, S., Lócsi, L., Schipp, F.: Rational function systems in ECG processing. In: Moreno-Díaz, R., Pichler, F., Quesada-Arencibia, A. (eds.) EUROCAST 2011. LNCS, vol. 6927, pp. 88–95. Springer, Heidelberg (2012). https://doi.org/10.1007/978-3-642-27549-4_12
5. Fridli, S., Schipp, F.: Biorthogonal systems to rational functions. Ann. Univ. Sci. Bp. Sect. Comp. **35**, 95–105 (2011)
6. Fridli, S., Kovács, P., Lócsi, L., Schipp, F.: Rational modeling of multi-lead QRS complexes in ECG signals. Ann. Univ. Sci. Bp. Sect. Comp. **37**, 145–155 (2012)
7. Gilián, Z., Kovács, P., Samiee, K.: Rhythm-based accuracy improvement of heart beat detection algorithms. In: Computing in Cardiology Conference, pp. 269–272 (2014)
8. Goldberger, A.L., et al.: PhysioBank, PhysioToolkit, and PhysioNet: components of a new research resource for complex physiologic signals. Circulation **101**(23), 215–220 (2000). http://circ.ahajournals.org/cgi/content/full/101/23/e215
9. Heuberger, P.S.C., Van den Hof, P.M.J., Wahlberg, B. (eds.): Modelling and Identification with Rational Orthogonal Basis Functions. Springer-Verlag, London Limited, London (2005)
10. Jiang, X., Zhang, L.Q., Zhao, Q.B., Albayrak, S.: ECG arrhythmias recognition system based on independent component analysis feature extraction. In: Proceedings IEEE Region 10 Conference, pp. 1–4. (2006). https://doi.org/10.1109/tencon.2006.343781
11. Kovács, P., Lócsi, L.: RAIT: the rational approximation and interpolation toolbox for Matlab, with experiments on ECG signals. Int. J. Adv. Telecommun. Electech. Sign. Syst. **1**(2–3), 67–75 (2012). https://doi.org/10.11601/ijates.v1i2-3.18
12. Lagerholm, M., Peterson, C., Braccini, G., Edenbrandt, L., Sornmo, L.: Clustering ECG complexes using Hermite functions and self-organizing maps. IEEE Trans. Biomed. Eng. **47**(7), 838–848 (2000). https://doi.org/10.1109/10.846677
13. Lócsi, L.: Approximating poles of complex rational functions. Acta Univ. Sapientiae-Math. **1**(2), 169–182 (2009)
14. Luz, E.J.S., Schwartz, W.R., Cámara-Cháveza, G., Menotti, D.: ECG-based heart-beat classification for arrhythmia detection: a survey. Comput. Methods Programs Biomed. **127**, 144–164 (2016). https://doi.org/10.1016/j.cmpb.2015.12.008
15. Moody, G.B., Mark, R.G.: The impact of the MIT-BIH arrhythmia database. IEEE Eng. Med. Biol. Mag. **20**(3), 45–50 (2001). https://doi.org/10.1109/51.932724
16. Nelder, J.A., Mead, R.: A simplex method for function minimization. Comput. J. **7**(4), 308–313 (1965). https://doi.org/10.1093/comjnl/7.4.308
17. Osowski, S., Hoa, L.T., Markiewic, T.: Support vector machine-based expert system for reliable heartbeat recognition. IEEE Trans. Biomed. Eng. **51**(4), 582–589 (2004). https://doi.org/10.1109/TBME.2004.824138
18. Prasad, G.K., Sahambi, J.S.: Classification of ECG arrhythmias using multi-resolution analysis and neural networks. In: Proceeding of Conference Convergent Technology Asia-Pacific Region, pp. 227–231 (2003). https://doi.org/10.1109/TENCON.2003.1273320
19. Robert, K., Colleen, E.C.: Basis and Treatment of Cardiac Arrhythmias, 1st edn. Springer, New York (2006). https://doi.org/10.1007/3-540-29715-4
20. Rodriguez, J., Goni, A., Illarramendi, A.: Real-time classification of ECGs on a PDA. IEEE Trans. Inf. Techol. Biomed. **9**(1), 23–34 (2005). https://doi.org/10.1109/TITB.2004.838369

21. Sansone, M., Fusco, R., Pepino, A., Sansone, C.: Electrocardiogram pattern recognition and analysis based on artificial neural networks and support vector machines: a review. J. Healthc. Eng. **4**(4), 465–504 (2013). https://doi.org/10.1260/2040-2295.4.4.465
22. Ye, C., Kumar, B.V., Coimbra, M.T.: Heartbeat classification using morphological and dynamic features of ECG signals. IEEE Trans. Biomed. Eng. **59**(10), 2930–2941 (2012). https://doi.org/10.1109/TBME.2012.2213253
23. Zhang, D.: Wavelet approach for ECG baseline wander correction and noise reduction. In: Proceedings IEEE International Conference Engineering Medicine Biology Society, pp. 1212–1215 (2005). https://doi.org/10.1109/IEMBS.2005.1616642

Rational Variable Projection Methods in ECG Signal Processing

Péter Kovács[(⊠)] [iD]

Department of Numerical Analysis, Eötvös L. University,
Pázmány Péter stny. 1/C, Budapest 1117, Hungary
kovika@inf.elte.hu

Abstract. In this paper we develop an adaptive electrocardiogram (ECG) model based on rational functions. We approximate the original signal by the partial sums of the corresponding Malmquist–Takenaka–Fourier series. Our aim in the construction of the model was twofold. Namely, besides good approximation an equally important point was to have direct connection with medical features. To this order, we consider the rational optimization problem as a special variable projection method. Based on the natural segmentation of a heartbeat into P, QRS, T waves, we use three complex parameters, i.e. the poles of the rational functions. For the optimization of the parameters, we apply constrained optimization. As a result every pole corresponds to one of these waves. We developed and tested our method by using the MIT-BIH Arrhythmia Database.

Keywords: Malmquist–Takenaka system · Constrained optimization
ECG modeling

1 Introduction

Certain problems in signal processing like compression, feature extraction, classification, etc., can be interpreted as approximation and optimization problems. To this end, an appropriate function space $\mathcal{H}$ with norm $\|\cdot\|$ is chosen first to the given problem and the signals are considered as members of this space. In order to simplify the representation, linear models are preferred. Then, the signals $f \in \mathcal{H}$ are usually modeled by linear combinations of the elements of a specific function system $\{\Phi_k \mid 0 \leq k < n\} \subset \mathcal{H}$ in such a way that the approximation error is minimized:

$$\text{dist}(f, \mathcal{S}) := \min_{g \in \mathcal{S}} \|f - g\| = \|f - \tilde{f}\|, \tag{1}$$

where $\mathcal{S} = \text{span}\{\Phi_k \mid 0 \leq k < n\}$. For instance, in [3,4], the ECG signals were considered as elements of $\mathcal{H} = H^2(\mathbb{D})$, the Hardy space on the open unit

P. Kovács—This research was supported by the Hungarian Scientific Research Funds (OTKA) No K115804.

R. Moreno-Díaz et al. (Eds.): EUROCAST 2017, Part II, LNCS 10672, pp. 196–203, 2018.
https://doi.org/10.1007/978-3-319-74727-9_23

disk $\mathbb{D}$, while the function system was the so-called Malmquist–Takenaka (MT) system:

$$\Phi_k(\mathbf{a}; z) = \frac{\sqrt{1 - |a_k|^2}}{1 - \overline{a}_k z} \prod_{j=0}^{k-1} \frac{z - a_j}{1 - \overline{a}_j z} \qquad (z \in \overline{\mathbb{D}}, 0 \le k < n), \qquad (2)$$

where $\mathbf{a} := (a_0, a_1, \ldots, a_{n-1}) \in \mathbb{D}^n$ denotes the inverse pole vector of these rational functions. One can show that for any $f \in H^2(\mathbb{D})$ the radial limit function $f(e^{it}) := \lim_{r \to 1-0} f(re^{it})$ also exists, which belongs to $L^2(\mathbb{T})$ (see e.g., [12]). In this case the norm in Eq. (1) can be written as $\|f\| = \|f\|_{L^2(\mathbb{T})}$, where $\mathbb{T}$ denotes the unit circle (or torus) and $\|.\|_{L^p(\mathbb{T})}$ is the usual Lebesgue norm. Therefore one can define a scalar product in $H^2(\mathbb{D})$ by

$$\langle f, g \rangle := \frac{1}{2\pi} \int_{-\pi}^{\pi} f(e^{it}) \overline{g}(e^{it}) \, dt \qquad (f, g \in H^2(\mathbb{D})). \qquad (3)$$

The MT system is an orthonormal function system with respect to this scalar product, hence $\langle \Phi_i(\mathbf{a}; \cdot), \Phi_j(\mathbf{a}; \cdot) \rangle = \delta_{ij}$ $(0 \le i, j < n)$ for any $\mathbf{a} \in \mathbb{D}^n$. Thanks to this property and the fact that $(H^2(\mathbb{D}), \langle \cdot, \cdot \rangle)$ is a Hilbert space, the orthogonal projection to $S(\mathbf{a}) := \text{span}\{\Phi_k(\mathbf{a}; \cdot) \mid 0 \le k < n\}$ gives the solution to the minimization problem in Eq. (1). The orthogonal projection can be easily calculated as

$$S(\mathbf{a}) \ni \widetilde{f}(\mathbf{a}; z) := \sum_{k=0}^{n-1} \langle f(\cdot), \Phi_k(\mathbf{a}; \cdot) \rangle \Phi_k(\mathbf{a}; z), \qquad (4)$$

where $\mathbf{a} \in \mathbb{D}^n$ is a given inverse pole vector. Note that the subspace $S(\mathbf{a})$ depends on the parameter vector $\mathbf{a}$ (cf. Eq. (1)), thus the rational approximation $\widetilde{f}(\mathbf{a}; z)$ is an adaptive signal model.

Proper selection of the features is of key importance in data reduction problems such as signal compression, smoothing and classification. Hence the parameter vector $\mathbf{a}$ must be adequately chosen. To this end the following optimization problem is considered:

$$\min_{\mathbf{a} \in \mathbb{D}^n} r_2(\mathbf{a}) := \min_{\mathbf{a} \in \mathbb{D}^n} \|f(\cdot) - \widetilde{f}(\mathbf{a}; \cdot)\|, \qquad (5)$$

where $r_2(\mathbf{a})$ is the so-called variable projection functional [6].

We will apply this adaptive model for ECG signals $f \in H^2(\mathbb{D})$. The optimal inverse pole vector $\mathbf{a} \in \mathbb{D}^n$, which minimizes $r_2(\mathbf{a})$, is determined by global and local optimization techniques. Additionally, in order to restrict the search space of the optimization we will develop constraints related to medical features of the electrocardiogram. The performance of these algorithms are demonstrated on real ECG signals of the PhysioNet MIT-BIH database [5]. Note that our case study includes only ECGs, however, the proposed algorithm can be applied to other signals as well (see e.g., [11]).

2   Optimization

For a given vector of inverse poles $\mathbf{a} \in \mathbb{D}^n$ the MT system is completely defined by Eq. (2). These inverse poles can also be repeated $m_0, m_1, \ldots, m_{n-1}$ number of times. In what follows, we denote the vector of different inverse poles by $\mathbf{a} \in \mathbb{D}^n$ and the corresponding multiplicities by $\mathbf{m} \in \mathbb{N}_+^n$. The vector $\mathbf{a} \in \mathbb{D}^n$ along with $\mathbf{m} \in \mathbb{N}_+^n$ define the following inverse pole vector:

$$\mathbf{b} := \big(\underbrace{a_0, \ldots, a_0}_{m_0}, \ldots, \underbrace{a_{n-1}, \ldots, a_{n-1}}_{m_{n-1}}\big) \in \mathbb{D}^N, \tag{6}$$

where $N = m_0 + m_1 + \ldots + m_{n-1}$. Hereinafter, we will consider the MT system $\{\Phi_k(\mathbf{a}, \mathbf{m}; \cdot) \,|\, 0 \leq k < n\} := \{\Phi_k(\mathbf{b}; \cdot) \,|\, 0 \leq k < n\}$.

Following [4], we choose three different inverse poles $\mathbf{a} = (a_0, a_1, a_2)$, which are related to the QRS complex and the T, P waves of the heartbeats. Using the fact that the QRS complex is the most significant wave among those three, we choose m_0 to be greater than m_1 and m_2 (see e.g., Table 1). According to these heuristics, we need to find the optimal positions of the inverse poles a_0, a_1 and a_2. To this end, we minimize the variable projection functional in Eq. (5). We apply the trust-region [9] and the particle swarm optimization (PSO) [7] techniques to this optimization problem. In the former case, the cost function is locally approximated by a quadratic model, which is used to predict the next step toward the minimum. In our experiments we utilized the lsqnonlin routine, which is the MATLAB implementation of trust-region-reflective algorithm. PSO is a stochastic search method that evaluates the cost function at several points in an iteration. Thanks to these trials we can explore the main characteristics of the cost function. Thus, the PSO has less chance for being trapped in local optima. Here, we used the hyperbolic variation of the PSO algorithm (HPSO) that keeps the candidate poles inside $\mathbb{D}$ via the Poincaré disk model [8].

Note that minimizing r_2 in Eq. (5) is already a constrained optimization problem, since all the inverse poles should lie inside $\mathbb{D}$ (the torus $\mathbb{T}$ is not included). In order to get a compact search space we derive further restrictions for these parameters based on medical features of the ECG.

3   Constraints for Modeling ECGs

In a cardiac cycle, the heart produces a signal that contains four entities called P, T, U waves and the QRS complex. Generally, the amplitude of the U wave is not significant, hence we will consider the P, T waves and the QRS complex only. The shapes of these waves are quite similar to those of the individual members of the MT system, which gives one of the main motivations of using rational functions in ECG signal processing (see e.g., [4]). In Fig. 1, we restricted the first element of the MT system to the unit circle and displayed the real part of $\Phi_0(\mathbf{a}, \mathbf{m}; e^{it})$ ($t \in [-\pi, \pi)$), where $\mathbf{a} = a_0 = 0.5 + 0i$ and $\mathbf{m} = m_0 = 1$. Notice that this function gives the basic shape of the approximated P, QRS and

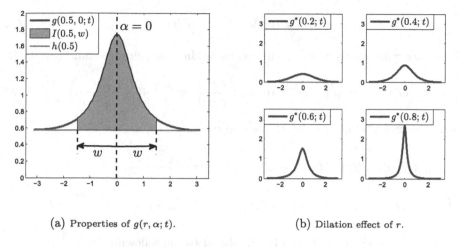

(a) Properties of $g(r, \alpha; t)$.   (b) Dilation effect of r.

Fig. 1. Real part of the first element of the MT system at $\mathbb{T}$.

$\mathbb{T}$ curves. Therefore, we will use this property to derive the constraints for the inverse poles.

Let us consider the simple case, when we approximate only the QRS complex of a heartbeat. To this end, we need to optimize only one inverse pole $a_0 \in \mathbb{D}$, which is repeated $m_0 \in \mathbb{N}_+$ number of times. If $a_0 = re^{i\alpha}$ ($r \in [0, 1)$, $\alpha \in [-\pi, \pi)$), the real part of $\Phi_0(a_0, m_0; e^{it})$ at $t \in [-\pi, \pi)$ (cf. Eq. (2)) is as follows:

$$g(r, \alpha; t) := \operatorname{Re} \Phi_0(a_0, m_0; e^{it}) = \sqrt{1 - r^2} \cdot \frac{1 - r\cos(t - \alpha)}{1 - 2r\cos(t - \alpha) + r^2} \quad (t \in [-\pi, \pi)).$$

The angle α is interpreted as a translation parameter that provides a good time localization property. Namely, the optimization of the angle of the inverse pole is equivalent to finding the position of the waves. In order to predict the onset and the offset of the QRS complex we apply the well-known Pan–Tompkins algorithm [10]. These points can be naturally represented on the interval $[-\pi, \pi]$ and are denoted by QRS_{on} and QRS_{off}. Then, the constraints for the angles can be defined as follows:

$$-\pi \leq \alpha_P \leq QRS_{on}, \quad QRS_{on} \leq \alpha_{QRS} \leq QRS_{off}, \quad QRS_{off} \leq \alpha_T \leq \pi.$$

These constraints are demonstrated in Fig. 2(b). Before turning to the restriction of the radius r, we note that the translation in time does not change the shape of the curve $g(r, \alpha; t)$. Thus, for the sake of simplicity we will consider the case $\alpha = 0$ in the rest of this section.

The radius r of the inverse pole is a kind of dilation parameter, which is shown in Fig. 1(b). It is easy to see that $g(r, 0; t) \geq h(r) = \sqrt{1 - r^2}/(1 + r)$. Therefore, we will examine the function $g^*(r; t) := g(r, 0; t) - h(r)$ instead. We set the constraints $r_{lb} \leq r \leq r_{ub}$ for the dilation r such that p percentage of the overall area, i.e. the integral of $g^*(r; t)$ should be concentrated into the interval $[-w, w]$:

$$p \cdot \int_{-\pi}^{\pi} g^*(r;t)dt \le \int_{-w}^{w} g^*(r;t)dt. \tag{7}$$

These integrals can be easily calculated by taking the primitive function of g^*:

$$G^*(r;t) := \sqrt{1-r^2} \cdot \left(\arctan\left(\frac{r-1}{r+1} \cot \frac{t}{2} \right) + \frac{t}{2} - \frac{t}{1+r} \right). \tag{8}$$

Using the fact that g^* is an even function we get:

$$\int_{-\pi}^{\pi} g^*(r;t)dt = 2 \cdot (G^*(r;\pi) - G^*(r;0)) = 2\pi r \cdot \frac{\sqrt{1-r^2}}{1+r},$$

$$\int_{-w}^{w} g^*(r;t)dt = 2 \cdot (G^*(r;w) - G^*(r;0)) = 2G^*(r;w) + \pi\sqrt{1-r^2} =: I(r,w).$$

Hence, the inequality in (7) can be simplified to the following form:

$$0 \le \arctan\left(\frac{r-1}{r+1} \cot \frac{w}{2} \right) + \frac{w}{2} - \frac{w}{1+r} + \frac{\pi}{2} - p \cdot \frac{\pi r}{1+r} =: u(p,w;r). \tag{9}$$

The trivial solution of this inequality is $r = 0$ for which $u(p,w;0) = 0$. We will show that the function $u(p,w;r)$ $(p \in (0,1)$ $w \in (0,\pi))$ has at most one root for $r \in (0,1)$. To this end, let us consider the derivative of the right hand side with respect to r:

$$\frac{\partial u}{\partial r} = \frac{1}{(1+r)^2}\left(w - p\pi + \frac{2\cot\frac{w}{2}}{1 + \left(\frac{r-1}{r+1}\cot\frac{w}{2}\right)^2} \right). \tag{10}$$

If $p\pi \le w$, the derivative $\partial u/\partial r$ is positive and so (9) holds for every $r \in [0,1]$. In case $p\pi > w$, the derivative $\partial u/\partial r$ has at most one root in $(0,1)$. Indeed, we can express r in the form

$$r = \frac{1-c}{1+c}, \qquad \text{where} \qquad c = \tan\frac{w}{2} \cdot \sqrt{\frac{2\cot\frac{w}{2}}{p\pi - w} - 1}. \tag{11}$$

If the square root in the definition of c exists and $c \in (0,1)$, then $\frac{1-c}{1+c}$ is the unique root of $\partial u/\partial r$ in the interval $(0,1)$. Hence, for fixed $p \in (0,1)$ and $w \in (0,\pi)$, the function $u(p,w;r)$ $(r \in (0,1))$ has at most one extremum. On one hand, if the value at this point is positive, then $u(p,w;0) = 0$ and $u(p,w;1) = (1-p)\pi/2 > 0$ imply that the inequality in (9) holds for every $r \in [0,1]$. The same is true, when the extremum does not exists. On the other hand, a negative minimum along with the continuity of u and $u(p,w;1) > 0$ provide a unique root r_w, and (9) holds for every $r \in [r_w, 1]$. In order to find r_w we applied the MATLAB fzero routine, which was initialized by r0=1.

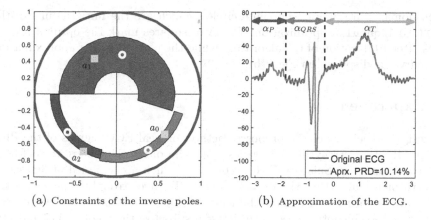

(a) Constraints of the inverse poles.   (b) Approximation of the ECG.

Fig. 2. Constrained optimization of an ECG signal. (Color figure online)

Now, we can derive the heuristics for the radius of the inverse poles corresponding to P, T waves and the QRS complex. Namely, we demand that 70% of the overall integral of g^* should be concentrated into the interval $[-w, w]$. To this end, we set $p = 0.7$, while the length of the interval should be equal to the average duration of the P, T waves and the QRS complex. According to Table 3.1 in [2], we recall the average length of the QRS complex, which is 100 ms $\pm$ 20 ms. Then, we set $w_{lb} = 120/2$ ms, $w_{ub} = 80/2$ ms and compute the nontrivial solutions r_{lb}, r_{ub} for Eq. (9). These values defines the lower and upper bounds for the radius of the inverse pole corresponding to the QRS complex: $r_{QRS} \in [r_{lb}, r_{ub}] \subseteq [0, 1]$. The average duration of the P wave is 110 ms $\pm$ 20 ms, thus we can repeat our calculations by setting $w_{lb} = 130/2$ ms, $w_{ub} = 90/2$ ms. The onset of the QRS complex and the offset of the T wave defines the QT interval. The duration of the QT segment varies with the heart rate, which can be predicted by the length of the RR intervals. The corrected QT interval is called QTc $= QT/\sqrt{RR}$. The average length of the QTc interval is 400 ms $\pm$ 40 ms. In order to estimate the average maximum/minimum width of the T wave we apply the following formulas:

$$\max T \text{ width} = \sqrt{RR} \cdot \max QTc - \min ST - \min QRS,$$

$$\min T \text{ width} = \sqrt{RR} \cdot \min QTc - \max ST - \max QRS.$$

where ST denotes the average length of the ST segment: 70 ms $\pm$ 10 ms. The constraints for the inverse pole of the T wave are $w_{lb} = \max T$ width$/2$ ms, $w_{ub} = \min T$ width$/2$ ms, which depend on the heart rate. Note that for the MT system, the variable projection functional $r_2(\mathbf{a})$ is a continuous function. In addition, we can restrict the original search space $\mathbb{D}^n$ to a compact domain by using our constraints. Hence, the solution to the new constrained optimization problem exists. One can see an example in Fig. 2(a), where we displayed the constraints corresponding to the ECG in Fig. 2(b). The middle points of the red, blue and green domains are marked by white circles, which are the initial points of the MATLAB lsqnonlin optimization. The cyan squares mark the optimal inverse poles after 10 iteration of the algorithm, while the corresponding approximation (red curve) is displayed in Fig. 2(b).

4   Experiments

We test the efficiency of the proposed method on real ECG signals of the PhysioNet MIT-BIH Arrhythmia Database [5]. The dataset includes 48 half-hour long two-channel ambulatory ECG recordings, which are sampled at 360 Hz. It is worth mentioning that the duration of the T wave strongly depends on the heart rate. In order to test how the lengths of the heart beats affect the constraints and so the optimization, we choose a subset of the records: 117 and 119. The latter one is ideal for testing this phenomenon, since it contains extremely varying periods (see e.g., [1]).

In the experiment, we segmented the first channel of the whole records into more than 3500 heart beats. Then, we applied different approaches to solve the optimization problem in Eq. (5). On one hand, we executed the HPSO for optimizing the inverse poles in the whole unit disk. Recall that the HPSO does not need any constraints to keep the inverse poles inside $\mathbb{D}$ [8]. Here, the size of the swarm and the number of iteration were equal to 30. On the other hand, we restricted the search space and applied the constraints in Sect. 3 along with 10 iteration of the lsqnonlin MATLAB routine. We evaluated the results by the percent root mean square difference (PRD), which is a conventional measure of the numerical distortion in ECG signal processing. The PRD of the approximation $\tilde{f}(\mathbf{a}; \cdot)$ in Eq. (4) can be defined as

$$\mathrm{PRD} := \frac{\left\| f(\cdot) - \tilde{f}(\mathbf{a}; \cdot) \right\|}{\left\| f(\cdot) - \overline{f} \right\|} \times 100,$$

where $\overline{f}$ denotes the mean of f. Note that we used Intel Core i5-4570 3.2GHz CPUs for testing. Table 1 shows the average execution time and the PRDs for various multiplicities $\mathbf{m}$ of each record. The result of the HPSO can be used as a reference to the proposed constrained optimization algorithm. Although the HPSO produced better PRDs, the difference is less than 2.6% and 1.6% for the records 117 and 119, respectively. In addition, the price of robustness is paid in the speed of the HPSO, which is about 10 times slower than the lsqnonlin method.

We implemented the proposed algorithm in MATLAB. The procedures that construct the appropriate constraints for an ECG, the optimization and the experiments are available at the website: http://numanal.inf.elte.hu/~kovi/docs/pubs/ConstrainedVarPro.rar.

Table 1. PRDs for various inverse pole configurations and optimization methods.

PRDs (%) of the constrained lsqnonlin								Average (sec) execution time
$m =$ (3, 2, 2)	(4, 2, 2)	(4, 3, 3)	(5, 4, 4)	(6, 4, 4)	(7, 5, 5)	(8, 4, 4)	(8, 6, 6)	
117 16.93	15.81	11.30	8.80	8.72	7.69	8.58	7.30	355
119 12.24	10.15	8.49	5.94	5.42	4.60	4.98	4.01	445
PRDs (%) of the unconstrained HPSO								Average (sec) execution time
$m =$ (3, 2, 2)	(4, 2, 2)	(4, 3, 3)	(5, 4, 4)	(6, 4, 4)	(7, 5, 5)	(8, 4, 4)	(8, 6, 6)	
117 14.76	13.24	9.59	7.95	7.66	7.11	7.28	6.72	3907
119 12.02	9.10	6.96	4.57	4.18	3.44	3.70	3.02	4284

References

1. Chou, H.H., Chen, Y.J., Shiau, Y.C., Kuo, T.S.: An effective and efficient compression algorithm for ECG signals with irregular periods. IEEE Trans. Biomed. Eng. **53**(6), 1198–1205 (2006)
2. Clifford, G.D., Azuaje, F., McSharry, P.: Advanced Methods And Tools for ECG Data Analysis. Artech House, Massachusetts (2006)
3. Fridli, S., Kovács, P., Lócsi, L., Schipp, F.: Rational modeling of multi-lead QRS complexes in ECG signals. Ann. Univ. Sci. Budapest. Sect. Comp. **37**, 145–155 (2012)
4. Fridli, S., Lócsi, L., Schipp, F.: Rational function systems in ECG processing. In: Moreno-Díaz, R., Pichler, F., Quesada-Arencibia, A. (eds.) EUROCAST 2011. LNCS, vol. 6927, pp. 88–95. Springer, Heidelberg (2012). https://doi.org/10.1007/978-3-642-27549-4_12
5. Goldberger, A.L., Amaral, L.A.N., Glass, L., Hausdorff, J.M., Ivanov, P.C., Mark, R.G., Mietus, J.E., Moody, G.B., Peng, C.K., Stanley, H.E.: PhysioBank, PhysioToolkit, and PhysioNet: components of a new research resource for complex physiologic signals. Circulation **101**(23), 215–220 (2000)
6. Golub, G.H., Pereyra, V.: The differentiation of pseudo-inverses and nonlinear least squares problems whose variables separate. SIAM J. Num. Anal. (SINUM) **10**(2), 413–432 (1973)
7. Kennedy, J., Eberhart, R.C.: Particle swarm optimization. In: Proceedings of IEEE International Conference on Neural Networks, vol. 4, pp. 1942–1948 (1995)
8. Kovács, P., Kiranyaz, S., Gabbouj, M.: Hyperbolic particle swarm optimization with application in rational identification. In: Proceedings of the 21st European Signal Processing Conference (EUSIPCO), pp. 1–5 (2013)
9. Nocedal, J., Wright, S.J.: Numerical Optimization. Springer Science+Business Media, LLC, New York (2006). https://doi.org/10.1007/978-0-387-40065-5
10. Pan, J., Tompkins, W.J.: A real-time QRS detection algorithm. IEEE Trans. Biomed. Eng. **32**(3), 230–236 (1985)
11. Samiee, K., Kovács, P., Gabbouj, M.: Epileptic seizure detection in long-term EEG records using sparse rational decomposition and local Gabor binary patterns feature extraction. Knowl.-Based Syst. **118**, 228–240 (2017)
12. Zygmund, A.: Trigonometric Series I.–II. Cambridge University Press, Cambridge (1959)

Stochastic Computing Using Droplet-Based Microfluidics

Werner Haselmayr[1](✉), Andreas Grimmer[2], and Robert Wille[2]

[1] Institute for Communications Engineering and RF-Systems,
Johannes Kepler University Linz, Linz, Austria
werner.haselmayr@jku.at
[2] Institute for Integrated Circuits, Johannes Kepler University Linz, Linz, Austria
{andreas.grimmer,robert.wille}@jku.at

Abstract. In this work, we consider the realization of stochastic computing in the microfluidic domain. To this end, we exploit the fact that both, the bit streams and the operations required for stochastic computing can be realized in microfluidic systems through droplet streams and microfluidic gates. Simulating the trajectory of the individual droplets through the microfluidic gates confirmed the validity of our approach.

Keywords: Droplet-based microfluidics · Microfluidic computing
Microfluidic gates · Stochastic computing

1 Introduction

Droplet-based microfluidic systems refer to systems, where tiny volumes of fluids, so-called *droplets*, flow in channels of micrometer scale [1]. Currently, such systems are frequently used as platform for the realization of *Labs-on-Chip* (LoC) devices, where droplets contain biological/chemical samples and undergo several processes to execute certain laboratory experiments, e.g., DNA sequencing, cell analysis or drug discovery [1,2]. But beyond that, droplet-based microfluidics recently also found interest in domains such as information transmission and simple computing.

For example, a simple droplet-based communication system was proposed for the first time in [3]. This idea was later extended in [4] by introducing different methodologies for information encoding using droplets, e.g., the presence/absence of droplets or the distance between two consecutive droplets. With respect to using microfluidic systems for computing, initial work has been conducted in [5,6]. Here, the presence/absence of a droplet is used to represented Boolean values, while their flow through a dedicated microfluidic system realizes the desired Boolean functions.

In this work, we further extend these concepts in order to realize *stochastic computing* (SC, [7]) in the microfluidic domain. In SC, real numbers between 0 and 1 are represented by a stochastic bit stream which allows to realize usually

© Springer International Publishing AG 2018
R. Moreno-Díaz et al. (Eds.): EUROCAST 2017, Part II, LNCS 10672, pp. 204–211, 2018.
https://doi.org/10.1007/978-3-319-74727-9_24

complex arithmetic operations (such as a multiplication) through very simple logic gates (such as an AND gate) – yielding substantially more compact realizations of circuits. The fact that both concepts (the bit streams and the operations) can be realized in microfluidic systems (namely as droplet streams and microfluidic gates [5,6], respectively) motivates a more detailed consideration of SC using droplet-based microfluidic systems.

To this end, the remainder of this work provides the following contributions: First, we give a brief introduction to SC and review the principle of droplet-based microfludics as well as microfluidic gates in Sects. 2 and 3, respectively. Afterwards, we propose a concept for realizing stochastic arithmetic operations in droplet-based microfluidic systems by adopting the SC approach for the microfludic domain in Sect. 4 – including a validation of the proposed concepts through simulations based on the duality between microfluidic systems and time-varying electrical circuits. Finally, Sect. 5 concludes the paper and briefly discusses future work.

2   Stochastic Computing

In stochastic computing [7], a real number s in the unit range ($s \in [0, 1]$) is represented as a serial stochastic bit stream S. The desired number corresponds to the ratio of 1's included in the bit stream to the bit stream length, i.e., the probability for each bit in the stream to be 1 is given by $\Pr(S = 1) = s$. For example, the value $s = 7/10$ can be represented by a stochastic stream $S \hat{=} 1101011011$ with $\Pr(S = 1) = 7/10$. It is important to note that the positions of the 1's in the stream is not prescribed and, thus, many different streams for the same value exists.

The generation of a stochastic bit stream can be accomplished through a comparator as shown in Fig. 1. The comparator compares a w-bit random natural number R and the value $B = [s \times 2^w]$, with $[x]$ as the nearest integer function. The number R is drawn from an uniform distribution and the number B corresponds to the mapping of the desired real number s to an w-bit natural number. If $R < B$ the output of the comparator is 1 and, thus, the probability of a 1 appearing at the output of the comparator is given by $\Pr(R < B) = \Pr(S = 1) = [s \times 2^w]/2^w \approx s$.

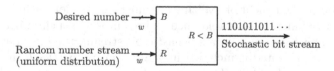

Fig. 1. Stochastic bit stream generation.

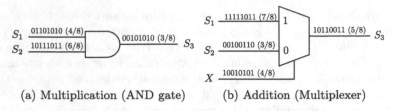

(a) Multiplication (AND gate)   (b) Addition (Multiplexer)

Fig. 2. Logic circuits for performing arithmetic operations in SC [7].

The main benefit of SC is that arithmetic operations can be performed with simple logic circuits. In the following, we briefly discuss the realization of multiplication and addition[1].

- **Multiplication:** A multiplication can be implemented by an AND gate as shown in Fig. 2(a). Let us consider two independent stochastic bit streams S_1 and S_2 at the input of an AND gate. According to the AND gate behavior the output stream S_3 is only 1, if both input streams are equal to 1. More formally this can be written as

$$s_3 = \Pr(S_3 = 1) = \Pr(S_1 = 1 \wedge S_2 = 1) = \Pr(S_1 = 1)\Pr(S_2 = 1) = s_1 s_2. \quad (1)$$

Thus, the AND gate can be used to compute the product of s_1 and s_2, which are represented by the input stochastic streams S_1 and S_2, respectively.

- **Addition:** A scaled addition can be implemented using a multiplexer as shown in Fig. 2(b). Let us consider two independent stochastic streams S_1 and S_2 at the input of the multiplexer and a stochastic stream X that selects the input to be forwarded to the output. If $X = 1$ or $X = 0$ the actual bit value of the output stream S_3 corresponds to the bit value of S_1 or S_2, respectively. More formally this can be written as

$$
\begin{aligned}
s_3 = \Pr(S_3 = 1) &= \Pr(X = 1)\Pr(S_1 = 1) + \Pr(X = 0)\Pr(S_2 = 1) \\
&= \Pr(X = 1)\Pr(S_1 = 1) + (1 - \Pr(X = 1))\Pr(S_2 = 1) \\
&= xs_1 + (1 - x)s_2. \quad (2)
\end{aligned}
$$

Thus, a multiplexer can be used to compute the scaled addition of s_1 and s_2, where s_1 and s_2 are weighted by x and $1 - x$.

In addition to the benefit of realizing arithmetic operations using simple logic gates, SC has an inherent fault tolerance and requires no synchronization among the streams. This is because, in a stochastic bit stream, the information on the number to be represented is included in the stream properties (number of 1's) and not in the individual bits. Thus, all bits have a similar weight and no higher order bits exists as in the conventional binary format. For example, a bit flip changes a stochastic stream from 000101010 to 100101010 which changes the corresponding number from 3/8 to 4/8 [7]. Thus, only a small error occurs due

---

[1] We refer to [8] for a description of the implementation for division and subtraction.

to the bit flip. In contrast, considering the number 3/8 in conventional binary format 0.011, a bit flip of a higher order bit causes a huge error, i.e., if 0.011 changes to 0.111 the numbers change from 3/8 to 7/8.

One major drawback of SC is that, for high accuracy, long stochastic bit streams are required – resulting in a high latency for the computation. For example, to increase the accuracy from 4 to 8 bits requires an increase of the stream length from 16 to 256 bits [7]. In a recent work, it is proposed to generate the stochastic streams deterministically rather than randomly (cf. Fig. 1), which leads to significantly lower latency while keeping the advantage of inherent fault tolerance [9].

Nevertheless, despite this drawback and motivated by the huge benefits discussed above, SC has been successfully used in various applications – including neural networks [8], control systems [10], image processing [11] and decoding of error-correcting codes [12,13].

3   Droplet-Based Microfluidics and Microfluidic Gates

In droplet-based microfludic systems [1], tiny volumes of fluids, so-called droplets, flow in closed microchannels, triggered by some external sources (e.g., pressure or syringe pump). Typically, the droplets are generated using a T-junction, where a fluid (dispersed phase) in form of droplets is dispersed into another immiscible fluid (continuous phase) acting as carrier fluid (cf. Fig. 3).

Originally, droplet-based microfluidic systems were used as a platform for the realization of LoCs, which execute certain laboratory experiments by including biological/chemical samples in the droplet [2]. But in the past few years, also several approaches for employing droplet-based microfluidics for information transmission or computing (microfludic gates) were proposed [3–6]. In this section, we briefly describe the principle of microfluidic gates [5,6] which will provide the basis for the proposed realization of SC in the microfluidic domain.

The working principle of microfluidic gates is based on two observations: First, droplets arriving at a junction flow along the channel with the lowest hydrodynamic resistance. Second, droplets increase the hydrodynamic resistance in a channel. For a rectangular channel with length L_c, width w_c and height h_c, the hydrodynamic resistance is given by [14]

$$R_c = \frac{\alpha \mu_c L_c}{w_c h_c^3},$$ (3)

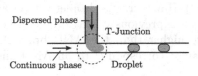

Fig. 3. Generation of droplets using a T-junction.

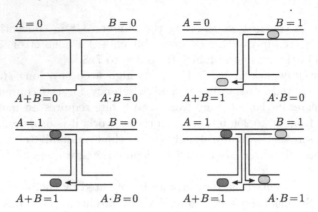

Fig. 4. Working principle of a microfluidic AND/OR gate [6].

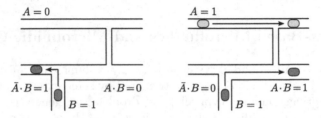

Fig. 5. Working principle of a microfluidic AND/NOT gate [6].

where μ_c denotes the dynamic viscosity of the carrier fluid and the dimensionless parameter α is given by $\alpha = 12[1 - 192h/(\pi^5 w_c)\tanh(\pi w_c/(2h_c))]^{-1}$.

The increase of the resistance due to a droplet in a channel is given by [15]

$$R_d = \frac{(\mu_d - \mu_c)\alpha L_d}{w_d h_d^3},\tag{4}$$

with L_d, w_d and h_d being the length, width and height of the droplet, respectively. Moreover, μ_d denotes the dynamic viscosity of the dispersed phase. Thus, the total hydrodynamic resistance of a channel which includes a droplet is given by $R = R_c + R_d$.

For the first time, microfluidic AND/OR and AND/NOT gates were proposed in [6] and are shown in Figs. 4 and 5, respectively. The channel dimensions of the AND/OR gate are chosen such that a droplet from input A or B arriving at the bottom junction flows into the OR branch $(A + B)$ due the lower hydrodynamic resistance (wider channel). However, if a droplet arrives at the junction and the OR branch is occupied by a previously sent droplet (increasing the resistance of the OR branch to be higher than the AND branch), it is directed to the AND branch. This behavior corresponds to an AND and OR gate behavior[2].

---

[2] To allow for a correct functionality for $A = 1$ and $B = 1$, the droplets must enter the gate with a slight time delay.

The channel dimensions of the NOT gate are set such that, if there is no droplet from input A, the droplet from input B flows into the branch $\bar{A} \cdot B$. By introducing a droplet into the input channel A, the flow towards the lower channel is reduced and the droplet from input B flows towards $A \cdot B$. Thus, by providing a droplet train for input B, the gate shown in Fig. 5 realizes a NOT gate.

4   Stochastic Computing Using Microfluidic Systems

Using microfluidic systems as reviewed in the previous section allows to realize SC. To this end, a serial stochastic bit stream as well as corresponding microfluidic gates need to be realized. The latter is already available as discussed before by means of Figs. 4 and 5. The stochastic bit stream can be realized by a droplet stream that can be generated by a T-junction as shown in Fig. 3. More precisely, the droplet transmission is divided into *time intervals* of duration T – generating a single or no droplet depending on whether a bit 1 or 0 occurs in the stochastic stream (cf. Fig. 6).

In the following, we validate the realization of SC in the microfludic domain through the example of a multiplication of two numbers. To this end, we evaluate the trajectory of individual droplets through a microfluidic system using the event-based simulator proposed in [15]. This simulator models microfluidic systems as time-varying electrical circuits.

We implemented an AND/OR gate as shown in Fig. 7 using the channel dimensions and fluid properties as specified in Table 1. We convert the numbers to be multiplied into two serial stochastic bit streams, which are used for the droplet generation. One bit stream represents the droplets which are injected into input A and the other represents the droplets which are injected into input B of the AND/OR gate. Depending on the value of the bits in the streams, a droplet is injected or not (presence/absence of a droplet for bit 1/0). More precisely, in case that both bit streams have length N, every time nT $(1 \leq n \leq N)$, a bit of both streams is injected in form of droplets. It is important to note that, in order to prevent droplets merging, the injection time of input A and B are slightly time-delayed.

Using these streams as input to the AND/OR gate, we can observe whether a droplet enters the OR or the AND branch. Note that the droplet stream exiting the AND branch represents the results of the multiplication.

As discussed in Sect. 2, a longer stochastic bit stream results in a higher accuracy. Thus, given a fixed amount of time, it is desirable to have a small time

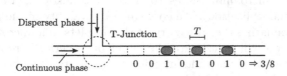

Fig. 6. Droplet stream representing the number 3/8.

Table 1. Simulation parameters.

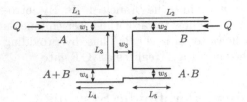

Fig. 7. Layout of AND/OR gate.

Flow rate Q	0.25 µl/s
Dyn. viscosity cont. phase μ_c	1.002 mPa · s
Dyn. viscosity disp. phase μ_d	5.511 mPa · s
Channel height h	70 µm
Channel lengths L_1, L_2	125 µm
Channel lengths L_3, L_4, L_5	62.5 µm
Channel widths w_1, w_2	25 µm
Channel width w_3	50 µm
Channel width w_4	30 µm
Channel width w_5	20 µm

interval T. However, in order to ensure a correct behavior of the AND/OR gate, T cannot be arbitrarily small. A correct behavior can only be ensured, when droplets injected at time nT do not influence droplets injected at time $(n + 1)T$ (e.g., by changing the resistances). In our simulations, we decreased the time interval T until droplets flow into wrong branches at the bottom junction and, thus, produced a wrong result. Considering the system specified in Fig. 7 and Table 1, the smallest droplet injection time interval which still guarantees a correct behavior has been determined as being $T = 2.9$ ms.

As an example, a video that shows the simulation result for the multiplication of 0.7×0.9 is available at www.jku.at/iic/eda/sc. This video confirms the working principle of the AND/OR gate (cf. Sect. 3). More precisely, it shows the injection of two bit streams of length $N = 10$. The droplets injected into input A have a probability $\Pr(A = 1) = 7/10$ and droplets injected into input B have a probability $\Pr(B = 1) = 9/10$. The result of the multiplication $\Pr(A = 1) \Pr(B = 1)$ is obtained by counting the number of droplets flowing into the AND branch (i.e. to the right bottom channel) and dividing this number by the stream length N. It is important to note that the result of the multiplication is 6/10, since we have only used stochastic bit streams of length $N = 10$. In order to improve the accuracy, the bit stream length must be increased as discussed in Sect. 2. We successfully conducted similar simulations using other inputs as well as other operations – confirming the validity of the proposed approach.

5   Conclusions

In this work, we proposed the realization of stochastic computing using droplet-based microfluidic systems. To this end, we represented stochastic bit streams as droplet streams (presence/absence of droplets for bit 1/0) and utilized existing realization of microfluidic gates to realize stochastic operations (e.g., an AND gate for a multiplication). We confirmed the validity of the proposed approach through evaluating the trajectory of the droplets in a microfluidic system. As future work, we will investigate the effect of imperfect droplet generation (e.g., different droplet volumes and injection time variation) and the implementation of more complex operations. Furthermore, we will consider ways to conduct corresponding simulations on discrete models such as proposed in [16].

References

1. Mark, D., Haeberle, S., Roth, G., von Stetten, F., Zengerle, R.: Microfluidic lab-on-a-chip platforms: requirements, characteristics and applications. Chem. Soc. Rev. **39**(3), 1153–1182 (2010)
2. Teh, S.-Y., Lin, R., Hung, L.-H., Lee, A.P.: Droplet microfluidics. Lab Chip **8**, 198–220 (2008)
3. Fuerstman, M.J., Garstecki, P., Whiteside, G.M.: Coding/decoding and reversibility of droplet trains in microfluidic networks. Science **315**(5813), 828–832 (2007)
4. Leo, E.D., Galluccio, L., Lombardo, A., Morabito, G.: Networked labs-on-a-chip (NLoC): introducing networking technologies in microfluidic systems. Nano Commun. Netw. **3**(4), 217–228 (2012)
5. Cheow, L.F., Yobas, L., Kwong, D.-L.: Digital microfluidics: droplet based logic gates. Appl. Phys. Lett. **90**(5), 054107-1–054107-3 (2007)
6. Prakash, M., Gershenfeld, N.: Microfluidic bubble logic. Science **315**(5813), 832–835 (2007)
7. Alaghi, A., Hayes, J.P.: Survey of stochastic computing. ACM Trans. Embed. Comput. Syst. **12**(2s), 92:1–92:19 (2013)
8. Brown, B., Card, H.: Stochastic neural computation I: computational elements. IEEE Trans. Comput. **50**(9), 891–905 (2001)
9. Jenson, D., Riedel, M.: A deterministic approach to stochastic computation. In: Proceedings of the IEEE/ACM International Conference on Computer-Aided Design, pp. 1–8 (2016)
10. Dinu, A., Cirstea, M.N., McCormick, M.: Stochastic implementation of motor controllers. In: Proceedings of the 2002 IEEE International Symposium on Electronics Industrial, pp. 639–644 (2002)
11. Hammadou, T., Nilson, M., Bermak, A., Ogunbona, P.: A 96×64 intelligent digital pixel array with extended binary stochastic arithmetic. In: Proceedings of the IEEE International Symposium on Circuits and Systems, pp. IV-772–IV-775 (2003)
12. Gaudet, V.C., Rapley, A.C.: Iterative decoding using stochastic computation. Electron. Lett. **39**(3), 299–301 (2003)
13. Tehrani, S.S., Gross, W.J., Mannor, S.: Stochastic decoding of LDPC codes. IEEE Commun. Lett. **10**(10), 716–718 (2006)
14. Fuerstman, M.J., Lai, A., Thurlow, M.E., Shevkoplyas, S.S., Stone, H.A., Whitesides, G.M.: The pressure drop along rectangular microchannels containing bubbles. Lab Chip **7**, 1479–1489 (2007)
15. Biral, A., Zordan, D., Zanella, A.: Modeling, simulation and experimentation of droplet-based microfluidic networks. Trans. Mol. Biol. Multi-scale Commun. **1**(2), 122–134 (2015)
16. Grimmer, A., Haselmayr, W., Springer, A., Wille, R.: A discrete model for networked labs-on-chips: linking the physical world to design automation. In: 2017 54th ACM/EDAC/IEEE Design Automation Conference (DAC), pp. 1–6, June 2017. https://doi.org/10.1145/3061639.3062186

Determination of Parasitic Capacitances in Inductive Components - A Comparison Between Analytic Calculation Methods and FEM-Simulation

Simon Merschak[1($\boxtimes$)], Mario Jungwirth[1], Daniel Hofinger[2], Alexander Eder[2], and Günter Ritzberger[2]

[1] University of Applied Sciences Upper Austria,
Campus Wels Stelzhamerstrasse 23, 4600 Wels, Austria
{simon.merschak,mario.jungwirth}@fh-wels.at
http://www.fh-ooe.at
[2] Research and Development, Fronius International GmbH,
Günter Fronius Strasse 1, 4600 Wels-Thalheim, Austria
{hofinger.daniel,eder.alexander,ritzberger.guenter}@fronius.com
http://www.fronius.at

Abstract. Flyback converters are commonly used in switching mode power supplies. An essential component of a flyback converter is the high-frequency transformer. In order to ensure a faultless operation of the converter, even at high operating frequencies, the parasitic capacitive effects in the transformer have to be considered during the engineering process. The negative effects of the parasitic capacitances on the currents in the converter windings increase with rising operation frequency [1]. The aim of this paper is to draw a comparison between analytic calculation methods and FEM-simulations by means of a simple sample transformer model. It is also verified that the dynamic component of the electric field can be neglected for the calculation of the parasitic capacitances for frequencies up to a few hundred kilohertz.

Keywords: Transformer parasitic capacitance · Flyback converter
Finite element method · Electrostatic simulation

1 Introduction

Parasitic capacitances appear between all conductors of a coil and between conductors, shields and the core. There are interwinding capacitances which occur between the conductors inside the winding and there are intrawinding capacitances between different coils. In Fig. 1 some parasitic capacitances are depicted. The size of the parasitic capacitances is dependent on wire geometry, coil size, insulation materials, coil form and winding style of the transformer. Parasitic capacitances can lead to oscillations of the currents in the transformer windings. In order to produce compact transformers, high operation frequequencies

© Springer International Publishing AG 2018
R. Moreno-Díaz et al. (Eds.): EUROCAST 2017, Part II, LNCS 10672, pp. 212–218, 2018.
https://doi.org/10.1007/978-3-319-74727-9_25

are required. But the negative effects of the parasitic capacitances on the currents in the transformer windings are increasing with rising operation frequency. Hence it is important to reduce the size of the parasitic capacitances.

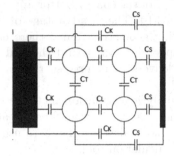

Fig. 1. Parasitic capacitances in the transformer windings

2   Methods of Calculation

There are two different approaches for the calculation of parasitic capacitances in transformers. The first one is based on easy to use equations which depend on a strong simplification of the complex task. The second one applies the finite element method to calculate a matrix of parasitic capacitances. The computational cost of the second approach is significantly higher [2]. Hence it is evaluated if the analytic approach is applicable for an automated calculation of the parasitic capacitances or if the finite element method has to be used.

2.1   Analytic Calculation Methods

Analytic approaches are fast calculation methods for the parasitic capacitances of specific transformer geometries. There are different analytic approaches which are all based on strong simplifications of the transformer geometry. Some of them are presented in [1,3]. For the analytic calculation only a few parameters of the windings are needed. Hence analytic equations are a good compromise between accuracy and calculation effort. For the analytic calculation of parasitic capacitances some assumptions have to be made:

- An electrostatic field between the windings is assumed. Eddy current fields are neglected.
- There is no distributed charge between the conductors. The charge is only on the conductor surface.
- A uniform voltage distribution along the winding is assumed.
- All windings of a coil have the same length.

The least complex model for the calculation of the parasitic coil-to-coil capacitance is the parallel plate capacitor model displayed in Fig. 2(a). In this model, the conductors of the coils are replaced by parallel equipotential surfaces. Each coil forms an equipotential surface and the parameters of the equation can be calculated from the geometry of the coil. A better approximation is the cylindrical capacitor model in Fig. 2(b). The disadvantage of these models is the high simplification of the winding geometry. Hence there can be a huge difference between the calculated capacitance value and the real value. [3] The basic-cell method presented in [1] delivers more precise results with higher calculation effort.

The main problem of analytic calculation methods is to find the right equation for each winding geometry and coil form. There is also not always an appropriate equation available. That makes it difficult to use the analytic calculation methods in an automated calculation tool.

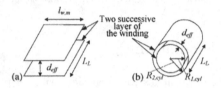

Fig. 2. (a) Parallel-plate capacitor model. (b) Cylindrical capacitor model. [3]

2.2 FEM Calculation Methods

A better method for the automated calculation of the parasitic capacitances is the finite element method. The FEM is a numerical method for solving complex problems of engineering and physics. The basic idea behind the FEM is that the examined structure is cut into a finite number of small elements. The small elements are connected at nodes. This process results in a set of simultaneous algebraic equations which can be solved for the electric potential of each node. Finally the electrostatic energy which is stored in the structure between the conductors of the transformer can be used to calculate equivalent lumped capacitances.

The FEM method can be structured in 5 steps [5]:

1. Discretisation: A complex problem is divided into small elements, connected with nodes. This meshing process is important for the accuracy of the results.
2. Element analysis: The electric potentials within the elements are expressed.
3. System analysis: The elements are coupled.
4. Boundary conditions: Boundary conditions are added to reduce the degrees of freedom.
5. Solving the linear set of equations.

The big advantage of the FEM is, that it can be used for complex geometries and complex physical processes. In contrast to the analytic equations, one calculation algorithm can be used for all different coil geometries and winding geometries. On the other hand there is a high computational effort and the solution is dependent on the density of the mesh during the discretisation process.

2.3   Complexity of the Simulation Model

In order to implement a fast FEM-simulation it is required to clarify if it is sufficient to calculate only the electrostatic field or if the eddy current field has to be considered too. Hence a simulation in the software Comsol Multiphysics ® is implemented. The electric field strength is calculated by use of Eq. (1).

$$E = -\nabla\phi - j \cdot \omega \cdot A \tag{1}$$

The first term of the equation represents the electrostatic energy and the second term represents the energy stored in the eddy current field. Two conductors coated with an insulation layer and surrounded by air are simulated at an operational frequency of 500 kHz.

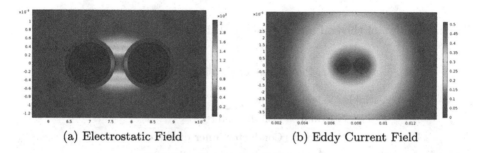

(a) Electrostatic Field   (b) Eddy Current Field

Fig. 3. Electrical field strength

Figure 3 depicts that the electrostatic field is concentrated between the two conductors. The electrostatic field is not dependent on the operation frequency. In contrast, the eddy current field is dependent on the operation frequency. The simulation shows that the electrostatic field is about a thousand times stronger than the eddy current field. So it is appropriate to neglect the eddy current field for the calculation of the parasitic capacitances.

3   Comparison of Results

Calculation and simulation of the turn-to-turn and layer-to-layer capacitances are conducted by use of simple coil models. The capacitances are calculated using analytical equations in Wolfram Mathematica and are compared to simulations performed in Comsol Multiphysics ®.

3.1   Turn-to-Turn Capacitance

The turn-to-turn capacitance is the capacitance between two adjacent conductors of a coil. The turn-to-turn capacitance of a simple single-layer air coil is calculated with different analytic equations and simulated in Comsol Multiphysics ®. The simulation model of the turn-to-turn capacitances is displayed in Fig. 4.

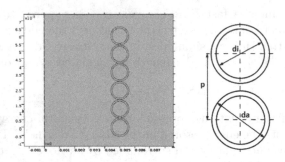

Fig. 4. Simulation model for turn-to-turn capacitance calculation

The parameters of the coil are depicted in Table 1.

Table 1. Parameters of the simulation model

Variable	Value	Name
D	10 mm	Coil diameter
di	1 mm	Conductor inner diameter
da	1.2 mm	Conductor outer diameter
ϵ_r	4	Relative permittivity of insulation
p	1.21 mm	Conductor center distance
d	1.2 mm	Coil center distance

The first simulation is carried out with no distance between the conductors (p = da). The used analytic Eqs. (2) and (3) are only valid for conductors which are in contact at the insulation surface.

$$C_{tt} = \epsilon_0 \cdot D \cdot \pi \cdot \left[\frac{\epsilon_r}{\ln \frac{da}{di}} + \cot \left(\frac{\Theta^*}{2} \right) - \cot \left(\frac{\pi}{12} \right) \right] \tag{2}$$

$$\Theta^* = \arccos \left(1 - \frac{\ln \frac{da}{di}}{\epsilon_r} \right) \tag{3}$$

The result of the calculation is 2,633 pF and the simulation result is 2,662 pF. In this case, the difference between the analytic calculation result and the simulation result is small and the analytic equations can be used for the fast estimation of the turn-to-turn capacitance.

If the distance between the conductors is inreased, other equations have to be used. Equations (4) and (5) take account of the distance between adjacent conductors [4].

$$C_{tt} = \cfrac{\pi^2 \cdot \epsilon_0 \cdot D_T}{\ln\left\{ \left[\cfrac{\frac{p}{2 \cdot di}}{\left(1 + \frac{t}{di}\right)^{1 - \frac{1}{\epsilon_r}}} \right] + \sqrt{\left(\cfrac{\frac{p}{2 \cdot di}}{\left(1 + \frac{t}{di}\right)^{1 - \frac{1}{\epsilon_r}}} \right)^2 - \left(1 + \frac{t}{di}\right)^{\frac{2}{\epsilon_r}}} \right\}} \quad (4)$$

$$t = \frac{da - di}{2} \quad (5)$$

The variable t is the insulation thickness of the conductor. For small air gaps p the equation works well but if the gap is increased, the difference between analytic calculation and simulation is getting bigger. For an air gap of $p = 0,2\,mm$ the analytic calculation delivers $C_{tt} = 1,42\,pF$ and the simulation result is $C_{tt} = 1,08\,pF$. The results doesn't match as well as before. The reason for this deviation is, that the equation only delivers exact results for a small range of input parameters like conductor diameter and insulation strength. Each equation is optimized for a defined coil form and winding geometry. There is no equation which fits for all winding geometries.

3.2   Coil-to-Coil Capacitance

The coil-to-coil capacitance is the capacitance between different coils of a transformer. The simulation model of the coil-to-coil capacitances consists of two coils and is displayed in Fig. 5. The parameter of this coils are equal to the parameters in Table 1. A simple analytic equation for the calculation of the coil-to-coil capacitance is based on the cylindrical capacitor model displayed in Fig. 2(b) [3].

$$C_{cc} = \epsilon_0 \cdot \epsilon_r \cdot \frac{2 \cdot \pi \cdot (n_t \cdot da)}{\ln(1 + \frac{d}{D})} \quad (6)$$

The variable n_t indicates the number of turns of the primary coil. The electrostatic simulation of this coils delivers a capacitance value of $C_{cc} = 17,01\,pF$. The analytic calculation result is $14.93\,pF$ and is therefore underestimating the simulation result by 12%.

There are far more analytic calculation methods for the calculation of the turn-to-turn and coil-to-coil capacitances. One example is the basic-cell method which is more precise but the computational cost is higher. Further information can be found in [3].

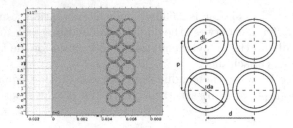

Fig. 5. Simulation model for coil-to-coil capacitance

4   Conclusion and Future Prospects

The parasitic turn-to-turn capacitance is calculated with two different equations. The first one is only valid for conductors which are in contact at the insulation surfaces. The calculated capacitance value is very close to the simulation result. If the distance between the conductors is increased, another equation has to be used and the conformity is not so good anymore. The difference is about 30% and can be explained by the limited range of validity for the input parameters of the analytic equations. The layer-to-layer capacitance is calculated by use of the cylindrical capacitor model. The result is underestimating the simulation result by 12%.

In summary it can be said, that analytic equations are suitable to estimate the parasitic capacitances of a specific transformer but they are not suitable for the automated calculation of them. The reason for this is that there are numerous analytic equations for different use cases and each equation is only valid for a specific range of input parameters. A FEM-based approch is prefered for the automated calculation of parasitic capacitances. Future steps will be to conduct a 3D simulation in the software ANSYS Maxwell to determine the influence of edge effects which are neglegted in the 2D simulation.

References

1. Massarini, A., Kazimierczuk, M.: Self-capacitance of inductors. IEEE Trans. Power Electron. **12**(4), 671–676 (1997)
2. Dalessandro, L., da Silveira Cavalcante, F., Kolar, J.: Self-capacitance of highvoltage transformers. IEEE Trans. Power Electron. **22**(5), 2081–2092 (2007)
3. Biela, J., Kolar, J.: Using transformer parasitics for resonant converters - a review of the calculation of the stray capacitance of transformers. IEEE Trans. Ind. Appl. **44**(1), 223–233 (2008)
4. Shadmand, M.B., Balog, R.S.: Determination of parasitic parameters in a high frequency magnetic to improve the manufacturability, performance, and efficiency of a PV inverter. In: 38th IEEE Photovoltaic Specialists Conference (PVSC), pp. 001368–001372 (2012)
5. Jin, J.: The Finite Element Method in Electromagnetics. Wiley, New York (2002)

Review of UHF-Based Signal Processing Approaches for Partial Discharge Detection

Benjamin Schubert[✉], Mauro Palo, and Thomas Schlechter

Global Energy Interconnection Research Institute Europe GmbH,
Markgrafenstr. 34, 10117 Berlin, Germany
{benjamin.schubert,palo,t.schlechter}@geiri.sgcc.com.cn

Abstract. Partial Discharge (PD) events are due to local defects in dielectrics and can cause damages to the electrical insulation and eventually to the whole power station. This paper reviews approaches describing procedures and numerical techniques for detecting, denoising, clustering, and classifying PDs in the ultra-high frequency range. For each method the mathematical background is recalled and one or few representative examples from selected papers are shortly described.

Keywords: Partial Discharge (PD) · Ultra-high frequency
Signal processing · Data analysis

1 Introduction

Partial Discharge is the effect of imperfections in gaseous, liquid, or solid dielectrics causing a local electrical field enhancement, which in turn may exceed the intrinsic field strength and thus induce an electron avalanche [2,8,12]. This can provoke a deterioration of the insulating material; hence, PDs are a common reason for failures in high voltage power stations [20].

Different types of PD sources have been distinguished, the most common being the corona, internal, and shallow discharge; they occur at the conducting heads of gas-insulated systems, within, and on the surface of the insulating material, respectively [12]. Effects of PD are noticeable as electromagnetic, acoustic, thermal, chemical, and electric disturbances. Moreover, different types of PD reflect into multiple recorded signals, what enables in principle the automatic detection and classification of PD events [23].

The electromagnetic effects of PDs extend up to the ultra high frequencies (UHF), ranging from 300 MHz up to 3 GHz (typical systems detect frequencies up to 1.5 GHz). The approaches focusing on this frequency range (UHF-based methods) have been diffusing in recent years among researchers and power plant operators due to its robustness against external noise [20,21]. UHF signals are mostly represented in time, phase, or spectral domain. In the first approach the PD waveform is analyzed, in the second the phase angle of the oscillating voltage at the occurrence time of PD is represented [11,23], while the representation in

© Springer International Publishing AG 2018
R. Moreno-Díaz et al. (Eds.): EUROCAST 2017, Part II, LNCS 10672, pp. 219–226, 2018.
https://doi.org/10.1007/978-3-319-74727-9_26

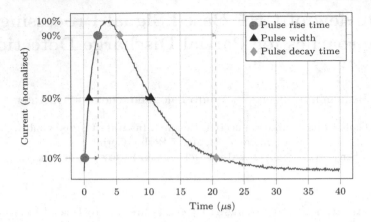

Fig. 1. Example of a PD pulse. Typically three times describe its shape: rise time, width and decay time. In this synthetic example, the PD pulse is modeled as $f(x) = A_0(e^{-\tau_1 t} - e^{-\tau_2 t})$ with τ_1 and τ_2 fixed as 0.25 µs and 0.5 µs, respectively. White noise with amplitude equal to 0.01% of the signal amplitude is superimposed. The resulting characteristic times are indicated by arrows in the figure.

the spectral domain looks at the frequency content [20]. In the time domain representation either the original oscillating waveform or its amplitude modulation is processed. In the latter case, three typical times describing the shape of the PD can be introduced [23], see Fig. 1.

As the fine characterization of PDs can potentially reduce the maintenance effort through real-time monitoring of critical events, many efforts have been made in recent years to optimize the detection, localization, and classification of PD events [15,17,21]. In the following we will describe a set of signal processing and data analysis techniques adopted for this aim. In detail, we will shortly review approaches used in denoising, detecting, reducing the dimensionality, and classifying PD events measured through its UHF signals.

2   Review of Methods

A PD monitoring system consists of three components, as schematically shown in Fig. 2. After recording the signals through a set of sensors, they are cleaned from noise, certain features are extracted, and the signal dimension is reduced by means of signal processing. The collected data is then analyzed in order to decide whether and what type of PD happened. Below some of the methods involved in these steps are described. For each method, one or more examples from selected papers are reported.

Energy-Based Detection. The authors in [26] proposed a PD detection method which uses only the received signal strength (RSSI) of the UHF signal radiated in consequence of the PD pulse. A couple of sensors measure the

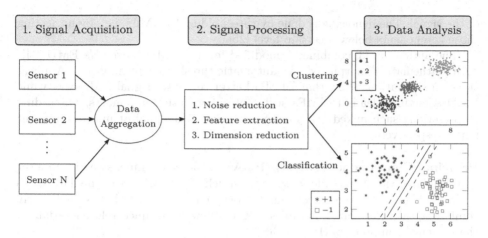

Fig. 2. The three components of a partial discharge monitoring system [23].

received power in four different frequency bands (where one of these channels covers the whole bandwidth) in the analog domain. Within each band, the power signal is integrated over a certain time, still in the analog domain. Afterwards, the information about the power within each band is sampled at a very low rate of several tens of MHz. The gained power distribution is then correlated with the phase relation of the 50 Hz AC high voltage. Based on the gained digitized phase relation and power distribution, a classification and characterization of the PD pulse is done. Compared to other approaches operating directly on the original time-domain signal, this method reduces the sampling rate requirements, since a great portion of work is done in the analog domain.

Discrete Wavelet Transform (DWT). As PD signals can be quite weak, their recordings may have a low signal-to-noise ratio (SNR). One approach for extracting the desired signal from the noisy background is based on the DWT, which is very well suited for extracting characteristic features from the underlying signal [14,18]. The main steps of wavelet-based denoising are: transformation of the signal into the wavelet space, thresholding of the coefficients, and back transformation. One of the major limitations of the wavelet transform (WT) is the proper selection of the mother wavelet, as it is difficult to catch all types of PD with a fixed mother wavelet [3]. This fact, however, can be exploited for the classification of multi-source PD patterns [2].

Empirical Mode Decomposition (EMD) is another technique utilized for noise reduction. Originally developed in [9] for analyzing nonlinear and non-stationary data, it was first used for denoising PD signals in [25]. Differently from the wavelet transform, in EMD selecting a certain mother function is not needed. EMD rather decomposes any given data into a collection of mono-component signals called intrinsic mode functions (IMFs), which are based on local properties of the signal. The denoising is done by thresholding the IMFs, i.e., by neglecting those components below a certain level [25].

The authors in [3] combined a modified form of EMD (*ensemble* EMD [24]) with mathematical morphology for automatic threshold determination. Applications on simulations show that EEMD better separates signal and noise, while EMD sometimes leads to IMFs still containing both components. Discarding or preserving such mixed IMFs distort the denoised signal or inject significant noise, respectively.

Principal Component Analysis (PCA) reduces the dimensionality of a set of interrelated variables while retaining as much of the variance of the data set as possible. This is achieved by rotating the original data space; that is, by linearly combining the interrelated variables to get a new set of uncorrelated variables, the principal components (PCs) [10].

Mathematically, PCA diagonalizes the covariance matrix:

$$S_{j,k} = \frac{1}{n-1} \sum_{i=1}^{n} (x_i^j - \hat{x}^j)(x_i^k - \hat{x}^k), \quad \boldsymbol{x}^{j,k} = \{x_1^{j,k}, x_2^{j,k}, \ldots, x_n^{j,k}\}, \quad (1)$$

where $\hat{x}^k$ is the mean value of the k-th vector.

The new set of coordinates are the eigenvectors of the covariance matrix. A number m ($m < n$) of PCs is selected, thus the original data is embedded in a space of lower dimension. The loss of variance from this dimension reduction is $R = \sum_{i=1}^{m} \lambda_i / \sum_{i=1}^{n} \lambda_i$, with λ_i being eigenvalues of $\boldsymbol{S}$ sorted in descending order.

In [16] PCA is applied as a pre-analysis for extracting representative attributes from PD simulated signals before the classification. The authors generate in laboratory five types of PD sources. For each source, 200 acquisitions are recorded. From this database a phase-resolved data set of 600 points is extracted by calculating three parameters (average and maximum PD pulse magnitude, and PD number count) in 200 equally spaced phase windows. Via PCA the number of significant features is reduced from 600 to 7.

In [7] a DWT is applied to PD pulses providing coefficients and energies of five decomposition levels. Each PD pulse is then defined by five detail coefficients plus one approximate coefficient from the final level of the decomposition levels. The six corresponding energies are adopted as features. PCA applied on the features of a set of experimental PD pulses reduces the dimensionality from six to three. The features are computed (in the time domain) from PD pulses that are mostly extracted from laboratory tests on either artificially defective conductor bars of AC rotating machines or during the accelerated aging of induction motor coils. The goal is the separation of multiple PD sources and of PD signal from noise.

In [1] the data set consists of 5000 recordings each of 5000 samples (sampling window of $2\,\mu$s). Measurements are acquired by a helical antenna placed close to a $400/275$-kV 500-MVA power transformer with a known internal defect. Before PCA, each signal is transformed in the frequency domain, thus halving the size of the samples (Nyquist theorem). Moreover, all signals $\boldsymbol{v}_i$ are normalized by the "range" method ($\overline{v}_i = \frac{v_i - \min(v_i)}{\max(v_i) - \min(v_i)}$), to ensure that all features have equal weight in the PCA. The number of PCs is fixed to six (cumulative variance is reduced by about 20%). In this way the dimension of the features data is reduced from 5000×2500 to 5000×6.

Support Vector Machine (SVM) is a classification technique also used for denoising and regression analysis. Based on training data with known class affiliation, SVM computes the hyperplane maximizing the margin between the clusters formed by data points of each class. The nearest data point of each class are called support vectors (see the lower right part in Fig. 2).

Mathematically the procedure can be formulated as an optimization problem. Given a set of training data $\{(x_i, y_i)\}_i$, $i \in \{1, \ldots, n\}$, $y \in \{-1, 1\}$, where vectors x_i are the data points and the corresponding class association is given by y_i, and a hyperplane defined by a vector w and a scalar b satisfying $w^\mathsf{T} x - b = 0$ for any x, the optimal hyperplane is found by

$$\min \| w \| \ \text{s.t.} \ y_i (w^\mathsf{T} x_i - b) \geq 1, \ \text{for} \ i \in \{1, \ldots, n\}. \tag{2}$$

Since the margin is proportional to $\frac{1}{\|w\|}$, it is maximized by minimizing $\|w\|$. If a linear hyperplane segregating two classes does not exist, the problem can be linearized by mapping it into a higher dimension in which the hyperplane is linear, what is referred to as kernel trick. Another possibility is to allow some overlapping of the classes with respect to a linear hyperplane by introducing a regularization parameter which penalizes data lying in the wrong cluster. In this way, a linear hyperplane can be found which separates the clusters. This problem is often less complex to solve than using the kernel trick [5, 19].

SVM has been utilized to identify different types of PD [5, 6]. In [13] SVM is compared with two other classification techniques, namely backpropagation NN (BPN) and self-organizing map (SOM). For two different types of input, SVM performs best in discriminating the three basic PD sources corona, surface, and internal discharge. The two input types are both based on phase-resolved patterns; while the first one uses the four statistical moments mean value, standard deviation, skewness, and kurtosis, the second one directly operates on the raw data and uses PCA for feature extraction and dimensionality reduction.

SVM is also utilized for denoising. The authors in [18] extract features using WT and apply an SVM classifier to separate the coefficients related to PD from those related to noise. This approach outperformed techniques based on linear filters and three other WT-based denoising procedures. This is also the case in [22], where the authors directly operate on the (simulated) PD signal and compare their approach with those based on Fourier and wavelet transforms.

Cumulative Energy Function (CE). The authors in [27] propose using the CE to separate PD signals of different types. CE calculates the cumulative power of a signal and normalize it to the signal's total power. This can be done in time (TCE) and frequency domain (FCE). The resulting function is then analyzed using the mathematical morphology gradient operation to extract certain features, such as rising steepness, width, sharpness, etc. The outcome is a multidimensional feature space which serves as input to various clustering approaches. Using the proposed techniques, different cases have been successfully investigated: (i) separation of PD and noise, (ii) separation of surface and internal discharge, and (iii) separate PD events at different voltages. Compared to the phase-resolved pattern analysis, the CE-based approach yields much clearer results, especially for the case of multiple PD types.

Neural Network (NN). An (artificial) NN is a computation model inspired by the structure of the brain. It is commonly described as a set of nodes V mimicking neurons that are interconnected through edges E (links). The nodes are normally decomposed into T disjoint layers. V_0, V_T, and $V_1, \ldots, V_{T-1}$ are the input, output, and hidden layers, respectively. In simple prediction problems the output layer contains a single neuron returning the network output [19].

A feedforward NN is described by a directed acyclic graph, $G = (V, E)$, and a weight function over the edges, $w : E \to \mathbb{R}$. Each single neuron is modeled as a simple scalar function (activation function), $\sigma : \mathbb{R} \to \mathbb{R}$. If we indicate by $v_{t,i}$ the i-th neuron of the t-th layer and by $o_{t,i}(\boldsymbol{x})$ the output of $v_{t,i}$ when the network is fed with the input vector $\boldsymbol{x}$, the input $a_{t+1,j}$ to the node $v_{t+1,j}$ and the output of $v_{t+1,j}$ are:

$$a_{t+1,j}(\boldsymbol{x}) = \sum_{r:(v_{t,r}, v_{t+1,j}) \in E} w(v_{t,r}, v_{t+1,j}) o_{t,r}(\boldsymbol{x}) \tag{3a}$$

$$o_{t+1,j}(\boldsymbol{x}) = \sigma(a_{t+1,j}(\boldsymbol{x})), \tag{3b}$$

where w is the weighting vector. Thus, once defined a NN by (V, E, σ, w), we obtain the function $h_{V,E,\sigma,w} : \mathbb{R} \to \mathbb{R}$. Normally the architecture of the NN (V, E, σ) is fixed and the weights over the edges w are the parameters specifying a hypothesis in the hypothesis class $\mathcal{H}$, with $\mathcal{H}_{V,E,\sigma} = \{h_{V,E,\sigma,w} : w : E \to \mathbb{R}\}$, where w are learned in the training phase (typically by backprojection method). These weights are then used for classifying new instances via Eq. (3).

In [4] a NN is applied to remove corona from internal PD. Signals are measured from a 300-kV gas-insulated switchgear section using a UHF coupler. Samples of PD, corona, and mixed signals are first decomposed by Wavelet-Packet Transform (a generalized DWT in which both detail and approximation coefficients are decomposed, leading to a full binary tree). For each tree node, three parameters are defined: $|\omega_{j,n}|^2$, $K_{j,n}$, $S_{j,n}$, $(j = 1, \ldots, 4)$, where $\omega_{j,n}$ are the decomposition coefficients, K is the kurtosis and S the skewness of the decomposed signals, thus forming three additional trees. Then, from each of these trees only two nodes are selected (following a criterion based on within- and between-class scatter), leading to a total of six selected parameters. These are finally the inputs of a three-layer NN with a backpropagation learning rule used as the classifier to discriminate among PD, corona, and mixed signals. This NN has thus six input, three hidden, and two output neurons for a three-class problem. A sigmoid-type activation function is used for both hidden and output layers. The authors use 40 PD, corona, and mixed events for training and 10 PD, 23 corona, and 13 mixed data sets for testing. Training of the NN converges after 15 epochs with almost no error, while the classification of the test shows an error of about 1%.

3 Conclusion

Currently, plenty of concurrent methods on PD evaluation in the UHF range are under investigation. Real world field tests will show, which method or combination of methods will be the most suitable. As a follow-up of the research presented in this paper, this aspect will be covered in a deeper sense.

Acknowledgment. This work was funded by the State Grid Corporation of China (SGCC) through the R&D project "Research of Key Technology of UHF Wireless Sensing based Substation Partial Discharge Monitoring and Location".

References

1. Babnik, T., Aggarwal, R.K., Moore, P.J.: Principal component and hierarchical cluster analyses as applied to transformer partial discharge data with particular reference to transformer condition monitoring. IEEE Trans. Power Deliv. **23**(4), 2008–2016 (2008)
2. Bartnikas, R.: Partial discharges. Their mechanism, detection and measurement. IEEE Trans. Dielectr. Electr. Insul. **9**(5), 763–808 (2002)
3. Chan, J., Ma, H., Saha, T., Ekanayake, C.: Self-adaptive partial discharge signal de-noising based on ensemble empirical mode decomposition and automatic morphological thresholding. IEEE Trans. Dielectr. Electr. Insul. **21**(1), 294–303 (2014)
4. Chang, C., Jin, J., Chang, C., Hoshino, T., Hanai, M., Kobayashi, N.: Separation of corona using wavelet packet transform and neural network for detection of partial discharge in gas-insulated substations. IEEE Trans. Power Deliv. **20**(2), 1363–1369 (2005)
5. Hao, L., Lewin, P.L.: Partial discharge source discrimination using a support vector machine. IEEE Trans. Dielectr. Electr. Insul. **17**(1), 189–197 (2010)
6. Hao, L., Lewin, P.L., Dodd, S.J.: Comparison of support vector machine based partial discharge identification parameters. In: Conference Record of 2006 IEEE International Symposium on Electrical Insulation, pp. 110–113 (2006)
7. Hao, L., Lewin, P.L., Hunter, J.A., Swaffield, D.J., Contin, A., Walton, C., Michel, M.: Discrimination of multiple PD sources using wavelet decomposition and principal component analysis. IEEE Trans. Dielectr. Electr. Insul. **18**(5), 1702–1711 (2011)
8. Hauschild, W., Lemke, E.: High-Voltage Test and Measuring Techniques. Springer, Heidelberg (2014). https://doi.org/10.1007/978-3-642-45352-6
9. Huang, N.E., Shen, Z., Long, S.R., Wu, M.C., Shih, H.H., Zheng, Q., Yen, N.C., Tung, C.C., Liu, H.H.: The empirical mode decomposition and the Hilbert spectrum for nonlinear and non-stationary time series analysis. In: Proceedings of the Royal Society of London A: Mathematical, Physical and Engineering Sciences, vol. 454, no. 1971, pp. 903–995 (1998)

10. Jolliffe, I.: Principal Component Analysis. Wiley Online Library, Hoboken (2002)
11. Judd, M.D., Yang, L., Hunter, I.B.B.: Partial discharge monitoring of power trans-formers using UHF sensors. Part I: sensors and signal interpretation. IEEE Electr. Insul. Mag. **21**(2), 5–14 (2005)
12. Küchler, A.: Hochspannungstechnik. VDI-Verlag, Düsseldorf (2009)
13. Lai, K.X., Phung, B.T., Blackburn, T.R.: Application of data mining on partial discharge part I: predictive modelling classification. IEEE Trans. Dielectr. Electr. Insul. **17**(3), 846–854 (2010)
14. Li-Xue, L., Cheng-Jun, H., Yi, Z., Xiu-Chen, J.: Partial discharge diagnosis on GIS based on envelope detection. WSEAS Trans. Syst. **7**(11), 1238–1247 (2008)
15. Ma, H., Chan, J.C., Saha, T.K., Ekanayake, C.: Pattern recognition techniques and their applications for automatic classification of artificial partial discharge sources. IEEE Trans. Dielectr. Electr. Insul. **20**(2), 468–478 (2013)
16. Ma, X., Zhou, C., Kemp, I.: Interpretation of wavelet analysis and its application in partial discharge detection. IEEE Trans. Dielectr. Electr. Insul. **9**(3), 446–457 (2002)
17. Markalous, S.M., Tenbohlen, S., Feser, K.: Detection and location of partial dis-charges in power transformers using acoustic and electromagnetic signals. IEEE Trans. Dielectr. Electr. Insul. **15**(6), 1576–1583 (2008)
18. de Oliveira Mota, H., da Rocha, L.C.D., de Moura Salles, T.C., Vasconcelos, F.H.: Partial discharge signal denoising with spatially adaptive wavelet thresholding and support vector machines. Electr. Power Syst. Res. **81**(2), 644–659 (2011)
19. Shalev-Shwartz, S., Ben-David, S.: Understanding Machine Learning: From Theory to Algorithms. Cambridge University Press, Cambridge (2014)
20. Siegel, M., Beltle, M., Tenbohlen, S.: Characterization of UHF PD sensors for power transformers using an oil-filled GTEM cell. IEEE Trans. Dielectr. Electr. Insul. **23**(3), 1580–1588 (2016)
21. Soomro, I.A., Ramdon, M.N.: Study on different techniques of partial discharge (PD) detection in power transformers winding: simulation between paper and EPOXY resin using UHF method. Int. J. Concept. Electr. Electron. Eng. **2**(1), 57–61 (2014)
22. Velayutham, M.R., Perumal, S., Basharan, V., Silluvairaj, W.I.M.: Support vector machine-based denoising technique for removal of white noise in partial discharge signal. Electr. Power Compon. Syst. **42**(14), 1611–1622 (2014)
23. Wu, M., Cao, H., Cao, J., Nguyen, H.L., Gomes, J.B., Krishnaswamy, S.P.: An overview of state-of-the-art partial discharge analysis techniques for condition mon-itoring. IEEE Electr. Insul. Mag. **31**(6), 22–35 (2015)
24. Wu, Z., Huang, N.E.: Ensemble empirical mode decomposition: a noise-assisted data analysis method. Adv. Adapt. Data Anal. **1**(01), 1–41 (2009)
25. Yong, Q., Cheng-Jun, H., Xiu-Chen, J.: Empirical mode decomposition based denoising of partial discharge signals. In: Proceedings of 5th WSEAS/IASME Inter-national Conference on Electric Power Systems, High Voltages, Electric Machines (2005)
26. Zhang, Y., Upton, D., Jaber, A., Ahmed, H., Saeed, B., Mather, P., Lazaridis, P., Mopty, A., Tachtatzis, C., Atkinson, R., Judd, M., de Fatima, M., Vieira, Q., Glover, I.: Radiometric wireless sensor network monitoring of partial discharge sources in electrical substations. Int. J. Distrib. Sens. Netw. **11**(9), 1–9 (2015)
27. Zhu, M.X., Zhang, J.N., Li, Y., Wei, Y.H., Xue, J.Y., Deng, J.B., Mu, H.B., Zhang, G.J., Shao, X.J.: Partial discharge signals separation using cumulative energy func-tion and mathematical morphology gradient. IEEE Trans. Dielectr. Electr. Insul. **23**(1), 482–493 (2016)

Algebraic and Combinatorial Methods
in Signal and Pattern Analysis

Gibbs Dyadic Differentiation on Groups - Evolution of the Concept

Radomir S. Stanković[1(✉)], Jaakko Astola[2], and Claudio Moraga[3]

[1] Department of Computer Science, Faculty of Electronic Engineering, Niš, Serbia
Radomir.Stankovic@gmail.com
[2] Department of Signal Processing, Tampere University of Technology,
Tampere, Finland
[3] Faculty of Computer Science, Technical University of Dortmund,
Dortmund, Germany

Abstract. Differential operators are usually used to determine the rate of change and the direction of change of a signal modeled by a function in some appropriately selected function space. Gibbs derivatives are introduced as operators permitting differentiation of piecewise constant functions. Being initially intended for applications in Walsh dyadic analysis, they are defined as operators having Walsh functions as eigenfunctions. This feature was used in different generalizations and extensions of the concept firstly defined for functions on finite dyadic groups. In this paper, we provide a brief overview of the evolution of this concept into a particlar class of differential operators for functions on various groups.

1 Introduction

The Walsh dyadic analysis emerged in late sixties and early seventies as a mathematical discipline aimed at providing a form of spectral Fourier-like analysis tailored for signals modeled by two-level piecewise constant functions and, therefore, making them compatible with digital computing devices based on elements with two stable states. At the same time, due to exactly this compatibility with devices based on binary arithmetic, dyadic analysis provided an answer to demands for a spectral analysis with simplified computations because the limited computing power of computing devices at that time.

The relentless complexity of contemporary and future digital systems, regains a new interest to Walsh dyadic analysis. Being primarily intended as a mathematical support in solving various signal processing and system design tasks, the dyadic analysis immediately required the concept of a derivative that will enable differentiation of piecewise constant functions.

The answer was provided by James Edmund Gibbs who introduced the concept that is now called the Gibbs dyadic derivative [7].

In a report from April 1970, in Sect. 4.1 of this report entitled Dyadic differentiation, Pichler extended this definition to real valued functions of a continuous non-negative real variable.

© Springer International Publishing AG 2018
R. Moreno-Díaz et al. (Eds.): EUROCAST 2017, Part II, LNCS 10672, pp. 229–237, 2018.
https://doi.org/10.1007/978-3-319-74727-9_27

Two papers [1] and [3], introducing the Butzer-Wagner derivative, which can be viewed as the extension of the concept of the dyadic differentiation to infinite dyadic group, alternatively the interval $[0, 1)$, were completed in preprint form in the autumn of 1971. See, also [2].

Since starting from the initial definitions, these operators were viewed as differential operators for functions on groups, and the same line of thinking was continued by Gibbs himself and his associates [9, 11], and have also been accepted by several other authors [19]. The scope of the definition was extended and generalized to various Abelian and finite and compact non-Abelian groups [17, 18], including also differentiation of different classes of signals and two-dimensional signals [20].

In this paper, we discuss evolution of the concept of the Gibbs dyadic differentiation into a variety of differential operators viewed uniformly as differential operators on groups.

As noticed above, Walsh dyadic analysis including the Gibbs dyadic differentiation appeared as an attempt to provide an answer to demands for a differential operator for piecewise constant functions used in solving certain mathematical and engineering problems.

1.1   Mathematical Problems

At the beginning of 20th century, a challenging problem was definition of an orthogonal system of functions such that series in terms of it will express the uniform convergence towards the given functions. This property was not possessed by any orthogonal set know up to that time, until Alfred Haar defined in 1909 in his PhD thesis defended in Göttingen a set of piecewise constant functions known now as Haar functions [12].

Another problem was related to the estimates of Lebesgue function and Lebesgue constant that quantify how much larger is the interpolation error compared to the smallest possible error, i.e., compared to the best polynomial approximation of the given function f. Hans Rademacher proposed certain estimates for the Lebesue function of a general orthonormal system and to prove that these are the best possible estimates introduced a set of piecewise constant functions, known as the Rademacher functions [16]. The manuscript was submitted on October 8, 1921. Rademacher functions are non complete, there exist other non-trivial functions orthogonal to all of them.

1.2   Engineering Problems

In late sixties, early seventies, there were developed many algorithms for different tasks in signal processing, including processing of audio, seismic, sonar, and radar signals. For example, concrete practical tasks to be solved include authentification, recognition, and voice tracking in the case of audio signals. Then, from the US side, another problem was spectral analysis of data from seismic sensing signals to observe the nuclear experiments performed in former USSR.

Many of these algorithms are based on Fourier analysis and, due to the limited computing power of digital devices available at that time, their application in practice was unfeasible. In particular, a fast algorithm to compute the Discrete Fourier transform (DFT) was highly needed as observed by members of the President Kenedy Scientific Advisory Committee around 1963. A member of this committee, Richard (Dick) Lawrence Garwin from Columbia University IBM Watson Research Center, New York City, served as a catalyst for the discovery and publication of the Cooley-Tukey Fast Fourier transform (FFT) called after the names of another member of the same committee, John Wilder Tukey, a Professor of Mathematics, Princeton University and Bell Labs, and, at that time, a young researcher James William Cooley at the IBM T. J. Watson Research Center [6].

2   Walsh Dyadic Analysis

Besides fast algorithms, as FFT for computing the DFT, an alternative approach to provide computationally efficient spectral analysis was based upon selecting different orthogonal systems of functions. In this setting, the Walsh functions, which can be viewed as a completing of the Rademacher system, thus, taking just two different values ±1 and related transform were a natural choice due to simplicity of computations and correspondence to binary encoded data to be processed [7,8].

Importance of the considered problems, as well as suitability of the Walsh functions as a subject providing foundations for highly acceptable solutions, is well recognized by pointing out that a series of workshops devoted exclusively to this subject was organized or supported by the US Navy Laboratory and some other important institutions and universities. For more details on the development and present state-of-art in Walsh dyadic analysis, we refer to [20,21].

3   Gibbs Dyadic Derivative

Since majority of intended applications was signal processing related, the lack of a differential operator that will permit differentiation of piecewise constant functions, like Rademacher, Walsh, and Haar functions, was apparent. Recall that the Newton-Leibniz derivative and many other related differentiation operators are defined to estimate the rate of change and the direction of change of a signal modeled by a function in a suitably selected function space. The task of finding an operator that will in the case of Walsh analysis play the role analogous to that of the Newton-Leibniz derivative in classical functional analysis was assigned by Ms. A. H. Gebbie, the Head of the National Physical laboratory, Middlesex, UK, to her associate James Edmund Gibbs. As witnessed by Mrs. Merion Gibbs, her husband Edmund provided the first completely formulated definition of a differentiator having the discrete Walsh functions as eigenfunctions in his personal diary on January 13, 1967.

Definition 1. *Consider the finite dyadic group* G_n *whose elements are binary* n-*tuples with modulo* 2 *addition* $\oplus$ *as the group operation that is in the context of switching theory also called logic EXOR. The set of elements of* G_n *can be isomorphically mapped to the set* $B_n = \{0, 1, \ldots, 2^n - 1\}$ *of non-negative integers smaller than* 2^n. *Denote by* L_n *the space of all bounded complex-valued functions* f *on* G_n *or* B_n.

For a function $f \in L_n$, *the Gibbs derivative* $f^{[1]}$ *is defined as*

$$f^{[1]} = -\frac{1}{2} \sum_{r=0}^{n-1} (f(x \oplus 2^r) - f(x)) 2^r, \quad x \in B_n.$$

The operator $D_n f = f^{[1]}$ *is called the Gibbs differentiator.*

Initially, Gibbs called this operator as the logical derivative, since the EXOR operation over binary representations of the argument x and 2^r is used.

This operator satisfies a property with resect to the Walsh transform analogous to the corresponding property of the Newton-Leibniz derivative and the Fourier transform. More precisely, in the classical analysis, the Fourier transform $F(w)$ of a function f and its derivative f' are related as $F'(w) = iwF(w)$. In the Walsh analysis, $S_{Df} = wS_f$, where S_f and S_{Df} are the Walsh transforms of f and its Gibbs dyadic derivative $f^{[1]}$.

From this property, it is possible to show another main property of the Gibbs derivative which is that the discrete Walsh functions are the 2^n solutions of the eigenvalue problem $D_n f = wf$. This property is used for certain generalizations and extensions of the concept of Gibbs dyadic derivative with the Butzer-Wagner derivative being the most widely known and probably most important and influential of them.

4   Butzer-Wagner Derivative

Butzer-Wagner derivative is an extension of Gibbs differentiation to functions defined on infinite dyadic group, which can be related to the unit interval $[0, 1)$ in the same way as G_n is related to B_n. In the case of dyadic rational numbers, we use their finite expressions.

Consider the orthonormal Walsh-Paley system $w_0(x), w_1(x), \ldots$, defined on the unit interval $[0, 1)$. The Walsh-Fourier transform pair is defined as

$$\hat{f}(k) = \int_0^1 f(u) w_k(u) du, \quad f(x) = \sum_{k=0}^{\infty} \hat{f}(k) w_k(x).$$

In spectral domain, the Butzer-Wagner derivative is defined as

$$D^{[k]} f(x) = \sum_{k=0}^{\infty} k^r \hat{f}(k) w_k(x).$$

This derivative has Walsh functions $w_k(x)$ as its eigenfunctions, i.e., $D^{[k]}w_k(x) = k^r w_k(x)$, for each $r \in N_0$ - the set of nonnegative integers. In the functional domain, the same definition reads as

$$D^{[1]}f(x) = \frac{1}{4}\sum_{j=1}^{\infty} 2^j (f(x) - f(x \oplus 2^{-j})),$$

where $x = \sum_{j=1}^{\infty} x_j 2^{-j}$, $y = \sum_{j=1}^{\infty} y_j 2^{-j}$, $x_j, y_j \in [0,1)$, $x \oplus y = \sum_{j=1}^{\infty} |x_j - y_j| 2^{-j}$.

Depending on the ordering of Walsh functions, distinguished are the Butzer-Wagner derivatives of the first and the second order corresponding to the Paley and Kaczmarz orderings, respectively. For a description of the development of the concept of Butzer-Wagner derivative, we refer to [5], where it is pointed out that this work have been done without knowledge of the related work by Franz Pichler reported in a technical report of the Dept. of Electrical Engineering, University of Maryland, USA [15].

In this report, thinking about the initial definition of the Gibbs derivative, in Sect. 4.1 entitled Dyadic differentiation Pichler wrote

Slight modifications must be made to obtain a theory of generalized differentiation defined for real valued functions of a continuous nonnegative real variable. Let f be such a function. To f we attach, if possible, a function $f^{[1]}$, given by

$$f^{[1]}(t) = \sum_{k=-\infty}^{\infty} (f(t) - f(t \oplus 2^{-k}))2^{k-2}.$$

If the function $f^{[1]}$ exists, we shall call it the first dyadic derivative of f.

Therefore, the term dyadic differentiation was coined by Pichler in 1970.

For the space $L^p(0,1)$ of functions of period 1 which are p-th power integrable $1 \le p < \infty$, in the norm $\|f\|_p = \left(\int_0^1 |f(x)|^p dx\right)^{1/p}$, the following definition is proposed.

Definition 2. *If for $f \in L_p(0,1)$, there exists $g \in L_p(0,1)$ such that*

$$\lim_{m\to\infty} \|\frac{1}{2}\sum_{j=0}^{m} 2^j (f(\cdot) - f(\cdot \oplus 2^{-j-1})) - g(\cdot)\| = 0,$$

then g is called the strong derivative of f denoted by $D^{[1]}f$.

It is explained by Butzer in [5] that *this definition differs from that given by Gibbs in [7] essentially in the sense that the factor 2^j is replaced by 2^{-j}, however, Gibbs definition is taken in the pointwise sense.*

Onneweer provided a definition of the dyadic derivative independent of the choice of enumeration of Walsh functions [14]. Further, by changing his original definition Onneweer [14] introduced a different derivative for which the additive characters of the local field are eigenfunctions, and the eigenvalues are equal to the norm of the character.

5   Dyadic Derivative on R_+

Besides the work by Pichler [15], the concept of dyadic differentiation was extended to functions defined on the positive part of the real axis by Butzer and Wagner [4].

Consider $R_+ = [0, \infty)$ and the space $L_1(R_+)$.

Definition 3. *If for $f \in L_1(R_+)$ there exists $g \in L^1(R_+)$, $m_1, m_2 \in P$, $P = N \cup \{0\}$, such that*

$$\lim_{m_1, m_2 \to \infty} \left\| \frac{1}{4} \sum_{j=-m_1}^{m_2} \frac{(f(\cdot) - f(\cdot \oplus 2^{-j}))}{2^{-j}} - g(\cdot) \right\| = 0,$$

then g is called the strong dyadic derivative of f and is denoted by $D^{[1]}f$.

If for $x \in R_+$, $\frac{1}{4} \sum_{j=-\infty}^{\infty} 2^j |f(x) - f(x \oplus 2^{-j})| = c$, then c is called the pointwise dyadic derivative of f at x and is denoted by $f^{[1]}$.

Designation *strong derivative* refers to the fact that the sum in question converges in the norm. If the sum converges in the pointwise sense, we speak on the *pointwise dyadic derivative*. For the dyadic group the derivative of Walsh function w_k, for $2^n \leq k < 2^{n+1}$ is $2^n w_k$.

6   Dyadic Derivative on Vilenkin Groups

Vilenkin group can be viewed as a generalization of the dyadic group, since consists of sequences $x = (x_0, x_1, \ldots)$, $x_k \in Z_k$, $Z_k = \{0, 1, \ldots, m_k\}$ - the finite cyclic group of order m_k. Thus, the Vilenkin group is $G = \times_{k=0}^{\infty} Z_k$.

Jeno Pál and Peter Simon generalized the concept of dyadic differentiation to Vilenkin groups. This definition can be expressed by using the following notations.

Consider the sequence $m = (m_k, k \in N)$ of positive integers such that $m_k \geq 2$. Define $M_0 = 1$, $M_{k+1} = m_k M_k$, $k \in N$. Then, $n = \sum_{k=0}^{\infty} n_k M_k$, for $0 \leq n_k < m_k$, $n_k \in N$. Define the functions $\phi_k^s(x) = \exp(2\pi i s x / m_k)$, $s \in \{0, 1, \ldots, m - 1\}$, $x \in Z_m$, $i^2 = -1$, which can be viewed as counterpart of Walsh functions for the dyadic group. Define $e_j = (0, 0, \ldots, 0, 1, 0, \ldots)$, i.e., all coordinates are 0 except the coordinate j, and $le_j = (0, 0, \ldots, 0, l, 0, \ldots)$. The generalized dyadic derivative is defined as

$$d_n f(x) = \sum_{j=0}^{n-1} M_j \sum_{k=0}^{m_j-1} \frac{k}{m_j} \sum_{l=0}^{m_j-1} \overline{\phi_j^k(l)} f(x + le_j).$$

Definition of dyadic differentiation is extended to locally compact Vilenkin group by Onneweer in 1979 [13,14]. For thus defined differential operator, the characters of the locally compact Vilenkin groups are the eigenfunctions.

7   Gibbs Derivatives on Abelian Groups

Already from the first definition of his logic derivative, Dr. Gibbs considered it as a differential operator on groups. By following this approach, the generalization of the concept to functions defined on Abelian groups was given by him and his associate B. Ireland in [9]. The way in which the extension is provided is probably the best expressed by quoting the following sentence from [9].

Among the basic properties of the differentiator so defined on any finite Abelian group is that each of its eigenmanifolds is spanned by a member of the associated system of the discrete Walsh-Lévy functions.

In this general setting, the Gibbs differentiator for complex-valued functions on each locally compact Abelian group is viewed as a linear operator having the characters of the group as its eigenfunctions. Gibbs and Ireland in 1971 pointed out that the transition from finite Abelian groups to compact infinite Abelian groups goes through viewing the compactness as a generalization of finiteness [9].

In a brief summary, when discussing this group-theoretic approach, the Gibbs differential operators were discussed on cyclic group of order 2 and its direct powers in [7], infinite dyadic group [1–3,15], cyclic groups of prime power orders and finite Abelian groups viewed as direct products of such cyclic groups [10,11], and Vilenkin groups as briefly mentioned above.

8   Gibbs Derivatives on Non-Abelian Groups

A generalization of the Gibbs differentiation to finite non-Abelian groups is done by replacing the role of group characters in the case of Abelian groups by the unitary irreducible representations of non-Abelian groups. For the case of finite non-Abelian groups, the definition of these operators can be briefly presented in the following way.

It is assumed that the group G of order g can be represented as the direct product of n not necessarily Abelian groups G_i of smaller orders g_i. Therefore,

$$G = \times_{i=1}^{n} G_i, \quad g = \prod_{i=1}^{n} g_i, \quad g_1 \leq g_2 \leq \cdots \leq g_n.$$

The dual object Γ for G consists of the unitary irreducible representations defined as group homomorphism $R : G \to GL(n, P)$, where $GL(n, P)$ is the General Linear Group whose support set consists of $(n \times n)$ matrices over a field P that can be either the field of complex numbers, real numbers, or a finite (Galois) field. Therefore,

$$\Gamma = \bigotimes_{i=1}^{n} \Gamma_i, \quad |\Gamma| = K = \prod_{i=1}^{n} K_i, \quad |\Gamma| < |G|.$$

With this notation, the generalized Fourier transform for functions on G into the field P, i.e., $f \in P(G)$, is defined by the Peter-Weyl theorem as

$$\mathbf{S}_f(w) = r_w g^{-1} \sum_{u=0}^{g-1} f(u) \mathbf{R}_w(u^{-1}), \quad f(x) = \sum_{w=0}^{K-1} Tr(\mathbf{S}_f(w) \mathbf{R}_w(x)),$$

where Tr denotes the trace of a matrix, and $\mathbf{R}_w(x)$ are matrix-valued elements of the unitary irreducible representations R_w of order r_w, with computations in P.

By the analogy to the definitions of the Gibbs derivative on other groups, the Gibbs derivative on finite non-Abelian groups in spectral domain is defined as

$$(Df)(x) = \sum_{w=0}^{K-1} wTr(\mathbf{S}_f(w)\mathbf{R}_w(x)).$$

In other words, this differential operator has unitary irreducible representations as eigenfunctions, with orders of representations as the corresponding eigenvalues.

References

1. Butzer, P.L., Wagner, H.J.: Approximation by Walsh polynomials and the concept of a derivative. In: Proceedings of Symposium on Applications of Walsh Functions, pp. 388–392, Washington D.C. (1972)
2. Butzer, P.L., Wagner, H.J.: On a Gibbs-type derivative in Walsh-Fourier analysis with applications. In: Proceedings of National Electronic Conference, vol. 27, pp. 393–398 (1972)
3. Butzer, P.L., Wagner, H.J.: Walsh-Fourier series and the concept of a derivative. Appl. Anal. **3**, 29–46 (1973)
4. Butzer, P.L., Wagner, H.J.: A calculus for Walsh functions defined on R+. In: Proceedings of Symposium on Applications of Walsh Functions, pp. 75–81, Washington D.C. (1973)
5. Butzer, P.L., Wagner, H.J.: Early contributions from the aachen school to dyadic Walsh analysis with applications to Dyadic PDEs and approximation theory. In: Stankovic, R., et al. (eds.) Dyadic Walsh Analysis from 1924 Onwards Walsh-Gibbs-Butzer Dyadic Differentiation in Science Volume 1 Foundations. ASMES, vol. 12, pp. 161–208. Atlantis Press, Paris (2015). https://doi.org/10.2991/978-94-6239-160-4_4
6. Cooley, J.W., Tukey, J.W.: On the origin and publication of the FFT paper. Curr. Contents ISI Phy. Chem. Earth Sci. **33**(51–52), 8–9 (1993)
7. Gibbs, J.E.: Walsh spectrometry, a form of spectral analysis well suited to binary digital computation, p. 24, National Physical Laboratory, Teddington, Middx, UK (1967)
8. Gibbs, J.E., Gebbie, H.A.: Application of Walsh functions to transform spectroscopy. Nature **224**(5223), 1012–1013 (1969). publication date 12/1969
9. Gibbs, J.E., Ireland, B.: Some generalizations of the logical derivative. DES report no. 8, National Physical Laboratory, p. 22+ii, August 1971
10. Gibbs, J.E., Ireland, B.: Walsh functions and differentiation. In: Proceedings of International Conference on Applications of Walsh Functions and Sequency Theory, pp. 147–176 (1974)
11. Gibbs, J.E., Simpson, J.: Differentiation on finite Abelian groups, p. 34, National Physical Laboratory, Teddington, Middx, UK (1974)
12. Haar, A.: Zur theorie der orthogonalen Funktionsysteme. Math. Annal. **69**, 331–371 (1910)

13. Onneweer, C.W.: Fractional differentiation on the group of integers of the p-adic or p-series field. Anal. Math. **3**, 119–130 (1977)
14. Onneweer, C.W.: On the definition of dyadic differentiation. Appl. Anal. **9**, 267–278 (1979)
15. Pichler, F.R.: Some aspects of a theory of correlation with respect to Walsh harmonic analysis, Technical report R-70-11. University of Maryland Technology Research (1970)
16. Rademacher, H.: Einige Sätze von allgemeinen Orthogonalfunktionen. Math. Ann. **87**, 122–138 (1922)
17. Stanković, R.S.: A note on differential operators on finite non-Abelian groups. Appl. Anal. **21**(1–2), 31–41 (1986)
18. Stanković, R.S.: Gibbs derivatives on finite non-Abelian groups. In: Butzer, P.L., Stanković, R.S. (eds.), Theory and Applications of Gibbs Derivatives: Proceedings of the First International Workshop on Gibbs Derivatives, 26–28 September 1989, Kupari-Dubrovnik, Yugoslavia, Matematički Institut, Beograd, pp. 269–297 (1990)
19. Stanković, R.S., Astola, J.T.: Gibbs Derivatives - the First Forty Years. TICSP Series #39. Tampere International Center for Signal Processing, Tampere, Finland (2008). ISBN 978-952- 15-1973-4, ISSN 1456-2774
20. Stanković, R.S., Butzer, P.L., Schipp, F., Wade, W.R., Su, W., Endow, Y., Fridli, S., Golubov, B.I., Pichler, F., Onneweer, K.C.W.: Dyadic Walsh Analysis from 1924 Onwards Walsh-Gibbs-Butzer Dyadic Differentiation in Science Volume 1, Foundations. A Monograph Based on Articles of the Founding Authors, Reproduced in Full. Atlantis Studies in Mathematics for Engineering and Science, Vol. 12. Atlantis Press, Springer (2015)
21. Stanković, R.S., Butzer, P.L., Schipp, F., Wade, W.R., Su, W., Endow, Y., Fridli, S., Golubov, B.I., Pichler, F., Onneweer, K.C.W.: Dyadic Walsh Analysis from 1924 Onwards, Walsh-Gibbs-Butzer Dyadic Differentiation in Science, Volume 2, Extensions and Generalizations. A Monograph Based on Articles of the Founding Authors, Reproduced in Full, Atlantis Studies in Mathematics for Engineering and Science. Atlantis Press (2015)

A Three-Level Hierarchy of Models for Lattices of Boolean Functions

Bernd Steinbach[1(✉)] and Christian Posthoff[2]

[1] Institute of Computer Science, Freiberg University of Mining and Technology, Bernhard-von-Cotta-Str. 2, 09596 Freiberg, Germany
`steinb@informatik.tu-freiberg.de`
[2] The University of the West Indies, St. Augustine Campus, Saint Augustine, Trinidad and Tobago
`christian@posthoff.de`

Abstract. The utilization of lattices of Boolean functions for the synthesis of circuits combines the benefits of more freedom for optimization with limited calculations on mark functions. We extend the known two-level hierarchy of lattices of Boolean functions by a third level. This new level increases the possibilities of synthesis approaches (An extended abstract of this paper was published in [4].).

Keywords: Lattice · Boolean function · Hierarchy · Mark function
Derivative operation · Independence function · Independence matrix
Rank

1 Introduction

A Boolean function is a mapping of the Boolean space $\mathbb{B}^n$ consisting of 2^n patterns $\mathbf{x}$ to the Boolean space $\mathbb{B}^1 = \mathbb{B} = \{0, 1\}$:

$$f: \quad \mathbb{B}^n \to \mathbb{B}.$$

Hence, a Boolean function specifies the output values for all 2^n input combinations. Very often not all combinations are needed as, e.g., in the case of the binary encoding of a decimal digit. The values of a subfunction within a circuit can also be arbitrarily chosen in many cases, e.g., for all patterns where one input of an AND-gate is equal to 0 the output of this AND-gate remains equal to 0 independent of the function values on the other inputs of this gate. Functions for which not all output values are specified are called *incompletely specified functions* (ISF). ISFs play a central role in the optimization of logic circuits as they represent the degrees of freedom for the assignment of a circuit structure [1].

From another point of view an ISF represents a set of Boolean functions from which an arbitrary one can be selected and realized in a circuit. An ISF with $|f_\varphi|$ don't cares specifies a set of $2^{|f_\varphi|}$ completely specified Boolean functions. This set of functions satisfies the rules of a Boolean lattice: commutativity, associativity, idempotence, absorption, distributivity, neutral elements, complement, and the laws of De Morgan.

R. Moreno-Díaz et al. (Eds.): EUROCAST 2017, Part II, LNCS 10672, pp. 238–245, 2018.
https://doi.org/10.1007/978-3-319-74727-9_28

2 Three Mark Functions of a First-Level Lattice

A benefit of a lattice is that an exponential number of Boolean functions can be described by two of the three mark functions $f_q(\mathbf{x})$, $f_r(\mathbf{x})$, and $f_\varphi(\mathbf{x})$. The set of all 2^n input patterns $\mathbf{x} = (x_1, x_2, \ldots, x_n)$ of an incompletely specified function can be divided into three disjoint sets:

- $\mathbf{x} \in$ don't-care-set $\Leftrightarrow f_\varphi(x_1, \ldots, x_n) = 1$
  $\Leftrightarrow$ it is allowed to choose the function value of $f(\mathbf{x})$
  without any restrictions,
- $\mathbf{x} \in$ ON-set $\Leftrightarrow f_q(x_1, \ldots, x_n) = 1$
  $\Leftrightarrow (f_\varphi(x_1, \ldots, x_n) = 0) \wedge (f(x_1, \ldots, x_n) = 1)$,
- $\mathbf{x} \in$ OFF-set $\Leftrightarrow f_r(x_1, \ldots, x_n) = 1$
  $\Leftrightarrow (f_\varphi(x_1, \ldots, x_n) = 0) \wedge (f(x_1, \ldots, x_n) = 0)$.

Example 1 (A First-Level Lattice). Assume that the variables x_i, $i = 0, \ldots, 3$, are used to encode a decimal digit d by $d = \sum_{i=0}^{3} 2^i * x_i$. The patterns $\mathbf{x}$ for the values d with $d = 10, \ldots, 15$ cannot appear. We use the symbol Φ in a Karnaugh-map to indicate that the associated function value can be arbitrarily chosen. The remaining 10 function values are specified in the left Karnaugh-map of Fig. 1. This Karnaugh-map can be interpreted as an incompletely specified function or, due to the six don't-cares, as a first-level lattice of 2^6 completely specified Boolean functions.

$x_2\,x_3$	$L^1 \langle f_q(\mathbf{x}), f_r(\mathbf{x}) \rangle$				$x_2\,x_3$	$f_\varphi(\mathbf{x})$				$x_2\,x_3$	$f_q(\mathbf{x})$				$x_2\,x_3$	$f_r(\mathbf{x})$			
0 0	1	1	1	0	0 0	0	0	0	0	0 0	1	1	1	0	0 0	0	0	0	1
0 1	1	Φ	Φ	1	0 1	0	1	1	0	0 1	1	0	0	1	0 1	0	0	0	0
1 1	Φ	Φ	Φ	Φ	1 1	1	1	1	1	1 1	0	0	0	0	1 1	0	0	0	0
1 0	0	1	1	1	1 0	0	0	0	0	1 0	0	1	1	1	1 0	1	0	0	0
	0	1	1	0 x_1		0	1	1	0 x_1		0	1	1	0 x_1		0	1	1	0 x_1
	0	0	1	1 x_0		0	0	1	1 x_0		0	0	1	1 x_0		0	0	1	1 x_0

Fig. 1. Karnaugh-maps of a first-level lattice and the associated mark functions.

The three mark functions $f_q(\mathbf{x})$, $f_r(\mathbf{x})$ and $f_\varphi(\mathbf{x})$ cover the whole Boolean space and are also mutually disjoint, so that one of these mark functions can be calculated based on the other two mark functions.

We prefer the mark functions $f_q(\mathbf{x})$ and $f_r(\mathbf{x})$ to specify the functions $f(\mathbf{x})$ belonging to a lattice; these functions must be greater than or equal to the ON-set function $f_q(\mathbf{x})$ and smaller than or equal to the complement of the OFF-set function $f_r(\mathbf{x})$:

$$f_q(\mathbf{x}) \leq f(\mathbf{x}) \leq \overline{f_r(\mathbf{x})}.$$

This inequality can be split into two inequalities and transformed into the single equation:

$$\overline{f(\mathbf{x})} \wedge f_q(\mathbf{x}) \vee f(\mathbf{x}) \wedge f_r(\mathbf{x}) = 0. \tag{1}$$

Hence, Equation (1) specifies by means of the mark functions $f_q(\mathbf{x})$ and $f_r(\mathbf{x})$ all functions $f(\mathbf{x})$ belonging to a lattice that is a first-level lattice in our hierarchy for which we use the short notation $L^1 \langle f_q(\mathbf{x}), f_r(\mathbf{x}) \rangle$.

The benefit of a first-level lattice is that not all functions of the lattice must be evaluated regarding a certain property, but it can be verified by means of the mark functions $f_q(\mathbf{x})$ and $f_r(\mathbf{x})$ whether the lattice contains at least one function that satisfies a wanted property. For instance, it is known from [2,5] that a lattice $L^1 \langle f_q(\mathbf{x}), f_r(\mathbf{x}) \rangle$ contains at least one function for which an OR-bi-decomposition with regard to $\mathbf{x}_a$ and $\mathbf{x}_b$ exists if it holds that

$$f_q(\mathbf{x}_a, \mathbf{x}_b, \mathbf{x}_c) \wedge \max_{\mathbf{x}_a}^{|\mathbf{x}_a|} f_r(\mathbf{x}_a, \mathbf{x}_b, \mathbf{x}_c) \wedge \max_{\mathbf{x}_b}^{|\mathbf{x}_b|} f_r(\mathbf{x}_a, \mathbf{x}_b, \mathbf{x}_c) = 0.$$

3   Second-Level Lattices of Boolean Functions

Derivative operations of the Boolean Differential Calculus [6,7] are very useful for the synthesis of circuits or the evaluation of properties of a given function [2,5]. There are three groups of derivative operations:

- vectorial derivative operations: $\frac{\partial f(\mathbf{x}_0, \mathbf{x}_1)}{\partial \mathbf{x}_0}, \min_{\mathbf{x}_0} f(\mathbf{x}_0, \mathbf{x}_1), \max_{\mathbf{x}_0} f(\mathbf{x}_0, \mathbf{x}_1),$
- single derivative operations: $\frac{\partial f(x_i, \mathbf{x}_1)}{\partial x_i}, \min_{x_i} f(x_i, \mathbf{x}_1), \max_{x_i} f(x_i, \mathbf{x}_1),$ and
- m-fold derivative operations: $\frac{\partial^m f(\mathbf{x}_0, \mathbf{x}_1)}{\partial \mathbf{x}_{01} \ldots \partial \mathbf{x}_{0m}}, \min_{\mathbf{x}_0}^m f(\mathbf{x}_0, \mathbf{x}_1),$
$\max_{\mathbf{x}_0}^m f(\mathbf{x}_0, \mathbf{x}_1), \Delta_{\mathbf{x}_0} f(\mathbf{x}_0, \mathbf{x}_1).$

Each derivative operation of a Boolean function results again in a Boolean function. This leads to the question how the set of functions can be described that has been calculated by one selected derivative operation for all functions of a first-level lattice. It has been shown in [3,6,7] that all three single derivative operations as well as all four m-fold derivative operations applied to all functions of a first-level lattice result again in a first-level lattice; formulas to calculate the mark functions of these new lattices are also given in [3,6,7]. However, it is not possible to express the results of any vectorial derivative operation of a first-level lattice again as a first-level lattice. The following example demonstrates the calculation of the vectorial derivative for all functions of a first-level lattice.

Example 2 (A Second-Level Lattice). A given fist-level lattice, shown in the top right Karnaugh-map of Fig. 2, has two don't cars and consequently describes four functions. The vectorial derivatives of these four functions are shown in the left four Karnaugh-maps of Fig. 2. It can be seen that these four functions specify a lattice: both the conjunction and the disjunction of any pair of these function result in one of these functions as emphasized by the thick blue arrows.

The Karnaugh-map on the bottom right of Fig. 2 indicates by the symbol Φ different values of the four vectorial derivatives. It seems that there is a contradiction between the four don't cares describing $2^4 = 16$ functions and the four functions used as source. These four functions satisfy the condition shown on the bottom right of Fig. 2. Hence, second-level lattices need in addition to the mark functions $f_q(\mathbf{x})$ and $f_r(\mathbf{x})$ the information which vectorial derivatives are equal to zero.

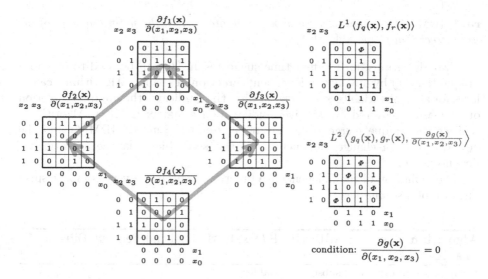

Fig. 2. Four functions of a second-level lattice. (Color figure online)

The functions of a second-level lattice can be independent of several directions of change; hence, all functions $f(\mathbf{x})$ of a second-level lattice satisfy

$$\overline{f(\mathbf{x})} \wedge f_q(\mathbf{x}) \vee f(\mathbf{x}) \wedge f_r(\mathbf{x}) \vee \bigvee_{i=1}^{k} \frac{\partial f(\mathbf{x})}{\partial \mathbf{x}_{0i}} = 0. \tag{2}$$

There are $2^n - 1$ different vectorial derivatives for Boolean functions of n variables, but only n of them are independent of each other. In order to specify the second-level lattices in a unique manner we define an independence matrix and the associated **rank**:

Definition 1 (Independence Matrix). *The **independence matrix** IDM(f) of a Boolean function $f(x_1, x_2, \ldots, x_n)$ is a Boolean matrix of n rows and n columns. The columns of the independence matrix are associated with the n variables of the Boolean space in the fixed order $(x_1, x_2, \ldots, x_n)$. The independence matrix has the shape of an echelon; all elements below the main diagonal are equal to 0. Values 1 of a row of the independence matrix indicate a set of variables for which the vectorial derivative of the function $f(x_1, x_2, \ldots, x_n)$ is equal to 0. The following rules ensure the uniqueness of the independence matrix:*

1. *Values 1 can only occur to the right of a value 1 in the main diagonal of the independence matrix.*
2. *All values above a value 1 in the main diagonal of the independence matrix are equal to 0.*

Definition 2 (Rank). *The **rank** of an independence matrix IDM(f) describes the number of independent directions of change of the Boolean function $f(\mathbf{x})$. The*

rank(IDM(f)) *is equal to the number of elements* 1 *in the main diagonal of the unique echelon shape of* IDM(f).

Two algorithms, for the first time suggested in [3], can be used to extend a unique IDM(f) by any new independent direction of change. Algorithm 1 calculates for a set of variables $\mathbf{x}_0$ the vector $\mathbf{s}_{min}$ that indicates the minimal direction of change not covered by the given independence matrix IDM(f). Algorithm 2 realizes the unique merge of the given independence matrix IDM(f) and the new direction of change all functions of the second-level lattice are also not depending on.

The definition of the independence function facilitates a very compact specification of a second-level lattice:

Algorithm 1. $\mathbf{s}_{min}$ = MIDC(IDM(f), $\mathbf{x}_0$): Minimal Independent Direction of Change

Input : $\mathbf{x}_0 \in \mathbf{x}$: evaluated subset of variables,
Input : IDM(f): unique independence matrix of n rows and n columns of $f(\mathbf{x})$
Output : $\mathbf{s}_{min}$: minimal direction of change

1: $j \leftarrow 1$
2: $\mathbf{s}_{min} \leftarrow BV(\mathbf{x}_0)$
3: **while** $j \leq n$ **do**
4:     **if** $(\mathbf{s}_{min}[j] = 1) \wedge (\text{IDM}(f)[j,j] = 1)$ **then**
5:         $\mathbf{s}_{min} \leftarrow \mathbf{s}_{min} \oplus \text{IDM}(f)[j]$
6:     **end if**
7:     $j \leftarrow j + 1$
8: **end while**

Algorithm 2. IDM(g) = UM(IDM(f), $\mathbf{x}_0$): Unique Merge

Input : $\mathbf{x}_0 \in \mathbf{x}$: subset of variables that satisfy $\frac{\partial f(\mathbf{x})}{\partial \mathbf{x}_0} = 0$, to merge with IDM($f$),
Input : IDM(f): unique independence matrix of n rows and n columns of $f(\mathbf{x})$
Output : IDM(g): unique independence matrix of the same size of $g(\mathbf{x})$ with $\frac{\partial g(\mathbf{x})}{\partial \mathbf{x}_0} = 0$

1: IDM(g) $\leftarrow$ IDM(f)
2: $\mathbf{s}_{min}$ = MIDC(IDM(f), $\mathbf{x}_0$)
3: **if** $\mathbf{s}_{min} > 0$ **then**
4:     $j \leftarrow$ IndexOfMostSignificantBit($\mathbf{s}_{min}$)
5:     $i \leftarrow 1$
6:     **while** $i < j$ **do**
7:         **if** IDM(g)$[i,j] = 1$ **then**
8:             IDM(g)$[i] \leftarrow$ IDM(g)$[i] \oplus \mathbf{s}_{min}$
9:         **end if**
10:         $i \leftarrow i + 1$
11:     **end while**
12:     IDM(g)$[j] \leftarrow \mathbf{s}_{min}$
13: **end if**

$$\begin{array}{c|c|c|c|c|} {}_i \diagdown {}^j & 1 & 2 & 3 & 4 \\ \hline 1 & 1 & 0 & 0 & 1 \\ \hline 2 & 0 & 1 & 0 & 0 \\ \hline 3 & 0 & 0 & 1 & 1 \\ \hline 4 & 0 & 0 & 0 & 0 \\ \hline \end{array}$$

$\text{IDM}(f)$

$$f^{id} = \frac{\partial f(\mathbf{x})}{\partial(x_1, x_4)} \vee \frac{\partial f(\mathbf{x})}{\partial x_2} \vee \frac{\partial f(\mathbf{x})}{\partial(x_3, x_4)}$$

$$\mathbf{rank}(\text{IDM}(f)) = 3$$

Fig. 3. Example of an independence matrix and the associated independence function.

Definition 3 (Independence Function). *The **independence function** $f^{id}(\mathbf{x})$ of a Boolean function corresponds to the independence matrix $\text{IDM}(f)$ such that*

$$f^{id}(\mathbf{x}) = \bigvee_{i=1}^{n} \frac{\partial f(\mathbf{x})}{\partial \mathbf{x}_{0i}}$$

where $\frac{\partial f(\mathbf{x})}{\partial \mathbf{x}_{0i}} = 0$ if all elements of the row i in $\text{IDM}(f)$ are equal to 0, and

$$x_j \in \mathbf{x}_{0i} \ \textit{if } \text{IDM}(f)[i, j] = 1.$$

Figure 3 shows an example of an independence function as well as the associated independence matrix and their rank.

A function $f(\mathbf{x})$ belongs to the lattice $L^2 \langle f_q(\mathbf{x}), f_r(\mathbf{x}), f^{id_2}(\mathbf{x}) \rangle$ if it satisfies (2) which can be described shorter using the independence function $f^{id_2}(\mathbf{x})$:

$$\overline{f(\mathbf{x})} \wedge f_q(\mathbf{x}) \vee f(\mathbf{x}) \wedge f_r(\mathbf{x}) \vee f^{id_2}(\mathbf{x}) = 0. \tag{3}$$

4 Third Level in the Hierarchy of Lattices

A lattice L^2, as described in (3), can be the result of a vectorial derivative of a lattice L^1. An implicit condition of a second-level lattice is that the mark functions $f_q(\mathbf{x})$ and $f_r(\mathbf{x})$ satisfy:

$$\bigvee_{i=1}^{k} \frac{\partial f_q(\mathbf{x})}{\partial \mathbf{x}_{0i}} \vee \bigvee_{i=1}^{k} \frac{\partial f_r(\mathbf{x})}{\partial \mathbf{x}_{0i}} = 0. \tag{4}$$

This condition is relaxed for third-level lattices of Boolean functions as shown in Example 3.

Example 3 (A Third-Level Lattice). It can easily be verified that the four functions in the left part of Fig. 4 establish a lattice of Boolean functions. Thin blue arrows connect pairs of identical function values belonging to the don't-care function $f_\varphi(\mathbf{x})$ for the direction of change (x_1, x_2, x_3). This lattice does not satisfy the rules of a second-level lattice due to the different values for the pairs of function values connected by a red dashed arrow with the same direction of change. Hence, it is an example of a third-level lattice.

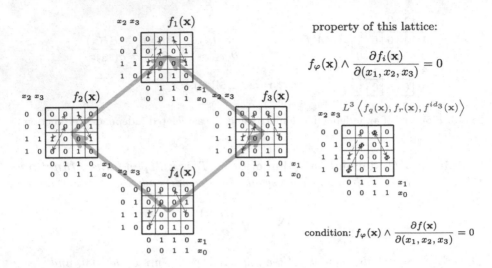

Fig. 4. Four functions of a third-level lattice. (Color figure online)

Lattices L^2 have the benefit that they describe only functions that are independent of certain directions of change. As shown in Example 3, there are lattices which do not satisfy this strong requirement for the whole Boolean space, but for the region of the don't-care-set. Lattices $L^3 \langle f_q(\mathbf{x}), f_r(\mathbf{x}), f^{id_2}(\mathbf{x}), f^{id_3}(\mathbf{x}) \rangle$ of this new third-level in the hierarchy of lattices are defined by:

$$\overline{f(\mathbf{x})} \wedge f_q(\mathbf{x}) \vee f(\mathbf{x}) \wedge f_r(\mathbf{x}) \vee \bigvee_{i=1}^{k} \frac{\partial f(\mathbf{x})}{\partial \mathbf{x}_{0i}} \vee f_\varphi(\mathbf{x}) \wedge \left(\bigvee_{j=1}^{l} \frac{\partial f(\mathbf{x})}{\partial \mathbf{x}_{0j}} \right) = 0, \quad (5)$$

where

$$f^{id_3}(\mathbf{x}) = \bigvee_{j=1}^{l} \frac{\partial f(\mathbf{x})}{\partial \mathbf{x}_{0j}}.$$

5   Hierarchy of Lattices of Boolean Functions

The hierarchy of lattices of Boolean functions, shown in Fig. 5, is determined by two influencing factors:

1. the independence functions, where a larger rank restricts the lattice to simpler functions, and
2. the mark functions $f_q(\mathbf{x})$ and $f_r(\mathbf{x})$, where a smaller number of function values 1 of $(f_q(\mathbf{x}) \vee f_r(\mathbf{x}))$ increases the number of functions belonging to the lattice.

Due to these influencing factors, the number function values 1 of the don't-care function is

$$|f_\varphi| = m * 2^{\mathrm{rank}(\mathrm{IDM}(f_\varphi))} \quad (6)$$

and the number of functions belonging to a lattice results in 2^m.

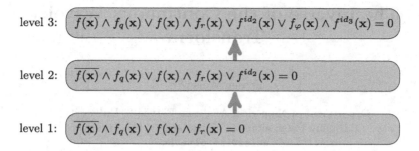

Fig. 5. Hierarchy of lattices of Boolean functions.

References

1. Brayton, R.K., Hachtel, G.D., McMullen, C.T., Sangiovanni-Vincentelli, A.: Logic Minimization Algorithms for VLSI Synthesis. Kluwer Academic Publishers, Dordrecht (1984)
2. Posthoff, C., Steinbach, B.: Logic Functions and Equations - Binary Models for Computer Science. Springer, Dordrecht (2004)
3. Steinbach, B.: Generalized lattices of Boolean functions utilized for derivative operations. In: Materiały konferencyjne KNWS 2013, Łagów, Poland, https://doi.org/10.13140/2.1.1874.3680, pp. 1–17 (2013)
4. Steinbach, B., Posthoff, C.: A hierarchy of models for lattices of Boolean functions. In: Quesada-Arencibia, A., Rodriguez, J.C., Moreno-Diaz Jr., R., Moreno-Diaz, R. (eds.) Computer Aided System Theory, Extended Abstracts. 16th International Conference on Computer Aided System Theory, EUROCAST 2017, pp. 213–214. IUCTC Universidad de Las Palmas, Grand Canaries (2017). ISBN 978-84-617-8087-7
5. Steinbach, B., Posthoff, C.: Boolean differential calculus - theory and applications. J. Comput. Theor. Nanosci. **7**(6), 933–981 (2010)
6. Steinbach, B., Posthoff, C.: Derivative operations for lattices of Boolean functions. In: Proceedings Reed-Muller Workshop 2013, Toyama, Japan, pp. 110–119 (2013). https://doi.org/10.13140/2.1.2398.6568
7. Steinbach, B., Posthoff, Ch.: Boolean Differential Calculus. Morgan & Claypool Publishers, San Rafael (2017). ISBN 9781627059220 (paperback). ISBN 9781627056175 (ebook). https://doi.org/10.2200/S00766ED1V01Y201704DCS052

The Inverse of the Continuous Wavelet Transform

Ferenc Weisz[✉]

Department of Numerical Analysis, Eötvös L. University,
Pázmány Péter sétány 1/C, Budapest 1117, Hungary
weisz@inf.elte.hu

Abstract. In this paper we summarize some recent results about the convergence of the inverse of the continuous wavelet transform.

1 Introduction

Wavelet analysis has established itself in the last 20–30 years as a fertile branch of analysis, and has a lot of links to real world applications, e.g. to image and signal analysis and to storage of fingerprints. There are many interesting books in this area, e.g. Daubechies [5], Hernandez and Weiss [8], Walnut [15], moreover several papers have appeared [1,9–11,20–22]. The continuous wavelet transform of f with respect to a wavelet g is defined by

$$W_g f(x,s) = \langle f, T_x D_s g \rangle \qquad (x \in \mathbb{R}, s \in \mathbb{R}, s \neq 0),$$

where D_s is the dilation operator and T_x the translation operator. Under some conditions on g and γ the inversion formula holds for all $f \in L_2(\mathbb{R})$:

$$\int_{\mathbb{R}} \int_{\mathbb{R}} W_g f(x,s) T_x D_s \gamma \, \frac{dxds}{s^2} = C_{g,\gamma} f,$$

where the equality is understood in a vector-valued weak sense (see Daubechies [5] and Gröchenig [7]). The convergence of this integral is an important problem. In fact, there are several results on the convergence of the inverse continuous or discrete wavelet transform (see e.g. [1,9–11,16,18,20–22]). In this paper we summarize the results about the convergence of

$$\lim_{S \to 0, T \to \infty} \int_{S \leq |s| \leq T} \int_{\mathbb{R}} W_g f(x,s) T_x D_s \gamma \, \frac{dxds}{s^2}$$

including the case when $T = \infty$.

F. Weisz—This research was supported by the Hungarian Scientific Research Funds (OTKA) No. K115804.

R. Moreno-Díaz et al. (Eds.): EUROCAST 2017, Part II, LNCS 10672, pp. 246–253, 2018.
https://doi.org/10.1007/978-3-319-74727-9_29

2 Wiener Amalgam Spaces

The space $L_p(\mathbb{R})$ is equipped with the norm

$$\|f\|_p := \begin{cases} \left(\int_{\mathbb{R}} |f|^p \, d\lambda \right)^{1/p}, & 0 < p < \infty; \\ \sup_{\mathbb{R}} |f|, & p = \infty. \end{cases}$$

These spaces can be generalized as follows. A measurable function f belongs to the *Wiener amalgam space* $W(L_p, \ell_q)(\mathbb{R})$ $(1 \leq p, q \leq \infty)$ if

$$\|f\|_{W(L_p,\ell_q)} := \left(\sum_{k \in \mathbb{Z}} \|f(\cdot + k)\|_{L_p[0,1)}^q \right)^{1/q} < \infty,$$

with the obvious modification for $q = \infty$. It is easy to see that $W(L_p, \ell_p)(\mathbb{R}) = L_p(\mathbb{R})$ and

$$W(L_\infty, \ell_1)(\mathbb{R}) \subset L_p(\mathbb{R}) \subset W(L_1, \ell_\infty)(\mathbb{R}) \qquad (1 \leq p \leq \infty).$$

3 Continuous Wavelet Transform

The *translation* and *dilation* of a function f are defined, respectively, by

$$T_x f(t) := f(t - x), \qquad D_s f(t) := |s|^{-1/2} f(s^{-1} t),$$

where $t, x \in \mathbb{R}, s \in \mathbb{R}, s \neq 0$. The *continuous wavelet transform* of f with respect to a *wavelet* g is defined by

$$W_g f(x, s) := |s|^{-1/2} \int_{\mathbb{R}} f(t) \overline{g(s^{-1}(t - x))} \, dt = \langle f, T_x D_s g \rangle,$$

$(x \in \mathbb{R}, s \in \mathbb{R}, s \neq 0)$ when the integral does exist. Plancherel's theorem is well-known for Fourier transforms: if $f, g \in L_2(\mathbb{R})$, then

$$\|f\|_2 = \left\| \widehat{f} \right\|_2 \quad \text{and} \quad \langle f, g \rangle = \left\langle \widehat{f}, \widehat{g} \right\rangle,$$

where the *Fourier transform* of $f \in L_1(\mathbb{R})$ is given by

$$\widehat{f}(x) = \int_{\mathbb{R}} f(u) e^{-2\pi \imath x u} \, du \qquad (x \in \mathbb{R}, \imath = \sqrt{-1}).$$

Now we present the analogues of these results for continuous wavelet transforms.

Theorem 1. *Suppose that* $g \in L_2(\mathbb{R})$ *and*

$$C_g := \int_{\mathbb{R}} |\widehat{g}(s)|^2 \frac{ds}{|s|} < \infty.$$

If $f \in L_2(\mathbb{R})$, *then*

$$\int_{\mathbb{R}} \int_{\mathbb{R}} |W_g f(x, s)|^2 \frac{dx \, ds}{s^2} = C_g \|f\|_2^2.$$

The proofs of this theorem is usually uncomplete in the literature (see e.g. [4,5,7]). The proofs are correct if $g \in L_1(\mathbb{R}) \cap L_2(\mathbb{R})$. The non-trivial extension to all $g \in L_2(\mathbb{R})$ is usually missing (see [19]). More generally, suppose that $\mathbb{X}$ and $\mathbb{Y}$ are two Banach spaces, with $\mathbb{X}_0 \subset \mathbb{X}$ is dense in $\mathbb{X}$. If a linear operator ℓ is defined on $\mathbb{X}$ and $\ell : \mathbb{X}_0 \to \mathbb{Y}$ is bounded, then ℓ is not necessarily bounded from $\mathbb{X}$ to $\mathbb{Y}$ (see Bownik [2]). Of course, the unique extension of $\ell\big|_{\mathbb{X}_0}$ is bounded from $\mathbb{X}$ to $\mathbb{Y}$. However, it is not sure that the extension is equal to ℓ on the whole $\mathbb{X}$.

We can easily give conditions such that C_g is finite.

Proposition 1. *If $g \in L_2(\mathbb{R})$, $\widehat{g}(0) = 0$ and $\int_{\mathbb{R}} (1 + |x|)\, |g(x)|\, dx < \infty$, then C_g is finite.*

Now we formulate Theorem 1 for the scalar product of two wavelet transforms.

Theorem 2. *Suppose that $g_1, g_2 \in L_1(\mathbb{R}) \cap L_2(\mathbb{R})$ and*

$$C_{g_1,g_2} := \int_{\mathbb{R}} \overline{\widehat{g_1}(s)}\widehat{g_2}(s)\, \frac{ds}{|s|}$$

is a finite number. If $f_1, f_2 \in L_2(\mathbb{R})$, then

$$\int_{\mathbb{R}} \left(\int_{\mathbb{R}} W_{g_1} f_1(x,s)\overline{W_{g_2} f_2(x,s)}\, dx \right) \frac{ds}{s^2} = C_{g_1,g_2} \langle f_1, f_2 \rangle.$$

The preceding result is stated often for all $g_1, g_2 \in L_2(\mathbb{R})$ and without the bracket. However, in this case we have to suppose that $C_{g_1} < \infty$ and $C_{g_2} < \infty$.

Theorem 3. *Suppose that $g_1, g_2 \in L_2(\mathbb{R})$, $C_{g_1} < \infty$ and $C_{g_2} < \infty$. If $f_1, f_2 \in L_2(\mathbb{R})$, then*

$$\int_{\mathbb{R}} \int_{\mathbb{R}} W_{g_1} f_1(x,s)\overline{W_{g_2} f_2(x,s)}\, \frac{dx\, ds}{s^2} = C_{g_1,g_2} \langle f_1, f_2 \rangle.$$

Of course, the finiteness of C_{g_1} and C_{g_2} implies the finiteness of C_{g_1,g_2}.

4   Inversion Formulas

Let us recall some results for the inverse Fourier transform. Suppose first that $f \in L_p(\mathbb{R})$ for some $1 \leq p \leq 2$. Then the *Fourier inversion formula*

$$f(x) = \int_{\mathbb{R}} \widehat{f}(u)e^{2\pi \imath xu}\, du \qquad (x \in \mathbb{R})$$

holds if $\widehat{f} \in L_1(\mathbb{R})$. This motivates the definition of the *Dirichlet integral* $\tau_T f$, which is given by

$$\tau_T f(x) := \int_{-T}^{T} \widehat{f}(u)e^{2\pi \imath xu}\, du,$$

where $f \in L_p(\mathbb{R})$ $(1 \le p \le 2)$ and $T \in \mathbb{R}_+$. It is known that for $f \in L_p(\mathbb{R})$, $1 < p < \infty$,

$$\lim_{T \to \infty} \tau_T f = f \qquad \text{a.e. and in the } L_p(\mathbb{R})\text{-norm.}$$

The norm convergence is due to Riesz [12] and the almost everywhere convergence to Carleson [3] (see also Grafakos [6]). If $p > 2$, then we suppose that $f \in L_p(\mathbb{R}) \cap L_2(\mathbb{R})$ or we extend the integral in $\tau_T f$ to every $f \in L_p(\mathbb{R})$.

In this section, we formulate the analogous results for the inverse wavelet transforms. The *inverse wavelet transform* is given by

$$\int_{\mathbb{R}} \int_{\mathbb{R}} W_g f(x,s) T_x \mathcal{D}_s g \, \frac{dx \, ds}{s^2}.$$

The basic question is, how can we define the integral of the vector-valued function

$$(x,s) \mapsto W_g f(x,s) T_x \mathcal{D}_s g(\cdot) \, ?$$

An obvious interpretation is that for each fixed t, we consider the pointwise integral

$$\int_{\mathbb{R}} \int_{\mathbb{R}} W_g f(x,s) T_x \mathcal{D}_s g(t) \, \frac{dx \, ds}{s^2}.$$

However, this integral is not well defined for general functions $f, g \in L_2(\mathbb{R})$. Our pointwise inversion formula reads as follows.

Theorem 4. *If $g, \gamma \in L_2(\mathbb{R})$, $C_g < \infty$, $C_\gamma < \infty$, $f, \widehat{f} \in L_1(\mathbb{R})$, then for almost every $t \in \mathbb{R}$,*

$$\int_{\mathbb{R}} \left(\int_{\mathbb{R}} W_g f(x,s) T_x \mathcal{D}_s \gamma(t) \, dx \right) \frac{ds}{s^2} = C_{g,\gamma} f(t).$$

Notice that if $f, \widehat{f} \in L_1(\mathbb{R})$, then $f \in L_2(\mathbb{R})$. Indeed,

$$f = \left(\widehat{f} \right)^{\vee} \in L_1(\mathbb{R}) \cap L_\infty(\mathbb{R}) \subset L_2(\mathbb{R}).$$

Another interpretation of the integral in the inverse wavelet transform is the following definition. We say that the *vector-valued weak integral*

$$\int_{\mathbb{R}} \int_{\mathbb{R}} W_g f(x,s) T_x \mathcal{D}_s g \, \frac{dx \, ds}{s^2}$$

is equal to the function $f \in L_2(\mathbb{R})$,

$$f = \int_{\mathbb{R}} \int_{\mathbb{R}} W_g f(x,s) T_x \mathcal{D}_s g \, \frac{dx \, ds}{s^2}$$

if for all $h \in L_2(\mathbb{R})$,

$$\langle f, h \rangle = \int_{\mathbb{R}} \int_{\mathbb{R}} W_g f(x,s) \, \langle T_x \mathcal{D}_s g, h \rangle \, \frac{dx \, ds}{s^2}.$$

The next two results were proved for all $f, g, \gamma \in L_2(\mathbb{R})$ e.g. in Daubechies [5] or Gröchenig [7].

Theorem 5. *If $g, \gamma \in L_2(\mathbb{R})$, $C_g < \infty$ and $C_\gamma < \infty$, then*

$$\int_{\mathbb{R}} \int_{\mathbb{R}} W_g f(x, s) T_x \mathcal{D}_s \gamma \, \frac{dx \, ds}{s^2} = C_{g,\gamma} f$$

for all $f \in L_2(\mathbb{R})$, where the integral is understood in a vector-valued weak sense.

Theorem 6. *Assume that $f, g, \gamma \in L_2(\mathbb{R})$, $C_g < \infty$ and $C_\gamma < \infty$. If $0 < S < T \leq \infty$ and $0 < B < \infty$, then*

$$\lim_{S \to 0, T \to \infty} \int_{S \leq |s| \leq T} \int_{|x| \leq B} W_g f(x, s) T_x \mathcal{D}_s \gamma \, \frac{dx \, ds}{s^2} = C_{g,\gamma} f$$

in the L_2-norm.

In the last theorem the integral can be understood pointwise as well as in the vector-valued weak sense. Now let $B = \infty$ and, similar to $\tau_T f$, introduce the operators with the pointwise integrals

$$\rho_{S,T} f := \int_{S \leq |s| \leq T} \int_{\mathbb{R}} W_g f(x, s) T_x \mathcal{D}_s \gamma \, \frac{dx \, ds}{s^2}$$

and

$$\rho_S f := \int_{S \leq |s|} \int_{\mathbb{R}} W_g f(x, s) T_x \mathcal{D}_s \gamma \, \frac{dx \, ds}{s^2}.$$

One can show stronger convergence results for these operators as $S \to 0$ and $T \to \infty$. The next theorem is due to Rao et al. [11].

Theorem 7. *Assume that $g, \gamma \in L_2(\mathbb{R}) \cap L_1(\mathbb{R})$ with $C_g < \infty$ and $C_\gamma < \infty$. If $f \in L_p(\mathbb{R})$ for some $1 < p < \infty$, then*

$$\lim_{S \to 0, T \to \infty} \rho_{S,T} f = C_{g,\gamma} f$$

and

$$\lim_{S \to 0} \rho_S f = C_{g,\gamma} f,$$

both a.e. and in the L_p-norm.

Supposing more conditions about g and γ, Li and Sun [10] extended this result as follows. We say that h is a log-majorant function if h is positive, decreasing as a function on $(0, \infty)$ and

$$h(|\cdot|) \ln(2 + |\cdot|) \in L_1(\mathbb{R}).$$

It is easy to see that in this case $h \in L_2(\mathbb{R}) \cap L_1(\mathbb{R})$ and even $h \in W(L_\infty, \ell_1)(\mathbb{R})$. Let

$$C'_{g,\gamma} := -\int_{\mathbb{R}} (g^* * \gamma)(x) \ln |x| \, dx,$$

where

$$g^*(y) := \overline{g(-y)}.$$

Theorem 8. *Suppose that the functions g and γ have log-majorants and*

$$\int_{\mathbb{R}} (g^* * \gamma)(x)\, dx = 0.$$

(i) If $1 < p < \infty$ and $f \in L_p(\mathbb{R})$, then

$$\lim_{S \to 0, T \to \infty} \rho_{S,T} f = C'_{g,\gamma} f \qquad in\ the\ L_p\text{-}norm.$$

(ii) If $1 \le p < \infty$ and $f \in L_p(\mathbb{R})$, then

$$\lim_{S \to 0} \rho_S f = C'_{g,\gamma} f \qquad in\ the\ L_p\text{-}norm.$$

Of course, under some conditions, $C_{g,\gamma}$ coincides with $C'_{g,\gamma}$ (see Rubin and Shamir [13] and Saeki [14]). By the theorem about the maximal function (see e.g. Grafakos [6]),

$$\lim_{h \to 0} \frac{1}{2h} \int_{-h}^{h} f(x - s)\, ds = f(x)$$

for a.e. $x \in \mathbb{R}$. Then

$$\lim_{h \to 0} \frac{1}{2h} \int_{-h}^{h} f(x - s) - f(x)\, ds = 0.$$

A point $x \in \mathbb{R}$ is called a Lebesgue point of f if the following stronger condition holds:

$$\lim_{h \to 0} \frac{1}{2h} \int_{-h}^{h} |f(x - s) - f(x)|\, ds = 0.$$

Theorem 9. *Almost every point $x \in \mathbb{R}$ is a Lebesgue point of $f \in W(L_1, \ell_\infty)(\mathbb{R})$.*

The following theorem is due again to Li and Sun [10].

Theorem 10. *Suppose that the functions g and γ have log-majorants and*

$$\int_{\mathbb{R}} (g^* * \gamma)(x)\, dx = 0.$$

If $1 \le p < \infty$ and $f \in L_p(\mathbb{R})$, then

$$\lim_{S \to 0, T \to \infty} \rho_{S,T} f(x) = C'_{g,\gamma} f(x)$$

and

$$\lim_{S \to 0} \rho_S f(x) = C'_{g,\gamma} f(x)$$

for all Lebesgue-points of f.

In the next two theorems due to the author [17], we do not suppose that $g, \gamma \in W(L_\infty, \ell_1)(\mathbb{R})$, instead we suppose more smoothness and that $g, \gamma \in L_1(\mathbb{R}) \cap L_2(\mathbb{R})$, and we show similar results for larger function spaces, namely for the Wiener amalgam spaces.

Theorem 11. *Assume that $g, \gamma \in L_1(\mathbb{R}) \cap L_2(\mathbb{R})$, $\widehat{g}$ and $\widehat{\gamma}$ are differentiable, $\widehat{g}'$ and $\widehat{\gamma}'$ are bounded and*

$$|\widehat{g}(r)|, |\widehat{\gamma}(r)| \le Cr^\alpha \qquad (0 < r \le 1)$$

for some $\alpha > 0$.

(i) If $1 \le p < \infty$, $1 < q < \infty$ and $f \in W(L_p, \ell_q)(\mathbb{R})$, then

$$\lim_{S \to 0, T \to \infty} \rho_{S,T} f = C_{g,\gamma} f \qquad \text{in the } W(L_p, \ell_q)\text{-norm.}$$

(ii) If $1 \le q < \infty$ and $f \in W(L_1, \ell_q)(\mathbb{R})$, then

$$\lim_{S \to 0, T \to \infty} \rho_{S,T} f(x) = C_{g,\gamma} f(x)$$

for all Lebesgue-points of f.

Theorem 12. *In addition to the conditions of Theorem 11, assume that $\gamma \in W(L_\infty, \ell_1)(\mathbb{R})$.*

(i) If $1 \le p, q < \infty$ and $f \in W(L_p, \ell_q)(\mathbb{R})$, then

$$\lim_{S \to 0} \rho_S f = C_{g,\gamma} f \qquad \text{in the } W(L_p, \ell_q)\text{-norm.}$$

(ii) If $1 \le q < \infty$ and $f \in W(L_1, \ell_q)(\mathbb{R})$, then

$$\lim_{S \to 0} \rho_S f(x) = C_{g,\gamma} f(x)$$

for all Lebesgue-points of f.

Note that, for all $1 \le p, q < \infty$,

$$W(L_1, \ell_q)(\mathbb{R}) \supset W(L_p, \ell_q)(\mathbb{R}), L_q(\mathbb{R}).$$

As we can see in the next result, under some weak conditions, Theorems 11 and 12 can be applied.

Theorem 13. *If $g \in L_1(\mathbb{R}) \cap L_2(\mathbb{R})$, $\widehat{g}(0) = 0$ and*

$$\int_{\mathbb{R}} |x|^\alpha |g(x)| \, dx < \infty$$

for some $0 < \alpha \le 1$, then

$$|\widehat{g}(r)| \le Cr^\alpha \qquad (0 < r \le 1).$$

Moreover, $\widehat{g}'$ is bounded for $\alpha = 1$.

References

1. Ashurov, R.: Convergence of the continuous wavelet transforms on the entire Lebesgue set of L_p-functions. Int. J. Wavelets Multiresolut. Inf. Process. **9**, 675–683 (2011)
2. Bownik, M.: Boundedness of operators on Hardy spaces via atomic decompositions. Proc. Am. Math. Soc. **133**, 3535–3542 (2005)
3. Carleson, L.: On convergence and growth of partial sums of Fourier series. Acta Math. **116**, 135–157 (1966)
4. Chui, C.K.: An Introduction to Wavelets. Academic Press, Boston (1992)
5. Daubechies, I.: Ten Lectures on Wavelets. SIAM, Philadelphia (1992)
6. Grafakos, L.: Classical and Modern Fourier Analysis. Pearson Education, New Jersey (2004)
7. Gröchenig, K.: Foundations of Time-Frequency Analysis. Birkhäuser, Boston (2001)
8. Hernández, E., Weiss, G.: A First Course on Wavelets. CRC Press, Boca Raton (1996)
9. Kelly, S.E., Kon, M.A., Raphael, L.A.: Local convergence for wavelet expansions. J. Func. Anal. **126**, 102–138 (1994)
10. Li, K., Sun, W.: Pointwise convergence of the Calderon reproducing formula. J. Fourier Anal. Appl. **18**, 439–455 (2012)
11. Rao, M., Sikic, H., Song, R.: Application of Carleson's theorem to wavelet inversion. Control Cybern. **23**, 761–771 (1994)
12. Riesz, M.: Sur la sommation des séries de Fourier. Acta Sci. Math. (Szeged) **1**, 104–113 (1923)
13. Rubin, B., Shamir, E.: Carlderon's reproducing formula and singular integral operators on a real line. Integr. Equat. Oper. Theory **21**, 78–92 (1995)
14. Saeki, S.: On the reproducing formula of Calderon. J. Fourier Anal. Appl. **2**, 15–28 (1995)
15. Walnut, D.F.: An Introduction to Wavelet Analysis. Birkhäuser, Basel (2002)
16. Weisz, F.: Inversion formulas for the continuous wavelet transform. Acta Math. Hung. **138**, 237–258 (2013)
17. Weisz, F.: Convergence of the inverse continuous wavelet transform in Wiener amalgam spaces. Analysis **35**, 33–46 (2015)
18. Weisz, F.: Inverse continuous wavelet transform in Pringsheim's sense on Wiener amalgam spaces. Acta Math. Hung. **145**, 392–415 (2015)
19. Weisz, F., Dutta, H., Rhoades, B.: Multi-dimensional summability theory and continuous wavelet transform. Current Topics in Summability Theory and Applications, pp. 241–311. Springer, Singapore (2016). https://doi.org/10.1007/978-981-10-0913-6_6
20. Wilson, M.: Weighted Littlewood-Paley Theory and Exponential-Square Integrability. Lecture Notes in Mathematics. Springer, Berlin (2008). https://doi.org/10.1007/978-3-540-74587-7
21. Wilson, M.: How fast and in what sense(s) does the Calderon reproducing formula converge? J. Fourier Anal. Appl. **16**, 768–785 (2010)
22. Zayed, A.: Pointwise convergence of a class of non-orthogonal wavelet expansions. Proc. Am. Math. Soc. **128**, 3629–3637 (2000)

The Reed-Muller-Fourier Transform Applied to Pattern Analysis

Claudio Moraga[1](✉) and Radomir S. Stanković[2]

[1] Faculty of Computer Science, TU Dortmund University, 44221 Dortmund, Germany
Claudio.Moraga@tu-dortmund.de
[2] Department of Computer Science, Faculty of Electronic Engineering,
University of Niš, 18000 Niš, Serbia
Radomir.Stankovic@gmail.com

Abstract. This paper introduces the analysis of pattern properties by means of the two-sided Reed-Muller-Fourier transform. Patterns are modelled as matrices of pixels and an integer coding for the colors is chosen. Work is done in the ring $(Z_p, \oplus, \cdot)$, where $p > 2$ is not necessarily a prime. It is shown that the transform preserves the (diagonal) symmetry of patterns, is compatible with different operations on patterns, and allows detecting and localizing noise pixels in a pattern. Finally, it is shown that there are patterns which are fixed points of the transform.

1 Introduction

The Reed-Muller-Fourier transform (RMF), defined in the ring $(Z_p, \oplus, \cdot)$, p an integer larger than 2, was introduced in [1] aiming to unify relevant characteristics of the Reed-Muller transform (RM) and the Discrete Fourier transform (DFT), to be applied in the non-binary integer domain allowing to obtain polynomial expressions for multiple-valued functions. In the binary case, the RM transform has a self inverse matrix representation, which is lower triangular, and exhibits a Kronecker product structure [2–4]. When the RM transform was extended to the non-binary domain [5], it retained the property of realizing a bijection in the set of functions for a given valuedness and arity, but it lost the lower triangular structure and its self inversion. The DFT, however is lower triangular in all integer domains. To obtain the desired combination of properties for the RMF transform, the Gibbs algebra defined in terms of the Gibbs multiplication for the Instant Fourier Transform [6] was selected and extended to the non-binary domain in [1].

In the context of patterns, let Γ be a finite ordered set of colors with cardinality p, and let $\beta : \Gamma \to Z_p$ be a bijection assigning an element of Z_p to each color in such a way that the ordering of the colors is preserved. Pixels are atoms of a picture and carry a single color. (The size of a pixel is defined according to requirements of geometric and chromatic resolution for a pattern under consideration.) A pattern is an array of pixels. In this paper, a pattern is also represented as a matrix with entries from Z_p obtained by applying β to every

© Springer International Publishing AG 2018
R. Moreno-Díaz et al. (Eds.): EUROCAST 2017, Part II, LNCS 10672, pp. 254–261, 2018.
https://doi.org/10.1007/978-3-319-74727-9_30

pixel of a picture. Operations among patterns are conducted on the corresponding numerical matrices in the ring $(Z_p, \oplus, \cdot)$. Pattern attributes associated to a matrix are understood as attributes of the pattern represented by such a matrix.

2 Formalisms

In what follows, some properties of patterns will be studied in a transform domain. In this paper, we select the Reed-Muller-Fourier (RMF) transform [1,2,7]. It is known that this transform matrix is lower triangular, self-inverse and has a Kronecker product structure [8]. Moreover, this transform is based on the Gibbs convolutional product [4]. Figure 1 shows the basic transform matrices for $p = 4, 5$, and 6 as will be used in this paper.

$$
\mathbf{R}_3(1) = \begin{bmatrix} 3 & 0 & 0 & 0 \\ 3 & 1 & 0 & 0 \\ 3 & 2 & 3 & 0 \\ 3 & 3 & 1 & 1 \end{bmatrix} \qquad
\mathbf{R}_4(1) = \begin{bmatrix} 4 & 0 & 0 & 0 & 0 \\ 4 & 1 & 0 & 0 & 0 \\ 4 & 2 & 4 & 0 & 0 \\ 4 & 3 & 2 & 1 & 0 \\ 4 & 4 & 4 & 4 & 4 \end{bmatrix} \qquad
\mathbf{R}_5(1) = \begin{bmatrix} 5 & 0 & 0 & 0 & 0 & 0 \\ 5 & 1 & 0 & 0 & 0 & 0 \\ 5 & 2 & 5 & 0 & 0 & 0 \\ 5 & 3 & 3 & 1 & 0 & 0 \\ 5 & 4 & 0 & 4 & 5 & 0 \\ 5 & 5 & 2 & 4 & 1 & 1 \end{bmatrix}.
$$

$$p = 4 \qquad\qquad p = 5 \qquad\qquad p = 6$$

Fig. 1. Basic RMF transform matrices for and $p = 4, 5$, and 6.

The following notation will be used in the rest of the paper: $\mathbf{A}(n)$ will denote a $(p^n \times p^n)$ matrix (and pattern). $\mathbf{A}(n, m)$ will denote a $(p^n \times p^m)$ matrix (and pattern). For the RMF-transform matrix, the notation $\mathbf{R}_p(n)$ will be used. (If a related statement is valid for all p, the index p may be omitted.)

For a given p and n, the RMF-transform matrix is defined as the n-th Kronecker power of the basic RMF-transform matrix $\mathbf{R}_p(1)$. Therefore,

$$\mathbf{R}_p(n) = \bigotimes_{i=1}^{n} \mathbf{R}_p(1).$$

$H(n)$ will represent a matrix with all entries equal to 1 (H for "high") and $L(n)$ will denote a matrix with all entries equal to 0 (L for "low"). Finally, $C(n)$ will represent a matrix with all entries equal to 0, except for the element at the left upper corner, where the entry equals 1 (leading to the name "corner"). For the analysis of patterns, a two-sided RMF-transform will be used (see definition below), a transformed pattern will be called "spectrum" and will be identified by Σ. Unless otherwise specified, all operations will be done in the ring $(Z_p, \oplus, \cdot)$.

3   Analysis of Patterns by Using the RMF-Transform

In this section, we will present some considerations related to the application of the RMF-transform to analysis of patterns.

Definition 1. *Given a pattern $A(n, m)$, its spectrum $\Sigma_A(n, m)$, is calculated as follows*

$$\Sigma_A(n, m) = \mathbf{R}_p(n) \cdot \mathbf{A}(n, m) \cdot (\mathbf{R}_p(m))^T, \tag{1}$$

where the superindex T denotes the transposition of a matrix.

Lemma 1. *The inverse RMF transform recovers a pattern from its spectrum as follows*

$$\mathbf{A}(n, m) = \mathbf{R}_p(n) \cdot \Sigma_A(n, m) \cdot (\mathbf{R}_p(m))^T, \tag{2}$$

Proof: Since $R_p(n)$ is its own inverse, the assertion follows by applying $R_p(n)$ and $(R_p(m))^T$ at both sides of Eq. (1).

Lemma 2. *For all p the RMF spectrum of a square symmetric pattern is symmetric.*

Proof: If $A(n)$ is symmetric, it holds that $A(n) = (A(n))^T$. Then,

$$\begin{aligned}\Sigma_A(n) &= \mathbf{R}(n) \cdot \mathbf{A}(n) \cdot (\mathbf{R}(n))^T = \mathbf{R}(n) \cdot (\mathbf{A}(n))^T \cdot (\mathbf{R}(n))^T \\ &= \langle \mathbf{R}(n) \cdot \mathbf{A}(n) \cdot (\mathbf{R}(n))^T \rangle^T = (\Sigma_A(n))^T.\end{aligned}$$

Lemma 3. *If $\mathbf{Q}(n, m) = \mathbf{A}(n, m) \oplus \mathbf{B}(n, m)$ then $\Sigma_Q(n, m) = \Sigma_A(n, m) \oplus \Sigma_B(n, m)$.*

Proof:

$$\begin{aligned}\Sigma_Q(n, m) &= \mathbf{R}(n) \cdot \mathbf{Q}(n, m) \cdot (\mathbf{R}(m))^T \\ &= \mathbf{R}(n) \cdot (\mathbf{A}(n, m) \oplus \mathbf{B}(n, m)) \cdot (\mathbf{R}(m))^T \\ &= \mathbf{R}(n) \cdot \mathbf{A}(n, m) \cdot (\mathbf{R}(m))^T \oplus \mathbf{R}(n) \cdot \mathbf{B}(n, m) \cdot (\mathbf{R}(m))^T \\ &= \Sigma_A(n, m) \oplus \Sigma_B(n, m).\end{aligned}$$

See examples in Fig. 2.

Lemma 4. *The spectrum of the Kronecker product [3, 4] of two patterns equals the Kronecker product of the respective spectra. If $\mathbf{Q}(n + r, m + s) = \mathbf{A}(n, m) \otimes \mathbf{B}(r, s)$ then $\Sigma_Q(n + r, m + s) = \Sigma_A(n, m) \otimes \Sigma_B(r, s)$.*

Proof:

$$\begin{aligned}\Sigma_Q(n + r, m + s) &= \mathbf{R}(n + r)\mathbf{Q}(n + r, m + s)(\mathbf{R}(m + s))^T \\ &= (\mathbf{R}(n) \otimes \mathbf{R}(r))(\mathbf{A}(n, m) \otimes \mathbf{B}(r, s))(\mathbf{R}(m) \otimes \mathbf{R}(s))^T.\end{aligned}$$

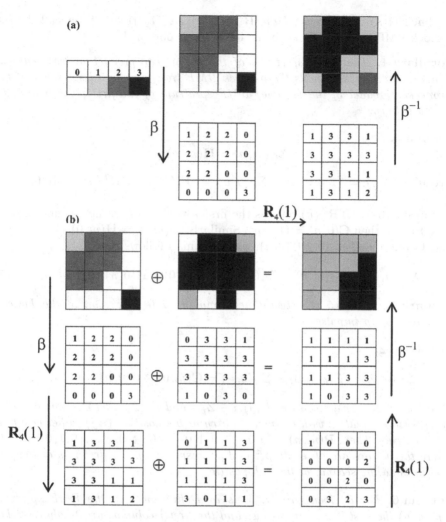

Fig. 2. Example of Lemmas 2 and 3 with $p = 4$ and $n = 1$. (a) The spectrum of a symmetric pattern is symmetric, (b) the spectrum of the sum of two patterns equals the sum of their spectra.

With the compatibility theorem between Kronecker and matrix products [3],

$$\Sigma_Q(n+r, m+s) = (\mathbf{R}(n)\mathbf{A}(n,m)\mathbf{R}(m)^T) \otimes (\mathbf{R}(r)\mathbf{B}(r,s)(\mathbf{R}(s)^T)$$
$$= \Sigma_A(n,m) \otimes \Sigma_B(r,s). \tag{3}$$

Notice that Eq. (3) may also be written as

$$\Sigma_Q(n+r, m+s) = (\mathbf{I}(n)\Sigma_A(n,m)\mathbf{I}(m)) \otimes (\mathbf{R}(r)\mathbf{B}(r,s)(\mathbf{R}(s)^T)$$
$$= (\mathbf{I}(n) \otimes \mathbf{R}(r))(\Sigma_A(n,m) \otimes \mathbf{B}(r,s))(\mathbf{I}(m) \otimes (\mathbf{R}(s)^T))$$
$$= (\mathbf{I}(n) \otimes \mathbf{R}(r))(\Sigma_A(n,m) \otimes \mathbf{B}(r,s))(\mathbf{I}(m) \otimes (\mathbf{R}(s))^T. \tag{4}$$

Since $(\mathbf{I}(n) \otimes \mathbf{R}(r)) = Diag(\mathbf{R}(r), \mathbf{R}(r), \ldots, \mathbf{R}(r))$ let it be called "the n-Block RMF transform". Then the following is obtained:

Corollary 1. *The RMF spectrum of the Kronecker product of two patterns* $\mathbf{A}(n, m)$ *and* $\mathbf{B}(r, s)$, *equals the n-Block RMF transform based spectrum of the Kronecker product of the (simple) RMF spectrum of the first pattern and the second pattern.*

Lemma 5.
$$\Sigma_C(m, n) = \mathbf{H}(m, n).$$

Proof: $\mathbf{C}(0, n) = [1, 0, \ldots, 0]$. $\Sigma_C(0, n) = \mathbf{R}(0)\mathbf{C}(0, n)\mathbf{R}^T(n) = [1, 0, \ldots, 0]$ $\mathbf{R}^T(n)$.

Since $[1, 0, \ldots, 0]\mathbf{R}^T(n)$ returns the first row of $\mathbf{R}t(n)$, which equals $\mathbf{H}(0, n)$ -(see Fig. 2)- then $\mathbf{C}(0, n) = \mathbf{H}(0, n)$. Similarly, $\mathbf{C}(m, 0) = \mathbf{H}(m, 0)$. Considering that $\mathbf{C}(m, n) = \mathbf{C}(m, 0) \otimes \mathbf{C}(0, n)$, with Lemma 4 follows that:

$$\Sigma_C(m, n) = \Sigma_C(m, 0) \otimes \Sigma_C(0, n) = \mathbf{H}(m, 0) \otimes \mathbf{H}(0, n) = \mathbf{H}(m, n). \tag{5}$$

Remark 1. *Lemma 5 is a two-dimensional discrete extension of the Fourier transform of an impulse.*

Corollary 2.

$$\Sigma_H(m, n) = \Sigma_H(m, 0) \otimes \Sigma_H(0, n) = \mathbf{C}(m, 0) \otimes \mathbf{C}(0, n) = \mathbf{C}(m, n).$$

Definition 2. *For a given pair* $(i, j), i \in Z_{p^m}$ *and* $j \in Z_{p^n}$, *let* $\mathbf{P}(m, n)$ *denote a "perturbation" matrix with a single 1 entry at the position* (i, j), *otherwise having 0 entries. Then,* $\mathbf{P}(m, n) = [0, 0, \ldots, 0, 1, 0, \ldots, 0] \otimes [0, 0, \ldots, 0, 1, 0, \ldots, 0]^T$, *where the vectors are of length p^m and p^n respectively, the first 1 is at the j-th position and the second, at the i-th position.*

Lemma 6. *For a given pair* (i, j) *as in Definition 2, in the RMF spectrum* $\Sigma_P(m, n)$ *the first i rows are 0-rows and the first j columns are 0-columns. The entry at the position* (i, j) *has the absolute value 1 and the remaining entries are mostly non-zero entries.*

Proof: Let $\mathbf{P}(0, n) = [0, 0, \ldots, 0, 1, 0, \ldots, 0]$, with the 1 at the j-th position. Then,

$$\begin{aligned}
\Sigma_P(0, n) &= \mathbf{R}(0) \cdot [0, 0, \ldots, 0, 1, 0, \ldots, 0] \cdot \mathbf{R}^T(n) \\
&= [1] \cdot [0, 0, \ldots, 0, 1, 0, \ldots, 0] \cdot \mathbf{R}^T(n) \\
&= [0, 0, \ldots, 0, 1, 0, \ldots, 0]\mathbf{R}^T(n).
\end{aligned}$$

It may be seen that the product $[0, 0, \ldots, 0, 1, 0, \ldots, 0] \cdot \mathbf{R}^T(n)$ extracts the j-th row of $\mathbf{R}^T(n)$. Since $\mathbf{R}^T(n)$ is upper triangular -(recall Fig. 2)- the j-th row has a prefix of j 0s and the first non-zero entry equals $(-1)^{j-1} \bmod p$.

Similarly, $\mathbf{P}(m,0) = [0,0,\ldots,0,1,0,\ldots,0]^T$, with the 1 at the i-th position. Then,

$$\mathbf{P}(m,0) = \mathbf{R}(m) \cdot [0,0,\ldots,0,1,0,\ldots,0]^T \cdot \mathbf{R}^T(0)$$
$$= \mathbf{R}(m) \cdot [0,0,\ldots,0,1,0,\ldots,0]^T.$$

The product $\mathbf{R}(m)[0,0,\ldots,0,1,0,\ldots,0]^T$ extracts the i-th column of $\mathbf{R}(m)$. Since $\mathbf{R}(m)$ is lower triangular, its i-th column has a prefix of i 0s and the first non-zero entry equals $(-1)^{i-1} \bmod p$.

It may be seen that $\mathbf{P}(0,n) \otimes \mathbf{P}(m,0)$ will have the first i rows and the first j columns with 0 entries. Moreover, at (i,j) the entry has magnitude 1 mod p.

Example 1. *For $p = 6$, Fig. 3 shows $\mathbf{P}(1)$ with $(i,j) = (4,2)$.*

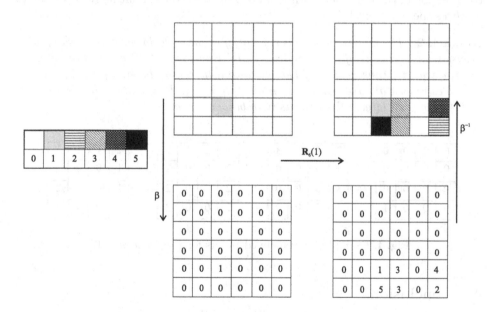

Fig. 3. Example of the effect of a perturbation matrix.

It may be seen that the "left upper non-zero" pixel in the spectrum indicates the position of the perturbing pixel. This property may be used to detect and localize a noise pixel.

Definition 3. *A mosaic is the Kronecker product of $\mathbf{H}(n,m)$ and a basic pattern $\mathbf{B}(r,s)$.*

Example 2. *Let $p = 4$ and (for space limitations) let $n = 1$ for $\mathbf{H}$. Then define $\mathbf{M}(1+r,1+s) = \mathbf{H}(1) \otimes \mathbf{B}(r,s)$. From Lemma 4 and Corollary 2 follows that $\Sigma_M(1+r,1+s) = \Sigma_H(1) \otimes \Sigma_B(r,s) = C(1) \otimes \Sigma_B(r,s)$.*

$$M(1+r,1+s) = \begin{bmatrix} \mathbf{B}(r,s) & \mathbf{B}(r,s) & \mathbf{B}(r,s) & \mathbf{B}(r,s) \\ \mathbf{B}(r,s) & \mathbf{B}(r,s) & \mathbf{B}(r,s) & \mathbf{B}(r,s) \\ \mathbf{B}(r,s) & \mathbf{B}(r,s) & \mathbf{B}(r,s) & \mathbf{B}(r,s) \\ \mathbf{B}(r,s) & \mathbf{B}(r,s) & \mathbf{B}(r,s) & \mathbf{B}(r,s) \end{bmatrix},$$

$$\Sigma_M(1+r,1+s) = \begin{bmatrix} \mathbf{\Sigma_B(r,s)} & \mathbf{L}(r,s) & \mathbf{L}(r,s) & \mathbf{L}(r,s) \\ \mathbf{L}(r,s) & \mathbf{L}(r,s) & \mathbf{L}(r,s) & \mathbf{L}(r,s) \\ \mathbf{L}(r,s) & \mathbf{L}(r,s) & \mathbf{L}(r,s) & \mathbf{L}(r,s) \\ \mathbf{L}(r,s) & \mathbf{L}(r,s) & \mathbf{L}(r,s) & \mathbf{L}(r,s) \end{bmatrix}.$$

It becomes apparent that if a random noise pixel is added to the mosaic during the building process, with Lemma 3, the 0-region of the spectrum will be clearly "contaminated", thus *detecting* the presence of a noise pixel. Moreover the position of the upper left corner of the non-zero contaminating region clearly *localizes* the noise pixel.

Remark 2. *The matrix which has a 1-entry at the right lower corner being otherwise 0 is a fixpoint of the two-sided RMF transform. Notice that this may be interpreted as a consequence of Lemma 6 and, therefore, is valid for all p. Additional examples of fixed points are shown in Fig. 4. Recall that with Lemmas 3 and 4 new (and larger) fixed points may be generated.*

0	0	0	0
0	0	0	0
0	0	0	0
0	0	0	1

2	3	3	3
3	0	0	0
3	0	0	0
3	0	0	0

0	2	2	2
2	0	0	2
2	0	0	2
2	0	0	3

2	1	1	1
1	0	0	2
1	0	0	2
1	0	0	3

Fig. 4. Examples of some fixed points for $p = 4$ and $n = 1$.

4 Conclusions

For the first time the RMF transform has been applied to analyze properties of patterns. Possibly the most relevant result refers to the possibility of both detecting and localizing noise pixels in patterns.

References

1. Stanković, R.S.: Some remarks on Fourier transforms and differential operators for digital functions. In: Proceeding of the 22nd International Symposium on Multiple-Valued Logic, Sendai, Japan, pp. 365–370 (1992). https://doi.org/10.1109/ISMVL. 1992.186818
2. Karpovsky, M.G., Stanković, R.S., Astola, J.T.: Spectral Logic and Its Application for the Design of Digital Devices. Wiley, Hoboken (2008)

3. Horn, R.A., Johnson, C.R.: Topics in Matrix Analysis. Cambridge University Press, New York (1991)
4. Graham, A.: Kronecker Products and Matrix Calculus with Applications. Ellis Horwood Ltd., Chichester (1981)
5. Green, D.H., Taylor, I.S.: Multiple-valued switching circuit design by means of generalized Reed-Muller expansions. Dig. Process. **2**, 63–81 (1976)
6. Gibbs, J.E.: Instant Fourier transform. Electron. Lett. **13**(5), 122–123 (1977)
7. Stanković, R.S.: The Reed-Muller-Fourier transform—computing methods and factorizations. In: Seising, R., Allende-Cid, H. (eds.) Claudio Moraga: A Passion for Multi-Valued Logic and Soft Computing. SFSC, vol. 349, pp. 121–151. Springer, Cham (2017). https://doi.org/10.1007/978-3-319-48317-7_9
8. Stanković, R.S., Stanković, M., Moraga, C.: The Pascal triangle (1654), the Reed-Muller-Fourier transform (1992), and the Discrete Pascal transform (2005). In: Proceedings of the 46th International Symposium on Multiple-Valued Logic, pp. 229–234. IEEE Press, New York (2016). https://doi.org/10.1109/ISMVL.2016.24

Some Spectral Invariant Operations for Functions with Disjoint Products in the Polynomial Form

Milena Stanković[1($\boxtimes$)], Claudio Moraga[2], and Radomir S. Stanković[1]

[1] Department of Computer Science, Faculty of Electronic Engineering, Niš, Serbia
`milena.stankovic@elfak.ni.ac.rs`
[2] Technical University of Dortmund, 44221 Dortmund, Germany

Abstract. It has long been known that some transformations of a binary function produce a permutation of some coefficients in the Walsh-Hadamard spectrum or just change the sign of some coefficients. Those operations are known as *spectral invariant operations*. In this paper some new spectral invariant operations are defined for the functions representable by disjoint quadratic polynomial forms. It is shown that these new invariant operations are useful for characterization of the bent functions.

Keywords: Invariant operations · Walsh-Hadamard spectrum
Bent functions

1 Introduction

Spectral invariant operations for binary functions are defined as operations that do not change the absolute values of Walsh-Hadamard spectral coefficients. There are five such invariant operations considered in the literature: complement of the function, complement of the input variable, permutation of two input variables, linear translation and disjoint linear translation, [1,2]. Their application results in sign changes and permutation of certain subsets of spectral coefficients.

From the other side, it is well known that bent functions that are important for cryptography are characterized through their Wash-Hadamard spectra. Bent functions are defined as functions with flat spectra, i.e., for bent functions all the Walsh-Hadamard coefficients have te same absolute values $2^{\frac{n}{2}}$, where n is the number of variables. Also, simplest bent functions have disjoint quadratic polynomial forms. Considering the properties of the invariant operations it follows that it will be possible to generate all bent functions with given number of input variables, starting from disjoint quadratic forms, and by applying some invariant operations. However, the known invariant operations cannot increase the degree (the number of variables in the product terms) in the polynomial form of the function. Since it is proved that polynomial form of bent functions may have

© Springer International Publishing AG 2018
R. Moreno-Díaz et al. (Eds.): EUROCAST 2017, Part II, LNCS 10672, pp. 262–269, 2018.
https://doi.org/10.1007/978-3-319-74727-9_31

terms of degree until $\frac{n}{2}$, it is possible to conclude that some additional invariant operations must exist. In this paper, such new spectral invariant operations will be defined and explored.

2   Walsh Hadamard Spectrum

In this section, we present the definition of the Walsh-Hadamard spectrum and five known invariant operations.

Definition 1 *Walsh-Hadamard spectrum*
The Walsh-Hadamard spectrum $\mathbf{S}_f(n)$ of the binary function $f(x_1, \cdots, x_n)$ of n variables is defined as:

$$\mathbf{S}_f(n) = \mathbf{W}(1)^{\otimes n}\mathbf{F}(n), \quad \mathbf{W}(1) = \begin{bmatrix} 1 & 1 \\ 1 & -1 \end{bmatrix}, \tag{1}$$

and $\mathbf{F}(n)$ is the truth vector of the function f in the $(0,1) \to (1,-1)$ encoding.

$\mathbf{S}_f(n)$ is a vector of 2^n coefficients. Here we will use a notation of the coefficients with binary subscripts $S_{b_1, b_2, \cdots, b_n}$, where $b_i \in (0,1), i = 1, 2, \cdots, n$:

$$\mathbf{S}_f(n) = [S_{00...0}, S_{00...1}, \cdots, S_{11...1}].$$

2.1   Spectral Invariant Operations

It is well known that some operations in the original domain result in the permutation of the coefficients in the spectral domain or with the change of the sign of some coefficients. There are five such invariant operations.

Definition 2 *(Complement of the function)*
By complement of the function $f(x_1, \cdots, x_n)$ a new function

$$g(x_1, \cdots, x_n) = \bar{f}(x_1, \cdots, x_n)$$

is generated. This operation produces the change of signs of the all coefficients:

$$\mathbf{S}_g = -\mathbf{S}_f.$$

Example 1. *Consider the function $f(x_1, x_2, x_3) = x_1 \oplus x_2 x_3$ and the function $f_1(x_1, x_2, x_3) = \bar{f}(x_1, x_2, x_3)$.*
The spectrum of the function f is $\mathbf{S} = [0, 0, 0, 0, 4, 4, 4, -4]$ while the spectrum of the function f_1 is $\mathbf{S}_1 = [0, 0, 0, 0, -4, -4, -4, 4]$. Note that all coefficients are multiplied by -1.

Definition 3 *(Complement of the input variable)*
By complement of the input variable x_j of the function $f(x_1, \cdots, x_j, \cdots, x_n)$ a new function $g(x_1, \cdots, x_j, \cdots, x_n) = f(x_1, \cdots, \bar{x}_j, \cdots, x_n)$ is generated. This operation produces the change of the signs of the following coefficients in the spectral domain:

$$S_{g_{b_1, \cdots, b_j=1, \cdots, b_n}} = -S_{f_{b_1, \cdots, b_j=1, \cdots, b_n}}.$$

Example 2. *Consider the function* $f(x_1, x_2, x_3) = x_1 \oplus x_2 x_3$ *and the function*

$$f_2(x_1, x_2, x_3) = f(x_1, \bar{x}_2, x_3).$$

The function f_2 *is obtained from function* f *by complementing the input variable* x_2. *The spectrum of the function* f_2 *is* $\mathbf{S}_2 = [0, 0, 0, 0, 4, 4, -4, 4]$. *The signs of following coefficients is changed:* $S_{2b_1,1,b_3} = -S_{b_1,1,b_3}$, *or*

$$S_{2010} = -S_{010}, \quad S_{2011} = -S_{011}, \quad S_{2110} = -S_{110}, \quad S_{2111} = -S_{111}.$$

Definition 4 *(Permutation of two input variables)*
By permutation of the input variables x_{j_1} *and* x_{j_2} *of the function* $f(x_1, \cdots, x_{j_1}, x_{j_2}, \cdots, x_n)$ *a new function*

$$g(x_1, \cdots, x_{j_1}, x_{j_2}, \cdots, x_n) = f(x_1, \cdots, x_{j_2}, x_{j_1}, \cdots, x_n)$$

is generated. This operation produces the permutation of the following coefficients in the spectral domain:

$$S_{g_{b_1, \cdots, b_{j_1}=0, b_{j_2}=1, \cdots, b_n}} \leftrightarrow S_{f_{b_1, \cdots, b_{j_1}=1, b_{j_2}=0, \cdots, b_n}}, \; and$$

$$S_{g_{b_1, \cdots, b_{j_1}=1, b_{j_1}=0, \cdots, b_n}} \leftrightarrow S_{f_{b_1, \cdots, b_{j_1}=0, b_{j_2}=1, \cdots, b_n}},$$

Example 3. *Consider the function* $f(x_1, x_2, x_3) = x_1 \oplus x_2 x_3$ *and the function* $f_3(x_1, x_2, x_3) = x_2 \oplus x_1 x_3$. *Function* f_3 *is optained by permutation of the input variables* x_1 *and* x_2 *in the function* f. *The spectrum of the function* f_3 *is* $\mathbf{S}_3 = [0, 0, 4, 4, 0, 0, 4, -4]$. *It is visible that*

$$S_{3010} = S_{100}, \quad S_{3011} = S_{101}, \quad S_{3100} = S_{010}, \quad S_{3101} = S_{011}.$$

Definition 5 *(Linear translation)*
By adding of the input variable x_j *to the function* $f(x_1, \cdots, x_j, \cdots, x_n)$ *a new function*

$$g(x_1, \cdots, x_j, \cdots, x_n) = x_j \oplus f(x_1, \cdots, x_j, \cdots, x_n)$$

is generated. This operation produces the permutation of the following coefficients in the spectral domain:

$$S_{g_{b_1, \cdots, b_j=1, \cdots, b_n}} \leftrightarrow S_{f_{b_1, \cdots, b_j=0, \cdots, i_n}},$$

$$S_{g_{b_1, \cdots, b_j=0, \cdots, b_n}} \leftrightarrow S_{f_{b_1, \cdots, b_j=1, \cdots, b_n}}.$$

Example 4. *Consider the function* $f(x_1, x_2, x_3) = x_1 \oplus x_2 x_3$ *and the function* $f_4(x_1, x_2, x_3) = x_2 \oplus x_1 \oplus x_2 x_3$. *The function* f_4 *is optained by adding the variable* x_2 *to the function* f. *The spectrum of the function* f_4 *is* $\mathbf{S}_4 = [0, 0, 0, 0, 4, -4, 4, 4]$ *wich means that*

$$S_{4b_1,0,b_3} = S_{b_1,1,b_3}, \quad S_{4b_1,1,b_3} = S_{b_1,0,b_3}.$$

Definition 6 *(Disjoint linear translation)*
By replacing of the input variable x_j with $x_j \oplus x_k$ into function f a new function

$$g(x_1, \cdots, x_j, \cdots, x_n) = f(x_1, \cdots, x_j \oplus x_k, \cdots, x_n)$$

is generated. This operation produces the permutation of the following pairs of spectral coefficients:

$$S_{g\, b_1, b_2, \cdots, b_j = 1, b_k = 1, \cdots, b_n} \leftrightarrow S_{f\, b_1, b_2, \cdots, b_j = 1, ib_k = 0, \cdots, b_n}$$

Example 5. *Consider the function $f(x_1, x_2, x_3) = x_1 \oplus x_2 x_3$ and the function $f_5(x_1, x_2, x_3) = x_1 \oplus x_1 x_2 \oplus x_2 x_3$. The function f_5 is optained by replacing x_3 with $x_1 \oplus x_3$. The spectrum of the function f_5 is $\mathbf{S}_5 = [0, 4, 0, -4, 4, 0, 4, 0]$, and following pairs of spectral coefficients will be permuted: $S_{50, b_2, 1} \leftrightarrow S_{1, b_2, 0}$, $S_{51, b_2, 0} \leftrightarrow S_{0, b_2, 1}$.*

Also, it exists a generalization of the invariant operations for Multi-valued functions, [3, 5].

3   New Operations in the Spectral Domain

In this section, new spectral invariant operation will be defined first for functions with an odd number of input variables, and then for functions with an even number of variables.

Theorem 1 *(Odd number of variables).*
Modification of a binary function f with an odd number of input variables $n = 2k + 1$ $(k \geq 2)$, which has a sum of disjoint products of two variables in its polynomial form: $f(x_1, \cdots, x_n) = x_{i_1} x_{i_2} \oplus x_{i_3} x_{i_4}$, where $(i_1, i_2) \cap (i_3, i_4) = \phi$, into a function

$$g(x_1, \cdots, x_n) = f(x_1, \cdots, x_n) \oplus x_{j_1} x_{j_2} x_{j_3}$$

where $j_1 \in (i_1, i_2)$, $j_2 \in (i_3, i_4)$ and $j_3 \notin (i_1, i_2, i_3, i_4)$ results in the following relations of the pairs of spectral coefficients:

$$S_{g\, b_1, \cdots, b_{j_4} = 1, b_{j_5} = 1, b_{j_3} = 0, \cdots, b_n} \leftrightarrow S_{f\, b_1, \cdots, b_{j_4} = 1, b_{j_5} = 1, b_{j_3} = 1, \cdots, b_n},$$

$$S_{g\, b_1, \cdots, b_{j_4} = 1, b_{j_5} = 1, b_{j_3} = 1, \cdots, b_n} \leftrightarrow S_{f\, b_1, \cdots, b_{j_4} = 1, b_{j_5} = 1, b_{j_3} = 0, \cdots, b_n},$$

where $j_4 = \{i_1, i_2\} \setminus j_1$ and $j_5 = \{i_3, i_4\} \setminus j_2$. All coefficients with $b_{j_4} = b_{j_5} = 1$, and $b_{j_3} = 0$ in the subscripts are permuted with the coefficients with $b_{j_4} = b_{j_5} = 1$, and $b_{j_3} = 1$ in the subscripts.

Proof: Consider the function $f(x_1, \cdots, x_n)$ with an odd number of input variables n, which is given by the disjoint polynomial form $f(x_1, \cdots, x_n) = x_{i_1} x_{i_2} \oplus x_{i_3} x_{i_4}$. By adding the product of variables $x_{j_1} x_{j_2} x_{j_3}$ where $j_1 \in (i_1, i_2)$, $j_2 \in (i_3, i_4)$ and $j_3 \notin (i_1, i_2, i_3, i_4)$ the function $g(x_1, \cdots, x_n) = x_{i_1} x_{i_2} \oplus x_{i_3} x_{i_4} \oplus x_{j_1} x_{j_2} x_{j_3}$ is generated.

To simplify the proof, consider the case when $j_1 = i_1, j_2 = i_3$, and $j_3 = i_5$. With this operation only the values of the function f for $x_{i_1} = 1, x_{i_3} = 1$, and $x_{i_5} = 1$ will be complemented. Since $f(x_1, \cdots, x_n) = x_{i_1}x_{i_2} \oplus x_{i_3}x_{i_4}$, the values of the function in those points will be: $f(x_1, \cdots, x_n) = 1 \cdot x_{i_2} \oplus 1 \cdot x_{i_4} = x_{i_2} \oplus x_{i_4}$, and the values of the function $g(x_1, \cdots, x_n) = x_{i_1}x_{i_2} \oplus x_{i_3}x_{i_4} \oplus x_{i_1}x_{i_3}x_{i_5} = 1 \cdot x_{i_2} \oplus 1 \cdot x_{i_4} \oplus 1 = x_{i_2} \oplus x_{i_4} \oplus 1$, as it is shown in Table 1.

In Table 1 the first five columns show the values of the input variables $x_{i_1}, x_{i_2}, x_{i_3}, x_{i_4}$, and x_{i_5}. In the columns f, g, F, and G the values of the functions f and g in $(0, 1)$ and $(1, -1)$ encoding are shown. In the next two columns the values of the Walsh functions through which the coeficients $S_{b_{i_2}=1,b_{i_4}=1}$ and $S_{b_{i_2}=1,b_{i_4}=1,b_{i_5}=1}$ are calculated, and finally in the last two columns the values of $S_{b_{i_2}=1,b_{i_4}=1}$ for the function g and the values of $S_{b_{i_2}=1,b_{i_4}=1,b_{i_5}=1}$ for the function f are shown.

The values in the column f are in correlation with the values of the Walsh function which is used for calculation of the coefficients with $b_{i_2} = b_{i_4} = 1$, and $b_{i_5} = 0$ in the subscript, while the values of the column g are in correlation with the values of the Walsh function used for calculation of the coefficient with $b_{i_2} = b_{i_4} = 1$, and $b_{i_5} = 1$ in the subscript. From that follows that the changes in the function f will have influence only on the coefficients having $b_{i_2} = b_{i_4} = 1$, and $b_{i_5} = 0$ in the subscripts, and for function g they will have the same values as the coefficients of the function f having $b_{i_2} = b_{i_4} = 1$, and $b_{i_5} = 1$. Also, the coefficients of the function g having $b_{i_2} = b_{i_4} = 1$, and $b_{i_5} = 1$ in the subscripts will have the same values as the coefficients of the function f having $b_{i_2} = b_{i_4} = 1$, and $b_{i_5} = 0$ in the subscripts.

Table 1. Proof

x_{i_1}	x_{i_2}	x_{i_3}	x_{i_4}	x_{i_5}	f	g	F	G	$X_{i_2}X_{i_4}$	$X_{i_2}X_{i_4}X_{i_5}$	$S_{f_{b_{i_2}=1,b_{i_4}=1,b_{i_5}=0}}$	$S_{g_{b_{i_2}=1,b_{i_4}=1,b_{i_5}=1}}$
1	0	1	0	1	0	1	1	-1	1	-1	1	1
1	0	1	1	1	1	0	-1	1	-1	1	1	1
1	1	1	0	1	1	0	-1	1	-1	1	1	1
1	1	1	1	1	0	1	1	-1	1	-1	1	1

Example 6. *The function $f_6(x_1, \cdots, , x_5) = x_1x_2 \oplus x_3x_4$ has the spectrum*

$$\mathbf{S}_6 = [8, 0, 8, 0, 8, 0, -8, 0, 8, 0, 8, 0, 8, 0, -8, 0, 8, 0, 8, 0, 8, 0, -8, 0,$$
$$-8, 0, -8, 0, -8, 0, 8, 0]^T,$$

while the function $f_7(x_1, \cdots, x_5) = f_6(x_1, \cdots, x_5) \oplus x_1x_3x_5$, the spectrum

$$\mathbf{S}_7 = [8, 0, 8, 0, 8, 0, -8, 0, 8, 0, 0, 8, 8, 0, 0, -8, 8, 0, 8, 0, 8, 0, -8, 0,$$
$$-8, 0, 0, -8, -8, 0, 0, 8]^T.$$

In this case the following relations of the pairs of coefficients exist:

$$S_{7_{b_1,b_2=1,b_3,b_4=1,b_5=0}} \leftrightarrow S_{6_{b_1,b_2=1,b_3,b_4=1,b_5=1}} \text{ for } b_1, b_3 \in (0, 1).$$

Also, adding of product of tree variables to the existing function $f(x_1, \cdots, x_n)$ with an odd number of input variables $n = 2 * k + 1$, it is possible when f has polynomial form with sum of disjoint products of two variables and some additional part (some terms are products of bigger number of variables). For example, if $f(x_1, \cdots, x_n) = x_{i_1} x_{i_2} \oplus x_{i_3} x_{i_4} \oplus h(x_1, \cdots, x_n)$, where $(i_1, i_2) \cap (i_3, i_4) = \phi$ by adding the product $x_{j_1} x_{j_2} x_{j_3}$ where $j_1 \in (i_1, i_2)$, $j_2 \in (i_3, i_4)$ and $j_3 \notin (i_1, i_2, i_3, i_4)$ the function $g(x_1, \cdots, x_n) = f(x_1, \cdots, x_n) \oplus x_{j_1} x_{j_2} x_{j_3}$ will be generated. This operation in the spectral domain results in the interchange of some pairs of spectral coefficients. However, the relation between subscripts in this case is more complex and depends of the part which is added to the disjoint products in the function f.

Example 7. *Consider function* $f_8(x_1, x_2, x_3, x_4, x_5) = x_1 x_2 \oplus x_3 x_4 \oplus x_1 x_2 x_5$ *with the spectrum:* $\mathbf{S}_8 = [12, -4, 12, -4, 12, -4, -12, 4, 4, 4, 4, 4, 4, 4, -4, -4, 4, 4, 4, 4, 4, 4, -4, -4, -4, -4, -4, -4, -4, -4, 4, 4]^T$.
And the function $f_9(x_1, x_2, x_3, x_4, x_5) = x_1 x_2 \oplus x_3 x_4 \oplus x_1 x_2 x_5 \oplus x_1 x_3 x_5$ *with the spectrum:* $\mathbf{S}_9 = [12, -4, 4, 4, 12, -4, -4, -4, 4, 4, 4, 4, 4, 4, -4, -4, 4, 4, 12, -4, 4, 4, -12, 4, -4, -4, -4, -4, -4, -4, 4, 4]^T$.
Following relations between coefficients exist:
$S_{9\,00010} \leftrightarrow S_{8\,10010}$, $S_{9\,00011} \leftrightarrow S_{8\,10011}$, $S_{9\,00110} \leftrightarrow S_{8\,10110}$, $S_{9\,00111} \leftrightarrow S_{8\,10111}$.

In analogy to Theorem 1 in the cases when the given function $f(x_1, \cdots, x_n)$ with an odd number of input variables $n = 2k + 1$, $k \geq 3$, has a polynomial form with m, $m \leq k$ disjoint product of variables it is possible to add the product of $m + 1$ variables and to preserve the values of the spectral coefficient with some permutations.

Example 8. *Consider the function* $f_{10}(x_1, \cdots, x_7) = x_1 x_2 \oplus x_3 x_4 \oplus x_5 x_6$ *and the function* $f_{11}(x_1, \cdots, x_7) = x_1 x_2 \oplus x_3 x_4 \oplus x_5 x_6 \oplus x_1 x_3 x_5 x_7$, *with the spectrum:*
$\mathbf{S}_{10} = [16, 0, 16, 0, 16, 0, -16, 0, 16, 0, 16, 0, 16, 0, -16, 0, 16, 0, 16, 0, 16, 0, -16, 0,$
$-16, 0, -16, 0, -16, 0, 16, 0, 16, 0, 16, 0, 16, 0, -16, 0, 16, 0, 16, 0, 16, 0, -16,$
$0, 16, 0, 16, 0, 16, 0, -16, 0, -16, 0, -16, 0, -16, 0, 16, 0, 16, 0, 16, 0, 16, 0,$
$-16, 0, 16, 0, 16, 0, 16, 0, -16, 0, 16, 0, 16, 0, 16, 0, -16, 0, -16, 0, -16, 0,$
$-16, 0, 16, 0, -16, 0, -16, 0, -16, 0, 16, 0, -16, 0, -16, 0, -16, 0, 16, 0, -16,$
$0, -16, 0, -16, 0, 16, 0, 16, 0, 16, 0, 16, 0, -16, 0]^T$,
while the spectrum of the function f_{11} *is:*
$\mathbf{S}_{11} = [16, 0, 16, 0, 16, 0, -16, 0, 16, 0, 16, 0, 16, 0, -16, 0, 16, 0, 16, 0, 16, 0, -16, 0,$
$-16, 0, -16, 0, -16, 0, 16, 0, 16, 0, 16, 0, 16, 0, -16, 0, 16, 0, 16, 0, 16, 0, -16,$
$0, 16, 0, 16, 0, 16, 0, -16, 0, -16, 0, -16, 0, -16, 0, 16, 0, 16, 0, 16, 0, 16, 0,$
$-16, 0, 16, 0, 16, 0, 16, 0, -16, 0, 16, 0, 16, 0, 16, 0, -16, 0, -16, 0, -16, 0,$
$-16, 0, 16, 0, -16, 0, -16, 0, -16, 0, 16, 0, -16, 0, -16, 0, -16, 0, 16, 0, -16,$
$0, -16, 0, -16, 0, 16, 0, 16, 0, 16, 0, 16, 0, -16, 0,]^T$.
Following pair of coefficients (in decimal notation) are related:
$S_{11\,42} \leftrightarrow S_{10\,43}$, $S_{11\,46} \leftrightarrow S_{10\,47}$, $S_{11\,58} \leftrightarrow S_{10\,59}$, $S_{11\,62} \leftrightarrow S_{10\,63}$,
$S_{11\,106} \leftrightarrow S_{10\,107}$, $S_{11\,110} \leftrightarrow S_{10\,111}$, $S_{11\,122} \leftrightarrow S_{10\,123}$, $S_{11\,126} \leftrightarrow S_{10\,127}$.

Theorem 2 *(Even number of input variables).*

Modification of a binary function $f(x_1, \cdots, x_n)$ with an even number of input variables $n = 2k$ ($k \geq 3$), witch has a sum of disjoint products of two variables in its polynomial expression:

$$f(x_1, \cdots, x_n) = x_{i_1} x_{i_2} \oplus x_{i_3} x_{i_4} \oplus x_{i_5} x_{i_6},$$

where $(i_1, i_2) \cap (i_3, i_4) \cap (i_5, i_6) = \phi$, into a function

$$g(x_1, \cdots, x_n) = f(x_1, \cdots, x_n) \oplus x_{j_1} x_{j_2} x_{j_3},$$

where $j_1 \in (i_1, i_2)$, $j_2 \in (i_3, i_4)$ and $j_3 \in (i_5, i_6,)$; results in the interchange of the following pairs of spectral coefficients:

$$S_{g\ldots, b_{j_4}=1, b_{j_5}=1, b_{j_3}=1, \ldots} \leftrightarrow S_{f\ldots, b_{j_4}=1, b_{j_5}=1, b_{j_3}=0, \ldots}.$$

Example 9. *Consider the function $f_{12}(x_1, x_2, x_3, x_4, x_5, x_6) = x_1 x_2 \oplus x_3 x_4 \oplus x_5 x_6$ with the spectrum:* $\mathbf{S}_{12} = [8, 8, 8, -8, 8, 8, 8, 8, -8, 8, 8, 8, 8, -8, -8, -8, -8, 8, 8, 8, 8, -8, 8, 8, 8, 8, -8, 8, 8, 8, 8, -8, -8, -8, -8, 8, 8, 8, 8, -8, 8, 8, 8, 8, -8, 8, 8, 8, 8, -8, -8, -8, -8, 8, -8, -8, -8, 8, -8, -8, -8, 8, -8, -8, -8, 8, 8, 8, 8, 8, -8]^T$,
and the function $f_{13}(x_1, x_2, x_3, x_4, x_5, x_6) = x_1 x_2 \oplus x_3 x_4 \oplus x_5 x_6 \oplus x_1 x_3 x_5$, with the spectrum: $\mathbf{S}_{13} = [8, 8, 8, -8, 8, 8, 8, 8, -8, 8, 8, 8, 8, -8, -8, -8, -8, 8, 8, 8, 8, -8, 8, -8, 8, 8, 8, 8, 8, -8, -8, 8, -8, -8, 8, 8, 8, 8, -8, 8, 8, 8, 8, -8, 8, 8, 8, 8, -8, -8, -8, -8, 8, -8, -8, -8, 8, -8, -8, -8, 8, -8, 8, 8]^T$.
In this case the following pairs of coefficients are permuted:
$S_{1320} \leftrightarrow S_{1222}, \quad S_{1321} \leftrightarrow S_{1223}, \quad S_{1328} \leftrightarrow S_{1230}, \quad S_{1329} \leftrightarrow S_{1231},$
$S_{1352} \leftrightarrow S_{1254}, \quad S_{1353} \leftrightarrow S_{1255}, \quad S_{1360} \leftrightarrow S_{1262}, \quad S_{1361} \leftrightarrow S_{1263}.$

It is possible to use this invariant operation for larger number of variables. Also, we will show that adding of product to the polynomial forms with disjoint products is possible also for small number of variables. However in this case this operation is covered by the already known invariant operations.

Example 10. *Consider the function $f_{14}(x_1, x_2, x_3) = x_1 x_2$ and the function $f_{15}(x_1, x_2, x_3) = x_1 x_2 \oplus x_1 x_3$. The function f_{15} is generated from f_{14} according to Theorem 1. However, it is possible to generate f_{15} from the function f_{14} by use of Disjoint spectral translation given by Definition 4, by replacing x_2 by $x_2 \oplus x_3$.*

4   Invariant Operation and Bent Functions

Note that function f_{12} from the Example 9 has disjoint homogeneus polynomial form, with disjoint products of pairs of variables where all variables of the function are included. This function has a flat spectrum, absolute values of all coefficients are equal, which means that f_{12} is a bent function. From Theorem 2 follows that with this invariant operation it is possible to generate from bent function with disjoint products of pairs of variables new bent function with cubic terms, as in case of the function f_{13}. By use of other invariant operations it is possible to generate from function f_{13} complex bent functions with large number of quadratic and cubic terms. Examples of such bent functions are shown in [4].

Example 11. *By replasing the variable x_1 with $x_1 \oplus x_4$ from function f_{13} the following function will be generated:*

$$f_{16}(x_1, x_2, x_3, x_4, x_5, x_6) = x_1 x_2 \oplus x_4 x_2 \oplus x_3 x_4 \oplus x_5 x_6 \oplus x_1 x_3 x_5 \oplus x_1 x_3 x_5.$$

The spectrum of function f_{16} remains flat and this functions is bent.

By replacing the variable x_3 with $x_3 \oplus x_6$ from function f_{16} the following function will be generated:

$$f_{17}(x_1, x_2, x_3, x_4, x_5, x_6) = x_1 x_2 \oplus x_4 x_2 \oplus x_3 x_4 \oplus x_4 x_4 \oplus x_5 x_6 \oplus x_1 x_3 x_5$$
$$\oplus x_1 x_5 x_6 \oplus x_3 x_4 x_5 \oplus x_4 x_5 x_6.$$

In comparison with f_{16} the function f_{17} has bigger number of quadratic and cubic terms in the polynomial form, but still remains bent.

5   Conclusions

Some new spectral invariant operations are defined for functions which have disjoint quadratic polynomial forms. The effect of these operations is visible for functions with $n \geq 5$ input variables. As result of applications of these new operations only the values of some subsets of coefficients are permuted, like in case of the spectral invariant operations which were known until now. It is also shown that the new invariant operations are useful for generation of bent functions. The properties of the existing and new spectral invariant operations has important consequences on bent functions. A function obtained by application of one or more spectral invariant operations to a bent function is also bent function.

References

1. Edwards, C.R.: The application of the Rademacher-Walsh transform to Boolean function classification and threshold logic synthesis. Trans. IEEE **C24**, 48–62 (1975)
2. Hurst, S.L.: The Logical Processing of Digital Signals, Crane. Russak & Company Inc./Edward Arnold, New York, London (1978)
3. Karpovsky, M., Stankovic, R.S., Astola, J.: Spectral Logic and Its Applications for the Design of Digital Devices. Wiley, Hoboken (2008)
4. Qu, C., Seberry, J., Pieprzyk, J.: A construction of hyperbent functions with polynomial trace form. Discrete Appl. Math. **102**, 133–139 (2000)
5. Stankovic, S., Stankovic, M., Astola, J.: Representation of multiple-valued functions wuth flat Vilenkin-Chrestenson spectra by decision diagram. J. Multiple-Valued Logic Soft Comput. **23**(5–6), 485–501 (2014)

Computer Vision, Deep learning and Applications

DetectionEvaluationJ: A Tool to Evaluate Object Detection Algorithms

C. Domínguez, M. García, J. Heras$^{(\boxtimes)}$, A. Inés, E. Mata, and V. Pascual

Department of Mathematics and Computer Science, University of La Rioja,
La Rioja, Spain
{cesar.dominguez,manuel.garcia,jonathan.heras,adrian.ines,eloy.mata,
vico.pascual}@unirioja.es

Abstract. Object detection is an area of computer vision with applications in several contexts such as biomedicine and security; and it is currently growing thanks to the availability of datasets of images, and the use of deep learning techniques. In order to apply object detection algorithms is instrumental to know the quality of the regions detected by them; however, such an evaluation is usually performed using ad-hoc tools for each concrete problem; and, up to the best of our knowledge, it does not exist a simple and generic tool to conduct this task. In this paper, we present DetectionEvaluationJ an open-source tool that has been designed to evaluate the goodness of object detection algorithms in any context and using several metrics. This tool is independent from the programming language employed to implement the detection algorithms and also from the concrete problem where such algorithms are applied.

1 Introduction

Object detection algorithms are applied in diverse computer vision applications; for instance, surveillance [25], traffic monitoring [10], or melanoma detection [2]. Recent advances in this area have been leaded by the availability of open datasets (e.g. the PASCAL VOC 2012 [3], the MS COCO datasets [13], or the ILSVRC competition [19]) and the application of deep learning techniques [18,21].

In order to evaluate the quality of object detection algorithms, the regions of interest (ROIs) located by such algorithms are compared against the regions manually annotated by experts (such regions are known as the *gold standard* or *ground truth*) using different metrics. Some of the most widely employed measures in this context are the area of intersection-over-union between two detections [3], or pixel-level specificity, precision, and recall [23].

Measuring the quality of object detection algorithms manually is not sensible since it is a time-consuming process – especially when dealing with hundreds of images, or when many algorithms are compared. Developing ad-hoc tools for concrete scenarios is not a solution either; it means reinventing the wheel several

Partially supported by Ministerio de Industria, Economía y Competitividad, project MTM2014-54151-P.

R. Moreno-Díaz et al. (Eds.): EUROCAST 2017, Part II, LNCS 10672, pp. 273–280, 2018.
https://doi.org/10.1007/978-3-319-74727-9_32

times, and usually this approach depends on a concrete programming language; hence, it is difficult to compare the results obtained with algorithms implemented in different programming languages. Therefore, a simple and generic tool to automatically evaluate object detection algorithms, independently of the concrete problem and the programming language employed to implement them, could be helpful; however, and up to the best of our knowledge, such a tool does not exist.

In this work, we have filled this gap by developing *DetectionEvaluationJ*, an open-source tool that has been designed to evaluate the goodness of object detection algorithms using several metrics, and that is independent to the concrete problem where the detection algorithms are applied, and does not depend either on the programming language employed to implement the detection algorithms.

2   DetectionEvaluationJ

DetectionEvaluationJ has been developed as a plugin of ImageJ [20] — an image-analysis tool that has been successfully employed to deal with many problems in life sciences [1,5,8,14]. In order to evaluate an object detection algorithm, DetectionEvaluationJ takes as input a set of images, the gold standard associated with such images, and the detected regions obtained by the algorithm; and, it generates, as output, a report that summarises the quality of the detection based on several available measures. The workflow of DetectionEvaluationJ is depicted in Fig. 1. The rest of this section is devoted to explain the main features of DetectionEvaluationJ.

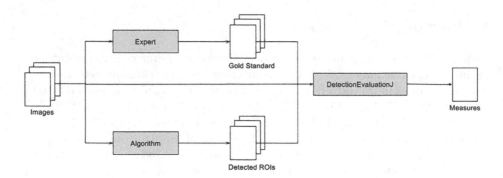

Fig. 1. Workflow of DetectionEvaluationJ

2.1   DetectionEvaluationJ's Input

As we have previously mentioned, the first component of DetectionEvaluationJ's input is the set of images, $\mathcal{S}$, where the detection algorithm will be employed to detect the objects. DetectionEvaluationJ supports the most common standard image-formats including tiff, jpeg, png, gif, and bmp.

In addition to the set of images $\mathcal{S}$, DetectionEvaluationJ takes as input the gold standard, $\mathcal{G}$, associated with such images, and the regions, $\mathcal{D}$, obtained by the detection algorithm. DetectionEvaluationJ can handle different kinds of regions (including rectangles, circles, polygons, points, and other geometrical figures) both for the gold standard and the detected regions; but, all the regions in $\mathcal{G}$ and $\mathcal{D}$ must have the same type to compare them properly.

The sets $\mathcal{D}$ and $\mathcal{G}$ can be loaded in DetectionEvaluationJ using a new format called ROIXML. The ROIXML format is based on the XML format and is, therefore, independent of any particular computer system and extensible for future needs. The structure of XML files following the ROIXML format is fixed by an XML schema [4], that not only determines the structure of XML files but also specifies and restricts the content of their elements — see Fig. 2 for a summary of the XML Schema of ROIXML, and the project webpage for the whole schema. This schema has been developed taking into account the information that is needed to encode different kinds of ROIs. The ROIXML format simplifies interoperability since it allows users from systems like OpenCV [11] or Matlab [15] to generate files that can be read by DetectionEvaluationJ — the programs to generate files in this format using OpenCV are available in the project webpage.

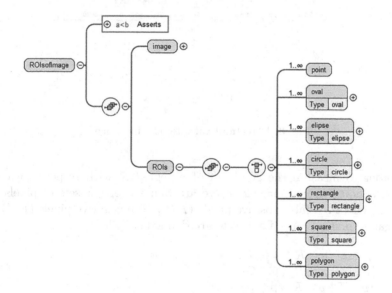

Fig. 2. XML Schema of ROIXML

Generating an XML file following the ROIXML format from the set of regions obtained by an object detection algorithm is a simple task — due to the fact that there are many programming tools that simplify this process. However, the XML language is not human-oriented, and, therefore, creating directly the gold standard using an XML format is not sensible since this task is conducted by human experts in the area wherein the images are obtained. We tackle this problem in DetectionEvaluationJ by using ImageJ's features for ROI management. DetectionEvaluationJ integrates an enhanced version of ImageJ's "ROI manager", a tool that allows the users to handle (add, remove, and modify) multiple selections (ROIs) from different locations on an image, and, thanks to the improvement introduced in DetectionEvaluationJ, save the annotations in the ROIXML format for further usage.

2.2   Interface and Measures Provided by DetectionEvaluationJ

Using the interface shown in Fig. 3, the DetectionEvaluationJ's users can load the gold standard and the detected regions; and, subsequently, measure how good are the detected regions regarding the gold standard.

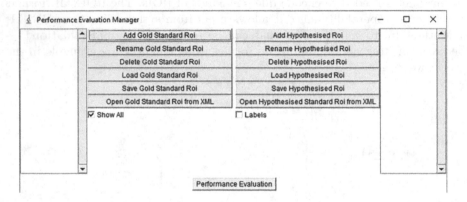

Fig. 3. DetectionEvaluationJ's interface

In order to measure the goodness of a region $\mathcal{R}$ with respect to a gold-standard region $\mathcal{G}$ in an image $\mathcal{I}$, there are four important sets of pixels: true positive pixels (TP), false positive pixels (FP), false negative pixels (FN), and true negative pixels (TN). These sets are defined as follows:

- $TP = \{p \in \mathcal{I} \mid p \in \mathcal{R} \land p \in \mathcal{G}\}.$
- $FP = \{p \in \mathcal{I} \mid p \in \mathcal{R} \land p \notin \mathcal{G}\}.$
- $FN = \{p \in \mathcal{I} \mid p \notin \mathcal{R} \land p \in \mathcal{G}\}.$
- $TN = \{p \in \mathcal{I} \mid p \notin \mathcal{R} \land p \notin \mathcal{G}\}.$

From these sets, several metrics can be defined. In the case of DetectionEvaluationJ, we conducted a thorough review of articles related to object detection, and spotted the measures listed in Table 1. All the measures provided in Table 1 are implemented in DetectionEvaluationJ, and this tool has been designed to incorporate easily new measures in the future.

Table 1. Measures provided by DetectionEvaluationJ

Measure	Equation	Also known as	Reference
Positive	$P = TP + FN$	-	-
Negative	$N = TN + FP$	-	-
Accuracy	$ACC = \frac{TP+TN}{P+N}$	Rand index	[7]
Precision	$PR = \frac{TP}{TP+FP}$	-	[7]
Recall	$Rec = \frac{TP}{P}$	Sensitivity, True positive rate	[7]
Fallout	$Fall = \frac{FP}{N}$	False positive rate	[7]
Specificity	$Spe = \frac{TN}{N}$	True negative rate	[7]
False negative rate	$FNR = \frac{FN}{P}$	-	[17]
Negative predictive value	$NPV = \frac{TN}{FN+TN}$	-	[6]
False discovery rate	$FDR = \frac{FP}{TP+FP}$	-	[6]
LR+	$LR+ = \frac{Rec}{1-Spe}$	Positive likelihood	[6]
LR-	$LR- = \frac{1-Rec}{Spe}$	Negative likelihood	[6]
F-measure ($\alpha = 0.5, 1, 2$)	$F(\alpha) = \frac{1}{\alpha\frac{1}{PR} + (1-\alpha)\frac{1}{Rec}}$	$F(1)$ = Dice coefficient	[7]
Intersection over union	$IOU = \frac{TP}{TP+FN+FP}$	Jaccard index	[7]
Fowlkes-mallows index	$G = \sqrt{PR * Rec}$	G-measure	[16]
Matthews correlation coefficient	$MCC = \frac{TP \times TN - FP \times FN}{\sqrt{P \times N \times (TP+FP) \times (FN+TN)}}$	-	[16]
Youden's J statistic	$J = Rec + Spe - 1$	Youden's index, Informedness	[6]
Markedness	$M = PR + NPV - 1$	Yule coefficient	[17]
Diagnostic odds ratio	$DOR = \frac{LR+}{LR-}$	-	[6]
Balanced accuracy	$BACC = \frac{Rec+Spe}{2}$	Galanced classification rate	[24]

3   Discussion and Conclusions

As we have previously explained, the main application of DetectionEvaluationJ consists in measuring the goodness of a detection algorithm compared to a gold-standard. In addition, this tool can be employed to compare the effectiveness of several algorithms; in particular, the users can load the regions detected by many algorithms and compare them by analysing the tables containing the measures previously presented and the ROC space [12], see Fig. 4. More generally, this tool can be applied to compare two sets of regions obtained by any means. For instance, this can be applied to study inter-rater agreement among experts [22] or to compare manual drawings against expected results.

DetectionEvaluationJ is a free and open-source tool that allows the evaluation of object detection algorithms, and that is independent from the programming language employed to implement the detection algorithms and also from the concrete problem where such algorithms are applied. This tool has been already successfully employed to evaluate algorithms for detecting halos in antibiogram images [1] and for detecting DNA bands in gel images [9].

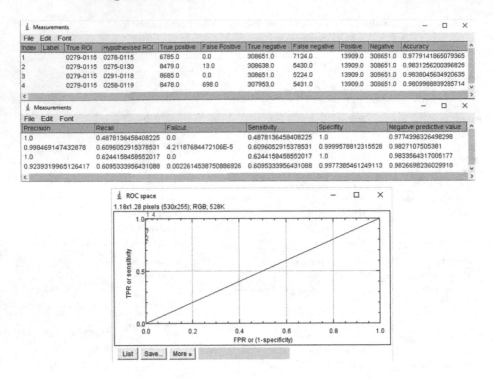

Fig. 4. Top and Middle. Tables produced by DetectionEvaluationJ containing the measures for several algorithms (each row corresponds to the results associated with an algorithm). **Bottom.** ROC Space generated by DetectionEvaluationJ.

4   Availability and Requirements

- Project name: DetectionEvaluationJ.
- Project home page: https://joheras.github.io/DetectionEvaluationJ/.
- Operating system(s): platform independent.
- Programming language: Java.
- License: GNU GPL 3.0.
- Any restrictions to use by non-academics: None.
- Dependencies: ImageJ.

The project home page contains the installation instructions and usage examples.

References

1. Alonso, A., et al.: AntibiogramJ: a tool for analysing images from disk diffusion tests. Comput. Methods Program. Biomed. **143**, 159–169 (2017)
2. Codella, N., et al.: Deep learning ensembles for melanoma recognition in dermoscopy images. IBM J. Res. Dev. **61**(4), 5:1–5:15 (2017)
3. Everingham, M., et al.: The PASCAL visual object classes challenge 2012 (VOC2012) results (2012). http://www.pascal-network.org/challenges/VOC/voc2012/workshop/index.html
4. Evjen, B., et al.: Professional XML. Wiley Publishing Inc., Hoboken (2007)
5. Ghasemian, F., et al.: An efficient method for automatic morphological abnormality detection from human sperm images. Comput. Methods Program. Biomed. **122**(3), 409–420 (2015)
6. Glas, A.S., et al.: The diagnostic odds ratio: a single indicator of test performance. J. Clin. Epidemiol. **56**, 1129–1135 (2003)
7. Gutman, D., et al.: Skin lesion analysis toward melanoma detection: a challenge at the international symposium on biomedical imaging (ISBI) 2016. arXiv preprint arXiv:1605.01397 (2016)
8. Heras, J., et al.: GelJ - a tool for analyzing DNA fingerprint gel images. BMC Bioinf. **16**, 270 (2015)
9. Heras, J., et al.: Surveying and benchmarking techniques to analyse DNA gel fingerprint images. Brief. Bioinf. **17**(6), 912–925 (2015)
10. Huang, S.C., Chen, B.H.: Highly accurate moving object detection in variable bit rate video-based traffic monitoring systems. IEEE Trans. Neural Netw. Learn. Syst. **24**(12), 1920–1931 (2013)
11. Kaehler, A., Bradski, G.: Learning OpenCV 3. O'Reilly Media, Sebastopol (2015)
12. Lasko, T.A., et al.: The use of receiver operating characteristic curves in biomedical informatics. J. Biomed. Inf. **38**(5), 404–415 (2005)
13. Lin, T.-Y., Maire, M., Belongie, S., Hays, J., Perona, P., Ramanan, D., Dollár, P., Zitnick, C.L.: Microsoft COCO: common objects in context. In: Fleet, D., Pajdla, T., Schiele, B., Tuytelaars, T. (eds.) ECCV 2014. LNCS, vol. 8693, pp. 740–755. Springer, Cham (2014). https://doi.org/10.1007/978-3-319-10602-1_48
14. Mata, G., et al.: SynapCountJ: a validated tool for analyzing synaptic densities in neurons. Commun. Comput. Inf. Sci. **690**, 41–55 (2017)
15. MathWorks: Matlab version 9.0.0.341360 (R2016a). The MathWorks Inc., Natick, Massachusetts (2016)
16. Orlando, J.I., et al.: A discriminatively trained fully connected conditional random field model for blood blessed segmentation in fundus images. IEEE Trans. Biomed. Eng. **64**(1), 16–27 (2015)
17. Powers, D.M.W.: Evaluation: from precision, recall and F-factor to ROC, informedness, markedness and correlation. Int. J. Mach. Learn. Technol. **2**(1), 37–63 (2011)
18. Redmon, J., Farhadi, A.: YOLO9000: better, faster, stronger. arXiv preprint arXiv:1612.08242 (2016)
19. Russakovsky, O., et al.: ImageNet large scale visual recognition challenge. Int. J. Comput. Vis. (IJCV) **115**(3), 211–252 (2015)
20. Schneider, C.A., Rasband, W.S., Eliceiri, K.W.: NIH Image to ImageJ: 25 years of image analysis. Nature Methods **9**(7), 671–675 (2012)
21. Shaoqing, R., et al.: Faster R-CNN: towards real-time object detection with region proposal networks. IEEE Trans. Pattern Anal. Mach. Intell. **39**(6), 1137–1149 (2017)

22. Silva, J.S., et al.: Algorithm versus physicians variability evaluation in the cardiac chambers extraction. IEEE Trans. Inf. Technol. Biomed. **16**(5), 835–841 (2012)
23. Wolf, C., Jolion, J.M.: Object count/area graphs for the evaluation of object detection and segmentation algorithms. Int. J. Doc. Anal. Recogn. **8**, 280–296 (2006)
24. Zalama, E., et al.: Road crack detection using visual features extracted by gabor filters. Comput. Aided Civil Infrastruct. Eng. **29**, 342–358 (2014)
25. Zhai, M., et al.: Object detection in surveillance video from dense trajectories. In: 14th IAPR International Conference on Machine Vision Applications. IEEE (2015)

Evaluation of Whole-Image Descriptors
for Metric Localization

Manuel Lopez-Antequera[1,2]([✉]), Javier Gonzalez-Jimenez[1],
and Nicolai Petkov[2]

[1] MAPIR-UMA Group, Instituto de Investigación Biomédica de Málaga (IBIMA),
University of Malaga, Málaga, Spain
{mlopezantequera,javiergonzalez}@uma.es
[2] Johann Bernoulli Institute of Mathematics and Computing Science,
University of Groningen, Groningen, The Netherlands
n.petkov@rug.nl
http://mapir.isa.uma.es, http://www.cs.rug.nl/is

Abstract. Appearance-based localization attempts to recover the position (and orientation) of a camera based on the images that it captured and a previously stored collection of images. Recent advances in image representations extracted using convolutional neural networks for the task of place recognition have produced whole-image descriptors which are robust to imaging conditions, including small viewpoint changes. In previous work, we have used these descriptors to perform localization by performing descriptor interpolation to compare the appearance of the image that is currently captured with the *expected* appearance at a candidate location. In this work, we directly study the behaviour of recently developed whole-image descriptors for this application.

1 Introduction and Related Work

Image-based localization is often regarded as a nearest-neighbor search, also known as place recognition, where the pose space is represented by a set of discrete locations. Although efficient solutions exist for this discretization, the general case of localizing in continuous space without the use of local keypoints remains a challenge. Describing images using holistic (whole-image) descriptors is the standard practise for topological localization (*"in which place is the image?"*). However, holistic descriptors can also be used for metric localization (*"what is the pose where the image has been taken from?"*).

We study the general case of metric localization (Fig. 3), in which there are no restrictions on the position of the camera. Instead of selecting the best location from a grid, graph or collection of known images, any point in the pose space can be selected as the estimate.

Performing continuous metric localization without local keypoints is challenging, however, it has shown promising results in previous work: in [3], the authors use global Fourier descriptors on omnidirectional images, performing regression

© Springer International Publishing AG 2018
R. Moreno-Díaz et al. (Eds.): EUROCAST 2017, Part II, LNCS 10672, pp. 281–288, 2018.
https://doi.org/10.1007/978-3-319-74727-9_33

through a Gaussian process and representing the location of the camera by a particle filter. Recently, we [5] performed localization through appearance regression, using a convolutional neural network-based descriptor trained for place recognition as the image representation. These results indicate that metric localization through appearance regression is feasible and merits further research.

In this paper, we analyze the current state of the art in convolutional neural network-based holistic descriptors with respect to their applicability for metric localization through appearance regression and discuss the desired characteristics of a whole-image descriptor for their use in such systems. First, we will describe the components of such a localization system in order to establish a framework. We then examine the behaviour of several state-of-the-art convolutional neural network-based holistic descriptors in search for proportionality and smoothness in the change of their values with respect to camera motion. Finally, we leverage recent advances in image synthesis to invert the representations extracted using these descriptors, in order to understand what visual characteristics are being modelled by them.

Our findings confirm that descriptors trained to perform place recognition are better suited for image-based localization than generic internal representations of networks trained to perform object recognition, while also being less demanding on memory and compute time.

2   Metric Localization with Whole-Image Descriptors

Whole-image descriptors are normally used to generate place recognition candidates for loop closure in SLAM systems, that is, to detect when a vehicle revisits a place in order to close the SLAM loop and correct any drift that has been accumulated through visual odometry (Fig. 1).

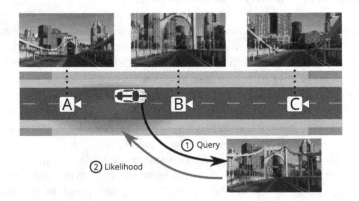

Fig. 1. The appearance of the query image suggests that it is located somewhere between images A and B. State of the art holistic descriptors present smooth changes in their values with changes in camera pose and could be used to localize on a finer grain than nearest-neighbor approaches.

In previous work [5] we explored the possibility of using these whole-image descriptors for the full visual localization pipeline. A system was developed to obtain the location and orientation of a moving camera, based solely on the sequence of captured images. The main concept that allows the system to work is the evaluation of the similarity between the appearance of the currently captured image and images which are previously captured from many locations in the environment. Since space is continuous and images cannot be captured at every possible location, a need for *interpolating* image appearance or image descriptors arises. In [5], we used Gaussian Processes to perform this interpolation, but the general concept doesn't depend on this: There is a need to interpolate or predict the descriptor that would be extracted at any location. In this work, we study a set of commonly available CNN based descriptors with respect to some suitable properties for this task.

3   Experiments

We attempt to shed light into the behaviour of holistic descriptors. Even though their usefulness for localization has been studied previously, some assumptions and properties have not been individually studied.

3.1   Interpolation Using Different *Baselines*

In this experiment we study some properties of the interpolation of the appearance of images over a set of images captured by a front-facing camera on a moving vehicle, a sample of which is shown in Fig. 2. We do this to evaluate the behaviour of the different feature extractors when the pose of the camera changes. First, we extract whole-image descriptors for all images in the sequence, using different state of the art convolutional neural networks [1,2,4]. Then, we

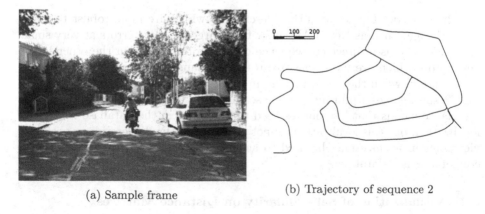

(a) Sample frame   (b) Trajectory of sequence 2

Fig. 2. We used sequence 2 from the KITTI dataset for the experiments

subsample the sequence, separating it into a training set and a test set. We perform linear interpolation of the descriptors over the vehicle position (in meters, over the trajectory of the sequence) and compare the error when using different descriptors.

The error is computed as the L2 norm of the distance between the original descriptor extracted from the images and the interpolated descriptor. As different descriptors have different dimensionality and a different scale in descriptor space, direct comparison is not possible. We normalize each curve by dividing by the maximum error obtained at very large baselines, when interpolation is no longer feasible.

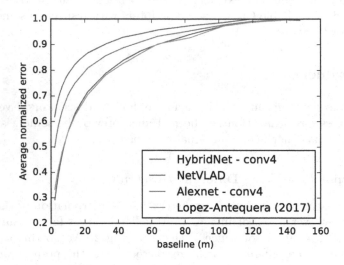

Fig. 3. Normalized error when performing linear interpolation as a function of the average separation between data points (baseline).

The resulting Fig. 3 shows that descriptors which are more robust to viewpoint changes such as NetVLAD suffer from larger relative errors at very small baselines. This is a direct consequence of that robustness. On the other hand, descriptors which are more sensitive to these changes also encode information that is useful when the camera moves in small increments. As a take away message, a suitable descriptor must be chosen for each application. Larger viewpoint invariance means that less images need to be stored in the database to achieve localization, but this localization cannot be as fine-grained. Descriptors with less viewpoint invariance may be used to achieve finer grained localization, at the cost of a denser database.

3.2   Visualization of Self-similarity on Distance Matrices

We select the first 200 images of the second sequence of the KITTI odometry dataset and extract whole-image descriptors for each frame. We then calculate the pairwise euclidean distances for every pair and plot them in distance matrices (see Fig. 4). By observing the thickness and smoothness of the main diagonal, we can extract insight about the viewpoint invariance that each descriptor presents.

NetVLAD descriptors change drastically when the image is not exactly the same (there is a very sharp diagonal), but then present distance changes proportional to camera motion. The descriptors extracted from Alexnet behave differently, depending on the pooling operation that is performed: when maxpooling, the diagonal is sharp and there is little viewpoint invariance. Average pooling allows the descriptor distance to change more smoothly.

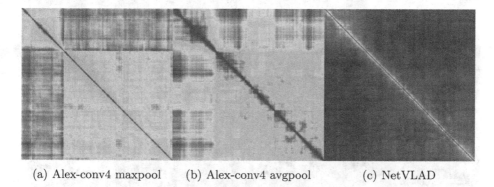

  (a) Alex-conv4 maxpool   (b) Alex-conv4 avgpool   (c) NetVLAD

Fig. 4. Confusion matrices corresponding to the first 200 frames of the KITTI-odometry-02 sequence. Again, we can observe another manifestation of the viewpoint invariances offered by the different descriptors.

3.3   Evaluation of Descriptor Smoothness w.r.t Pose Changes

Both of our previous experiments were limited to one dimensional movement, simplifying the problem. Here we attempt to visualize the influence of camera translation and rotation separately.

Ideally, whole-image descriptors should present smooth change as the camera moves around the scene if they are to be interpolated. In [5] we use a Gaussian Process to interpolate the appearance of image descriptors over translation and rotation (yaw motion) of the camera. To study if descriptors indeed change smoothly when the camera moves, we have extracted whole-image descriptors from all of the KITTI sequences. We then calculate the descriptor distance between every pair of images in the dataset, and plot the average descriptor distance on a polar plot where distance and rotation are kept separate (see Fig. 5).

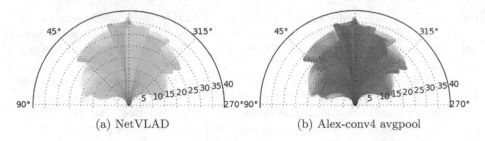

(a) NetVLAD   (b) Alex-conv4 avgpool

Fig. 5. Descriptor distances plotted against rotation and translation

Fig. 6. Test sample (first column) and images synthesized to yield the same descriptor as the test sample. Caffenet (conv5) is shown on top, and Hybridnet (conv5) on the bottom.

3.4 Visualization of Whole-Image Descriptors Through Reconstruction

We attempt to visualize what different convolutional neural network-based holistic descriptors are representing in order to gain some insight about what image features are relevant for place recognition. To do so, we leverage recent work [6] in which a deep generator network is trained to synthesize images. This network can be then included in an optimization process so that the generated image produces a descriptor that is similar to a desired descriptor.

More formally, we first extract a whole-image descriptor $d_x = f(x)$ from the CNN that we attempt to visualize. The generative model produces another image $y = g(l)$ where l is the latent vector (activations from the fc6 layer in this case). This generated image y also produces a whole-image descriptor $d_y = f(y)$. As all of the chain is made up of differentiable operations, it can be optimized with gradient descent to minimize $||d_y - d_x||$. The end result is an image that is completely synthetic but produces a descriptor which is similar to that of the original image.

We performed this operation on a selection of convolutional neural networks and input images, over several random initializations of the latent vector, obtaining several synthetic images which can be seen in Fig. 6. Although these experiments are merely qualitative, we can observe how Alexnet, trained for general object recognition, is representing finer details which are not so relevant por place recognition. In contrast, in the images reconstructed from Hybridnet [2] (a network trained to perform place recognition), we can observe how some details are blurred out, such as street markings or cars, which should be ignored when performing place recognition.

Acknowledgments. This work has been supported by the Spanish Government, the "European Regional Development Fund ERDF" under contract DPI2014-55826-R, and the EU-H2020 project MOVECARE.

References

1. Arandjelović, R., Gronat, P., Torii, A., Pajdla, T., Sivic, J.: NetVLAD: CNN architecture for weakly supervised place recognition (2015). arXiv http://arxiv.org/abs/1511.07247
2. Chen, Z., Jacobson, A., Sunderhauf, N., Upcroft, B., Liu, L., Shen, C., Reid, I., Milford, M.: Deep learning features at scale for visual place recognition, January 2017. arXiv http://arxiv.org/abs/1701.05105
3. Huhle, B., Schairer, T., Schilling, A., Straßer, W.: Learning to localize with Gaussian process regression on omnidirectional image data. In: 2010 IEEE/RSJ International Conference on Intelligent Robots and Systems (IROS) (2010)
4. Lopez-Antequera, M., Gomez-Ojeda, R., Petkov, N., Gonzalez-Jimenez, J.: Appearance-invariant place recognition by discriminatively training a convolutional neural network. Pattern Recogn. Lett. (2017). http://www.sciencedirect.com/science/article/pii/S0167865517301381

5. Lopez-Antequera, M., Petkov, N., Gonzalez-Jimenez, J.: Image-based localization using Gaussian processes. In: 2016 International Conference on Indoor Positioning and Indoor Navigation (IPIN) (2016)
6. Nguyen, A., Dosovitskiy, A., Yosinski, J., Brox, T., Clune, J.: Synthesizing the preferred inputs for neurons in neural networks via deep generator networks. In: Advances in Neural Information Processing Systems 29 (2016)

Filtering and Segmentation of Retinal OCT Images

Miguel Alemán-Flores[1](✉) and Rafael Alemán-Flores[2]

[1] Departamento de Informática y Sistemas,
Universidad de Las Palmas de Gran Canaria, Las Palmas, Spain
miguel.aleman@ulpgc.es
[2] Departamento de Morfología, Universidad de Las Palmas de Gran Canaria,
Las Palmas, Spain
raleman@dmor.ulpgc.es

Abstract. This work presents a method for the segmentation of optical coherence tomography images of the retina. Before segmenting the tomography, anisotropic diffusion is applied to reduce noise, but preserve the relevant edges. Afterward, the intensity profile of the images is analyzed to extract an initial approximation for the segmentation of three bands within the retina. Finally, a combination of attraction and regularization terms is used to refine the segmentation by fitting the limits of the bands to the highest gradients and smoothing their shapes to make them more regular. From the bands extracted in the different slices of the tomography, a three-dimensional reconstruction is performed for a better visualization of the results.

1 Introduction

The field of medical images includes a wide range of modalities, which are used for the diagnosis and monitoring of different kinds of pathologies. They share some common problems, such as the presence of noise, blurred edges, low contrast or heterogeneities. OCT (Optical Coherence Tomography) is an image modality that has important applications in several medical fields, such as ophthalmology, dermatology or cardiology. In this work, we focus on retinal optical coherence tomography, which provides a series of cross-sectional images of the retina. Some relevant pathologies which can be analyzed by means of the retinal OCT are age-related macular degeneration, macular edema, macular hole, epiretinal membrane or central serous chorioretinopathy. As we deal with tomographies, we need to cope with the combination of two- and three-dimensional visualizations and large amounts of data. Most OCT units allow some automatic processing of the acquired images. Nevertheless, the high rate of noise and the heterogeneity of the retinal layers limit the possibilities of an automatic or semi-automatic analysis of the images.

Between the internal limiting membrane, which is the innermost part of the retina, and Bruch's membrane, which is the limit between the retina and the

© Springer International Publishing AG 2018
R. Moreno-Díaz et al. (Eds.): EUROCAST 2017, Part II, LNCS 10672, pp. 289–296, 2018.
https://doi.org/10.1007/978-3-319-74727-9_34

choroid, several layers can be found (for example, the retinal nerve fiber layer, the ganglion cell layer, the inner and outer plexiform layers, the inner and outer nuclear layers, or the retinal pigment epithelium). Taking into account the intensity ranges, these layers can be observed as three bands. The lower band is brighter than the rest of the layers, the upper band is wider, but not so bright as the previous one, and the intermediate band is relatively dark. In this work, we intend to automatically extract these three bands. In Fig. 1 we can see these bands in two different OCT scans (one with macular edema, and one where the typical shape of the macula can be noticed).

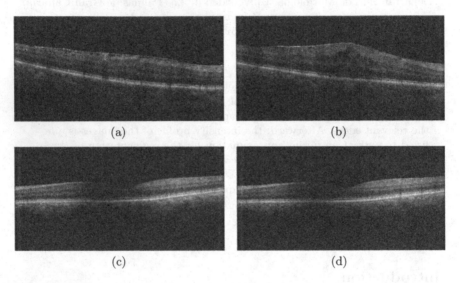

(a)   (b)

(c)   (d)

Fig. 1. Sample slices of two retinal OCT scans: (a) and (b) correspond to an OCT scan of a macular edema, while (c) and (d) illustrate a case with the typical shape of the macula.

These images present a significant amount of noise, mainly due to the acquisition process. For this reason, the first step in the processing of the images is a filtering stage. Since we want to preserve the limits of the regions, we apply an anisotropic implementation of Perona-Malik approach [1]. In this implementation, the diffusion is performed with the neighboring voxels according to the magnitude of the gradient, in such a way that those voxels where the gradient is lower contribute more to the diffusion process. Once the slices have been filtered, we apply an adaptation of the active contours technique to extract the bands described above by means of what we call active bands. In our proposal we combine three terms: an expansion term to obtain the initial approximation, a regularization term to smooth the limits of the bands, and an attraction term to adjust the limits to the highest gradients.

From the contours of the bands that have been extracted in the different slices of the tomography, a three-dimensional reconstruction of the retina can be obtained, so that it is possible to study its shape, extract measurements, or perform any other further analysis.

2   Anisotropic Filtering

With the aim of reducing noise, but preserving the edges, we first apply an implementation of Perona-Malik filter [1]:

$$u_t = div\left(k\left(\|\nabla u\|\right)\nabla u\right), \tag{1}$$

where we use:

$$k\left(x\right) = e^{-\beta x}. \tag{2}$$

As we deal with a series of uniformly spaced images, when the distance is short we can apply the filter in three dimensions. In such case, the noise reduction process also takes into account the neighbors in the previous and next images in the OCT, i.e., the values at the same position in the neighboring images. Since the distance between two consecutive images may not be the same as the distance between the pixels within an image, different weights can be assigned to the neighbors in the different coordinates.

Depending on the similarity of the various elements in the 3D image, their contrast and texture, the value of β in (2) can be adapted, as well as the number of iterations in the following discrete approach:

$$u_{i,j,k}^{n+1} = u_{i,j,k}^n + \frac{dt}{2\left(dh\right)^2}M\left(u_{i,j,k}^n\right), \tag{3}$$

where $M\left(u_{i,j,k}^n\right)$ is the result of convolving at each point (i,j,k) in the iteration n with the $3\times 3\times 3$ mask whose coefficients are:

$$C_{i+a,j,k} = k_{i+a,j,k} + k_{i,j,k}$$
$$C_{i,j+a,k} = k_{i,j+a,k} + k_{i,j,k}$$
$$C_{i,j,k+a} = k_{i,j,k+a} + k_{i,j,k}$$
$$C_{i,j,k} = -k_{i+1,j,k} - k_{i-1,j,k} - k_{i,j+1,k}$$
$$-k_{i,j-1,k} - k_{i,j,k+1} - k_{i,j,k-1} - 6k_{i,j,k} \tag{4}$$

and $a \in \{-1, 1\}$. The values of $k_{i,j,k}$ are obtained from (2) as follows:

$$k_{i,j,k} = e^{-\beta\|\nabla u\|_{i,j,k}}. \tag{5}$$

An increase of β preserves more edges, but also noise. Therefore, its value must be adapted to the amount of noise present in the image and the relevance of the edges to be considered. This approach results in an anisotropic implementation, which preserves the most relevant edges and whose results can be observed in Fig. 2.

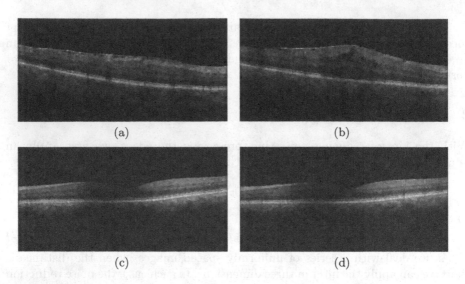

(a)   (b)

(c)   (d)

Fig. 2. Result of the anisotropic filtering process for the slices in Fig. 1.

3   Active Contours

Geodesic active contours, also known as snakes, are based on the minimization of the following energy with respect to the contour C:

$$E_{gac}(C) = \int_C g_\sigma(C(s))\, ds, \qquad (6)$$

where $C : [0, L] \to \mathbb{R}^2$, $L > 0$, is a rectifiable curve parameterized by arc-length s, and ds denotes the arc-length element. The function $g_\sigma(x, y)$ is used to stop the evolution of the snake when it approaches the edges. It is a smooth decreasing function of the modulus of the gradient of a regularized version of the image $I(x, y)$ on which the segmentation is performed, and acts as an edge detector [2]. The level set formulation of the geometric curve evolution is given by:

$$\frac{\partial u}{\partial t} = \|\nabla u\| \operatorname{div}\left(g_\sigma(I)\frac{\nabla u}{\|\nabla u\|}\right). \qquad (7)$$

If we expand this equation, we obtain the following expression, in which the first term controls the smoothness of the contour and the second one makes the contour evolve toward the highest gradients:

$$\frac{\partial u}{\partial t} = g_\sigma(I)\|\nabla u\| \operatorname{div}\left(\frac{\nabla u}{\|\nabla u\|}\right) + \lambda \nabla u \nabla g_\sigma(I). \qquad (8)$$

The parameter $\lambda > 0$ is introduced to balance the contribution of both terms. If we increase the value of λ, the attraction term will have a higher contribution and the contour will try to fit to the highest gradients in the current configuration. On the other hand, decreasing its value will round the contour, making it tend to a more regular outline. When an initial approximation is available, geodesic active contours permit to improve the pre-segmentation, since the contour adapts to the minimum of the energy in (6). This pre-segmentation must be relatively close to the real contour of the region to segment. Otherwise, the effect of the second term in (8) will not be enough to overcome the regularizing effect of the first term and, instead of approaching the real edges, the snake will be rounded and will tend to reduce. Active contours have previously been used for the segmentation of different types of medical images (see for instance [3]).

4   Active Bands

The extraction of the initial approximation is one of the most important drawbacks of active contours. Manual delimitation is extremely time-consuming, even more when dealing with 3D images, as in our case. The use of region-growing algorithms is too risky when the limits are not clearly defined. This is the reason why we use a combination of different morphological operators. Instead of adopting the classical level-set approach with a range of values and a limit to separate the inner and outer regions, we work with an upper and a lower limit of the retina and two internal limits to separate the three bands.

A classical approach when applying morphological operators to obtain an initial approximation consists in using a balloon force which expands the region from the initial seed while the magnitude of the gradient is lower than a certain threshold [4,5]. This can be expressed with the following PDE:

$$\frac{\partial u}{\partial t} = g\left(I\right) v \|\nabla u\|, \tag{9}$$

where $g\left(I\right)$ is a stopping function. This PDE can be used for both, expanding or contracting contours, depending on the sign of v.

However, we try to find the limits of three bands in every slice of the OCT scan instead of expanding in all directions. Therefore, we proceed as follows: for each column of each filtered image in the series, we extract the intensity profile, i.e., the evolution of the intensity as we move from top to bottom. We apply a Gaussian convolution to the profile of the filtered image in order to obtain a more suitable profile. As observed in Fig. 3, there are two peaks which correspond to the upper and lower bright bands. In this smoothed profile, we extract the two most significant local maxima, which allow us to identify the locations the first and third bands. From these local maxima, we expand up- and downward until the intensity has decreased a given percentage form the corresponding maximum. This way, we obtain an approximation for the limits of the first and third bands and, indirectly, for the second one.

However, this initial approach does not provide completely satisfactory limits. Therefore, two more terms are introduced. One of them is a regularizing term, which aims at smoothing the edges and filling the holes of the segmentation,

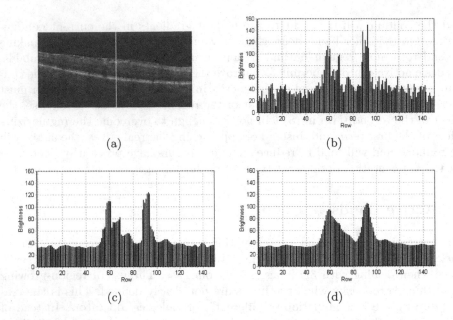

Fig. 3. Analysis of the intensity profile: (a) sample column to analyze the profile, (b) profile for the original image, (c) profile for the filtered image, (d) smoothed profile for the filtered image.

avoiding an extremely irregular contour. In the classical snakes, this is obtained by controlling the curvature of the contour as follows:

$$\frac{\partial u}{\partial t} = g\left(I\right)\left\|\bigtriangledown u\right\| \left(div\left(\frac{\bigtriangledown u}{\left\|\bigtriangledown u\right\|} \right) \right). \tag{10}$$

In our case, we rely on statistical filters. Since the initial process is applied column by column, there may be significant oscillations within a given image and also between neighboring images, as observed in Fig. 4. For each position which has been extracted, the median of the values for the same limit (upper or lower limit of a certain band) in a neighborhood of the same image allows eliminating the possible outliers. On the other hand, the mean of the limits in the neighborhood generates a smoother shape. Finally, the median across the images allows correcting the loss of track when shadows or other artifacts avoid finding the limits.

Finally, the third term is an attraction term, similar to that described in Sect. 3 (second term in (8)):

$$\frac{\partial u}{\partial t} = \bigtriangledown g\left(I\right) \bigtriangledown u. \tag{11}$$

In our adaptation to active bands, the attraction term moves the limits toward the higher gradients. That is to say, they are shifted upward or downward

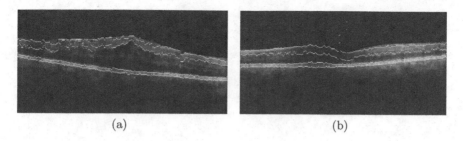

(a)   (b)

Fig. 4. Examples of initial approximations when only the expansion term is used.

when the magnitude of the gradient increases in those directions. The second and third terms are applied alternatively to balance the attraction with the regularization.

Since the process is performed column by column, we are able to cope with local variations of intensities, shadowing or other factors. Moreover, the statistical values allow correcting the false results which may appear. Figure 5 illustrates the results of these three terms when searching for the limits of the three bands.

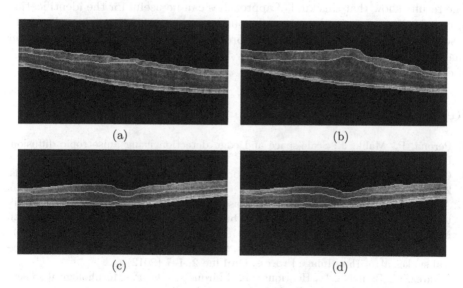

Fig. 5. Limits of the three bands for the slices in Fig. 1 when the expansion, regularization and attraction terms are combined.

Once the bands have been segmented, the limits of the three bands in the series of slices are connected to generate an elevation field and visualize the results in three dimensions. Figure 6 shows two examples of this kind of reconstruction.

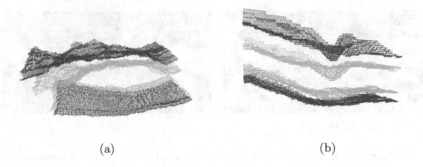

(a)   (b)

Fig. 6. Three-dimensional reconstruction of the bands for two retinal OCT scans.

5   Conclusion

This work introduces a new approach for the segmentation of retinal OCT images. Inspired in the active contour technique, we have developed a novel method, in which the terms of expansion, regularization and attraction have been adapted to the extraction of the limits of bands, instead closed contours. The results show that this kind of approaches can be useful for the identification of certain structures in the retina. In our case, we can extract three bands to described the shape of the retina and provide some information for its further analysis. This segmentation can be very helpful in the diagnosis or monitoring of certain pathologies.

References

1. Perona, P., Malik, J.: Scale-space and edge detection using anisotropic diffusion. IEEE Trans. Pattern Anal. Mach. Intell. **12**(7), 629–639 (1990)
2. Caselles, V., Kimmel, R., Sapiro, G.: Geodesic active contours. Int. J. Comput. Vis. **22**(1), 61–79 (1997)
3. Alemán-Flores, M., Álvarez-León, L., Caselles, V.: Texture-oriented anisotropic filtering and geodesic active contours in breast tumor ultrasound segmentation. J. Math. Imaging Vis. **28**(1), 81–97 (2007)
4. Alvarez, L., Baumela, L., Márquez-Neila, P., Henríquez, P.: A real time morphological snakes algorithm. Image Process. On Line **2**, 1–7 (2012)
5. Alvarez, L., Baumela, L., Henríquez, P., Márquez-Neila, P.: Morphological snakes. In: Proceedings CVPR, pp. 2197–2202 (2010)

Towards Egocentric Sentiment Analysis

Estefania Talavera[1,2](✉), Petia Radeva[1], and Nicolai Petkov[1,2]

[1] University of Barcelona, Barcelona, Spain
`petia.ivanova@ub.edu`
[2] University of Groningen, Groningen, The Netherlands
{`e.talavera.martinez,n.petkov`}`@rug.nl`

Abstract. The availability and use of egocentric data are rapidly increasing due to the growing use of wearable cameras. Our aim is to study the effect (positive, neutral or negative) of egocentric images or events on an observer. Given egocentric photostreams capturing the wearer's days, we propose a method that aims to assign sentiment to events extracted from egocentric photostreams. Such moments can be candidates to retrieve according to their possibility of representing a positive experience for the camera's wearer. The proposed approach obtained a classification accuracy of 75% on the test set, with deviation of 8%. Our model makes a step forward opening the door to sentiment recognition in egocentric photostreams.

Keywords: Egocentric images · Moment retrieval · Sentiment analysis

1 Introduction

Lifelogging describes an egocentric vision of the experiences of a person. Nowadays, the use of small wearable cameras, which capture images in certain intervals, is increasing considerably. Such images provide an overview of the daily activities of a person that can be interpreted as a visual log of the day. This information can be used to examine a person's pattern of behaviour; daily habits, such as eating habits, social interactions, indoor or outdoor activities, are recorded in such images. Although our mood is influenced by the environment and social context that surrounds us, egocentric data do not always catch our attention or induce the same emotion when retrieved. We consider that the creation of a diary of positive moments in an electronic way, by combining several cues, will help to improve the inner perception of the user's own life. Therefore, in this work we seek for *positive moments* that can raise the user's positiveness.

Several experiments have been conducted in that direction, in [14] the authors presented a survey of positive psychology strategies that demonstrated to be potentially effective as tools for the treatment of depression. As an example, in [1] the authors suggested to the participants activities such as *walking in a park, visiting a friend*, or *going to the Student Union and saying hello to someone*, that resulted in participants' suffering to decline or disappear. Their study gives ideas

© Springer International Publishing AG 2018
R. Moreno-Díaz et al. (Eds.): EUROCAST 2017, Part II, LNCS 10672, pp. 297–305, 2018.
https://doi.org/10.1007/978-3-319-74727-9_35

about the type of moments that can retrieve positive feelings; nature, landscapes, friends, or smiling people staring at us, among others.

Sentiment analysis from images is a novel research field. Given the challenge of image sentiment recognition and the ambiguity of the problem, we analyse images sentiment assigning a discrete ternary sentiment value (positive (1), neutral (0) or negative (−1) value), similar to [18]. In the literature, sentiment recognition from images has been approached based on different types of data features. Attributes such as facial expressions are used for sentiment prediction in [8,20]. The combination of visual and textual information from the images [16,17] appeared due to the wide use of online social media and microblogs. In such websites, images are posted with short descriptive comments by the user. Audio features were also included in the models presented by [11,13].

Despite several works having approached the understanding of how people can be affected seeing images, the field of sentiment analysis from images has not been yet settled. The labelling of the images is not yet established, leading to different questions when addressing sentiment recognition from images. As examples, we can find available datasets labelled as: amusement, anger, awe, disgust, excitement, fear, sad [10,19], Positive/Negative [2]-Twitter, Positive/Negative/Neutral [5], with values of Pleasure, Arousal and Dominance [7], with sentiment values from −2 to 2 [2], where the extremes correspond to Negative and Positive sentiments, respectively. Despite the above mentioned works, to the best of our knowledge, none of them has dealt with the sentiment recognition from egocentric photostreams.

Recently, with the outstanding performance of the Convolutional Neural Networks (CNN), several approaches on sentiment analysis have relied on supervised learning through deep learning techniques, such as [3,8,9,19]. One of the more remarkable approaches, in [2], introduced a Visual Sentiment Ontology (VSO), based on the Plutchik's wheel of emotions [12], and a visual concept detector called SentiBank. The VSO is built by 3022 semantic concepts called Adjective Noun Pairs (ANP) represented by images from the social net Flickr with them as tags. The ANPs are composed by a pair of a noun and an adjective, with a sentiment value associated between [−2 : 2]. They defend that an object, according to its appearance has a different sentiment value associated to it, like 'lonely boat' (−1.43) and 'traditional boat' (1,37), or 'noisy bird' (−1) and 'cute bird' (1,37). They proposed the semantic concepts baseline classification based on visual features (RGB, SIFT, LBP, etc.) extracted from the images. In [4], a Deep Neural Network named DeepSentiBank was trained on Caffe for the VSO semantic concepts classification. The authors relied on the concepts with higher number of images and with a classification accuracy associated. From the original 3022 concepts in [2] they select 2089 in [4].

To the best of our knowledge, previous to our work [15] there was no approaches addressing sentiment recognition from egocentric photostreams. We propose to analyse the output of the DeepSentiBank per image as semantic representation. We defined a classification model where the one-vs-all SVM classifiers were trained and evaluated with the features describing semantic and global information from the images.

In this work, we approach the same problem from a different perspective. We analyse the relation to each other semantic concepts extracted from images that belong to the same scene. A scene is described as a group of sequential images related between them and describing the same event. Our contribution is an analytic tool for positive emotion retrieval seeking for events that best represent a positive moment to be retrieved within the whole set of a day photostream. We focus on the event's sentiment description where we are observers without inner information about the event, i.e. from an objective point of view of the moment under analysis.

The rest of the paper is organized as follows. In Sect. 2, we describe the sentiment analysis method and the features selection procedure. In Sect. 3, we describe the proposed dataset, while in Sect. 3, we describe the experimental setup, the evaluation, and discuss our findings. Finally, Sect. 4 draws conclusions and outlines future lines of work.

2 Method

Given an egocentric photostream, we propose scene emotion analysis seeking for events that represent and can retrieve a positive feeling from the user. We apply event-based analysis since single egocentric images cannot capture the whole essence of the situation. By combining information from several images that represent the same scene, we get closer to a better understanding of the event.

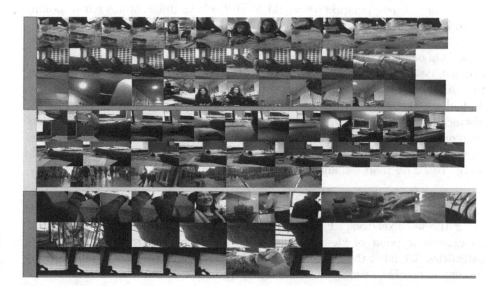

Fig. 1. Example of events extracted from the photostream by applying the temporal segmentation method introduced in [6]. Each row of the figure represents an event of the day. Image events labelled as Positive (green), Neutral (yellow) and Negative (red). (Color figure online)

2.1 Temporal Segmentation

We apply temporal segmentation on the egocentric photostreams using the proposed method in [6]. The clustering procedure is performed on an image representation that combines visual features extracted by a CNN with semantic features in terms of visual concepts extracted by Imagga's auto-tagging technology[1]. In Fig. 1 we present some examples of events extracted from the dataset, we introduce below.

2.2 Event's Sentiment Recognition

The model relies on semantic concepts extracted from the images to infer the event sentiment associated. However, it relies not only on the semantic concepts extracted by the net with their sentiment associated, but also on how those semantic concepts can be interpreted by the user. We apply the DeepSentiBank Convolutional Neural Network [4] to extract the images semantic information since it is the only introduced model that extract semantic concepts (ANPs) with sentiment values associated. Given an image, the output of the network is a 2089-D feature vector, where the values correspond to the ANPs likelihood in the image.

Besides taking into account the sentiment associated to the ANPs, the influence of the common concepts within an event are also analysed. We categorize the noun into Positive, Neutral or Negative. There is a wide range of semantic concepts within the ontology, but many of them seem to repeat concepts that even from the user perspective would be difficult to differentiate when looking at an image; such as "girl" from "woman" or "lady".

When facing our egocentric images challenge, the VSO presents several drawbacks. On one hand, this tool is trained to recognise up to 2089 concepts, which can not describe all possible scenarios. On the other hand, despite including that big amount of concepts, many of them categorize objects into categories difficult to visually interpret or differ by the human eye. Examples can be the distinction between 'child', 'children', 'boy', or 'kid' from an image. In order to overcome this problem, we generate a parallel ontology with what we consider an egocentric view of the concepts, i.e., we cluster the concepts a person would merge based on their semantic.

Egocentric analysis of the VSO: We cluster the semantic concepts based on the similarities between the noun components of the ANPs, which are computed using the wordNet tool[2]. Following what would be considered as similar from an egocentric point of view, we manually refine the resulted clusters into 44 categories. We label the clusters as Positive, Neutral or Negative. In Table 1 we present some of the egosemantic clusters.

---

[1] http://www.imagga.com/solutions/auto-tagging.html.
[2] http://wordnet.princeton.edu.

Table 1. Examples of clustered concepts based on their semantic similarity, initially grouped following the distance computed by the WordNet tool.

Positive			Neutral			Negative		
petals	christmas	award	car	study	bible	tumb	bug	nightmare
rose	winter	present	cars	science	book	tumbstone	bugs	accident
flora	snow	honor	machine	history	card	monument	insect	shadows
park	santa	gift	vehicle	economy	stiletto	grave	worm	noise
yard	sketch	heroes	rally	market	sins	memorial	cockroach	scream
plant	cartoon	dolls	train	industry	record	stone	decay	night
garden	drawing	dolls	competition	statue	paper	graveyard	garbage	darkness
	comics	toy	race	sculpture	poem	cementery	trash	shadow
	illustration	toys	control	museum	interview	grief	shit	
	humor	lego	metal			pain		

2.3 Sentiment Model

Given an event, the event's sentiment analysis model (see Fig. 2) performs as follows:

1. Given the ego-photostream we apply the temporal segmentation, analyse events with a minimum of 6 images, i.e. that last for at least 3 min.
2. Extract the ANPs of each event frame and rank them by their probability $(Prob_{ANP_j})$ of describing an image.
3. Select the top-5 ANPs per image, since we consider that those are the concepts with higher relevance, thus better capturing the image's information. After this step the model ends up with a total of M semantic concepts per event, where $\{M = \text{Number of images} \times 5\}$.
4. Cluster the M semantic concepts based on their Wordnet-based nouns semantic distances. As a result, we have clusters of concepts with semantic similarity. For the event sentiment computation (S_{event}), focus on the largest cluster.
5. Finally, fuse the sentiment associated to the ANPs and noun's cluster following the Eq. (1):

$$S_{event} = \sum_j (\alpha * S_{ANP_j} + \beta * S_{Noun_j}), j = 1 : N_{ANP}, \tag{1}$$

where $S_{ANP_j} = (S_{ANP_j}^{VSO} * Prob_{ANP_j})$, $S_{ANP_j}^{VSO}$ is the ANP's sentiment given by the VSO and S_{Noun} is the label of the noun, α and β are the contributions (%) of the ANPs and the nouns. Take into account the probability associated to the ANPs aiming to penalize the ANPs with low relation to the image content.

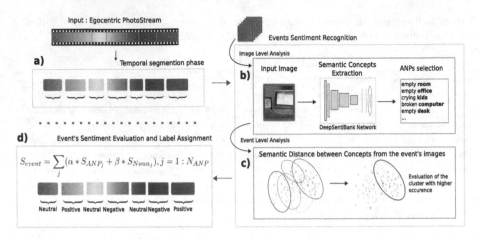

Fig. 2. Sketch of the proposed method. First, a temporal segmentation is applied over the egocentric photostream (a). Later, semantic concepts are extracted from the images using the DeepSentiBank [4] (b). The semantic concepts with higher occurrence are selected as event descriptors (c). Finally, the ternary output is obtained by merging the sentiment values associated to the event's semantic concepts (d).

3   Experiments Setup

3.1   Dataset

We collected a dataset of 4495 egocentric pictures, which we call UBRUG-Senti. The user was asked to wear the Narrative Clip Camera[3] fixed to his/her chest during several hours every day and was asked to continue with his/her normal life. Since the camera is attached to the chest, the frames vary following the user's movement and describe the user's view of his/her daily indoor/outdoor activities. It involves challenging backgrounds due to the scene variation, handled objects appearing and disappearing during images sequences, and the movement of the user. The camera takes a picture every 30 s, hence each day around 1500 images are collected for processing. The images have a resolution of 5 MP and JPG format.

After the temporal clustering [6], we obtained a dataset composed of 4495 images grouped in a total of 98 events. The events were manually labelled based on how the user felt while reviewing them. The labels assigned were *Positive* (36), *Negative* (43) and *Neutral* (19). Some examples are given in Fig. 1.

3.2   Experiments

During the experimental phase, we evaluated the contributions of ANPs and nouns by defining different combinations of α and β. We performed a balanced 5-fold cross validation. For each of the folds, we used 80% of the total of events

---

[3] http://getnarrative.com/.

Table 2. Parameter-selection results

	Accuracy			F-Score		
	beta = 0.2	beta = 0.5	beta = 0.8	beta = 0.2	beta = 0.5	beta = 0.8
	alpha = 0.8	alpha = 0.5	alpha = 0.2	alpha = 0.8	alpha = 0.5	alpha = 0.2
Ours	0.60	0.63	**0.73**	0.35	0.43	**0.59**
Evaluating 3 Clusters	0.68	0.66	0.68	0.48	0.45	0.48
Evaluating with weights	0.65	0.65	0.66	0.41	0.43	0.47

per label of our dataset and compute the best pair of α and β values. This is a parameters selection process that is later re-evaluated in a test phase with a different set of events.

Validation: To evaluate the effectiveness of the scene detection approach, we use the *Accuracy*, as the rate of correct results, and the *F-Score* (F1). The F1 is defined as : $F1 = 2(RP)/(R+P)$, where P is the precision ($P = TP/(TP+FP)$, R is the recall ($R = TP/(TP + FN)$) and TP, FP and FN respectively are the number of true positives, false positives and false negatives of the event's sentiment label correctly identified.

Results: Tables 2 and 3 present the results achieved by the proposed method at image and event level, respectively. The model achieves an average training accuracy of $73 \pm 3.8\%$ and F-score of $59 \pm 5.4\%$ and test accuracy of $75 \pm 8.2\%$ and F-score of $61 \pm 13.2\%$, when $\alpha = 0.8$ and $\beta = 0.2$, i.e. when the ANP information is considered; although the major contribution comes from the noun sentiment associated. As expected, neutral events are the most challenging ones to classify.

Table 3. Test set results

	Accuracy	F-Score
	beta = 0.8	
	alpha = 0.2	
Ours	**0.75 ± 0.08**	**0.60 ± 0.13**
Evaluating 3 Clusters	0.69 ± 0.1	0.50 ± 0.15
Evaluating with weights	0.74 ± 0.1	0.58 ± 0.15

In order to contextualize our results, we fine-tune the well-known *GoogleNet* deep convolutional neural network [9] to classify into Positive, Neutral and Negative. We use 80%, 10% and 10% of the dataset for training, validation and testing respectively. The network achieves an accuracy of **55%**.

From the results we can conclude that the application of the DeepSentiBank presents drawbacks when applied to egocentric photostreams. To begin with and as commented before, the 2089 ANPs not necessarily have the power to represent what the image captured about the scene, taking into account the difficulty to detect them automatically (Mean average accuracy of the net $\sim$25%). Moreover, the ANPs present the limitation that they are classified strictly into Negative or Positive concepts. Thus, moments from our daily routine, which are often considered as neutral, are difficult to recognize.

4 Conclusions

We present a new model for positive moments recognition from our digital memory, composed by images recorded by the Narrative wearable camera. It analyses semantic concepts called ANPs extracted from the images. These semantic concepts have a sentiment value associated and describe the appearance of concepts in the images. The sentiment prediction tool is based on new semantic distance of ANPs and fusion of ANPs and nouns sentiments extracted from egocentric photostreams. The proposed approach obtained a classification accuracy of 75% on the test set, with deviation of 8%. Future experiments will address the generalization of the model over datasets collected by other wearable cameras, as well as recorded by different users. Analysing the results obtained, we conclude that the polarity of the ANPs makes it difficult to classify 'Neutral' events. However, most of our daily life is composed by neutral events, which can be considered as routine. Thus, in future lines we will address the routine recognition and retrieval.

Acknowledgements. This work was partially founded by Ministerio de Ciencia e Innovación of the Gobierno de España, through the research project TIN2015-66951-C2. SGR 1219, CERCA, *ICREA Academia 2014* and Grant 20141510 (Marató TV3). The funders had no role in the study design, data collection, analysis, and preparation of the manuscript.

References

1. Beck, J., et al.: Stimulating therapeutic change with interpretations: a comparison of positive and negative connotation. Couns. Psychol. **29**, 551 (1986)
2. Borth, D., Ji, R., Chen, T., Breuel, T., Chang, S.-F.: Large-scale visual sentiment ontology and detectors using adjective noun pairs, pp. 223–232. ACM (2013)
3. Campos, V., et al.: Diving deep into sentiment: understanding fine-tuned CNNs for visual sentiment prediction. In: ASM, pp. 57–62 (2015)
4. Chen, T., Borth, D., Darrell, T., Chang, S.-F.: DeepSentiBank: Visual Sentiment Concept Classification with Deep Convolutional Neural Networks, p. 7 (2014)
5. Dan-Glauser, E.S., Scherer, K.R.: The Geneva affective picture database (GAPED): a new 730-picture database focusing on valence and normative significance. Behav. Res. Methods **43**(2), 468–477 (2011)
6. Dimiccoli, M., Talavera, E., Nikolov, S.G., Radeva, P.: SR-Clustering: Semantic Regularized Clustering for Egocentric Photo Streams Segmentation (2015)
7. Lang, P., Bradley, M., Cuthbert, B.: International affective picture system (IAPS): technical manual and affective ratings. In: NIMH, pp. 39–58 (1997)
8. Levi, G., Hassner, T.: Emotion recognition in the wild via convolutional neural networks and mapped binary patterns. In: ICMI, pp. 503–510 (2015)
9. Ma, M., Fan, H., Kitani, K.M.: Going deeper into first-person activity recognition. In: CVPR (2016)
10. Machajdik, J., Hanbury, A.: Affecitve image classification using features inspired by psychology and art theory. In: ICM, pp. 83–92 (2010)
11. Nojavanasghar, B., et al.: EmoReact: a multimodal approach and dataset for recognizing emotional responses in children. In: ICMI 2016, pp. 137–144 (2016)

12. Plutchik, R.: Emotion: A Psychoevolutionary Synthesis. Harper & Row, New York City (1980)
13. Poria, S., et al.: Fusing audio, visual and textual clues for sentiment analysis from multimodal content. Neurocomputing **174**, 50–59 (2014)
14. Santos, V., et al.: The role of positive emotion and contributions of positive psychology in depression treatment: systematic review. Clin. Pract. Epidemiol. Ment. Health: CP & EMH **9**, 221–237 (2013)
15. Talavera, E., Strisciuglio, N., Petkov, N., Radeva, P.: Sentiment recognition in egocentric photostreams. In: Alexandre, L.A., Salvador Sánchez, J., Rodrigues, J.M.F. (eds.) IbPRIA 2017. LNCS, vol. 10255, pp. 471–479. Springer, Cham (2017). https://doi.org/10.1007/978-3-319-58838-4_52
16. Wang, M., Cao, D., Li, L., Li, S., Ji, R.: Microblog sentiment analysis based on cross-media bag-of-words model. In: ICIMCS, pp. 76–80 (2014)
17. You, Q., et al.: Cross-modality consistent regression for joint visual-textual sentiment analysis of social multimedia. In: WSDM, pp. 13–22 (2016)
18. You, Q., et al.: Robust image sentiment analysis using progressively trained and domain transferred deep networks. In: AAAI, pp. 381–388 (2015)
19. You, Q., Luo, J., Jin, H., Yang, J.: Building a large scale dataset for image emotion recognition: the fine print and the benchmark, CoRR (2016)
20. Yuan, J., et al.: Sentribute: image sentiment analysis from a mid-level perspective categories and subject descriptors. In: WISDOM, pp. 101–108 (2013)

Interactive Three-Dimensional Visualization System of the Vascular Structure in OCT Retinal Images

Joaquim de Moura[✉], Jorge Novo, Marcos Ortega, Noelia Barreira, and Manuel G. Penedo

Department of Computing, University of A Coruña, A Coruña, Spain
{joaquim.demoura,jnovo,mortega,nbarreira,mgpenedo}@udc.es

Abstract. This paper proposes an automated tool for the 3D visualization of the retinal arterio-venular tree using Optical Coherence Tomography (OCT) images. The methodology takes advantage of different image processing techniques that initially segments the vessel tree and estimates its corresponding calibers. Then, the depths for the entire vessel tree are also calculated. With all this information, the 3D reconstruction of the vessel tree is achieved, interpolating with B-splines all the segments, obtaining a smooth representation that facilitates its inspection. This model allows the visualization and manipulation of the 3D vessel tree by means of graphical affine transformations, including translation, scaling and rotation. Thus, the method offers a complete and comfortable visualization of the 3D real layout of the vasculature that permits to proceed with more reliable diagnostic processes involving the retinal microcirculation analysis.

Keywords: Computer-aided diagnosis · Vascular structure
Retinal imaging · Optical Coherence Tomography

1   Introduction

Computer-aided diagnosis (CAD) systems has become one of the major research subjects in medical imaging [1]. These systems facilitate the work of clinical experts in the different diagnostic processes, facilitating and simplifying their work. Optical Coherence Tomography (OCT) is a standard imaging technique in ophthalmology that can provide non-invasive medical images with high resolution [2]. These images provide relevant medical information about the measures of the biological tissues such as retinal layers [3] and other structures [4]. Ophthalmologists use OCT scans for the analysis of the vascular tree and produce a diagnosis in different diseases like diabetes [5], hypertension [6] or arteriosclerosis [7]. Therefore, the use of automatic tools for the 3D visualization of the vessel tree is relevant as they facilitate the specialists' work, increasing their productivity and helping to establish preventive and therapeutic strategies.

© Springer International Publishing AG 2018
R. Moreno-Díaz et al. (Eds.): EUROCAST 2017, Part II, LNCS 10672, pp. 306–313, 2018.
https://doi.org/10.1007/978-3-319-74727-9_36

In the state-of-the-art, we can find many approaches that faced the retinal analysis in classical retinographies. Hence, different methods were proposed for the extraction and representation of the retinal vessel tree. As reference, Zhang et al. [8] based their proposal on the application of adaptive thresholds for the localization of the vascular structures. Mendonça and Campilho [9] employed a methodology that combines the detection of centerlines with the subsequent application of region growing to achieve the final vessel segmentation. In the case of Leandro et al. [10], an approach was implemented based on the continuous wavelet transform using the Morlet wavelet, integrating the information over multiple classification scales. Espona et al. [11] proposed a methodology based on the use of deformable contour models, incorporating domain specific knowledge such as topological properties of the blood vessels to identify them.

Only a small number of works have appeared that use OCT images to deal with the issue of the vasculature segmentation. Additionally, these few proposals consist of limited methodologies that still offer 2D representations of the retinal vasculature. Niemeijer et al. [12] used a 2D projection of the vessel pattern to obtain a high contrast between the vessel silhouettes and the retinal background. Guimarães et al. [13] employed the OCT fundus images to locate the depth of the vessels, enclosed in the study of abnormal retinal vascular patterns. Despite that, most of these works do not pay special attention to the visualization of the segmentation results, key issue that can facilitate significantly the doctor's work, specially in cases like this that involves a three-dimensional visualization.

We propose, in this work, an automated tool for the three-dimensional visualization of the retinal arterio-venular tree using OCT images. The method uses the 3D vessel coordinates as well as the corresponding calibers to render a comfortable three-dimensional visualization of the vascular structure. This interactive system provides useful information to the doctors that can be of a great utility to obtain accurate diagnosis in a large variability of pathologies.

2   Methodology

Our methodology receive, as input, a set of OCT images. These images are complemented with the corresponding near-infrared reflectance retinography of the eye fundus that is provided in combination with the OCT sections. These sections represent, in a cross-sectional view, the biological tissues such as retinal layers and other structures. Figure 1 includes an illustrative example of an OCT image.

The proposed visualization system uses the (x, y, z) coordinates and the calibers, d of the entire vasculature. The extraction of this information is organized in a set of progressive stages [14]. Firstly, the arteriovenous tree is extracted in the near-infrared reflectance retinography. Subsequently, their calibers and depth are estimated at the all the positions of the vessel structure. Using all this information, the three-dimensional reconstruction of the vessel tree is achieved interpolating with B-splines all the segments, producing a smooth representation. Following sections explain each step in more detail.

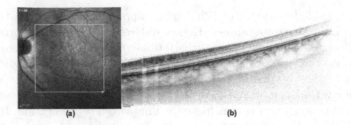

Fig. 1. Example of OCT image. (a) Near-infrared reflectance retinography. (b) OCT section.

2.1 Vessel Tree Extraction and Depth Estimation

Firstly, we segment the vessels in the near-infrared reflectance retinography to obtain the (x, y) coordinates and the vessel caliber d. The retinal arteriovenous tree is extracted by means of well-established image processing techniques. The vessels can be thought as creases (ridges or valleys), where the level curves are used to calculate the crest and valley lines. Then, a method of thinning is applied to obtain the representation of each vascular segment where all the vessels are represented by one-pixel width segments, that is, their coordinates (x, y). The vascular caliber d is obtained by means of the calculation of the distance between the edges (limits of the vessel) of the crease image of each vessel coordinate. Figure 2(a) illustrates an example of the vessel tree extraction.

Once the retinal arteriovenous tree is estimated in the near-infrared reflectance retinography, we can obtain the corresponding vessel depth z in the associated OCT sections, for each coordinate (x, y). To achieve this, the vascular profiles are identified in the OCT images using two stages: (1) mapping of the (x, y) coordinates in the OCT sections and (2) depth vessel identification. Firstly, we identify the positions of the vessels in the OCT sections by the intersection of the section and the vessel tree in the near-infrared reflectance

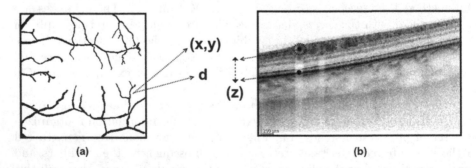

Fig. 2. Example of vessel tree extraction and depth estimation. (a) Vessel tree extraction, where (x, y) are the vessel coordinates and d is the corresponding vessel caliber. (b) Depth vascular estimation, z, in the OCT section.

retinography. This intersection identify the mapping region of the OCT section where the vessels are located. Subsequently, the vascular depth, z, is calculated in the mapped areas by the distance between the vascular profile (darkest spot in the mapping region) and the Retinal Pigment Epithelium (RPE) layer of the retina. Figure 2(b) shows an example of estimation of the vascular depth.

2.2   3D Vasculature Reconstruction

Next, we perform the three-dimensional reconstruction of the vascular structure. To achieve this, we use the information that was obtained in the previous phases: the spacial coordinates (x, y, z) and the vessel calibers, d. In a three-dimensional cartesian coordinate system ($\mathbb{R}^3$), each point P is represented by three real numbers, coordinates (x, y, z), indicating the positions of the perpendicular projections from the point to three fixed, perpendicular, graduated lines, called the axes which intersect at the origin O, which is the point with coordinates $(0, 0, 0)$. In this work, each vessel structure is represented as a segment S, where each point P_i of the vascular segment S is represented by its three-dimensional cartesian coordinates (x, y, z) and the vessel calibers d.

Firstly, we construct the vascular segment S from the points P_i representing the vascular structure with coordinates (x, y, z). For this, we a use B-spline $S(u)$, a function that has minimal support with respect to a given degree, smoothness, and domain partition, defined by:

$$S(u) = \sum_{i=0}^{n} B_{i,m}(u)P_i \qquad 2 \leq m \leq n+1, \tag{1}$$

where P_i is the ith control point of the $(n+1)^{th}$ control point of the curve and $B_{i,m}$ are the B-spline blending functions (also called the B-spline basis functions), which are basically polynomials of degree $m-1$. In this work, the basis function $B_{i,m}(u)$ is defined by the recursion formula of Cox-de Boor [15] using an order value $m = 2$. An example of this process is shown in Fig. 3.

Once all the curves $S(u)$ are obtained for all the points P_i that represent the vascular coordinates, we can perform the three-dimensional reconstruction of the

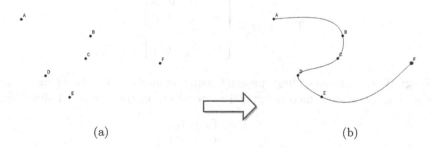

(a)   (b)

Fig. 3. Example of the three-dimensional representation process. (a) Set of points (x, y, z) of the plane. (b) Interpolation with B-spline curves between the set of points.

arterio-venular tree. To achieve this, we use the vascular calibers, d, associated to the points P_i. The vessels are reconstructed as tubular shapes (following the tubular vessel structure) centering on the B-spline curve $S(u)$ with a diameter size equivalent to the caliber d. Subsequently, a post-processing is applied offering a smooth representation of the vessel structure and, therefore, minimizing abrupt transitions between consecutive coordinates of the vessel. Figure 4 illustrate this three-dimensional representation process over a curve.

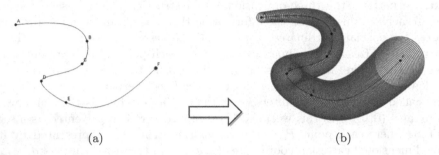

(a)   (b)

Fig. 4. Example of the three-dimensional representation process. (a) Interpolation with B-spline curves between points. (b) Three-dimensional tube along a spline.

This model allows the visualization and manipulation of the three-dimensional vessel tree by means of graphical affine transformations, including translation, scaling and rotation. Under the use of clinicians, these operations are applied over all the identified vessel coordinates, rendering the new set of coordinates and the corresponding calibers, for the user visualization. The application of any affine transformation is indicated by the user interactively with the system.

Translation: To achieve the translation operation within a three-dimensional space, we use of a matrix T (Eq. 2), where D_x, D_y and D_z represent the coordinates to move respectively in the x, y and z directions.

$$T(x, y, z) = \begin{bmatrix} 1 & 0 & 0 & 0 \\ 0 & 1 & 0 & 0 \\ 0 & 0 & 1 & 0 \\ D_x & D_y & D_z & 1 \end{bmatrix} \tag{2}$$

Scaling: We obtain the scaling operation using a matrix S (Eq. 3), where k, l and m represent the scaling factors applied respectively in the x, y and z directions.

$$S(k, l, m) = \begin{bmatrix} k & 0 & 0 & 0 \\ 0 & l & 0 & 0 \\ 0 & 0 & m & 0 \\ 0 & 0 & 0 & 1 \end{bmatrix} \tag{3}$$

Rotation: To perform the rotation operation on the x, y or z axes, we use the following matrices R_x, R_y and R_z (Eqs. 4, 5 and 6, respectively), where α denotes the angle of rotation.

$$R_x(\alpha) = \begin{bmatrix} 1 & 0 & 0 & 0 \\ 0 & cos(\alpha) & sin(\alpha) & 0 \\ 0 & -sin(\alpha) & cos(\alpha) & 0 \\ 0 & 0 & 0 & 1 \end{bmatrix} \tag{4}$$

$$R_y(\alpha) = \begin{bmatrix} cos(\alpha) & 0 & -sin(\alpha) & 0 \\ 0 & 1 & 0 & 0 \\ sin(\alpha) & 0 & cos(\alpha) & 0 \\ 0 & 0 & 0 & 1 \end{bmatrix} \tag{5}$$

$$R_z(\alpha) = \begin{bmatrix} cos(\alpha) & sin(\alpha) & 0 & 0 \\ -sin(\alpha) & cos(\alpha) & 0 & 0 \\ 0 & 0 & 1 & 0 \\ 0 & 0 & 0 & 1 \end{bmatrix} \tag{6}$$

3   Experimental Results

The proposed method was tested using a dataset of 392 OCT retinal images that were taken with a confocal scanning laser ophthalmoscope, Spectralis®OCT (Heidelberg Engineering), that offers the near-infrared reflectance retinography combined with the corresponding OCT sections. These sections were obtained from both left and right eyes, all centered on the macula, with a resolution of 1520×496 pixels. In order to test the performance of our system, this automated tool has been evaluated by an expert who has validated its functionality and usefulness. Figure 5 illustrates the proposed methodology, where an example can be observed with the main graphical transformations that this automatic visualization tool allows.

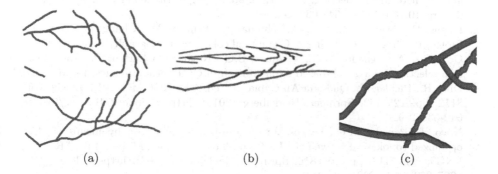

(a)   (b)   (c)

Fig. 5. Example of interactive three-dimensional visualization of the vessel tree. (a) Initial visualization. (b) Rotation operation. (c) Scaling operation.

4   Discussion and Conclusions

In this paper, we presented a new interactive three-dimensional visualization system of the vascular structure in OCT retinal images. Our proposal offers a complete set of information for a three-dimensional analysis of the arterial vessel tree, aiding the clinical experts to identify the vascular alterations that may lead to the early detection of various types of pathologies, such as diabetes, hypertension or arteriosclerosis.

The proposed methodology exploits different techniques to identify the vessel structure and estimate their calibers, using the near-infrared reflectance retinography images. Subsequently, the depth of the vascular profiles is obtained in the OCT sections. The three-dimensional reconstruction is performed using B-splines to interpolate all the vessel points, obtaining a smooth representation of the vascular structure. This automatic tool allows the clinical experts to visualize and manipulate the retinal vessel tree by means of the main graphical transformations, such as translation, scaling and rotation. The implemented tool was tested by an expert clinician, validating its well-functioning as well as stating its utility in the analysis of the retinal vasculature for the early diagnosis of different diseases.

Acknowledgments. This work is supported by the Instituto de Salud Carlos III, Government of Spain and FEDER funds of the European Union through the PI14/02161 and the DTS15/00153 research projects and by the Ministerio de Economía y Competitividad, Government of Spain through the DPI2015-69948-R research project. Also, this work has received financial support from the European Union (European Regional Development Fund - ERDF) and the Xunta de Galicia, Centro singular de investigación de Galicia accreditation 2016–2019, Ref. ED431G/01; and Grupos de Referencia Competitiva, Ref. ED431C 2016-047.

References

1. Doi, K.: Computer-aided diagnosis in medical imaging: achievements and challenges. In: Dössel, O., Schlegel, W.C. (eds.) World Congress on Medical Physics and Biomedical Engineering. IFMBE, p. 96. Springer, Heidelberg (2009). https://doi.org/10.1007/978-3-642-03904-1_26
2. Huang, D., Swanson, E., Lin, C., Schuman, J., Stinson, W., Chang, W., Hee, M., Flotte, T., Gregory, K., Puliafito C.: Optical coherence tomography. Science (1991)
3. González, A., Penedo, M.G., Vázquez, S.G., Novo, J., Charlón, P.: Cost function selection for a graph-based segmentation in OCT retinal images. In: Moreno-Díaz, R., Pichler, F., Quesada-Arencibia, A. (eds.) EUROCAST 2013. LNCS, vol. 8112, pp. 125–132. Springer, Heidelberg (2013). https://doi.org/10.1007/978-3-642-53862-9_17
4. Novo, J., Penedo, M.G., Santos, J.: Optic disc segmentation by means of GA-optimized topological active nets. In: Campilho, A., Kamel, M. (eds.) ICIAR 2008. LNCS, vol. 5112, pp. 807–816. Springer, Heidelberg (2008). https://doi.org/10.1007/978-3-540-69812-8_80

5. Klein, R., Sharrett, A.R., Klein, B.E., Chambless, L.E., Cooper, L.S., Hubbard, L.D., Evans, G.: Are retinal arteriolar abnormalities related to atherosclerosis? The atherosclerosis risk in communities study. Arterioscler. Thromb. Vasc. Biol. **20**, 1644–1650 (2000)
6. Won, T.Y., Mitchell, P.: Hypertensive retinopathy. N. Engl. J. Med. **351**, 2310–2317 (2004)
7. Nguyen, T., Wong, T.Y.: Retinal vascular changes and diabetic retinopathy. Curr. Diab. Rep. **113**(9), 277–283 (2009)
8. Zhang, Y., Hsu, W., Lee, M.L.: Detection of retinal blood vessels based on nonlinear projections. J. Sig. Process. Syst. **55**(1–3), 103 (2009)
9. Mendonça, A.M., Campilho, A.: Segmentation of retinal blood vessels by combining the detection of centerlines and morphological reconstruction. IEEE Trans. Med. Imaging **25**(9), 1200–1213 (2006)
10. Leandro, J., Cesar, J., Jelinek, H.: Blood vessels segmentation in retina: preliminary assessment of the mathematical morphology and of the wavelet transform techniques. In: Computer Graphics and Image Processing, pp. 84–90 (2001)
11. Espona, L., Carreira, M., Penedo, M., Ortega, M.: Retinal vessel tree segmentation using a deformable contour model. In: IEEE Transactions on Neural Networks, pp. 1–4 (2008)
12. Niemeijer, M., Garvin, M., Ginneken, B., Sonka, M., Abramo, M.: Vessel segmentation in 3D spectral OCT scans of the retina. Med. Imaging, vol. 6914, p. 69141R (2008)
13. Guimarães, P., Rodrigues, P., Bernardes, R., Serranho, P.: 3D blood vessels segmentation from optical coherence tomography. Acta Ophthalmol. **90**(s249) (2012)
14. de Moura, J., Novo, J., Ortega, M., Charlón, P.: 3D retinal vessel tree segmentation and reconstruction with OCT images. In: Campilho, A., Karray, F. (eds.) ICIAR 2016. LNCS, vol. 9730, pp. 716–726. Springer, Cham (2016). https://doi.org/10.1007/978-3-319-41501-7_80
15. Biswas, S., Lovell, B.C.: Bezier and Splines in Image Processing and Machine Vision. Springer Science and Business Media, London (2008). https://doi.org/10.1007/978-1-84628-957-6

Deep Reinforcement Learning in Serious Games: Analysis and Design of Deep Neural Network Architectures

Aline Dobrovsky$^{(\boxtimes)}$, Cezary W. Wilczak, Paul Hahn, Marko Hofmann, and Uwe M. Borghoff

Fakultät für Informatik, Universität der Bundeswehr München,
85577 Neubiberg, Germany
{aline.dobrovsky,cezary.wilczak,paul.hahn,marko.hofmann,
uwe.borghoff}@unibw.de

Abstract. Serious games present a noteworthy research area for artificial intelligence, where automated adaptation and reasonable NPC behaviour present essential challenges. Deep reinforcement learning has already been successfully applied to game-playing. We aim to expand and improve the application of deep learning methods in SGs through investigating their architectural properties and respective application scenarios. In this paper, we examine promising architectures and conduct first experiments concerning CNN design and analysis for game-playing. Although precise statements about the applicability of different architectures are not yet possible, our findings allow for concluding some general recommendations for the choice of DL architectures in different scenarios. Furthermore, we point out promising prospects for further research.

Keywords: Deep learning · Serious games
Convolutional neural networks · Neural network visualization

1 Introduction

Serious Games (SGs) rank among the most important future e-learning trends; they attain enhanced public acceptance and importance [3]. Although more frequently used in recruitment and training, their production is still effortful and expensive. The generation of human behaviour for non-player characters (NPCs), player identification, adaptivity to the player, content generation and general game playing remain prevalent challenges [1,10]. SGs can profit from the application of machine learning methods to create diverse behaviour and from automated adaptation to increase learning effectiveness. Deep learning (DL) methods and deep reinforcement learning (DRL) in particular demonstrated to be an opportunity for application. Considerable results have already been achieved by this general method. DL means machine learning methods using multiple processing layers, usually deep neural networks (DNN). DRL means the combination of reinforcement learning and DL. A famous example is deep Q-learning

R. Moreno-Díaz et al. (Eds.): EUROCAST 2017, Part II, LNCS 10672, pp. 314–321, 2018.
https://doi.org/10.1007/978-3-319-74727-9_37

for learning to play Atari games [13], where a convolutional neural network was trained with a variant of Q-learning on different games and partially outperformed human game players.

A general requirement for the application of DL in games is the demand for a cost-effective development process due to limited time and capacities. Furthermore, AI behaviour and its development have to be accessible to, comprehensible for and influenceable by different domain experts. Game AI has to face very different, prevalent challenges, depending on game and application scenario, including action-state space representation (from symbolic to raw data) and size, observability, non-determinism and credit assignment. Exemplary application areas are NPC control an the generation of human behaviour, human player replacement in multi-player games, automated adaptation of game content and progress, game-testing, etc.

We aim to expand and improve the application of DL in games through offering a novel framework including interactive learning. An overview of the original conceptual framework and its components is described in [2], although it has been fundamentally extended by adding the aspect of automated game adaptation by now. In this paper, we attempt to examine the usability of DL methods in SGs through literature research and first experiments. We investigate DNN architectures and show important factors considering their application.

2 Deep Learning Architectures

The term "architecture" summarizes all characteristics that define the structure of the deep neural network; including general structure, types of layers and connections, topology of different layers, weights and activation functions. This problem is also strongly related to hyperparameter optimization. In this work, however, we mean to rather explore general properties and aim to find general recommendations for using DL in SGs. Our methodical approach comprised literature analysis of different architectures, identification of further research prospects and experimental investigation of specific aspects. This section presents selected architectures with regard to existing promising practical applications of DL in different domains. The approaches listed here should serve as inspirational examples for related application possibilities in SGs.

2.1 Investigation of Exemplary Architectures

Feedforward architectures are already extensively used for classification and clustering tasks. Feedforward NNs are actually function approximators, and, according to the universal approximation theorem, a network of appropriate size can achieve a desired degree of accuracy [4]. Especially convolutional neural networks (CNNs) have recently attracted attention due to their successes in practical applications. CNNs consist of several layers of convolutions and in each layer, filters are applied to data and their respective values are trained. The structure is hierarchical: every layer can detect distinctive features of the input data on another abstraction level. Besides, spare connections and parameter sharing improve the training time. Because of their structure, CNNs are principally designed for 2D data, but can be adapted to other dimensions as well. Applications include computer vision and acoustic modelling for automated speech recognition; examples for 1D, 2D and 3D image and audio data can be found in ([4], p. 349). A commendable example for supervised learning is given by the GoogLeNet Inception architecture [14] that won the ILSVRC (image classification challenge) 2014. The handcrafted architecture comprises 27 layers, whereby inception modules count as one layer. A main goal was to increase depth and width of the network while trying to keep an acceptable computational effort.

CNNs can be used in the area of unsupervised learning as well, e.g. for unsupervised visual representation learning in videos [16]. The key issue of this work is that the otherwise necessary supervisional feedback can be substituted by using visual tracking, assuming that connected patches in tracks should have similar visual representations. As architecture, a siamese-triplet network with a ranking loss function is used. Every triplet uses the same architecture as Alexnet [8], an image classification architecture that won the ILSVRC 2012. CNN architectures have already been successfully used in games in combination with reinforcement learning (RL). In 2015, Google presented an approach of using a combination of RL and DNN for playing Atari video games. Their input state representation was a simplified version of the raw pixels of the game screen and their output controller moves. They tested their approach on 49 Atari games. The experiments showed that their variant of deep-Q-Learning even reached human level performance on over more than half of the games [13]. Investigating this success, the architecture exploits the CNN 2D image recognition potential. Their method performed best in quick-moving games but poorer than human players in long-horizon games where planning and a specific sequence of actions is important.

In contrast to feedforward techniques, recurrent NN architectures possess an internal memory through directed, cyclic connections. Their structure is better suited for processing sequences of arbitrary length, context modelling and time dependencies. A RNN with hidden-hidden recurrent connections actually corresponds to a universal turing machine. As architecture, the long short-term memory (LSTM) model is commonly used, because it avoids the vanishing gradient problem. LSTM proved good at handwriting recognition, speech recognition, machine translation, image captioning and parsing [4]. A third, emerging, type of architecture is provided by deep generative models. In contrast to the former mentioned discriminative models, a generative architecture can be used for probability distribution representation and sample generation. Particularly generative adversarial networks (GANs) have recently shown impressive capabilities. A GAN architecture consists of a generator and discriminator component, whereby the discriminator evaluates content by estimating if its original data or created by the generator. GANs have been used, for example, to generate contextually matching image samples with mandated properties or text to image synthesis (cf. [5]).

Altogether, we found very few concrete game applications of DL; the major activities remain in academic research. However, the mentioned approach of Atari game-playing could also be used in SGs, in restricted scenarios, as NPC control or human replacement, and also for game testing purposes. An effective DRL application to a more complex game could be achieved through exploiting additional symbolic game information for training, as has been shown in a 3D shooter [9]. Another possible DL application area is demonstrated by [12], where stacked denoising autoencoders are used for player goal recognition, which is a player-modeling task. Moreover, in [11], CNNs are used to recognise emotions through face images for player affect analysis.

2.2 Conclusions and Further Research Prospects

Our investigations indicate that, at the present time, suggesting concrete architectures for specific tasks is unfeasible. The suitability of an architecture is dependent on a variety of different criteria, e.g. amount and type of training data, reward function, computational and time limits, convergence properties and comprehensibility of learning behaviour. For the time being, deciding for a concrete architecture remains a trial and error process (cf. [4] p. 192). Though, we can still derive general ideas of suitable application areas for the different DNN types. CNNs are obviously suitable for all image recognition related tasks and present a good starting point, because efficient exemplary implementations and sophisticated frameworks are available. A RL approach is commonly used in games because of the lack of labelled training data. Although the learning efficiency continually increases through refining architectures and learning algorithms, long training times and uncertain results are still to be expected. Application opportunities in (visually) complex game scenarios are supported by input preprocessing or additional game information. For preprocessing, DL techniques like autoencoders can be considered. Furthermore, non-visual game information can be transformed into visual representation. We assume that e.g. simplified top-down views on strategy game scenarios are interpretable by a CNN. This can be extended by 2D presentations of context information, like in the upcoming Google Startcraft2 DL project [15]. The use of RNN in games seems to be still at the beginning. Applications for translation and language modelling are quite conceivable. Additionally, generating NPC behaviour in combination with CNN as shown in [9] is encouraging. Generative models seem promising for application in the prevalently challenging area of automated (procedural) content generation.

3 Experiments

Within the scope of two bachelor theses [6,17] we examined the DQN approach of [13] in video games concerning architectural structure and analysis possibilities of CNNs. We used Tensorflow on a desktop PC on the Atari games Ms PacMan and Breakout. As input, the NN received a scaled, grayscale screenshot of the game and had to compute actions as Atari controller moves.

3.1   Architecture Comparison

We compared two architectures during 2×2 days of training to identify the influence of an additional pooling layer. Figure 1(c) shows the architecture *Arch2* with an extra pooling layer. In contrast, *Arch1* didn't have the 3rd layer and was therefore $12.25\times$ bigger than *Arch2*. During the first two days of training, both architectures' scores rose to 300 game reward points on average and stagnated subsequently (Fig. 1(a)). This performance corresponds roughly to a human beginner. The maximum value, measured each 100 episodes, was regularly between 1000 and 2000, with spikes of 3870 (*Arch1*) and 4370 (*Arch2*). During the same time (2 runs in 4 days) and with the same configuration, *Arch1* trained 23 million timesteps and *Arch2* 29 million. In the test runs after 4 days of training (5 × 100 episodes without learning), *Arch1* reached a maximum episode mean value of 3380, and *Arch2* of 2250; with episode means in general between 200 and 400 (Fig. 1(b)).

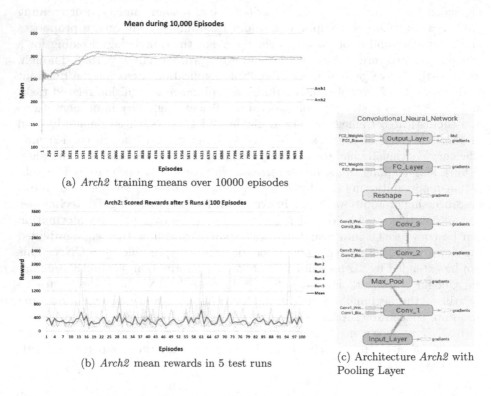

(a) *Arch2* training means over 10000 episodes

(b) *Arch2* mean rewards in 5 test runs

(c) Architecture *Arch2* with Pooling Layer

Fig. 1. Architecture comparison

3.2   Architecture Analysis

For architecture analysis, we examined different visualization methods and applied the method of Harley (cf. [7]) to a network trained on Breakout. Visualization can reveal shortcomings of architecture choice and training process.

For example, noisy filters could mean an unconverged network or a wrongly adjusted learning rate. By contrast, clean structures are an indication for effective training. The layer activation visualization comprehensibly shows activated neurons and reveals adaptation to specific features, e.g. detection of particular game objects.

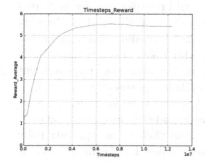

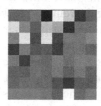

Fig. 2. Average game reward during 12 million timesteps

Fig. 3. Exemplary layer 1 filter at the beginning of training

Fig. 4. Same filter after 2 million timesteps

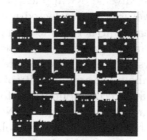

Fig. 5. Game situation in Breakout

Fig. 6. Activation maps of layer 1 at the beginning of training

Fig. 7. Activation maps of layer 1 after 2 million timesteps

As in the first experiment, learning stagnated relatively early in training (Fig. 2). Figures 3 and 4 show the first of four 8×8 filters of the first layer before and during the training. Figures 6 and 7 show the 32 activation maps (due to 32 filters) of the first convolutional layer in the game state of Fig. 5. The filter weights exhibit emerging of a clear structure and the feature maps show that important game features, the paddle and the ball, are recognized.

3.3   Experiment Conclusions

As already mentioned, CNN architectures are a mighty tool, but their design relies on assumptions and a process of trial and error. A lot of configuration options are available that heavily influence the learning performance (from general architecture design to specific parameters and different learning algorithms and reward function). We assume the latter and small computing capacity to be the reason why we couldn't achieve Google's performance. Additionally, we imagine Ms Pacman to be a challenging object of investigation because of contextual game-state change (ghost vulnerability isn't clearly graphically represented). Regarding architectures, bigger models could possibly convey more distinct state features in games but would also require more training than the simplistic approach. An encouraging finding is that our additional pooling layer had no negative influence on performance but allowed for faster training time. DNN largely remain blackboxes but visualization can offer a significant improvement in interpreting their behaviour. Visualization entails the potential of a powerful developer tool; it can help to optimize nets and reduce time for fault diagnostics. Comprehensive toolboxes are already available and can potentially be used as an essential part of neural network training. Nonetheless, interpretation by experienced persons is still necessary to detect anomalies. Furthermore, various other not visualisable possible sources of error exist.

4   Conclusions and Future Work

We aimed to inquire the usability of DL in SGs and investigated if general recommendations for applying DL in diverse SG application scenarios can be derived dependent on architecture. We presented selected architectures and illustrated their apparent and also imaginable further application possibilities. In our experiments, we examined architecture design and comparison and presented visualization as a method for NN analysis. Although we have to conclude initially that we can't offer precise statements about the applicability of different architectures, we nonetheless found promising first recommendations for further research. Architectures like CNNs, LSTMs, autoencoders and GANs already exhibit amazing successes and could probably be used for NPC control, player modelling or content creation. Some game adaptations could also improve the applicability of DL, e.g. creating a visual state representation or using preprocessing autoencoders for CNN application. Although DL methods aren't popular in the SG scene yet, we are convinced that our findings can inspire further research for beneficial applications in games.

References

1. Brisson, A., Pereira, G., Prada, R., Paiva, A., Louchart, S., Suttie, N., Lim, T., Lopes, R., Bidarra, R., Bellotti, F., et al.: Artificial intelligence and personalization opportunities for serious games. In: Eighth AIIDE Conference, vol. 2012, pp. 51–57 (2012)
2. Dobrovsky, A., Borghoff, U.M., Hofmann, M.: Applying and augmenting deep reinforcement learning in serious games through interaction. Periodica Polytech. Electr. Eng. Comput. Sci. **61**(2), 198–208 (2017). https://doi.org/10.3311/PPee.10313. ISSN 2064-5279

3. Doujak, G.: Serious Games und Digital Game Based Learning. Spielebasierte E-Learning Trends der Zukunft. GRIN Verlag, Munich (2015)
4. Goodfellow, I., Bengio, Y., Courville, A.: Deep Learning. Adaptive Computation and Machine Learning Series. MIT Press, Cambrige (2017)
5. Goodfellow, I.J.: NIPS 2016 tutorial: generative adversarial networks. CoRR abs/1701.00160 (2017). http://arxiv.org/abs/1701.00160
6. Hahn, P.: Visualisierung von Convolutional Neural Networks. Bachelorarbeit, Universität der Bundeswehr München (2017)
7. Harley, A.W.: An interactive node-link visualization of convolutional neural networks. In: Bebis, G., et al. (eds.) ISVC 2015. LNCS, vol. 9474, pp. 867–877. Springer, Cham (2015). https://doi.org/10.1007/978-3-319-27857-5_77
8. Krizhevsky, A., Sutskever, I., Hinton, G.E.: Imagenet classification with deep convolutional neural networks. In: Pereira, F., Burges, C.J.C., Bottou, L., Weinberger, K.Q. (eds.) Advances in Neural Information Processing Systems 25, pp. 1097–1105. Curran Associates, Inc. (2012)
9. Lample, G., Chaplot, D.S.: Playing FPS games with deep reinforcement learning. arXiv preprint arXiv:1609.05521 (2016)
10. Lara-Cabrera, R., Nogueira-Collazo, M., Cotta, C., Fernández-Leiva, A.J.: Game artificial intelligence: challenges for the scientific community. In: Proceedings 2st Congreso de la Sociedad Española para las Ciencias del Videojuego Barcelona, Spain (2015)
11. Martinez, H.P., Bengio, Y., Yannakakis, G.N.: Learning deep physiological models of affect. IEEE Comput. Intell. Mag. 8(2), 20–33 (2013)
12. Min, W., Ha, E., Rowe, J.P., Mott, B.W., Lester, J.C.: Deep learning-based goal recognition in open-ended digital games. In: AIIDE. Citeseer (2014)
13. Mnih, V., Kavukcuoglu, K., Silver, D., Rusu, A.A., Veness, J., Bellemare, M.G., Graves, A., Riedmiller, M., Fidjeland, A.K., Ostrovski, G., et al.: Human-level control through deep reinforcement learning. Nature 518(7540), 529–533 (2015)
14. Szegedy, C., Liu, W., Jia, Y., Sermanet, P., Reed, S., Anguelov, D., Erhan, D., Vanhoucke, V., Rabinovich, A.: Going deeper with convolutions. In: Proceedings of the IEEE Conference on Computer Vision and Pattern Recognition, pp. 1–9 (2015)
15. Vinyals, O.: DeepMind and Blizzard to release Starcraft II as an AI research environment (2016). https://deepmind.com/blog/deepmind-and-blizzard-release-starcraft-ii-ai-research-environment/. Accessed 23 May 2017
16. Wang, X., Gupta, A.: Unsupervised learning of visual representations using videos. In: Proceedings of the IEEE International Conference on Computer Vision, pp. 2794–2802 (2015)
17. Wilczak, C.: Convolutional Neural Networks: Betrachtung und Vergleich verschiedener Architekturen und ihrer Merkmale in Computerspielen. Bachelorarbeit, Universität der Bundeswehr München (2017)

Who is Really Talking? A Visual-Based Speaker Diarization Strategy

Pedro A. Marín-Reyes[1(✉)], Javier Lorenzo-Navarro[1],
Modesto Castrillón-Santana[1], and Elena Sánchez-Nielsen[2]

[1] Instituto Universitario SIANI, Universidad de las Palmas de Gran Canaria,
35017 Las Palmas, Spain
pedro.marin102@alu.ulpgc.es
[2] Departamento de Ingeniería Informática y de Sistemas, Universidad de la Laguna,
38271 Santa Cruz de Tenerife, Spain

Abstract. The speaker activity at the Canary Islands Parliament is recorded, and later manually annotated. This task can be modelled as a diarization problem, that is a way to automatically annotated who and when is speaking. In this paper, we propose the use of the visual cue to solve the diarization task. To perform this approach, it is mandatory to detect individuals, determine the one speaking, and extract features for matching. In order to test the performance of our proposal, we evaluate four different strategies based on the visual shot features.

Keywords: Visual diarization strategies · Local descriptors
Histogram distances · F-reid

1 Introduction

Speaker Diarization deals with annotating who and when a speaker is talking, it represents a challenge for the scientific community [1,2] that is mostly tackled using the audio cue. This problem can be tackled from a vision-based point of view, considering a re-identification process, i.e. detecting a speaker and checking whether he/she appears again.

A standard solution to audio-based speaker diarization is based on the procedure described by Tranter and Reynolds [3], being the approach adopted by the most recent literature. The purpose of speaker diarization is to split the audio recording of the different people interventions into segments. In this way, each segment represents a single speaker. After that, a clustering technique is used to group the different segments in order to include all the segments of one person in the same cluster. Different diarization scenarios have captured the attention of researchers, specially of those who investigate in the field of audio signals.

Ning et al. [4] have focused on Japanese Parliament sessions to the aim to solve speaker diarization, they segment the speech using Mel Frequency Cepstral Coefficient (MFCC) and Bayesian Information Criterion (BIC) as features.

R. Moreno-Díaz et al. (Eds.): EUROCAST 2017, Part II, LNCS 10672, pp. 322–329, 2018.
https://doi.org/10.1007/978-3-319-74727-9_38

Then, the Kullback-Leibler (KL) divergence is used at the clustering process as similarity measure between segments, obtaining the number of clusters by the value of the eigenvalues of the affinity matrix. These techniques are also used by Lupu et al. [5] in the Rumanian Parliament, using the system LIUM [6] to extract the audio of the sessions without taking into account the visual information of the videos.

To improve the results of only audio methodologies, Campr et al. [7] proposed the use of audio and visual information applied to Czech parliamentary recordings. Using Gaussian Mixture Model (GMM) to segment and detect in the audio any new speaker or recorded, for the latter also update the parameters of the corresponding GMM. After the face is detected and normalized, Local Binary Pattern features are extracted. For each group of consecutive faces, a cluster of key-faces are selected, to be later matched with different clusters, and they are compared among the different clusters. If the distance between two clusters is lower than a threshold, they are considered to be the same identity, otherwise it is a new person. After that, using the fusion of both diarization processes, the number of models is reduced with the audio-based diarization.

Furthermore, video processing can be used to detect the speakers, even without the audio information. Everingham et al. [8] propose a method to automatic annotation of film characters. To this purpose, both the subtitles and facial information are analized, where Scale-Invariant Feature Transform (SIFT) descriptor is used, and the clothing characterized by the YCbCr color histogram. In some cases, a person who is not speaking appears in the image, so, a speaker detector is implemented using the consecutive histogram differences of the mouth area. The matching process is done by a distance scheme of each character with the nearest representation of the face and clothing to assign an identity. Then, a Support Vector Machine (SVM) classifier is trained, one class with respect the others. Unlike the previous work, Sang and Xu [9] use scripts, instead of the subtitles to identify the name of the speaker. When all the faces are detected, they are grouped into several clusters using a clustering technique, matching the face identify using a graph fit, Error Correcting Graph Matching (ECGM).

The contributions of this paper are the following: (1) propose different strategies to assign the speaker ID from visual segments to an audio segment, (2) study different local descriptors to apply the above assignation, and (3) compare different distances to measure the similarity between descriptors in the problem of assign an ID.

2   Scenario

In this paper, we are focused on the diarization of parliamentary debates sessions using only video information. Specifically, this work is based on the Canary Islands Parliament in Spain. In this scenario, speaker interventions can be done from three different points: (1) at the presidential table where presidential deputies follow the guidelines to expose the topic during a predefined speaking time, (2) at the platform located, at front of the presidential table where the

Fig. 1. Different views of the Canary Islands Parliament.

deputies explain some topics, and (3) at the seats, the place where the deputies are sitting, in some cases they can stand up and intervene to answer another deputy. In those places, the interventions are recorded by a network of cameras distributed in the Parliament, which can do pan, tilt and zoom. Fig. 1 shows different images recorded in the Parliament. Those cameras are managed by a producer who decide the camera to focus the attention, which could lead to change the view during the intervention of a speaker, that situation increases the problem challenges involved in a vision-based system because the camera could be recording a person who is not talking.

3   Procedure

The speaker diarization problem is tackled based on a visual approach, using the face as the main source of information. For each detected face in each frame, the following processing is applied: Initially, the image is rotated till the position of the eyes is horizontal. Then, to generate the model, the faces have to satisfy the Biometric Keyframe condition [10], where the eyes and mouth distances match with a frontal pose.

Later, each key-face is transformed to grey-scale and the face or head shoulder (HS) pattern are obtained as a region of interest (ROI); the face pattern represents a ROI of the face area, and the HS pattern is composed by the face area adding the surrounding information as hair, clothing and background are included. Then, as features, local descriptors are extracted to obtain an histogram representation, using different grid size setups, because they have demonstrated good performance in facial analysis [11], an outline of the process is shown in Fig. 2. After the ROI is modelled, a matching stage is carried out by comparing the model against the database models, that is updated with each

Fig. 2. The image is normalized using face or HS pattern. Then, it is divided into 3×3 or 5×5 grids respectively where a local descriptor is applied to obtain the speaker model.

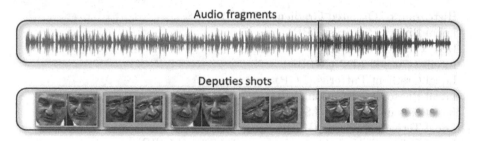

Fig. 3. Audio fragments can include different visual shots.

new identity found as the video is processed. The comparison is made using a histogram distance. The minimum distance model of the database is taken, and if this distance is larger than a fixed threshold, it is considered as a new speaker and added to the database, otherwise it is the same identity.

Besides, the parliamentary sessions, as well as other debates scenarios, we have to deal with shot changes while the intervener is talking as commented in Sect. 2. Therefore during a speaker intervention, i.e. to annotate his/her audio fragment, different deputies shots could appear, see an example in Fig. 3. The system has to give to the whole audio fragment corresponding to a single speaker the same ID through different visual shots. If the assignment is not correct, the diarization system will annotate with a wrong ID this shot or in worst case, the system creates a new ID, and for future comparisons the system will take into account the new false speaker. That increases the number of failures to the re-identification task. To solve the above problem, different strategies are proposed:

- **First Appearance (FA):** The person of the first shot detected by the system in the audio fragment is taken as representative.

- **Most Frequent (MF):** The person that the system detects a larger number of times in the audio fragment is taken as representative speaker shot.
- **Greatest Length (GL):** The person that the system detects as largest duration shot in the audio fragment is taken as representative speaker shot.
- **Greatest Total Length (GTL):** The person that the system detects as largest duration in the audio fragment is taken as representative speaker shot.

4 Experiments

The experiments have been performed using 29 videos[1] of different sessions of the Parliament. Those videos present different number of frames, shots and speakers; the mean duration of the videos is four hours. As local descriptors we have considered to test the following ones:

- Histogram of Oriented Gradients (HOG)
- Local Binary Patterns (LBP)
- LBP Uniform (LBPu2)
- Intensity based LBP (NILBP)
- Local Gradient Patterns (LGP)
- Local Phase Quantization (LPQ)
- Local Salient Patterns (LSP0)
- Local Ternary Patterns (LTP)
- Local Oriented Statistics Information Booster (LOSIB)
- LTP high (LTPh)
- LTP low (LTPl)
- Weber Local Descriptor (WLD)

Local descriptors are calculated in the mentioned ROI areas(face or HS) using a 5×5 and 3×3 grids. At the same time, the comparison of the models is computed with Canberra, Chebyshev, Cosine, Euclidean and kullback-Leibler divergence histogram measures. Besides, the different configurations are computed using the diarization strategies commented in the previous section.

To get an idea of the cost of the carried out experiments, 29 videos with two patterns with 2 grid configurations with 12 local descriptors with 5 different measures were processed, making a total of 6,960 experiments. Moreover, those experiments were validated by four diarization approximations, obtaining a total of 27,840 experiments.

4.1 Results

To evaluate the results, True Re-identification Rate (TRR) and True Distinction Rate (TDR) are taken from [12]. The TRR measure determinates how good is the system to re-identificate individuals, and the TDR represents the measure

---

[1] Videos available at http://www.parcan.es.

of how good is the system to distinguish between individuals. At the time, to evaluate the system, it could be the case that a system assigns only different IDs to the individuals detected, it will obtain 0% in TRR and 100% in TDR. We need to combine those values, at first, we will take the mean value. But, it will obtain a 50% being the worst system possible. To avoid this problem, it is taken the F_1 score, labelled as F_{reid}, that combines the TRR and TDR measures, as it is shown in Eq. 1.

$$F_{reid} = 2 \cdot \frac{TRR \cdot TDR}{TRR + TDR} \tag{1}$$

To focus in diarization methods we have calculate the F_{reid} mean value of all the videos processed. Although, the mean value of the different local descriptors and distance measures, see Table 1, where the Most Frequent approach matches the highest value independently of the kind of ROI and grid configuration. Taking into account this setup, the results improve 2.61% in the best case.

Table 1. Comparison of different patterns and number of grid respect different diarization approaches in term of the F_{reid} for the mean value of all the videos processed, descriptors and distances.

Strategy	Face		HS	
	3×3	5×5	3×3	5×5
FA	56.61	52.23	64.03	59.51
MF	56.70	54.91	**64.31**	59.57
GL	56.19	54.22	63.85	59.05
GTL	54.21	54.65	61.70	57.65

Table 2 shows the comparison of different local descriptors with different pattern and grid configuration. The best descriptor is Weber Local Descriptor that obtains an increment of 0.39% in relation to the second best descriptor, Histogram Oriented Gradients. The former obtains an improvement of 5.86% in relation to the worst descriptor for this configuration.

In relation to the histogram distance, Canberra is the best distance that matches the highest value, as we can see in Table 3. But in general, Kullback-Leibler divergence has a good behaviour for the different configurations and the difference between Canberra and this measure is insignificant.

Additionally, we highlight the use of HS and a 3×3 cells division, that obtains the best results for all the experiments. Specifically, the best configuration reported 74.09% with the Most Frequent approach, using a HS pattern with a 3×3 grid applying Weber Local Descriptor and comparing the models with a Canberra distance.

Table 2. Comparison of different patterns and number of grids respect different local descriptors in term of the F_{reid} for the mean value of all the videos processed, diarization approaches and distances.

Descriptor	Face		HS	
	3×3	5×5	3×3	5×5
HOG	56.81	55.09	65.86	63.51
LBP	55.03	53.17	62.10	58.36
LBPu2	55.95	55.52	63.93	58.97
LGP	53.52	51.72	64.58	59.41
LOSIB	49.56	47.57	64.72	60.65
LPQ	58.75	53.19	60.62	55.65
LSP0	55.49	54.70	59.25	53.75
LTPh	56.87	54.35	62.42	58.79
LTPl	56.40	54.05	63.29	57.02
LTP	56.97	54.65	62.75	59.77
NILBP	56.60	56.66	65.91	57.63
WLD	59.20	57.34	**66.25**	63.83

Table 3. Comparison of different patterns and number of grids respect different histogram distance measures in term of the F_{reid} for the mean value of all the videos processed, diarization approaches and descriptors.

Distance	Face		HS	
	3×3	5×5	3×3	5×5
Canberra	54.13	53.53	**65.86**	62.16
Chebyshev	52.84	47.25	55.76	46.81
Cosine	57.58	55.94	65.04	62.50
Euclidean	58.74	55.86	65.16	60.07
KL	56.35	57.43	65.54	63.18

5 Conclusion

This paper addresses four different strategies related to diarization problems, where these strategies avoid the annotation of false speaker in a vision-based context. Furthermore, the purpose of this article is to test various features related to computer vision to obtain a good configuration of parameters. So, Different local descriptors have been compared using HS and face patterns with two grid configurations, obtaining a general idea of their behaviour. Finally, multiple histogram measures have been compared, allowing us to know what configuration give us greater results for upcoming test.

In general, HS pattern matches the best results independently of the other parameters. In a same way, the 3×3 grid increases the performance of our diarization system. Moreover, Weber Local Descriptor is the best form to reduce the dimensionality of our problem, getting good results. At the time to compare the models, Canberra reports the best values. And last but not less important, to use a Most Frequent approach in a diarization system avoids the apparition of false speakers identification with an increment of 2.61% in terms of F_{reid} comparing the different diarization approaches using the HS pattern with a 3×3 grid.

Acknowledgement. This work is partially supported by Government of Spain through TIN2015-64395-R and by the Ministerio de Economía y Competitividad, Government of Spain and FEDER funds of the European Union through TIN2016-78919-R (MINECO/FEDER).

References

1. Miró, X.A., Bozonnet, S., Evans, N.W.D., Fredouille, C., Friedland, G., Vinyals, O.: Speaker diarization: a review of recent research. IEEE Trans. Audio Speech Lang. Process. **20**(2), 356–370 (2012)
2. Barra-Chicote, R., Pardo, J.M., Ferreiros, J., Montero, J.M.: Speaker diarization based on intensity channel contribution. IEEE Trans. Audio Speech Lang. Process. **19**(4), 754–761 (2011)
3. Tranter, S.E., Reynolds, D.A.: An overview of automatic speaker diarization systems. IEEE Trans. Audio Speech Lang. Process. **14**(5), 1557–1565 (2006)
4. Ning, H., Liu, M., Tang, H., Huang, T.: A spectral clustering approach to speaker diarization. In: Proceedings of ICSLP (2006)
5. Lupu, E., Apatean, A., Arsinte, R.: Speaker diarization experiments for Romanian parliamentary speech. In: 2015 International Symposium on Signals, Circuits and Systems (ISSCS), pp. 1–4, July 2015
6. Meignier, S., Merlin, T.: Lium spkdiarization: an open source toolkit for diarization. In: CMU SPUD Workshop, Dallas (Texas, USA), mars 2010
7. Campr, P., Kunešová, M., Vaněk, J., Čech, J., Psutka, J.: Audio-video speaker diarization for unsupervised speaker and face model creation. In: Sojka, P., Horák, A., Kopeček, I., Pala, K. (eds.) TSD 2014. LNCS (LNAI), vol. 8655, pp. 465–472. Springer, Cham (2014). https://doi.org/10.1007/978-3-319-10816-2_56
8. Everingham, M., Sivic, J., Zisserman, A.: Taking the bite out of automated naming of characters in TV video. Image Vis. Comput. **27**(5), 545–559 (2009)
9. Sang, J., Xu, C.: Robust face-name graph matching for movie character identification. IEEE Trans. Multimed. **14**(3), 586–596 (2012)
10. Marín-Reyes, P.A., Lorenzo-Navarro, J., Castrillón-Santana, M., Sánchez-Nielsen, E.: Shot classification and keyframe detection for vision based speakers diarization in parliamentary debates. In: Luaces, O., Gámez, J.A., Barrenechea, E., Troncoso, A., Galar, M., Quintián, H., Corchado, E. (eds.) CAEPIA 2016. LNCS (LNAI), vol. 9868, pp. 48–57. Springer, Cham (2016). https://doi.org/10.1007/978-3-319-44636-3_5
11. Castrillón-Santana, M., Lorenzo-Navarro, J., Ramón-Balmaseda, E.: Multi-scale score level fusion of local descriptors for gender classification in the wild. Multimed. Tools Appl. (2016, in press)
12. Cong, D.N.T., Khoudour, L., Achard, C., Meurie, C., Lezoray, O.: People re-identification by spectral classification of silhouettes. Sig. Process. **90**(8), 2362–2374 (2010). Special Section on Processing and Analysis of High-Dimensional Masses of Image and Signal Data

Detecting Hands in Egocentric Videos: Towards Action Recognition

Alejandro Cartas[1]([✉]), Mariella Dimiccoli[1,2], and Petia Radeva[1,2]

[1] University of Barcelona,
Gran Via de Les Corts Catalanes, 585, 08007 Barcelona, Spain
alejandro.cartas@ub.edu
[2] Computer Vision Centre,
Campus UAB, 08193 Cerdanyola del Valls, Barcelona, Spain

Abstract. Recently, there has been a growing interest in analyzing human daily activities from data collected by wearable cameras. Since the hands are involved in a vast set of daily tasks, detecting hands in egocentric images is an important step towards the recognition of a variety of egocentric actions. However, besides extreme illumination changes in egocentric images, hand detection is not a trivial task because of the intrinsic large variability of hand appearance. We propose a hand detector that exploits skin modeling for fast hand proposal generation and Convolutional Neural Networks for hand recognition. We tested our method on UNIGE-HANDS dataset and we showed that the proposed approach achieves competitive hand detection results.

Keywords: Ego-centric vision · First person vision · Hand-detection

1 Introduction

With the advances on wearable technologies in recent years, there has been a growing interest in analyzing data captured by wearable cameras [1]. In particular, due to the large number of potential applications, the analysis of human daily activities [2–5] has gained special attention. Daily activities are crucial to characterize human behavior, and enabling their automatic recognition would pave the road to novel applications in the field of Preventive Medicine, such as health monitoring [2,3], among others [6].

The hands are involved in a wide variety of daily tasks, such as typing on a self-phone keyboard, drinking coffee or riding a bike (see Fig. 1). Along with the objects being manipulated in a scene, the hands are often the main focus in the egocentric field of view. Consequently, their detection is a fundamental step towards action recognition. However, detecting hands in egocentric images is not a trivial task for three main reasons. First, the hands are intrinsically non-rigid and their shape appearance change continuously while manipulating objects. Second, the illumination conditions rapidly change in egocentric images as a consequence of the camera user movements across different locations. These

© Springer International Publishing AG 2018
R. Moreno-Díaz et al. (Eds.): EUROCAST 2017, Part II, LNCS 10672, pp. 330–338, 2018.
https://doi.org/10.1007/978-3-319-74727-9_39

Fig. 1. Examples of images showing actions involving hands. These pictures were captured by a chest-mounted wearable camera.

changes also affect the appearance of the hands and their recognition, as stated by Li and Kitani [7]. Third, the complexity of the method also depends on the camera used and its position on the body (head, shoulders, or chest). For instance, if the camera is worn on the chest, the focus of attention is lost and the location of hands in the field of view becomes more unpredictable. Available methods for detecting hands in egocentric images [7–9] are mostly based on hand-crafted features such as color histogram, texture and HOG in different color spaces.

The reminder of this paper is organized as follows. In the next Sect. 2 we review the state of the art on egocentric activity recognition and other works closely related to our. In Sect. 3, we introduce the proposed approach and in Sect. 4 we details the experiments performed. Finally, in section we draw some conclusions.

2   Related Work

In recent years, one of the first attempts to segment hands from egocentric images was proposed by Fathi et al. [8]. In order to determine regions containing hands and active objects, they modeled the background pixels using texture and boundary features. From the extracted foreground pixels, they distinguish between hands and objects using color histograms. Additionally, they introduced the Georgia Tech Ego-centric Activity (GTEA) dataset to test their model.

Li and Kitani [7] trained a pixel-level hand detector on images with more realistic egocentric characteristics such as motion, and extreme lighting and illumination changes. Their method combines superpixels with invariance descriptors, and color and texture features. They tested different combinations on the GTEA dataset and on their own proposed dataset, commonly referred as the zombie dataset. Although their results were better than other approaches, its method still failed when the hands were on dark or saturated regions. They extended their work by posing the detection problem as recommendation task using virtual probes [10]. Additionally, not only the hands are segmented by their method, but also the forearms.

Serra et al. [9] also proposed a hand segmentation that relies on the same combination of features HSV + LAB [7], but employed the Simple Linear Iterative Clustering (SLIC) algorithm for extracting superpixels. Moreover, they corrected segmentation problems by temporally smoothing the pixels and by joining segmented regions using a graph-based approach.

Betancourt et al. [11] proposed a two-stage hand detector using different color (RGB, HSV, LAB) and edge (HOG, GIST) features in addition with a classifier (SVM, random forests, decision trees). During the first stage, an image is divided using a grid in order to reduce the color features. In the second stage, the features are extracted and classified for each found region. The results on their own dataset indicate that the best performance is achieved combining HOG features and SVMs. In further work [12], they introduced the UNIGE-HANDS dataset and improved their detector to work on egocentric video sequences under the presence of image texture, color and luminosity variations. Specifically, they proposed a kalman filter that smooths the results the frame-by-frame classification results of SVMs.

More recently, a new egocentric dataset named EgoHands was introduced by Bambach et al. [13]. This dataset consists of videos where a pair of persons wear camera glasses in front of each other while playing a board game. Specifically, its purpose is to detect left and right hands and their respective owner at the pixel level. The pipeline of their approach is similar to R-CNN, but they provide a probabilistic region proposal and perform a pixel-level segmentation at the end of it. Besides, they performed an activity classification of the four board games played in the dataset using images containing only the detected hands, thus preserving the original location and sizes.

In this work, we propose a hand detector that exploits skin modeling for fast hand proposal generation and Convolutional Neural Networks for hand recognition. We tested our method on UNIGE-HANDS dataset [11] and we show that the proposed approach achieves competitive hand detection results.

3   Hand Detection

Our hand detector consists in a three-task architecture outlined in Fig. 2. We first detect regions containing skin pixels. Later, we generate a set of hand proposals using these regions. Finally, we classify the hand proposals using a Convolutional Neural Network (CNN).

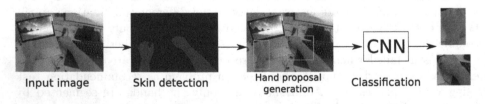

Input image   Skin detection   Hand proposal generation   Classification

Fig. 2. Outline of the proposed method for hand detection.

Skin detection. For this task, we use the pixel-level skin detection (PERPIX) method introduced in [7]. The PERPIX method models skin pixels by combining color (RGB, HSV, and LAB), texture (SIFT, ORB), and histogram features (Gabor filters).

Hand proposal generation. In order to generate hand proposals, we determine if each estimated skin-region in an image contains two pixel-connected arms. For instance, Fig. 3a shows a case where the arms are joined to each other and considered as one skin-region. First, we fit a straight line using the points from the boundary of the skin-region, as depicted in Fig. 3b. Next, if the mean squared error of the fit is greater than a fixed threshold, then the skin-region is considered as a two-arms region.

A two-arms region is split in two by applying a soft segmentation. The first step is to apply the k-means lines algorithm over the contour points of the skin blob. Since each line represent an arm, k is set to 2. Moreover, the calculated fit line at each iteration is the medial-axis line, obtained using orthogonal least squares. The second step is to perform a watershed transformation over the skin blob. The result of this operation are smaller sub-blobs that have soft boundaries, as seen on Fig. 3c. The last step is to assign each sub-blob to the closest line. This achieved by computing the smallest distance between the each sub-blob centroid and the lines, as shown on Fig. 3d.

After all resulting blobs are considered one-arm regions, then the hand proposals are extracted as follows. First, a rectangular convex-hull is calculated for each one-arm region. For example, extracted one-arm regions and their corresponding convex-hull are respectively shown in green and blue colors in Fig. 3d. Furthermore, in order to extract a hand from the convex-hull, we calculate a line

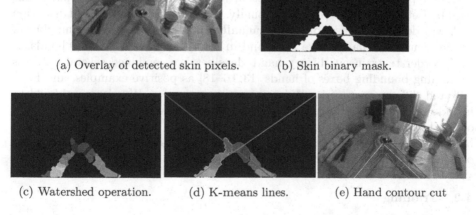

(a) Overlay of detected skin pixels.   (b) Skin binary mask.

(c) Watershed operation.   (d) K-means lines.   (e) Hand contour cut

Fig. 3. Example of a hand proposal generation over a skin region containing two pixel-connected arms. See text for detailed description. (Color figure online)

representing its wrist. We consider that a hand in the convex-hull is located in the side of the box closer to the center of the frame. As a result, we estimate the location of the *wrist* with respect to that side of the box. Second, a medial-axis line crossing the largest side of the convex-hull is computed. The *wrist* line perpendicularly intersects the medial-axis line and it is set at a fixed distance from the closest side to center of the frame. Figure 3d shows the medial-axis and the *wrist* lines in yellow and cyan colors, respectively. Finally, the hand proposal are obtained by cutting the one-arm regions using the *wrist* line. For instance, hand proposal boxes appear in red in Fig. 3d. More hand boxes can be proposed using different distances to *wrist*.

Hand recognition. To classify a hand proposal we created a binary classifier by fine-tuning the CaffeNet network [14] pre-trained on ImageNet [15].

4   Experimental Results

We describe the training and testing datasets in Sect. 4.1, and detail the skin and hand detection training in Sect. 4.2. We then present the experimental results on skin and hand detection tasks on Sect. 4.3.

4.1   Datasets

Our experiments were done using the UNIGE-HANDS dataset [12]. This dataset consists of 25 videos (292, 461 images) captured by a single person using a head-mounted camera. The labels provided indicate if arms appeared or not in each frame. The videos were filmed on 5 different settings: *office*, *street*, *bench*, *kitchen*, and *coffee bar*. Each setting has 4 training and 1 testing videos. Half of the training videos show the user arms, while the other half show only the setting. In the case of the testing videos, the user arms appears half of the time.

The reported results on skin detection were obtained on the same fixed-split used by Betancourt et al. [12]. Additionally, the evaluation on the hand detection task was done in a subset of 2,000 manually annotated images. The number of images containing hands were 1,000 and in total they were over 1,739 hands.

In order to train our binary hand classifier, we combined several datasets containing bounding boxes of hands [13, 16–18] as positive examples, and faces [19] and different categories [15] as negative examples. We also considered to include other hand datasets, but some of them considered the forearm as part of the hand [7, 8], or lack of hand annotations [4]. The number of images and bounding boxes by dataset are shown as a histogram in Fig. 4. The total number of images and bounding boxes is 761,946 and 872,414 respectively.

4.2   Training

We trained the PERPIX model using one training video for each setting category, i.e. we only used 5 videos for training. For each selected video, we uniformly

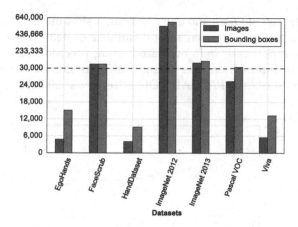

Fig. 4. Summary of the datasets used for training and validation. Histogram by dataset of the number of images and bounding boxes. Note the scale change on the vertical axis.

sampled 30 and 150 frames as the input for two training models. All the user skin regions in these frames were annotated and segmented.[1] The binary hand classifier network was created by fine-tunning the CaffeNet network [14]. It was fine-tuned for 20,000 iterations using Stochastic Gradient Descent, with a learning rate $\alpha = 0.001 = 10^{-3}$, a momentum $\mu = 0.9$, and weight decay equal to 5×10^{-4}.

4.3 Skin and Hand Detection

We made a skin detection performance comparison on the UNIGEN dataset with the HOG-SVM and DBN methods originally designed for it [11], as seen on Table 1. The PERPIX method offers competitive results using less training

Table 1. Skin-segmentation performance comparison. The HOG-SVM and DBN results correspond to [11] and the PERPIX results were obtained using the Per-pixel regression method [7] for 30 and 150 frames per settings video.

	True positive				True negative			
	HOG-SVM	DBN	PERPIX @30 frames	PERPIX @150 frames	HOG-SVM	DBN	PERPIX @30 frames	PERPIX @150 frames
Office	0.893	0.965	**0.973**	0.953	0.929	0.952	**0.986**	0.981
Street	0.756	0.834	0.872	**0.900**	0.867	**0.898**	0.586	0.574
Bench	0.765	0.882	0.773	**0.892**	0.965	**0.979**	0.954	0.948
Kitchen	0.627	0.606	**0.713**	0.628	0.777	**0.848**	0.789	0.830
Coffee bar	0.817	0.874	**0.996**	0.991	0.653	0.660	0.632	**0.688**
Total	0.764	0.820	0.862	**0.863**	0.837	**0.864**	0.799	0.815

---

[1] The annotations for skin detection training and hand detection evaluation are publicly available at http://gorayni.github.io.

data, specifically we only used 150 and 750 frames showing hands. Additionally, the results presented on [11] used a total number of 4439 frames. The results on the hand detection task were evaluated using precision/recall curves for 4 distinct values of intersection over union (IoU), as illustrated on Fig. 5. The average precision using the PASCAL VOC criteria was 20.01%.

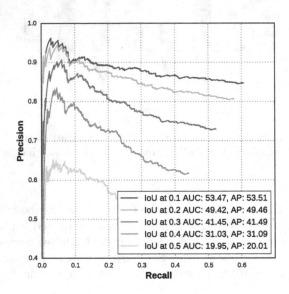

Fig. 5. Detection results on the UNIGE-HANDS test set for different values of the intersection over union (IoU) ratios.

5   Conclusions

We presented an egocentric hand detector method, which relies on skin modeling for fast hand proposal generation and a convolutional neural network for hand classification. We tested our method on the UNIGE-HANDS dataset and obtained an average precision of 0.216 when using the PASCAL VOC criteria. We showed that the proposed approach achieves competitive hand detection results. Future work will investigate how to incorporate hand detection to egocentric action recognition.

Acknowledgments. A.C. was supported by a doctoral fellowship from the Mexican Council of Science and Technology (CONACYT) (grant-no. 366596). This work was partially founded by TIN2015-66951-C2, SGR 1219, CERCA, *ICREA Academia'2014* and 20141510 (Marató TV3). M.D. is grateful to the NVIDIA donation program for its support with a GPU card.

References

1. Bolaños, M., Dimiccoli, M., Radeva, P.: Toward storytelling from visual lifelogging: an overview. IEEE Trans. Hum.-Mach. Syst. **47**, 77–90 (2017)
2. Karaman, S., Benois-Pineau, J., Mégret, R., Dovgalecs, V., Dartigues, J.F., Gaëstel, Y.: Human daily activities indexing in videos from wearable cameras for monitoring of patients with dementia diseases. In: 2010 20th International Conference on Pattern Recognition (ICPR), pp. 4113–4116. IEEE (2010)
3. Zariffa, J., Popovic, M.R.: Hand contour detection in wearable camera video using an adaptive histogram region of interest. J. Neuroeng. Rehabil. **10**, 1–10 (2013)
4. Rogez, G., Supancic, J.S., Ramanan, D.: Understanding everyday hands in action from RGB-D images. In: ICCV 2015 - IEEE International Conference on Computer Vision, Santiago, Chile (2015)
5. Cartas, A., Marín, J., Radeva, P., Dimiccoli, M.: Recognizing activities of daily living from egocentric images. In: Alexandre, L.A., Salvador Sánchez, J., Rodrigues, J.M.F. (eds.) IbPRIA 2017. LNCS, vol. 10255, pp. 87–95. Springer, Cham (2017). https://doi.org/10.1007/978-3-319-58838-4_10
6. Nguyen, T.H.C., Nebel, J.C., Florez-Revuelta, F., et al.: Recognition of activities of daily living with egocentric vision: a review. Sensors **16**, 72 (2016)
7. Li, C., Kitani, K.M.: Pixel-level hand detection in egocentric videos. In: Conference on Computer Vision and Pattern Recognition (CVPR), pp. 3570–3577. IEEE (2013)
8. Fathi, A., Ren, X., Rehg, J.M.: Learning to recognize objects in egocentric activities. In: 2011 IEEE Conference On Computer Vision and Pattern Recognition (CVPR), pp. 3281–3288. IEEE (2011)
9. Serra, G., Camurri, M., Baraldi, L., Benedetti, M., Cucchiara, R.: Hand segmentation for gesture recognition in EGO-vision. In: Proceedings of the 3rd ACM International Workshop on Interactive Multimedia on Mobile & #38; Portable Devices. IMMPD 2013, pp. 31–36. ACM, New York (2013)
10. Li, C., Kitani, K.: Model recommendation with virtual probes for egocentric hand detection. In: Proceedings of the IEEE International Conference on Computer Vision, pp. 2624–2631 (2013)
11. Betancourt, A., Lopez, M., Regazzoni, C.S., Rauterberg, M.: A sequential classifier for hand detection in the framework of egocentric vision. In: Conference on Computer Vision and Pattern Recognition, Columbus, Ohio, vol. 1. IEEE Computer Society (2014)
12. Betancourt, A., Morerio, P., Barakova, E.I., Marcenaro, L., Rauterberg, M., Regazzoni, C.S.: A dynamic approach and a new dataset for hand-detection in first person vision. In: Azzopardi, G., Petkov, N. (eds.) CAIP 2015. LNCS, vol. 9256, pp. 274–287. Springer, Cham (2015). https://doi.org/10.1007/978-3-319-23192-1_23
13. Bambach, S., Lee, S., Crandall, D., Yu, C.: Lending a hand: detecting hands and recognizing activities in complex egocentric interactions. In: 2015 IEEE International Conference on Computer Vision (ICCV). IEEE (2015)
14. Jia, Y., Shelhamer, E., Donahue, J., Karayev, S., Long, J., Girshick, R., Guadarrama, S., Darrell, T.: Caffe: convolutional architecture for fast feature embedding. arXiv preprint arXiv:1408.5093 (2014)
15. Russakovsky, O., Deng, J., Su, H., Krause, J., Satheesh, S., Ma, S., Huang, Z., Karpathy, A., Khosla, A., Bernstein, M., Berg, A.C., Fei-Fei, L.: ImageNet large scale visual recognition challenge. Int. J. Comput. Vis. (IJCV) **115**, 211–252 (2015)

16. Everingham, M., Eslami, S.M.A., Gool, L., Williams, C.K.I., Winn, J., Zisserman, A.: The pascal visual object classes challenge: a retrospective. Int. J. Comput. Vis. **111**, 98–136 (2014)
17. Eshed Ohn-Bar, S.M., Mogelmose, A., Trivedi, M.M.: Vision for intelligent vehicles and applications (VIVA) workshop and challenge. Workshop Chall. **13**, 17–30 (2015)
18. Mittal, A., Zisserman, A., Torr, P.H.S.: Hand detection using multiple proposals. In: British Machine Vision Conference (2011)
19. Ng, H.W., Winkler, S.: A data-driven approach to cleaning large face datasets. In: 2014 IEEE International Conference on Image Processing (ICIP), pp. 343–347 (2014)

Exploring Food Detection Using CNNs

Eduardo Aguilar[✉], Marc Bolaños, and Petia Radeva

Universitat de Barcelona and Computer Vision Center, Barcelona, Spain
{eduardo.aguilar,marc.bolanos,petia.ivanova}@ub.edu

Abstract. One of the most common critical factors directly related to the cause of a chronic disease is unhealthy diet consumption. Building an automatic system for food analysis could enable a better understanding of the nutritional information associated to the food consumed and thus, help taking corrective actions on our diet. The Computer Vision community has focused its efforts on several areas involved in visual food analysis such as: food detection, food recognition, food localization, portion estimation, among others. For food detection, the best results in the state of the art were obtained using Convolutional Neural Networks. However, the results of all different approaches were tested on different datasets and, therefore, are not directly comparable. This article proposes an overview of the last advances on food detection and an optimal model based on the GoogLeNet architecture, Principal Component Analysis, and a Support Vector Machine that outperforms the state of the art on two public food/non-food datasets.

Keywords: CNN · PCA · GoogLeNet · SVM · Food detection

1 Introduction

In the last decades, the amount of people with overweight and obesity is progressively increasing [1], whom generally maintain an excessive unhealthy diet consumption. Additionally to the physical and psychological consequences involved to their condition, these people are more prone to acquire chronic diseases such as heart diseases, respiratory diseases, and cancer [2]. Consequently, it is highly necessary to build tools that offer high accuracy in nutritional information estimation from ingested foods, and thus, improve the control of food consumption and treat people with nutritional problems.

Recently, the computer vision community has focused its efforts on several areas devoted to developing automated systems for visual food analysis, which usually involve using a food detection method [3–6]. These methods, also called food/non-food classification, have as purpose to determine the presence or absence of food in an image. Generally, they are applied as a pre-processing prior to food analysis, and can also be useful for selecting food images from huge datasets acquired from the WEB or from wearable devices.

© Springer International Publishing AG 2018
R. Moreno-Díaz et al. (Eds.): EUROCAST 2017, Part II, LNCS 10672, pp. 339–347, 2018.
https://doi.org/10.1007/978-3-319-74727-9_40

Food detection has been investigated in the literature in different works [3,6–9], where it has been proven that the best results obtained are based on Convolutional Neural Networks (CNN). The first method based using this technique was proposed by [3], which achieved a 93.8% using AlexNet model [10] on a dataset composed of 1,234 food images and 1,980 non-food images acquired from social media sources. They proved that using a CNN provided a 4% higher accuracy compared to using hand crafted features [7]. In [11], the authors improved the accuracy on this dataset to 99.1% using the NIN model [12]. In addition, they evaluated their model on other datasets, IFD and FCD, obtaining 95% and 96% of accuracy, respectively. Evaluation on a huge dataset with over 200,000 images constructed from Food101 [13] and ImageNet Challenge was done in [5], where the authors achieved 99.02% using an efficient CNN model based on inception module called GoogLeNet [14]. The same model was used in [4], the authors obtained 95.64% of accuracy on a dataset composed of Food101; food-related images extracted from the ImageNet Challenge dataset; and Pascal [15] (used as non-food images). Evaluation of different CNN models and settings was proposed by [9] on a dataset that we call RagusaDS. The authors obtained the best results using AlexNet for feature extraction and Binary SVM [16] for classification. In terms of accuracy, they achieved 94.86%. In [6], the authors apply fine-tuning on the last six layers of a GoogLeNet obtaining high accuracy, but tested their model on a balanced dataset of only 5,000 images (Food-5k). Since the proposed models were evaluated on different datasets, the results obtained are not directly comparable. Therefore, in order to compare our results with the state of the art, we selected the available datasets with more than 15,000 images.

Furthermore, we explored the food detection problem using the GoogLeNet, because this CNN model presented the best results in the classification of objects in the ILSVRC challenge [17]. In particular for food detection it has also presented good results on multiplies datasets with images acquired in different conditions [4–6]. Specifically, we propose a food detection model combining GoogLeNet for feature extraction, PCA [18] for feature selection and SVM for classification, which prove the best accuracy in the state of the art with respect to the previous works on the same datasets.

This article is organized as follows: in Sect. 2, we present our methodology. In Sect. 3, we present and discuss the datasets used and the results obtained. Finally, in Sect. 4, we present conclusions and future work.

2   Methodology for Food Detection

We propose a methodology for food detection, which involves the use of the GoogLeNet model for feature extraction, PCA for feature selection and SVM for classification. In Fig. 1, we show the pipeline of our food detection approach which will be explained below.

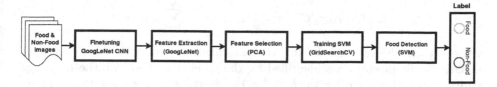

Fig. 1. Method overview for our food detection approach.

2.1 GoogLeNet for Feature Extraction

The first step in our methodology consists in training the GoogLeNet CNN model. For this purpose, we pre-train the GoogLeNet model on ImageNet [17], as a base model, and then we change the number of classes in the output layer for our binary classification problem (food/non-food). Then, GoogLeNet is fine-tuned on the last two layers until the accuracy on training set stops to increase, then we choose the model that gives the best accuracy on the validation set.

Once GoogLeNet is fine-tuned, we use the resulting model as a feature extractor. The feature vector for each image is extracted using the penultimate layer, with which a 1024-dimensional vector is obtained for each image. Then, we calculate a transformation that distributes normally the data through a Gaussian distribution function with zero mean and unit variance, by means of the feature vectors obtained from the training set. Finally, we normalize the data, in a range of $[-1, 1]$, by applying this transformation to each extracted feature vector.

2.2 PCA for Feature Selection

The following step in our methodology consists in reducing the dimensions of the feature vectors obtained in the previous steps by means of Principal Component Analysis (PCA) [18], which transforms the data to a new coordinate system leaving the greatest variance of the images in the first axes (principal components). We apply PCA on all feature vectors normalized from the training set and then the principal components are analyzed to select the first dimensions that retain the most discriminant information. To do this, we selected the features based on the Kaiser Criterion [19], which consists of retaining those components with eigenvalues greater than 1. The feature vectors reduced are used during the training of SVM and also during classification.

2.3 SVM for Classification

An SVM is a classification algorithm that optimizes a boundary $f(x) = Wx + b$ for maximizing the margin between two types of data, where the maximum margin is found solving a quadratic optimization problem. The dual SVM formulation is defined as follows:

$$\max_{\alpha} \quad \sum_{i=1}^{n} \alpha_i - \frac{1}{2} \sum_{i=1}^{n} \sum_{j=1}^{n} \alpha_i \alpha_j y_i y_j K(x_i, x_j)$$

$$s.t. \quad 0 \leq \alpha_i \leq C, \quad i = 1, 2, \ldots, n, \quad \sum_{i=1}^{n} \alpha_i yi = 0.$$

where x_i are the reduced feature vectors calculated from the n training set images, y_i the images class identified by the label -1 or 1, α_i are the Lagrange multipliers, and $K(x_i, x_j) = \tan h(\gamma x_i^T x_j)$ is the chosen kernel function, as in [9], for SVM classifier. The parameters C and γ are obtained by means of the Grid-SearchCV strategy, and the parameters α_i are adjusted during the optimization.

With the solution of the optimization problem, we compute the parameters of the objetive function f(x), where $W = \sum_{i=1}^{n} \alpha_i y_i x_i$, and b is the bias. Finally, the class is predicted using the $sgn(f(x))$ function, which returns +1 when $f(x)$ is positive and -1 when $f(x)$ is negative.

3 Experiments

3.1 Datasets

In this section, we present the selected datasets for the evaluation of the proposed model and comparison of the results. Both datasets, FCD and RagusaDS, were selected because they contain a significant amount of images, at least 15,000, and also they have free access to the images.

FCD was constructed from two public datasets widely used: Food-101 [13] and Caltech-256 [20] for food and non-food images, respectively (see Fig. 2). Food-101 is a dataset for food recognition, which contains 101 international food categories with 1,000 images each one. Caltech-256 contains 256 categories of objects with a total of 30,607 images, in which each object has a minimum of 80 images. For the construction of FCD, not all images of these datasets were considered. To balance the amount of food and non-food, we selected 250 images for each category in the Food-101. The selection was based on the color histogram of the images, keeping those with the highest color variance within the same category and thus, keeping the most highly variable set, obtaining a total of 25,250 food images. In Caltech-256, all images were selected except the food-related ones, resulting in 28,211 non-food images. To evaluate our approach, we used 64% of the images for training, 16% validation and 20% test.

Fig. 2. Example of images contained in the FCD dataset. Top row shows food images from Food101 and bottom row shows non-food images from Caltech256.

RagusaDS consists of three datasets acquired in different conditions: UNICT-FD889 [21] and Flickr-Food [8] for positive; and Flickr-NonFood [8] for negative samples. UNICT-FD889 is a dataset composed of 3,583 images of meals of 889 different dishes acquired from multiple perspectives with the same device in real-world scenarios, where images were acquired from a top view avoiding the presence of other objects. The Flickr images datasets were manually labeled as being food or non-food images. These datasets, which are called Flickr-Food and Flickr-NonFood, contain 4,805 images of food and 8,005 of non-food, respectively. Compared to UNICT-FD889, they contain less restricted images, and specifically for Flickr-Food the images can contain additional objects as well as food and were taken from different points of view. In total, the dataset contains 8,388 images of food and 8,005 of non-food (see Fig. 3). From the UNICT-FD889 dataset we split 80% of the data for training and 20% for validation. The first 3,583 images of Flickr-NonFood were also used for validation, and the remaining 4,422 as well as all images from Flickr-Food were used for testing.

Fig. 3. Example of images contained in the RagusaDS dataset. Top row shows food images from UNICT-FD889, middle row shows food images from Flickr-Food, and bottom row shows non-food images from Flickr-NonFood.

3.2   Experimental Setup

We used Caffe [22] for training our CNN model. We fine-tuned the last two layers of our model applying ten times the default learning rate. We set a learning rate of 1×10^3, with a decay of 0.96 every 5,000 iterations and a batch size of 32. We pre-processed the images by resizing them to 256×256 pixels, subtracted the training average of ImageNet and maintained the original color pixels scale. During training, data augmentation was applied by using horizontal mirroring and random crops of 224×224 pixels. During prediction, a center crop is applied.

The GoogLeNet was fine-tuned during 10 epochs, in the case of FCD, and during 40 epochs for RagusaDS. Training of the models was stopped when accuracy converged in the training set. The feature vector extracted from each image is reduced selecting the principal components based on the Kaiser Criterion, resulting in 186 dimensions on RagusaDS and 206 dimensions on FCD. As for the optimization of C and γ parameters, we applied a 3-fold cross validation. We defined a range of 14 values uniformly distributed on a base-10 logarithmic scale. In the case of the C parameter, we used a range from 1×10^{-4} to 1×10^2 and for γ parameter from 1×10^{-8} to 1×10^{-2}. Finally, the best parameters are used to train the SVM from scratch with all the training set.

3.3 Metrics

We used different metrics to evaluate the performance of our approach, namely: overall Accuracy (ACC), True Positive rate (TPr) and True Negative rate (TNr), which are defined as follows: $ACC = \frac{TP+TN}{T}$, where TP (True Positive) and TN (False Negative) are the amount of correctly classified images as Food and Non-Food, respectively; $TPr = \frac{TP}{TP+FN}$, where FN (False Negative) is the amount of misclassified images as Non-Food; $FNr = \frac{FN}{FP+TN}$, where FP (False Positive) is the amount of images misclassified as food.

3.4 Results

In this section, we present the results obtained during the experiments. In Table 1, the first two rows correspond to the state of the art algorithms that gave the best prediction on RagusaDS and FCD datasets, respectively. The last three methods are variations of our proposal, which is based on the GoogLeNet. The results show the ACC, the TPr and TNr obtained when evaluating each method on the FCD and RagusaDS datasets. In the case of the FCD, it can be seen that the model obtains a high precision in the global classification and maintains a slightly higher performance on TNr, which may be due to the small imbalance between food and non-food images of this dataset. On the other hand, for RagusaDS the difference between TPr and TNr is about 7% better for TNr. We believe that this occurs considering that food images used during training are very different from those used for evaluation and therefore the model is not able to recover enough discriminant information that allows to generalize over a sample acquired under different conditions. GoogLeNet + PCA-SVM is selected for the next experiment given that it achieved the best results on both datasets.

Table 1. Results obtained by models based on CNN on RagusaDS and FCD datasets on the food detection task. All results are reported in %.

	RagusaDS			FCD		
	ACC	TPr	TNr	ACC	TPr	TNr
AlexNet + SVM [9]	94.86	**94.28**	95.50	-	-	-
NIN [11]	-	-	-	96.4	96	97
GoogLeNet	94.66	91.53	98.06	98.87	98.48	**99.22**
GoogLeNet + SVM	94.95	91.53	**98.67**	98.96	**98.85**	99.06
GoogLeNet + PCA-SVM	**94.97**	91.57	**98.67**	**99.01**	**98.85**	99.15

Following, we trained the best model and evaluated its performance using RagusaDS and FCD datasets together, maintaining the same sets of training, validation and test, which we named RagusaDS + FCD. Table 2 shows the results obtained by training our approach using the training sets from RagusaDS + FCD

Table 2. Results obtained when GoogLeNet + PCA-SVM is trained on both datasets together (RagusaDS + FCD) and evaluated separately and jointly.

Test dataset	ACC	TPr	TNr
RagusaDS	95.78%	93.65%	98.10%
FCD	98.81%	98.60%	99.01%
RagusaDS + FCD	97.41%	96.19%	98.61%

and evaluating on the test sets from RagusaDS + FCD, RagusaDS and FCD. The results show that, when the model is trained on RagusaDS + FCD, it improves the classification significantly on RagusaDS although it presents a slight decrease on FCD. We believe that the improvement on RagusaDS is mainly due to an increase in the detection of food-related images. We deduce that by combining the training datasets, our method is able to extract features from various types of food acquired in different conditions, which allows to have a more robust classifier achieving a better generalization on the test set of RagusaDS dataset.

Some FPs FNs obtained in both datasets are shown in Fig. 4. Analyzing the FNs, we can observe that in the case of RagusaDS most errors occurred in images in which food was a liquid (drink, coffee, etc.). The reason for this is because the training set contains a wide variety of dishes but none of these correspond to beverages and therefore the classifier does not recognize them as food. In addition, other factors that influence classification are poorly labeled images such as food and also the cases where in the same image there are a lot of dishes. In the case of FCD, there are also some errors caused by wrong labels in both categories.

Fig. 4. FP (top) and FN (bottom) on RagusaDS (left) and FCD (right) datasets.

4   Conclusions

In this paper, we addressed the food detection problem and proposed a model that uses GoogLeNet for feature extraction, PCA for feature selection and SVM for classification. Furthermore, we applied a benchmark on the two more widely used publicly available datasets. From the results obtained, we observed that the best accuracy is achieved in both datasets with our proposed approach. Specifically, the improvement in the overall accuracy is more than 2% on FCD and about 1% for RagusaDS, when both datasets are combined for training and evaluated on the respective datasets. In addition, the overall accuracy when combining both datasets is 97.41%. As a conclusion, we explored the problem of food detection comparing the last works in the literature and our proposed approach provides an improvement on the state of art with respect to both public datasets. Moreover, models based on GoogLeNet, independently of the settings, gave the highest accuracy on the food detection problem. As future work, we will evaluate the performance of CNN-based models on larger datasets containing a much wider range of dishes and beverages such as food images and diversity of environments for non-food images.

Acknowledgement. This work was partially funded by TIN2015-66951-C2, SGR 1219, CERCA, *ICREA Academia'2014*, CONICYT Becas Chile, FPU15/01347 and Grant 20141510 (Marató TV3). The funders had no role in the study design, data collection, analysis, and preparation of the manuscript. We acknowledge Nvidia Corporation for the donation of a Titan X GPU.

References

1. Ng, M., et al.: Global, regional, and national prevalence of overweight and obesity in children and adults during 1980–2013: a systematic analysis for the global burden of disease study 2013. Lancet **384**, 766–781 (2014)
2. World Health Organization: Diet, nutrition and the prevention of chronic diseases. WHO Technical Report Series, vol. 916, p. 149 (2003)
3. Kagaya, H., Aizawa, K., Ogawa, M.: Food detection and recognition using convolutional neural network. In: ACM Multimedia, pp. 1085–1088 (2014)
4. Bolaños, M., Radeva, P.: Simultaneous food localization and recognition. In: ICPR (2016)
5. Myers, A., et al.: Im2Calories: towards an automated mobile vision food diary. In: ICCV (2015)
6. Singla, A., Yuan, L., Ebrahimi, T.: Food/non-food image classification and food categorization using pre-trained GoogLeNet model. In: Proceedings of the 2nd International Workshop on MADiMa (2016)
7. Kitamura, K., Yamasaki, T., Aizawa, K.: FoodLog. In: Proceedings of the ACM Multimedia 2009 Workshop on Multimedia for Cooking and Eating Activities (2009)
8. Farinella, G.M., Allegra, D., Stanco, F., Battiato, S.: On the exploitation of one class classification to distinguish food vs non-food images. In: Murino, V., Puppo, E., Sona, D., Cristani, M., Sansone, C. (eds.) ICIAP 2015. LNCS, vol. 9281, pp. 375–383. Springer, Cham (2015). https://doi.org/10.1007/978-3-319-23222-5_46
9. Ragusa, F., et al.: Food vs non-food classification. In: Proceedings of the 2nd International Workshop on MADiMa (2016)
10. Krizhevsky, A., Sutskever, I., Hinton, G.E.: ImageNet classification with deep convolutional neural networks. In: Advances in Neural Information Processing Systems, vol. 25, p. 19 (2012)
11. Kagaya, H., Aizawa, K.: Highly accurate food/non-food image classification based on a deep convolutional neural network. In: Murino, V., Puppo, E., Sona, D., Cristani, M., Sansone, C. (eds.) ICIAP 2015. LNCS, vol. 9281, pp. 350–357. Springer, Cham (2015). https://doi.org/10.1007/978-3-319-23222-5_43
12. Lin, M., Chen, Q., Yan, S.: Network in network. arXiv Preprint, p. 10 (2013)

13. Bossard, L., Guillaumin, M., Van Gool, L.: Food-101 – mining discriminative components with random forests. In: Fleet, D., Pajdla, T., Schiele, B., Tuytelaars, T. (eds.) ECCV 2014. LNCS, vol. 8694, pp. 446–461. Springer, Cham (2014). https://doi.org/10.1007/978-3-319-10599-4_29
14. Szegedy, C., et al.: Going deeper with convolutions. In: CVPR (2015)
15. Everingham, M., Van Gool, L., Williams, C.K.I., Winn, J., Zisserman, A.: The pascal visual object classes (VOC) challenge. Int. J. Comput. Vis. **88**, 303–338 (2010)
16. Cortes, C., Vapnik, V.: Support-vector networks. Mach. Learn. **20**, 273–297 (1995)
17. Russakovsky, O., et al.: ImageNet large scale visual recognition challenge. Int. J. Comput. Vis. **115**, 211–252 (2015)
18. Jollie, I.T.: Principal component analysis. J. Am. Statist. Assoc. **98**, 487 (2002)
19. Kaiser, H.F.: The application of electronic computers to factor analysis. Edu. Psychol. Measur. **20**, 141–151 (1960)
20. Griffin, G., Holub, A., Perona, P.: Caltech-256 object category dataset. Caltech mimeo **11**, 20 (2007)
21. Farinella, G.M., Allegra, D., Stanco, F.: A benchmark dataset to study the representation of food images. In: Agapito, L., Bronstein, M.M., Rother, C. (eds.) ECCV 2014. LNCS, vol. 8927, pp. 584–599. Springer, Cham (2015). https://doi.org/10.1007/978-3-319-16199-0_41
22. Jia, Y. et al.: Caffe: convolutional architecture for fast feature embedding. arXiv Preprint (2014)

Computer and Systems Based Methods and Electronic Technologies in Medicine

An Advanced Hardware Platform for Modern Hand-Prostheses

Peter Hegen[✉] and Klaus Buchenrieder

Universität der Bundeswehr München, Neubiberg, Germany
{peter.hegen,klaus.buchenrieder}@unibw.de

Abstract. While commercially available prostheses have seen mechanical improvements in recent years, new and improved myoelectric control schemes have been proposed in academia but have not made it into readily available devices. Conversely, current commercial prostheses only allow a limited number of analog-only input channels and can not be easily modified. However, research on myoelectric control schemes is frequently conducted using a higher number of channels and new control schemes necessitate the modification of the hand prostheses firmware.

In this contribution, we present new electronics and firmware for the commercial Steeper bebionic hand prosthesis. The firmware implements different control schemes for analog and digital sensors. To support both types of sensors, we bring forward a communication scheme for a combined interface, ensuring backwards compatibility.

Keywords: Electromyography · EMG · Prosthetic hands

1 Introduction

The human hand is a versatile tool that we rely on to perform everyday tasks. The loss of one or both hands is therefore a severe handicap which limits the ability to perform certain jobs. Fortunately, commercially available hand prostheses have seen steady improvements, especially in recent years. While some improvements have been achieved in the area of usability – like end-user accessible wireless configuration – most progress has been limited to the mechanics of the prostheses. Increasing the degrees of freedom in controlling fingers from one (cookie crusher model) to five or more (individual actuation of fingers, in other cases even parts of them) influenced the number of possible grips that can be performed with modern hand prostheses, both in research [3,7] and of commercial devices (e.g., Steeper bebionic or Touch Bionics i-limb).

Unlike the mechanics, the control scheme used in commercial prostheses has remained mostly the same since its introduction decades ago [5]. Conversely, academia has produced a number of proposals how to control hand prostheses more effectively or intuitively [9]. However, most of these proposals have been either tested only with prerecorded EMG signals and simple visualizations [1] or with custom prostheses that were purpose-built for these experiments [11]. Few

© Springer International Publishing AG 2018
R. Moreno-Díaz et al. (Eds.): EUROCAST 2017, Part II, LNCS 10672, pp. 351–358, 2018.
https://doi.org/10.1007/978-3-319-74727-9_41

concepts have been evaluated in the context of commercial prosthesis, e.g. [10], offering an upgrade path for prosthesis users.

This disconnect between research and the reality of users is likely caused by the comparatively high price of current commercial prostheses [4] and their closed nature. These can not be extended or reprogrammed easily. Therefore, the devices are limited to the input possibilities provided by the manufacturer. To avoid the latter issue, we have developed replacement electronics and firmware for the bebionic hand prosthesis. Replacing the electronics allowed us to add additional peripherals as well as sensors while keeping the excellent mechanics of the original prosthesis. Additionally, being able to customize the firmware allowed to implement features that would not have been possible to use with the original prosthesis. We will explain this with an example showing the simultaneous support for analog and digital sensors.

First, we will briefly describe the new electronics before explaining the structure of the firmware in greater detail, and how we support the use of both analog and digital sensors over the same interface.

2   Hardware

The bebionic hand prosthesis in its original form uses five microcontrollers for finger movements. Each of the four main digits – with the thumb being the remaining one – is directly controlled by its own microcontroller to track the finger position and to move it to a specified position upon request. Additionally, the processors monitor and limit the current consumption of the motor and return the current temperature of each board. While the microcontroller for the main digits only control the movement of their respective finger, the main microcontroller is responsible for the thumbs low level movement control as well as coordinating the movement of all fingers to form a single coherent grip pattern. Simultaneously, it also has to sample the analog input signals from the EMG sensors, detect movement patterns in it and facilitate user interaction through Bluetooth.

This hardware architecture was kept from the original hand prosthesis, as we initially meant to only replace the main microcontroller which defines the prosthesis' behaviour. Later, we also replaced the motor control boards of the index through pinky fingers (left photo of Fig. 1) in order to offload additional movement control tasks from the main controller. Accordingly, we ended up replacing all circuit boards inside the prosthesis, with only actuators like motors and buzzer remaining as original electrical parts inside the prosthesis. Due to choosing smaller components for replicating the existing functionality, board space was available for a USB debugging and programming port, a microSD card for long-term mass storage and an extensions header for I^2C based sensors, as can be seen in Fig. 1. We used the latter to connect a MPU6050 inertial measurement development board. As the orientation of the forearm influences the myographic signal generation [8] – e.g., due to decreased blood supply when holding it above the head or also muscle fatigue – it is important to record the orientation alongside the EMG signals to enable compensations for this effect. The offset compensation is not implemented yet, but we intend to address this issue in upcoming versions. Furthermore, we plan to integrate an active radio frequency identification (RFID) reader for communicating with passive tags. This feature, as introduced by Trachtenberg et al. [10], allows to switch grip patterns when the hand is in close proximity to such a passive RFID tag. While it lessens the mental burden upon the user in controlled/familiar environments, it is not universally applicable at all times as it requires prior setup for each specific situation.

Fig. 1. Prototype implementation of the new electronics for the hand prosthesis. Each of the four main digits (index, middle, ring and small finger) has its own motor control board (left) which offloads movement tasks from the main processor (right). The latter controls the motor for the thumb (not shown) and interfaces the peripherals, such as the Bluetooth module and SDcard shown in the right image.

3   Firmware

The firmware for our replacement boards is based on the FreeRTOS operating system. It was chosen in order to easily implement multiple functions running quasi-simultaneously. We will explain this aspect in further detail below. Additionally, the operating system ensures that time-critical parts of the code are prioritized over convenience functions. Especially the processing of input signals and control of the finger movements take priority to ensure that the prosthesis is operable at all times. Finally, separating the overall functionality into individual tasks allowed to modularize the code and provide synchronized access to peripherals. A single task is assigned exclusive control over each specific hardware functionality of the microcontroller. Using message queues, other tasks send requests and get synchronized such that hardware functions are not accessed interleaved by different parts of the code. The rational behind this design was to provide student contributors a simple interface for the hardware without a need to worry about racing conditions.

One such task is responsible for a major functionality of hand prostheses, acquiring and pre-processing the input signals. Concurrent commercial hand prostheses work by sampling the analog voltage of two connected EMG sensors, comparing their values to at least one user-specific threshold and finally executing a movement depending on which threshold was exceeded. Additionally, the most recent ones implement multiple grip patterns that the user can switch between by performing a specific gesture, e.g., forming a fist (co-contraction). In order to be backwards compatible to the original and other concurrent commercial hand prostheses, we implemented the same basic control scheme. The grip pattern implemented include basic traditional ones like holding a key or credit card but also ones based on feedback from an actual user of the bebionic prosthesis. Specifically, these grip pattern aim to support the usage of various tools needed to perform mechanical tasks, e.g., operating cordless power-tools.

Besides the classic thresholding scheme, we also implemented a simple linear classifier which operates on features computed from the input signal. The current firmware implements the root-mean-square (RMS), zero crossing (ZC) and waveform length (WL) features (refer to Phinyomark et al. [6] for an overview of EMG features). However, for current commercial prosthesis rectifying and sometimes also integrating sensors are used. Approximating the RMS computation in hardware, these sensors filter the huge parts of the frequency spectrum, rendering frequency-based features such as ZC useless. Additionally, the current de-facto standard connector of the bebionic and other hand prostheses (especially older ones) is limited to two uni-polar voltage inputs (sensors). Especially for research, a higher number of input channels would be highly desirable, e.g., when using sensors with multiple outputs [2]. Also, we wanted to support digital sensors as well as older analog ones. Therefore, we devised a handshake protocol to allow the usage of digital sensors over the same connection.

Consider the case of digital sensors connected to the analog input. Sending a digital signal right from the start would result in random or undefined behaviour of any legacy hand prosthesis connected to it. Therefore, it would need to emulate analog sensors, i.e., use a digital to analog converter (DAC) to output an analog signal, in order to provide backward compatibility. A prosthesis connected to such sensors, however, can not start sending digital information either, as that would short the sensors output to ground. A potential solution to that problem would be to configure both the users prosthesis and the digital sensors in their stump to expect a digital counterpart. Unfortunately, this has a significant drawback in that the user would be forced to use a hand prosthesis supporting the digital connection. User feedback suggests, that it is desirable to be able to connect older hand prosthesis models to the same stump as well. These models are (typically) mechanically and electrically less intricate, meaning these devices are less likely to fail and may allow greater strengths (e.g., when using only one bigger motor instead of several smaller ones in a similar form factor). Therefore, it would be best if both the digital sensors and the hand prosthesis would allow legacy counterparts at all times (after a power-cycle if necessary).

Therefore, we devised the handshake visualized schematically in Fig. 2. Upon power-up the digital sensors use their DAC to output an analog signal, proportional to the one measured, i.e., emulating the behaviour of purely analog sensors. Contrary to actual analog sensors, the DACs output is offset artificially from ground. These offsets are ignored during normal operation as they only require a shift of the thresholds used for classification. However, as the steady-state level is now significantly different from ground, a low-impedance load placed on the sensors output for a short time would be detectable by the sensor (1). A reasonable load needs to be selected in order not to overload the DAC of a digital sensor or the output amplifier of an analog one. Once the sensor detected an increased loading of its output for a specified time (e.g., several microseconds), it would switch off the DAC and send a digital message, e.g., an identifier of the supported digital protocol and its speed (2). Afterwards, it would re-enable the DAC output in case the previous loading was caused by some external interference and not by the hand prosthesis. In case the hand prosthesis supports the same digital protocol and speeds, it would acknowledge it by again loading the sensor output for a short time (3). Finally, both the sensor and the hand prosthesis switch to using digital communication thereafter (4).

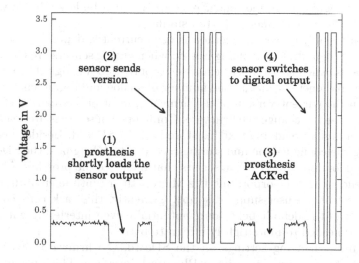

Fig. 2. Schematic visualization of the proposed handshake to carry digital communication over the analog interface of current prostheses. The prosthesis would initiate the handshake by briefly loading the input channel. Analog sensors should not be disturbed significantly if the loading is kept short enough, whereas digital sensors may switch off their DAC and continue in the shown sequence (1–4).

Note, that in the above description of the handshake, we did not specify a specific digital protocol. While a number of standard or custom protocols could be used, we suggest to use RS232-based serial communication. We tested this mode with another microcontroller acting as the digital sensors. It in turn sampled eight sensors, two of which were selected for the initial analog-only output. Once switched to the digital protocol, the microcontroller in the hand prosthesis and the external one communicated over a 2 MBaud serial connection. At this speed, up to 30 sensors sampled at 4 kHz with 16 bit per sample could be connected to the prosthesis. Additionally, this mode is implemented with push-pull output stages not suffering from external inferences as for example the I^2C bus. Indeed, in our test-setup, the hand prosthesis was connected to the external microcontroller through a pair of 30 cm wires plus cabling inside the prosthesis. Even in an open-bench setup with other electrical equipment nearby, we could not observe any dropped data (which was sent in frames and given a running number to detect missing/dropped data frames).

As the firmware for the prosthesis was kept as modular as possible, it can also be configured to use digital sensors with thresholding. In this case, only up to two sensor channels may be used, whereas arbitrary number of channels may be used with the linear classifier. In theory both methods could also be used cooperatively, i.e., a special gesture would allow to switch between these two systems at runtime. This might be useful to switch to the less intricate but more reliable thresholding when a safe operation is required (e.g., when driving a car). In less critical situations, the user could switch to the linear classifier which can recognize more movement patterns but can be less reliable when the signals for different movements are too similar.

In practise, only either one method is used currently, depending on the configuration. By default, thresholding is used when analog sensors are connected to the prosthesis, whereas the linear classifier is only used for digital input signals. While some parameters, such as the number of analog and digital channels used, are fixed in the current version of the firmware, most others can be configured without the need to change the firmware. Configuration settings are read at boot time from a human readable text file stored on the SDcard. Besides being used for holding the configuration and storing recorded EMG signals, the SDcard can also be used for supplying Python scripts that influence the prosthesis' behavior. An independent task incorporating the Micropython runtime environment continuously executes a user-supplied script. Currently, this task runs completely independent of the default behaviour, potentially even interfering with it. Further work is needed to more tightly integrate both.

Aside from changing settings on the SDcard, the firmware also allows to set parts of its configuration using a Bluetooth interface. This option is meant especially for settings that require live feedback, such as setting the thresholds for each channel. In order to determine a suitable threshold, it is preferable to be able to view the live EMG signals simultaneously. To accomplish both of these tasks at the same time, we developed a mobile companion application for the prosthesis, exemplarily shown in Fig. 3. Once connected to the prosthesis, it continuously plots a low-pass filtered version of the live EMG signals (Fig. 3b). Thresholds for each channel can be set independently in the same plot by dragging their visual representation to the desired value. Settings for individual thresholds can be locked in the settings tab of the application to prevent accidental changes. Finally, the application is also used for displaying debugging information (Fig. 3a) to help troubleshoot issues during development and operation.

 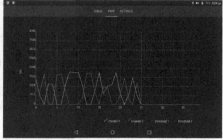

(a) error reports due to unplugged IMU (b) live EMG signal visualization

Fig. 3. Android application for diagnosing and configuring the prosthesis. The different tabs allow to view debug messages and status reports (a), visualize live EMG signals (b) and change settings (not shown). Thresholds can be set individually for each channel in the same plot used for the live EMG signals.

4 Conclusion

The hardware platform described in this paper aims to bridge the divide between mechanically proven commercial hand prosthesis designs that still rely on a decades-old control scheme and more current schemes/algorithms proposed in research. Using the bebionic hand prosthesis as a basis, it allows to implement various approaches that are not possible with the original one while keeping the form factor. As an example feature, we described the simultaneous support for analog and digital sensors over the same interface/connection while retaining backwards-compatibility to other sensors and hand prostheses.

While the current prototype achieves many of our initial goals, there is still room for improvement in some areas. Further integration of the electrical components of the system, extending the support for the Micropython interface of the prosthesis and implementation of several state of the art algorithms in the firmware are next steps we plan to pursue. The current separation of functionality into several individual printed circuit boards complicates the programming model and drastically increases the amount of manufacturing work compared to a more integrated version. Consequently, for the next iteration of the hardware, we plan to co-locate as much functionality as possible on the mainboard to support additional sensors. Concurrently, we plan to implement a suitable digital sensor/recording platform that is ready for integration into users shafts and implements the handshake method described in this work.

References

1. Attenberger, A., Buchenrieder, K.: Modeling and visualization of classification-based control schemes for upper limb prostheses. In: 2012 IEEE 19th International Conference and Workshops on Engineering of Computer-Based Systems, pp. 188–194, April 2012. https://doi.org/10.1109/ECBS.2012.32
2. Herrmann, S., Buchenrieder, K.: Fusion of myoelectric and near-infrared signals for prostheses control. In: Proceedings of the 4th International Convention on Rehabilitation Engineering & Assistive Technology, iCREATe 2010, pp. 54:154:4. Singapore Therapeutic, Assistive & Rehabilitative Technologies (START) Centre, Kaki Bukit TechPark II, Singapore (2010)
3. Krausz, N.E., Rorrer, R.A.L., Weir, R.F.ff.: Design and fabrication of a six degree-of-freedom open source hand. IEEE Trans. Neural Syst. Rehabil. Eng. **24**(5), 562–572 (2016). https://doi.org/10.1109/TNSRE.2015.2440177
4. McGimpsey, G., Bradford, T.C.: Limb prosthetics services and devices - critical unmet need: market analysis. White Paper, Bioengineering Institute Center for Neuroprosthetics, Worcester Polytechnic Institute, Worcester (2008)
5. McLean, L., Scott, R.N.: The early history of myoelectric control of prosthetic limbs (1945–1970). In: Muzumdar, A. (ed.) Powered Upper Limb Prostheses - Control, Implementation and Clinical Application, pp. 1–15. Springer, Heidelberg (2004). https://doi.org/10.1007/978-3-642-18812-1_1
6. Phinyomark, A., Quaine, F., Charbonnier, S., Serviere, C., Tarpin-Bernard, F., Laurillau, Y.: EMG feature evaluation for improving myoelectric pattern recognition robustness. Expert Syst. Appl. **40**(12), 4832–4840 (2013). https://doi.org/10.1016/j.eswa.2013.02.023
7. Resnik, L., Klinger, S.L., Etter, K.: The DEKA arm: its features, functionality, and evolution during the veterans affairs study to optimize the DEKA arm. Prosthet. Orthot. Int. **38**(6), 492–504 (2014). https://doi.org/10.1177/0309364613506913
8. Scott, S.H., Kalaska, J.F.: Reaching movements with similar hand paths but different arm orientations. I. Activity of individual cells in motor cortex. J. Neurophysiol. **77**(2), 826–852 (1997)
9. Singh, R.M., Chatterji, S., Kumar, A.: Trends and challenges in EMG based control scheme of exoskeleton robots—a review. Int. J. Sci. Eng. Res. **3**(8), 933–940 (2012)
10. Trachtenberg, M.S., Singhal, G., Kaliki, R., Smith, R.J., Thakor, N.V.: Radio frequency identification - an innovative solution to guide dexterous prosthetic hands. In: 2011 Annual International Conference of the IEEE Engineering in Medicine and Biology Society, pp. 3511–3514 (2011). https://doi.org/10.1109/IEMBS.2011.6090948
11. Wang, J., Ren, H., Chen, W., Zhang, P.: A portable artificial robotic hand controlled by EMG signal using ANN classifier. In: 2015 IEEE International Conference on Information and Automation, pp. 2709–2714, August 2015. https://doi.org/10.1109/ICInfA.2015.7279744

A Real-Time Classification System for Upper Limb Prosthesis Control in MATLAB

Andreas Attenberger[1](✉) and Sławomir Wojciechowski[2]

[1] FH Kufstein, University of Applied Sciences, Kufstein, Austria
andreas.attenberger@fh-kufstein.ac.at
[2] Wrocław University of Science and Technology, Wrocław, Poland
swojciechowski@pjwstk.edu.pl

Abstract. In this paper we present a MATLAB tool for processing both EMG and NIR sensor signals in real-time in order to provide a fully operational tool for clinical testing. After a short training phase, the decision tree classifier produces output for actuating a Michelangelo hand by Ottobock Healthcare. To validate the system design, it was tested with four probands performing wrist flexion, wrist extension and fist hand movement patterns. After a training phase, features were extracted in real-time from either the EMG or NIR sensor data for classification with the model created during the training phase. In this setup, NIR sensor data alone proved to be sufficient for distinguishing three hand movement patterns with two sensors. The classification accuracy is equal or better to standard EMG data recorded from the same sensor pick-up area on the forearm.

Keywords: Prosthesis control · Machine learning · EMG signal
NIR signal

1 Introduction and Related Work

Despite the significant amount of academic research on pattern recognition for hand prostheses, clinical application is still insufficient. The underlying control strategies of even modern myoelectric hand prostheses date back to the 1950s [1]. As a consequence, longterm acceptance by patients is significantly reduced [2]. Among the reasons are restricted functionality of these devices as well as a lack of clinical trials for classification-based control [3], with the bulk of existing research relying on pre-recorded myoelectric signals [2]. A possible solution for these problems is a system for testing classification-based control systems in real-time. Furthermore, the introduction of novel sensing technology like near-infrared (NIR) spectroscopy can further improve classification results [4].

In order to provide the possibility of real-time testing of upper limb prosthesis control systems, this paper brings forth a corresponding hardware and software system for processing myoelectric and near-infrared signals accordingly. All steps of the multi-stage classification process as proposed by Englehart et al. [5], including feature extraction, classifier training as well as the operational classifier phase for actuating the prosthesis, are provided.

© Springer International Publishing AG 2018
R. Moreno-Díaz et al. (Eds.): EUROCAST 2017, Part II, LNCS 10672, pp. 359–365, 2018.
https://doi.org/10.1007/978-3-319-74727-9_42

2 Method

In this section, a description of the system setup and the conducted experiment to validate the approach is given. First the hard- and software prototype for acquiring NIR signals along with EMG signals for comparison is presented. During live classification of the signals, a Michelangelo hand can be utilized to demonstrate prosthesis actuation. Finally, an experiment was conducted with four probands.

2.1 Hard- and Software Setup

The hardware setup consists of two custom-built sensors with combined EMG and NIR sensing capability connected to a PC equipped with a NI DAQ system as shown in Fig. 1. The combined sensor makes simultaneous acquisition of two-channel EMG and NIR signals from the same pick-up site possible, allowing direct comparison of these two different sensing methodologies. During near-infrared sensing, visible light with a wavelength close to infrared light is emitted by an LED [6]. In order to prevent tissue damage from the produced heat, the light needs to be pulsed. The output of the NIR LED is absorbed in the blood stream. As the amount of blood in the tissue is reduced during muscle contraction, these movements can be detected with photo-sensing through the varying amount of light which is not absorbed but reflected. While most frequency components of the EMG signal are present in the range from 0 to 500 Hz, the sampling rate for the NI USB-6229 DAQ was set to 4096 Hz to allow filtering of the NIR pulses as explained in the next section. MATLAB Version 7.12.0.635 (R2011a) and Simulink Version 7.7 (R2011a) were utilized for the software implementation of feature and classification methods with utilization of the MATLAB

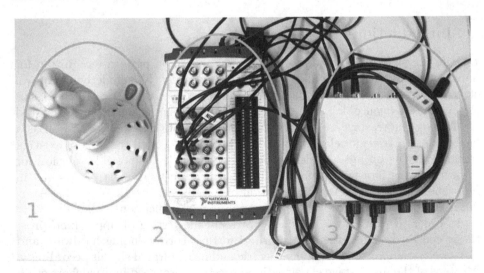

Fig. 1. The hardware setup for acquiring NIR and EMG signals with the (1) Michelangelo hand for prosthesis actuation (2) the data acquisition system and (3) the combined NIR-EMG sensors and amplifier.

Statistics toolbox. A graphical user interface designed in GUIDE (graphical user interface design environment) was created for user interaction during training and classification.

2.2 Feature Extraction

Different feature extraction methods exist for pre-processing the raw signal data before classifier training. In the experiment conducted, two basic feature algorithms were utilized to yield feature values denoting the average strength of the muscle contraction. In case of the myoelectric signals, the RMS signal was calculated for N samples $x_1, ..., x_n$ with the following formula [7]:

$$RMS = \sqrt{\frac{1}{N} \cdot \sum_{k=1}^{N} x_k^2}$$

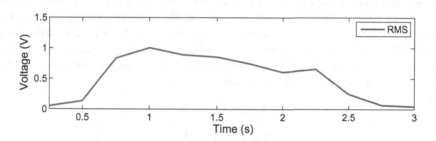

Fig. 2. The RMS feature calculated from a wrist extension EMG signal [8].

An example graph for the RMS feature values calculated from myoelectric signal data for a wrist extension is displayed in Fig. 2. In case of the near-infrared signal, the near-infrared signal feature (NIRS) is calculated from the raw pulsed sensor data with [9]:

$$\text{NIRS} = \text{Signal}(\overline{n}, \overline{e}) - \text{Offset}(\overline{n}, \overline{e}) \tag{1}$$

The vector $\overline{n} = (n_1, ..., n_k)$ denotes the samples recorded during the specified time window. Additionally $\overline{e} = (e_1, ..., e_k)$ contains samples corresponding to the on or off states of the LED. The function $Ena(e_i)$ determines if the LED was switched on or off in correspondence with pre-set voltage thresholds [6]). Signal and offset are finally calculated with the following formula:

$$\text{Signal}(\overline{n}, \overline{e}) = \frac{\sum_{i=1}^{k} n_i \cdot Ena(e_i)}{\sum_{i=1}^{k} Ena(e_i)} \tag{2}$$

$$\text{Offset}(\overline{n}, \overline{e}) = \frac{\sum_{i=1}^{k} n_i \cdot (1 - Ena(e_i))}{\sum_{i=1}^{k} (1 - Ena(e_i))} \tag{3}$$

The calculated feature data was subsequently input to the classifier algorithm for classifier training. A NIRS feature graph for a wrist extension movement is shown in Fig. 3.

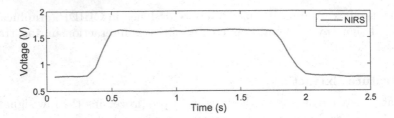

Fig. 3. The NIRS signal feature recorded during a wrist extension movement [8].

2.3 Classification

For real-time classification, we chose a simple decision tree classifier, which can be executed fast enough to deliver class labels for prosthesis actuation. Additionally, the decision tree classifier does not mandate parameter selection, which could be required when different probands are using the system. During decision tree classifier training, trees are constructed with subsets X_t for individual nodes T [10]. Along the tree every subset is further split up in two subsets (X_{tY}) and (X_{tN}), containing 'Yes'- or 'No'-answers to questions for each node T. The following conditions apply to the subsets:

$$X_{tY} \cap X_{tN} = \emptyset \, ; \tag{4}$$

$$X_{tY} \cup X_{tN} = X_t \, . \tag{5}$$

Training with feature data from known movement classes, a tree is constructed for the operational phase. During real-time classification, the data is acquired from the DAQ device and transmitted to the workspace with an event listener. Subsequently the decision tree classifier is executed with the trained model. Figure 4 shows the Simulink model in MATLAB for NIR signal acquisition.

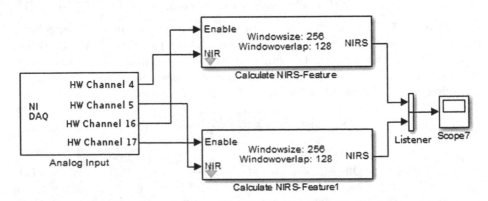

Fig. 4. The Simulink model utilized for NIRS feature extraction with a listener for MATLAB workspace inclusion, based on [11], p. 20.

3   Experimental Validation

The prototype was tested with four probands, recording two channels of EMG and NIR signals each from the same pick-up location with the custom-built combined sensors. Sensors were on the opposite sides of the forearm and fastened with an armband. Additionally, the NIR Sensors were affixed with adhesive tape and covered with aluminium foil in order to shield the sensors from ambient light

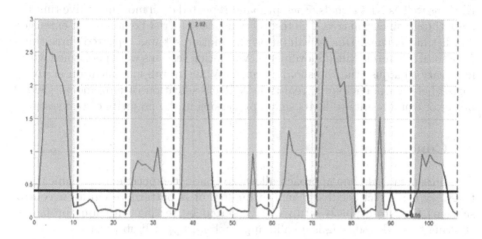

(a) First Sensor

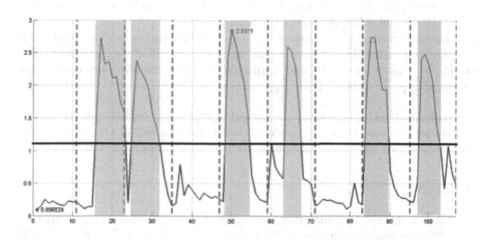

(b) Second Sensor

Fig. 5. A correctly applied threshold for RMS values of wrist extension, wrist flexion and fist movements [11].

sources. Before training, an adequate signal threshold had to be determined to mark the beginning and end of movements. The threshold was calculated from the mean of the RMS values during resting movements for each sensor respectively. The threshold varied between probands, with the measured value typically increased by 10 to 20% before training and validation. An example of a correctly applied threshold for the nine subsequent movements is displayed in Fig. 5. An incorrectly set threshold can lead to markedly reduced classification accuracy. Finding the correct signal threshold proved to be a challenge both for EMG as well as NIR signals. Each proband repeated each movement five times to train the classification tree. The test set consisted of wrist flexion, wrist extension and fist movements. During validation the motion set was repeated three times for a total of nine classified movements. Validation sessions were performed where necessary to adjust the threshold settings. While overall classification accuracy varied between participants, real-time NIR classification of movements yielded better or equal results to EMG signal processing in the majority of test sessions.

4   Conclusion

The presented real-time approach allows to conduct experiments in clinical settings, giving direct feedback on the quality of a pattern recognition control scheme. The prototypical setup was validated by testing with four probands, showing that NIR sensing is a valid approach for upper limb prosthesis control. Furthermore, it can potentially outperform myoelectrically controlled prosthetics. Threshold setting to distinguish resting movements from active motion is challenging for both signal types. This is one potential direction for future work. Furthermore, the approach should be extended with a higher number of tested grip patterns and different feature extraction methods for the two sensing methods. In this regard, combination of EMG and NIR sensing as well as other, more complex classifiers like support vector machines should be investigated. Integration of classification soft- and hardware into embedded devices could allow longterm in-place testing of new approaches.

Acknowledgment. The authors are grateful to Prof. Klaus Buchenrieder of the Universität der Bundeswehr for his support of their research as well as Otto Bock Healthcare for supplying the Michelangelo hand employed for testing the control scheme.

References

1. Jiang, N., Dosen, S., Müller, K.R., Farina, D.: Myoelectric control of artificial limbs - is there a need to change focus. IEEE Signal Process. Mag. **29**(5), 148–152 (2012)
2. Peerdeman, B., Boere, D., Witteveen, H.J.B., Huis in 't Veld, M.H.A., Hermens, H.J., Stramigioli, S., Rietman, J.S., Veltink, P.H., Misra, S.: Myoelectric forearm prostheses: state of the art from a user-centered perspective. J. Rehabil. Res. Develop. **48**(6), 719–738 (2011)

3. Scheme, E., Englehart, K.: Electromyogram pattern recognition for control of powered upper-limb prostheses: state of the art and challenges for clinical use. J. Rehabil. Res. Develop. **48**(6), 643 (2011)
4. Attenberger, A., Buchenrieder, K.: Modeling and visualization of classification-based control schemes for upper limb prostheses. In: Popovic, M., Schätz, B., Voss, S., (eds.) IEEE 19th International Conference and Workshops on Engineering of Computer-Based Systems, ECBS 2012, Novi Sad, Serbia, 11–13 Apr 2012, pp. 188–194. IEEE Computer Society (2012)
5. Englehart, K., Hudgins, B., Parker, P., Stevenson, M.: Classification of the myoelectric signal using time-frequency based representations. Med. Eng. Phys. **21**(6–7), 431–438 (1999)
6. Herrmann, S.: Direkte und proportionale Ansteuerung einzelner Finger von Handprothesen. Verlag Dr. Hut, Munich (2011)
7. Buchenrieder, K.: Dimensionality reduction for the control of powered upper limb prostheses. In: Proceedings of the 14th Annual IEEE International Conference and Workshops on the Engineering of Computer-Based Systems (ECBS 2007), pp. 327–333, March 2007
8. Attenberger, A.: Time analysis for improved upper limb movement classification: Dissertation. Universität der Bundeswehr München, Neubiberg (2016)
9. Herrmann, S., Attenberger, A., Buchenrieder, K.: Prostheses control with combined near-infrared and myoelectric signals. In: Moreno-Díaz, R., Pichler, F., Quesada-Arencibia, A. (eds.) EUROCAST 2011. LNCS, vol. 6928, pp. 601–608. Springer, Heidelberg (2012). https://doi.org/10.1007/978-3-642-27579-1_77
10. Theodoridis, S., Koutroumbas, K.: Pattern Recognition, 4th edn. Academic Press, Cambridge (2008)
11. Wojciechoswki, S.: A MATLAB classification system for upper limb prosthesis control. Master's thesis, Wrocław University of Science and Technology, Wrocław, Poland (2014)

The Metamodel of Heritage Preservation for Medical Big Data

Zenon Chaczko[1]([⊠]), Lucia Carrion Gordon[1], and Wojciech Bożejko[2]

[1] FEIT Faculty of Engineering and IT, UTS University of Technology,
Sydney, NSW, Australia
Zenon.Chaczko@uts.edu.au, Lucia.CarrionGordon@student.uts.edu.au
[2] Department of Automatics, Mechatronics and Control Systems,
Faculty of Electronics, Wrocław University of Science and Technology,
Wrocław, Poland
wojciech.bozejko@pwr.edu.pl

Abstract. At present the real challenge of Digital Data Preservation concerns methods of keeping all important attributes of the data and preserving their originality. The key is to keep the living part of the data. It is the essence of the Heritage concept. The Heritage is about the concrete data the concept gives the interconnection to other aspects of the reality. Nowadays the physical value and the aspects of items complete the relevance of information. But the question is what is heritage and which parameters defining the artifact or the information as a heritage? The context and the interpretation of data is the answer. The heritage term is defining as the crucial and central part of the presented research. Big data analytics in healthcare is evolving into a promising field for providing insight from very large data sets and improving outcomes while reducing costs.

Keywords: Data · Preservation · Digital · Heritage · Metamodel
Ontology

1 Introduction

There are two tendencies around the understanding of the management of the ideas. The ontology and the Epistemology of this study, centralized the future use and the Serendipity tendency of the item. However, in the perspective of the nonphysical items there is a World of Physical and Logical and how the Preservation need to look items and how will be the manifestation. Digital Preservation has evolved into a specialized, interdisciplinary research. Through the time the challenge in to jointly develop solutions. As the patterns and alternative solutions there are Information Retrieval and, Machine Learning or Software Engineering. The Digital Preservation show us the reality of the understanding of the World about the facility to have digital expressions rather than just physical. The Heritage of the collected information define the quality of the Data. This is specifically important in medical field. At this stage, the definition of Heritage involved

R. Moreno-Díaz et al. (Eds.): EUROCAST 2017, Part II, LNCS 10672, pp. 366–371, 2018.
https://doi.org/10.1007/978-3-319-74727-9_43

the presence not only the content. It is the express by itself the real meaning of the data. The perception of the importance and relevance of the information is measured through the definitions and the proposed Metamodel. Digital Data and Heritage Preservation concepts are related to medical data management, contextualization and storage. There are many issues and concerns around it. This research explores the precise definition, context and the need of patterns of heritage specifically in medicine. Patterns, the Metamodel and the Ontology of DHP.

2   Patterns, the Metamodel and the Ontology of DHP

One of the primarily concerns about the explosive amount of information and the complexity of the classification, is how to keep the principal characteristics data. A need to move away the traditional understanding of Heritage reflects the real meaning of the data. More artifacts and everyday life tendency is to have less physical representation in the World of Logic. The representation of the items refers to the tendency of more things nonphysical and ow through the Heritage it passes the attributes (Fig. 1).

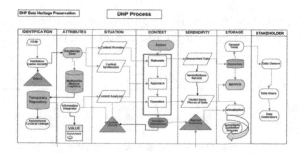

Fig. 1. DHP workflow model

The Metadata Model explores dynamic data representations and specifically the new relations, their origin and the mechanism(s) that generate these relations. The formatting of information provides the unique result as a digital age of the information. Other objective is the knowledge management and Ontology as techniques for analyzing information. The proposed metamodel aims to provide an alternative for the understanding of the Heritage Preservation concept that relates to important dimensions around the processed data and its origins. The different dimensions of the Digital Heritage Preservation capture the real significance of Data Heritage.

3   Value Based on Heritage

The search for values and meaning has become a pressing concern (see [1,2,7,16]).

Values are the subject of much discussion in contemporary society. In this postmodern, post-ideology, post-nation-state age, the search for values and meaning has become a pressing concern. In the field of cultural heritage conservation, values are critical to deciding what to conserve—what material goods will re p resent us and our past to future generations—as well as to determining how to conserve.

Discussions of values, of how social contexts shape heritage and conservation, and of the imperative of public participation are issues that challenge conventional notions of conservation professionals' responsibilities.

Values is most often used in one of two senses: first, as morals, principles, or other ideas that serve as guides to action (individual and collective); and second, in reference to the qualities and characteristics seen in things, in particular the positive characteristics (actual and potential).

Digital Preservation has evolved into a specialized, interdisciplinary research. Through the time the challenge in to jointly develop solutions. As the patterns and alternative solutions there are Information Retrieval and, Machine Learning or Software Engineering. The real fact of Digital Preservation show us the reality of the understanding of the World about the facility to hsve digital expressions rather than just physical. The Heritage of the collected information define the quality of the Data (Fig. 2).

At this stage, the definition of Heritage involves mostly the presence of the content. It is the expressed by itself the real meaning of the data. The perception

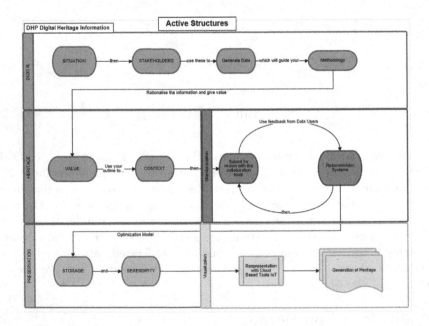

Fig. 2. Active structures

of the information relevance is measured using the definitions and matric of the proposed metamodel.

As concepts Digital Data and Heritage Preservation are related to data management, contextualization and storage. There are many issues and concerns around it. This research explores the precise definition, context and the need of patterns of heritage. The relations, interpretation and context give us the appropriate methods to keep information for a long term use. The management of massive amounts of critical data involves designing, modeling, processing and implementation of accurate systems. The methods to understand data have to consider two dimensions that this research has to focused on: access dimension and cognitive dimension. Both of these dimensions have relevance to get results because at the same time, ensure the correct data preservation.

Our cultural heritage, documents and artefacts increase regularly and place Data Management as a crucial issue. The first stage involves exploration and approaches based on review of recent advances. The second stage involves adaptation of architectural framework and development of software system architecture in order to build the system prototype. Increasing regulatory compliance mandates are forcing enterprises to seek new approaches to managing reference data. Sometimes the approach of tracking reference data in spreadsheets and doing manual reconciliation is both time consuming and prone to human error. As organizations merge and businesses evolve, reference data must be continually mapped and merged as applications are linked and integrated, accuracy and consistency, realize improved data quality, strategy lets organizations adapt reference data as the business evolves.

The massive amount of data and the growth of Big Data drive the society to preserve the information principally related with the lost of key information. The protagonism in the role of metadata and the requirement that data has to be keep in a long term open the alternative to focus on information management.

4 Big Data and Healthcare

The healthcare industry historically has generated large amounts of data, driven by record keeping, compliance & regulatory requirements, and patient care (see [9]). While most data is stored in hard copy form, the current trend is toward rapid digitization of these large amounts of data. Driven by mandatory requirements and the potential to improve the quality of healthcare delivery meanwhile reducing the costs, these massive quantities of data (known as 'big data') hold the promise of supporting a wide range of medical and healthcare functions, including among others clinical decision support, disease surveillance, and population health management [10–13].

For the purpose of big data analytics, this data has to be pooled. In the second component the data is in a 'raw' state and needs to be processed or transformed, at which point several options are available. A service-oriented architectural approach combined with web services (middleware) is one possibility [14] (Fig. 3).

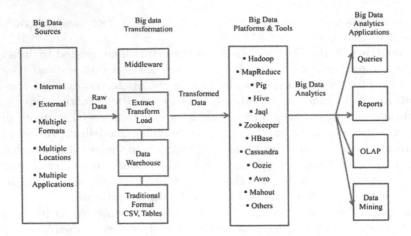

Fig. 3. An applied conceptual architecture of big data analytics [15]

5   Conclusion

Several challenges highlighted must be addressed. As big data analytics becomes more mainstream, issues such as guaranteeing privacy, safeguarding security, establishing standards and governance, and continually improving the tools and technologies will garner attention.

The use of tools and techniques like Steganography, the concepts of Software Architecture will have a real approach and meaningful characteristics for the relevance of the investigation.

The context, relation and situation of Heritage are impressive relevant in the research because it gives the sense of the future of the Knowledge in the World. Through medical process of Preservation will do a contribution for society advances.

The Business Process Management give us a good approach to the development of Performance and Data Preservation. Through process the increase of data can be justified.

The way to improve the understanding of the methodology, the information has to consider two dimensions: access dimension and cognitive dimension. Both of them have the level of importance in terms of the results.

While most platforms currently available are open source, the typical advantages and limitations of open source platforms apply. To succeed, big data analytics in healthcare needs to be packaged so it is menu-driven, user-friendly and transparent. Real-time big data analytics is a key requirement in healthcare.

References

1. Bożejko, W., Hejducki, Z., Wodecki, M.: Applying metaheuristic strategies in construction projects management. J. Civ. Eng. Manag. **18**(5), 621–630 (2012)
2. Bożejko, W., Pempera, J., Smutnicki, C.: Parallel tabu search algorithm for the hybrid flow shop problem. Comput. Ind. Eng. **65**, 466–474 (2013)

3. Nasir, S.A., Noor, N.L.M.: Integrating ontology-based approach in knowledge management system (KMS): construction of batik heritage ontology. In: Science and Social Research (CSSR), pp. 674–679 (2010)
4. Thalmann, S., Seeber, I., Maier, R., Peinl, R., Pawlowski, J.M., Hetmank, L., Kruse, P., Bick, M.: Ontology-based standardization on knowledge exchange in social knowledge management environments. In: Proceedings of the 12th International Conference on Knowledge Management and Knowledge Technologies, Graz, Austria, pp. 1–8 (2012)
5. Strodl, S., Becker, C., Neumayer, R., Rauber, A.: How to choose a digital preservation strategy: evaluating a preservation planning procedure. In: Proceedings of the 7th ACM/IEEE-CS Joint Conference on Digital Libraries, Vancouver, BC, Canada (2007)
6. Gordon, L., Chaczko, Z.: Digital patterns for heritage and data preservation standards. In: Borowik, G., Chaczko, Z., Jacak, W., Łuba, T. (eds.) Computational Intelligence and Efficiency in Engineering Systems. SCI, vol. 595, pp. 47–59. Springer, Cham (2015). https://doi.org/10.1007/978-3-319-15720-7_4
7. Challa, S., Gulrez, T., Chaczko, Z., Paranesha, T.N.: Opportunistic information fusion: a new paradigm for next generation networked sensing systems. In: 8th International Conference on Information Fusion, vol. 1, pp. 720–727 (2005)
8. McCay-Peet, L.: Investigating work-related serendipity, what influences it, and how it may be facilitated in digital environments. Ph.D. thesis, Dalhausie University (2014)
9. Raghupathi, W.: Data mining in health care. In: Kudyba, S. (ed.) Healthcare Informatics: Improving Efficiency and Productivity, pp. 211–223 (2010)
10. Burghard, C.: Big data and analytics key to accountable care success, White paper, IDC Health Insights, pp. 1–9 (2012)
11. Dembosky, A.: Data prescription for better healthcare. Financ. Times 11(12), 19 (2012)
12. Feldman, B., Martin, E.M., Skotnes, T.: Big Data in Healthcare Hype and Hope, October 2012. Dr. Bonnie 360 (2012)
13. Fernandes, L., O'Connor, M., Weaver, V.: Big data, bigger outcomes. J. AHIMA 83(10), 38–43 (2012)
14. Raghupathi, W., Kesh, S.: Interoperable electronic health records design: towards a service-oriented architecture. e-Serv. J. 5, 39–57 (2007)
15. Raghupathi, W., Raghupathi, V.: Big data analytics in healthcare: promise and potential. Health Inf. Sci. Syst. 2, 3 (2014)
16. Smutnicki, C., Bożejko, W.: Parallel and distributed metaheuristics. In: Moreno-Díaz, R., Pichler, F., Quesada-Arencibia, A. (eds.) EUROCAST 2015. LNCS, vol. 9520, pp. 72–79. Springer, Cham (2015). https://doi.org/10.1007/978-3-319-27340-2_10

Our Early Experience Concerning an Assessment of Laparoscopy Training Systems

Ryszard Klempous[1(✉)], Jerzy W. Rozenblit[2], Konrad Kluwak[1], Jan Nikodem[1],
Dariusz Patkowski[5], Sylwester Gerus[3], Mateusz Palczewski[4],
Zdzisław Kiełbowicz[5], and Andrzej Wytyczak-Partyka[1]

[1] Faculty of Electronics, Wrocław University of Science and Technology,
Wybrzeże Wyspiańskiego 27, 50-370 Wrocław, Poland
{ryszard.klempous,konrad.kluwak,jan.nikodem,
andrzej.wytyczak-partyka}@pwr.edu.pl
[2] Department of Electrical and Computer Engineering, Department of Surgery,
The University of Arizona, Tucson, USA
[3] Department of Paediatric Surgery, Wrocław Medical University, Wrocław, Poland
[4] Faculty of Veterinary Medicine, Wrocław University of Environmental and Life
Sciences, Wrocław, Poland
[5] Pediatric Surgery and Urology Department, Wrocław University of Medicine,
Wrocław, Poland

Abstract. This papers describes the use of professional, commercial laparoscopic simulators to create a database of training results achieved by several groups of participants. The aim of this project was to create a common parameter database for two devices (LapSim and eoSurgical) to evaluate the trainees' performance. In order to deepen the statistical analyses and to draw conclusions, it is useful to identify additional parameters concerning the experience for the subjects studied. Such parameter will help to further evaluate training effectiveness.

Keywords: Simulation laparoscopic training assessment
Virtual reality · Laparoscopic surgery training · Endoscopic techniques

1 Introduction

Over the last decade, the ViMed laboratory at Wroclaw University of Technology has built a training curriculum for laparoscopic surgeons [4,14–16]. Numerous training sessions have been conducted with surgical residents, students and specialist surgeons. Using a combination of Virtual Reality-based, commercial simulators and pelvi-trainers, a training program along with scoring methods has been developed.

Laparoscopic surgery (one of most common minimally invasive procedures) has been in development since the 1980s and is currently the gold-standard in many operations, and is being introduced in new procedures [3]. Along with the obvious benefits to the patient, it is very demanding on the operating surgeons

© Springer International Publishing AG 2018
R. Moreno-Díaz et al. (Eds.): EUROCAST 2017, Part II, LNCS 10672, pp. 372–379, 2018.
https://doi.org/10.1007/978-3-319-74727-9_44

and requires additional skills, which have not been developed as part of the classical surgical training in the past.

Several works have already confirmed the value of applying Virtual Reality (VR) based training at an early stage of laparoscopic education [5,6,9]. The work described in this paper relates to these findings.

2   Laparoscopy Training and Simulation Requirements

Laparoscopic surgery has far more advantages than conventional surgery when performed by professional surgeons. They minimize the complications associated with postoperative pain, blood loss, and scarring on large incisions. Its use allows the patient to return to home very quickly without any major complications. No sense of touch and limited movement in the work area require good hand eye coordination. This is a difficult technique that requires a lot of exercise to minimize the risk of surgery and prevent errors. The most important activities in a laparoscopic simulation laboratory consist of:

- Training using simulators.
- Development the techniques to improve outcomes.
- Enhancement of the assessment techniques.
- Development of devices that improve training outcomes.

Teaching on simulations and performed on laboratory models, offers trainees the opportunity to gain hands on experience in a non-patient based setting. New technologies, provide opportunities for introducing new, realistic training approaches. For instance, the effects of reconstruction algorithms for 3D scene based on images from the camera for operational training [7] are now research in the context of simulated medical environments. The key element of the system is a new approach to training [9], in which a 3D model of the operative field is the basis of the interaction between the trainees (veterinarians and surgeons) in the simulation system.

Using simulators, we can explore how to stimulate their effectiveness in training professionals and students and the effectiveness of the methodology used to perform the procedure. Considering the two phases of learning:

- basic skills learning,
- study of specific procedures,

we can distinguish the main activities of training for specialists as follows:

1. creating scenarios and roles,
2. training using simulators, without supervisor,
3. training with the supervision of a specialist,
4. review of obtained results and evaluation of training.

Concerning the last activity, the main features of the evaluation include:

- **WHAT** Score, Left/Right Instrument Path Length (m), Left/Right Instrument Angular Path (degrees), Left Instrument Outside View (number), Left Instrument Outside View (s),

- **WHERE** Left/Right Instrument Outside View (number), Left/Right Instrument Outside View (s),
- **WHEN** Total Time (s).

Comparative studies which investigate which instructional design features are especially important regarding to the specifications of learning objectives, learning groups and learning environments, are well justified.

For now, in the world of science, there are few publications that compare simulators for the acquiring basic skills of laparoscopic surgery. Because of that, the general aims and goals of our work was:

- What parameters are necessary to better analyze surgeon experience.
- How to incorporate all the training mechanisms required for the Fundamentals of Laparoscopic Surgery (FLS).
- How to design of metrics that will analyze better the performance of a trainee.
- How the simulators need to be modified for various levels of advancement of the student?

3   Specific Background

One of the areas of research during the period of the last ten years in the ViMed laboratory at Wroclaw University of Science and Technology has been the application of virtual methods for laparoscopic surgery training [7,10,14,16]. The number of surgical advanced procedures using endoscopic techniques is growing steadily. As mentioned in the section above, endoscopic surgery requires special skills which are difficult to acquire in an operating room [3,11]. The simulation teaching performed on laboratory models offers opportunity to provide hands on experience. In addition, discussion has been made on effects of reconstruction algorithms for 3D scene based on images from the camera for operational training [8]. A key element of the system is a new approach to training [13], in which a 3D model of the operative field is the basis of the interaction between the trainees (veterinarians and surgeons) and simulation system. Presentation and discussion of the outline of the 3D processing algorithm and the results of a test for a group of 16 veterinary physicians and 20 surgeons is established. The tests were conducted on LapSim (http://www.surgical-science.com/lapsim-the-proven-training-system/) and eoSim SurgTrac (http://www.eosurgical.com/) simulators. The number of parameters assessed were 22 and 21, respectively.

1. LapSim - 22 parameters: Start Time, Score, Status, Total Time (s), Left and Right Instrument: Misses (%), Path Length (m), Angular Path (degrees), Outside View (#, s); Tissue Damage (#), Maximum Damage (mm), Grasper Collided with Left Box (#), Left Box Lifted (#), Left Box Min Exposure Angle (degrees), Grasper Collided with Right Box (#), Right Box Lifted (#), Right Box Min Expo-sure Angle (degrees).
2. eoSurgical - 21 parameters: time, left and right hand appliance of: distance between(cm), acceleration(mm/s2), distance(m), handedness(%), off screen(%), smoothness(mm/s3), speed(mm/s).

The quality of the training evaluation was very important, from our experience in the areas of performance assessment in computerized surgical training systems [12], as well as the assessment of education process management [1,2], along with additional methods developed in other centers [6,9].

4   ViMed - Center for Virtual Medical Technologies

Laboratory Programming Interfaces and Modeling Lab at the Faculty of Electronics of Wroclaw University of Science and Technology had been in operation for over a decade. In 2012, the laboratory became the Center of Virtual Medical Technology (http://vimed.pwr.edu.pl/). Since then, we prepared work on the applications, the latest research in the field of information technology, automation and robotics in the medical field. The basic aim is to use virtual reality to conduct simulated laparoscopic surgical operations for training purposes. The current scope of activities of the laboratory is:

- Modeling and simulation of laparoscopic operations;
- Wireless Sensor Networks (WSN);
- Modeling and recognition of human gait; and
- International project teams "7/24".

Students in the Medical University of Wroclaw utilize the lab in semester-long classes hone their skills in the theory and practice of laparoscopic surgery.

The main goals of such training are as follows:

- to learn the basic skills on simple simulators with the measurement of progress
- to re-train practicing surgeons after an extended leave of absence

5   Experiments and Performance Exercise

The objective was to assess the effectiveness of the course conducted on an animal model to improve the manual skills of surgeons performing endoscopic procedures. The rating was carried out before the course and after its completion. The conducted research consisted of 3 experiments among 2 groups of trainees. In total, 32 people were involved: males and females, between 17 and 33 years of age.

The first experiment was conducted during laparoscopic training for expert surgeons that took place at the Wroclaw University of Environmental and Life Sciences in September 2016. A total of 16 people underwent the training, which consisted of lectures, training on live animals under anesthesia and training verification using the LapSim surgical simulator. In order to objectively verify the results of training, every trainee had executed a set of exercises in the simulator before and after the training on live animals. The exercise set was comprised of:

- Instrument Navigation - touching a simulated gallstone 5 times with each hand
- Clip Applying - stretch the vessel to apply a clips, put a clip to two highlighted areas and cut vessel with scissors
- Grasping - stretch the object and move the object to the target - 5 times for each hand.

The results of the experiment are recorded in Table 1.

Table 1. Laparoscopic training on LapSim simulator for 16 expert surgeons.

	[I N]		[L & G]		[C A]	
	Aver.	Var.	Aver.	Var.	Aver.	Var.
Score	77	15	74	15	67	27
Total time (s)	48	20	158	40	221	153
Left instrument path length (m)	1	0	3	1	2	2
Left instrument angular path (degrees)	182	60	622	206	483	343
Right instrument path length (m)	1	0	3	1	3	2
Right instrument angular path (degrees)	208	52	621	169	533	323
Left instrument outside view (#)	0	0	11	7	1	1
Left instrument outside view (s)	0	1	13	9	7	13
Right instrument outside view (#)	0	1	8	6	1	2
Right instrument outside view (s)	0	1	4	1	3	5

The second experiment has been conducted among a group of 16 students of the Wroclaw Medical University who had limited previous contact with surgery during theoretical courses and internships. Firstly, the group had been trained

Table 2. Laparoscopic training on eoSurogical simulator for 16 student surgeons.

n = 16	17 September 2017			20 September 2017		
	Aver.	Var.	Med.	Aver.	Var.	Med.
Total time (s)	175	75	173	74	25	67
Total distance between (cm)	6	1	6	6	1	6
Total acceleration (mm/s2)	1	1	1	1	0	1
Total distance (m)	5	7	2	1	0	1
Total speed (mm/s)	3	5	2	2	1	2
Left instrument acceleration (mm/s2)	1	0	1	1	0	1
Left instrument distance (m)	1	0	1	1	0	1
Left instrument handedness (%)	42	18	48	49	10	48
Left instrument outside View (%)	18	12	15	14	10	14
Left instrument speed (mm/s)	1	0	1	2	1	2
Right instrument acceleration (mm/s2)	1	2	1	1	0	1
Right instrument distance (m)	4	7	1	1	0	1
Right instrument handedness (%)	58	18	52	51	10	52
Right instrument outside View (%)	10	8	7	10	10	7
Right instrument speed (mm/s)	5	10	2	2	1	2

using the eoSurgical simulator and the Thread Transfer exercise, where a thread has to be passed through several loops using a laparoscopic instrument in minimal time. Table 2 gathers the results of the trainees in this group. Each of the trainees had repeated the exercise after three days.

After such preparation, the students were trained using the LapSim simulator and exercised using a set of tasks: grasping, clip application, and cutting. The results of this training session are gathered in Table 3.

Table 3. Laparoscopic training on LapSim simulator for 16 student surgeons.

	[Grasping]		[Cutting]		[Clip applying]	
	Aver.	Var.	Aver.	Var.	Aver.	Var.
Score	10	14	10	14	0	0
Total time (s)	189	78	189	78	225	127
Left instrument path length (m)	4	2	4	2	4	3
Left instrument angular path (degrees)	637	411	637	411	426	260
Right instrument path length (m)	2	1	2	1	4	3
Right instrument angular path (degrees)	539	303	539	303	609	513
Left instrument outside view (#)	2	1	2	1	5	6
Left instrument outside view (s)	1	1	1	1	5	7
Right instrument outside view (#)	1	1	1	1	3	4
Right instrument outside view (s)	1	1	1	1	12	21

6   Conclusions and Final Remarks

The results confirm that laparoscopy simulation training permits to learn the basic skills sufficiently in simple simulators and allows us to measure the progress. We also found that every laparoscopic surgeon after an extended leave should again test his or her skills in the simulator to determine the level of proficiency. The results show that the multiplicity of exercise has had a positive effect.

By contrast, reducing the time spent on exercises had a negative impact We are also expecting to increase significantly the size of our data base; this seems to be necessary to further develop our assessment system. The aim of the work was to create a common parameter database for different devices (here: LapSim and eoSurgical) to evaluate the exercise.

Our Vimed Laboratory is based on the experience of the ASTEC Laboratory (University of Arizona, Tucson, AZ, USA) and the Laboratorio de Simulacion y Formacion basada en Tecnologia (Universidad de Las Palmas de Gran Canaria, Spain) We plan to extend our cooperation with the Multimedia Network Lab (National Cheng-Kung University, Taiwan and the Remote Labs (University of Technology Sydney (Australia).

References

1. Chaczko, Z., Klempous, R., Nikodem, J., Rozenblit, J.: Assessment of education process management. In: 2016 IEEE 14th International Symposium on Applied Machine Intelligence and Informatics (SAMI), pp. 263–267. IEEE (2016)
2. Chaczko, Z., Dobler, H., Jacak, W., Klempous, R., Maciejewski, H., Nikodem, J., Nikodem, M., Rozenblit, J., Araujo, C.P.S., Sliwinski, P.: Assessment of the quality of teaching and learning based on data driven evaluation methods. In: 2006 7th International Conference on Information Technology Based Higher Education and Training, pp. nil21–nil36. IEEE (2006)
3. Esposito, C., Escolino, M., Miyano, G., Caione, P., Chiarenza, F., Riccipetitoni, G., Yamataka, A., Savanelli, A., Settimi, A., Varlet, F., et al.: A comparison between laparoscopic and retroperitoneoscopic approach for partial nephrectomy in children with duplex kidney: a multicentric survey. World J. Urol. **34**(7), 939–948 (2016)
4. Feng, C., Rozenblit, J.W., Hamilton, A.J.: A hybrid view in a laparoscopic surgery training system. In: Proceedings of the 14th Annual IEEE International Conference and Workshops on the Engineering of Computer-Based Systems, pp. 339–348 (2007)
5. Grantcharov, T.P., Kristiansen, V.B., Bendix, J., Bardram, L., Rosenberg, J., Funch-Jensen, P.: Randomized clinical trial of virtual reality simulation for laparoscopic skills training. Br. J. Surg. **91**(2), 146–150 (2004)
6. Jimbo, T., Ieiri, S., Obata, S., Uemura, M., Souzaki, R., Matsuoka, N., Katayama, T., Masumoto, K., Hashizume, M., Taguchi, T.: Effectiveness of short-term endoscopic surgical skill training for young pediatric surgeons: a validation study using the laparoscopic fundoplication simulator. Pediatr. Surg. Int. **31**(10), 963–969 (2015)
7. Klempous, R., Nikodem, J., Wytyczak-Partyka, A.: Application of simulation techniques in a virtual laparoscopic laboratory. In: Moreno-Díaz, R., Pichler, F., Quesada-Arencibia, A. (eds.) EUROCAST 2011. LNCS, vol. 6928, pp. 242–247. Springer, Heidelberg (2012). https://doi.org/10.1007/978-3-642-27579-1_31
8. Napalkova, L., Rozenblit, J.W., Hwang, G., Hamilton, A.J., Suantak, L.: An optimal motion planning method for computer-assisted surgical training. Appl. Soft Comput. **24**, 889–899 (2014)
9. Nasr, A., Gerstle, J.T., Carrillo, B., Azzie, G.: The pediatric laparoscopic surgery (PLS) simulator: methodology and results of further validation. J. Pediatr. Surg. **48**(10), 2075–2077 (2013)
10. Nikodem, J., Wytyczak-Partyka, A., Klempous, R.: Application of image processing and virtual reality technologies in simulation of laparoscopic procedures. In: Moreno-Díaz, R., Pichler, F., Quesada-Arencibia, A. (eds.) EUROCAST 2015. LNCS, vol. 9520, pp. 463–470. Springer, Cham (2015). https://doi.org/10.1007/978-3-319-27340-2_58
11. Patkowski, D., Chrzan, R., Wróbel, G., Sokół, A., Dobaczewski, G., Apoznański, W., Zaleska-Dorobisz, U., Czernik, J.: Laparoscopic splenectomy in children: experience in a single institution. J. Laparoendosc. Adv. Surg. Tech. **17**(2), 230–234 (2007)
12. Riojas, M., Feng, C., Hamilton, A., Rozenblit, J.: Knowledge elicitation for performance assessment in a computerized surgical training system. Appl. Soft Comput. **11**(4), 3697–3708 (2011)

13. Rozenblit, J.W., Feng, C., Riojas, M., Napalkova, L., Hamilton, A.J., Hong, M., Berthet-Rayne, P., Czapiewski, P., Hwang, G., Nikodem, J., et al.: The computer assisted surgical trainer: design, models, and implementation. In: Proceedings of the 2014 Summer Simulation Multiconference, pp. 211–220. Society for Computer Simulation International (2014)

14. Wytyczak-Partyka, A., Nikodem, J., Klempous, R., Rozenblit, J.: A novel interaction method for laparoscopic surgery training. In: 2008 Conference on Human System Interactions, pp. 858–861. IEEE (2008)

15. Wytyczak-Partyka, A., Nikodem, J., Klempous, R., Rozenblit, J., Feng, C.: Computer-guided laparoscopic training with application of a fuzzy expert system. Lecture Notes Artif. Intell. **5317**, 965–972 (2008)

16. Wytyczak-Partyka, A., Nikodem, J., Klempous, R., Rozenblit, J., Klempous, R., Rudas, I.: Safety oriented laparoscopic surgery training system. In: Moreno-Díaz, R., Pichler, F., Quesada-Arencibia, A. (eds.) EUROCAST 2009. LNCS, vol. 5717, pp. 889–896. Springer, Heidelberg (2009). https://doi.org/10.1007/978-3-642-04772-5_114

Developing Adaptive Visual Communication Interface to Improve Rehabilitation Services

Jan Nikodem[1]([⊠]), Paweł Szczęsny[1], and Konrad Kluwak[2]

[1] Department of Computer Engineering, Wrocław University of Science and Technology, Wrocław, Poland
jan.nikodem@pwr.edu.pl, pawelszczesny0@gmail.com
[2] Department of Control Systems and Mechatronics, Wrocław University of Technology, Wrocław, Poland
konrad.kluwak@pwr.edu.pl

Abstract. The contribution of this paper is to present a results of applications of image processing in developing adaptive visual communication interface (AVCI). In the proposed approach we extend the interface with additional visual modality directed from human to computer. In the AVCI for disabled people, the crucial is to provide computers some cognitive skills. For sake of this purpose we add adaptation as an additional functionality, which will be working in AVCI background, behind the desktop.

Keywords: Human-computer interactions · Adaptive interfaces
Rehabilitation services · Visual communication

1 Introduction

The development of human-computer interfaces (HCI) was driven by a new needs emerging in the field of human-machine communication, on the one hand, and by the perceptual capabilities of the entities between which this communication was carried out, on the other. The changes were incremental, but essential on both, the computer and human sides.

The first human-computer interface was a device which input and output was facilitated by a terminal equipped with a special keyboard for input and a row of lamps to show results [11]. Next HCI resembles a typewriter (it was call a teletypewriter TTY) using which a human could interact with a computer. TTY initiated a communication with the computer, based on the keyboard and commands entered using it. The typewriter supported both directions of information flow: a commands typed on typewriter was sending to the computer as well as its response was typing on typewriter too.

In general, the HCI communication space has always provided two-way flow of information [9]. The man-to-computer channel uses human skills (mainly manual, but sometimes also audible, mimic), while the computer-to-man channel utilize human perceptual capabilities (mainly visual, but sometimes audio descriptive or tactile features)[5].

© Springer International Publishing AG 2018
R. Moreno-Díaz et al. (Eds.): EUROCAST 2017, Part II, LNCS 10672, pp. 380–385, 2018.
https://doi.org/10.1007/978-3-319-74727-9_45

Over the years, HCI have been constantly changing their form and expand its functionality [4], but always as tools used by man, have had a great impact on human behavior and mind, and thus have a huge influence on society, economy and culture. One of the way HCI has impacted society is that it created specialized interfaces for people with disabilities or neurological disorders.

The proposed interface can perform as tools that support a specific therapy. Applications that allow to control the computer and are beneficial therapeutically can have wide usage. Mostly because of the fact, that society widely use computers. Why would users using a computer not have at the same time do some rehabilitation exercises? A web camera (ubiquitous in IT devices) can observe user behavior, and interface software by recognizing gestures or mimics can control the computer and concurrently record a progress in the exercise. An effective image processing software is available, hence appropriate control software and functionality of the new HCI interface can be perform by the person who design it.

Generally, this adaptive visual communication interface project is dedicated to disabled people, however, it can provide a framework for other interfaces with programmable adaptation to user behavior. Properly designed interface can take advantage of the user's interest surfing the Internet, to improve his skills in the area which is based on mimics, gestures and movement of body parts.

2   Human Centered Interfaces

An improvement of human-computer interfaces (HCI) should primarily base on users needs [3]. Of course all of the new ideas of adaptive visual communication interface (AVCI) need a strong sociological, technological and software background - because that's a manner how human cooperates with computer. Unconventional thinking of software developers and innovative approaches of technicians can help us to achieve it.

In the AVCI for disabled people, the crucial is to provide computers some cognitive skills. For sake of this purpose we want to add adaptation as an additional functionality, which will be working in AVCI background, behind the desktop.

Interface will be able to *get to know* users, using web camera and effective software for real time image processing. AVCI adaptation to a specific user will be within the predefined *window time* which duration is determined experimentally.

2.1   Asymmetry - Needs and Limits of Adaptability of AVCI

At first glance, the asymmetry in the AVCI interface is not unusual. The computer and the human are so different that interface between them cannot be symmetrical. This asymmetry matters, if it doesn't reflect the communication capabilities of both sides of communication channel.

Considering asymmetry of AVCI interfaces there are two aspects; asymmetry of roles and asymmetry of directions.

- *Asymmetry of roles*: In contemporary AVCI interfaces there is a ubiquitous dominance of the computer over the user. Concerning the machine-to-man communication, the interface software decides what information is passed to the human. What's more, the software dominates, even in opposite (human-to-machine) direction of communication, where one would expect that a person decide, what he wants to convey to machine. Contemporary touchscreen technology interfaces enable the user to pass commands to a computer, but eventually the software determines which buttons are visible on the screen and which ones are active. Hackers' actions show that modified/swapped interface software can even decide on what that activity will based on.
- *Directional asymmetry*: The AVCI interfaces now indicate a large asymmetry in the flow of information. The machine-to-man direction of communication channel is dominant in the interface. It is amazing that these trends can be observed also on the Internet - the man-to-Internet links are often slower than the corresponding links in opposite direction. Hence, the machine-to-man direction in AVCI, in more than 80% uses the visual channel and alternative sound but haptic channels are used significantly less often.

At present, the main symptom of asymmetry is the dominance of the computer, and thus the enormous amount of information transmitted in a very short time (very often on the threshold of human perception abilities).

The interface should possess a moderation in the transmission of information. There are a mechanisms like cookies etc. that do certain Internet-based selection, adaptive filtering to user requirements (as the software developer sees it), but they don't work efficiently and they are poorly performing in concrete cases.

It is evident (not only for me) that from the fact that in this year I booked a hotel in Provence, it doesn't follow that in the near future (or ever) I will again want to go there! How to persuade the browser that I don't want to repeatedly, negatively respond to such offers during the Internet surfing? Why, my multiple-fold negative decisions are not sufficient to stop that action?

Does the multiple-fold negative response to the browser's suggestion; *Google recommends Chrome browser. Do you want to try it now?*, should not suffice to block (at least for some time) a pop-up window, with a such proposal?

2.2   User-Specific AVCI for Disabled People and Rehabilitation

Generally, an interface is for the user and should serve him, not some *standard user*. People do not want to be treated like a standard. Users are different. Healthy people somehow deal with problems described in the end of previous section, but disabled and sick people are exposed to additional stress [1]. If you are healthy, you can imagine how big irritation is induce for healthy user, when he must repeatedly erase the pop-up window following the question *Do you really want to do it?*. It is much harder to imagine what kind of effort and stress is there for a user with motor disabilities or after amputation of upper limbs, whose the mouse replaces with a stick held in the mouth or fixed on the head, when he repeatedly cancels the pop-up window?

The interface developer should not assume that he knows which the interface activity will be accepted by the user as good [7]. What's more he is not able to catch this! A professional developer will provide a standard version of the AVCI interface, but he ensures the property of auto-tuning for this interface, during its exploitation. The AVCI interface should therefore be adaptable (personalized) to the needs of the user [10].

Compliance of HCI with adaptability is not a simple task [8]. It is necessary to address such development issues as the integration of multiple technologies, creation of a modular platform for AVCI, behavioral control, facial recognition, gestures, visual perception, object localization, access to databases and information from the Internet, memory and association of facts etc. We don't do it ourselves, we prefer to use libraries available in these areas.

However, there are also problems, that the software developer will not be able to solve without the cooperation with the occupational therapist and the patient as a interface user. Adaptation is a dynamic situation, constantly changing, difficult to predict, but in the process of rehabilitation we have a goal to which we go. So, some things we should anticipate because we would like to have an interface that will react both in cases where things go well, as well as when things go wrong.

3 Expanding of Human - Computer Visual Channel

In the proposed approach we are introducing an expanded visual channel directed from human to computer. In this area, we extend the interface with additional visual modality. We don't reduce the former interface and its device range (mouse, touch-pad, joystick, trackball) or the range of user functions available. Used Web browser does not require any intervention.

Proposed software (using XInput, HidApi), cooperates with the browser's interface control area in that way, that it collaborates in resource sharing mode with standard mouse input.

So user has a choice, he can use the mouse or our interface. We duplicate mouse actions such as moving the cursor to the right, to the left, up, down, as well as activities of left or right buttons. We do not support mouse actions such as mouse gestures, because for rehabilitated persons, they are often too sophisticated.

By using the camera we get an image of the body part of the user with the movement of which it will control the cursor. In our case this is the user's head, but it can be a limb, hand, forearm, frame, joint, knee, depending on the process of rehabilitation in which our interface is supposed to be helpful.

For a selected object (part of the human body) we define the landmarks, a constellation of points, which, as a general model of our object will serve us to detect object gestures. Having an image from the camera, we are detecting changes in the location of selected landmark points, and on the basis of these changes, we make control decisions.

The main task of building an interface is to write software that processes the image from the camera, so that the computer receives only a machine-readable descriptor of the object in the image (Fig. 1). In the image processing we use morphological operators (Sobel gradient, erosion, dilation, opening, closing etc.) based on different structuring elements. In the case of face detection, this leads to the Histogram Oriented Gradients (HOG) method [2]. In the case of the user's limbs, morphological operators are used to point out the skeleton (e.g. bwmorph (BW, 'skel', Inf) in MATLAB).

First, however, we must find the object in the image. As an object detector we use the HOG methods and linear support vector machine (SVM), which will allow us to indicate the position of the object on the screen [6]. We use the prepared dlib and OpenCV libraries and the sliding window combined with image pyramid ideas. The image pyramid concept allows us to find objects at different scales of an image. Starting from the bottom of pyramid, we have the image at its original size, but at each subsequent layer, the image size is reduced until a minimum size has been reached. Sliding window play an integral role in object detection, because is allow us to localize exactly where in an image an object resides.

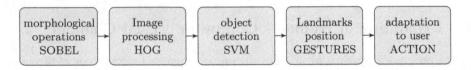

Fig. 1. An adaptive visual channel. From image processing to associative action.

When we know the position of the object on the screen, we span landmarks over it. As a result, we get constellations of points with calculated position of landmark points. By doing so, for subsequent frames of the video stream, we can keep track of the changes of these landmarks positions, and on the basis of those changes we classify the gestures that the user made.

Because of the adaptation purposes, for each gesture G associated with action a, we define the following structure

$$G^a = \{[g_1^a, g_2^a, g_3^a, ..., g_n^a]; MED, IQR, G_{min}, G_{max},\} \qquad (1)$$

where:

- $[g_1^a, g_2^a, g_3^a, ..., g_n^a]$ is n-element wide window of adaptation, remembering last n recent gestures G^a, in our program we set $n = 30$,
- MED, IQR - two variables; median and current acceptable range of changes, IQR - the interquartile range,
- G_{min}, G_{max} - static parameters, defined by the occupational therapist who set the limit of change (min, max) that the given gesture must match, to be interpreted by the interface as meaningful.

During the work, after each recognized gesture, the interface calculates the median and next determines the current permissible (in the vicinity of the median) range of gesture changes. For the purpose we use the interquartile range (IQR), as a measure of statistical dispersion, which is equal to the difference between 75-th and 25-th percentiles.

4   Conclusions

Tools requires human being with goals in mind, before they can be used to accomplish anything. We made an interface where computer using visual channel subsequently enucleating information from video stream. What's more, proposed human-computer communication channel was designed to realize adaptability to the user's disabilities. Our future work will deal with deriving optimal tuning of adaptation parameters.

References

1. Baltaci, S., Gokcay, D.: Stress detection in human-computer interaction: fusion of pupil dilation and facial temperature features. Int. J. Hum.-Comput. Interact. **32**(12) (2016). https://doi.org/10.1080/10447318.2016.1220069
2. Dalal, N., Triggs, B.: Histograms of oriented gradients for human detection. In: IEEE Compute Society Conference on Computer Vision and Pattern Recognition, vol. 1, pp. 886–893 (2005)
3. Evenson, S.: Design and HCI highlights. In: Presented at the HCIC 2005 Conference, Winter Park, Colorado, 6 February 2005
4. Grudin, J.: A moving target - the evolution of human-computer interaction. In: Human-Computer Interaction Handbook: Fundamentals, Evolving Technologies, and Emerging Applications, 3rd edn. Taylor & Francis Group (2012)
5. Light, A., Wakeman, I.: Beyond the interface: users perceptions of interaction and audience on websites. In: Clarke, D., Dix, A. (Eds.) Interfaces for the Active Web (Part 1), Special Issue. Interacting with Computers, vol. 13, No. 3, pp. 401–426 (2001)
6. Liu, Q., Lu, H., Metaxas, D.N.: Fundamentals in kernel discriminant analysis and feature selection for face recognition. In: Advanced Topics in Biometrics, pp. 129–148. World Scientific (2012). https://doi.org/10.1142/9789814287852_0006
7. Olson, J.S., Kellogg, W.A.: Ways of Knowing in HCI. Springer, New York (2014). https://doi.org/10.1007/978-1-4939-0378-8
8. Rogers, Y., Preece, J., Sharp, H.: Interaction Design: Beyond Human-Computer Interaction, 3rd edn. Wiley, Chichester (2011)
9. Sears, A., Jacko, J.A. (eds.): Human Computer Interaction Fundamentals. CRC Press/Taylor & Francis, Boca Raton (2009). ISBN 978-1-4200-8881-6
10. Vu, K.-P.L., Proctor, R.W.: Introduction to special issue: foundations of cognitive science for the design of human computer interactive systems. Int. J. Hum. Comput. Interact. **33**(1) (2017). https://doi.org/10.1080/10447318.2016.123948
11. Zuse, K.: The Computer - My Life. Springer, Heidelberg (1993). https://doi.org/10.1007/978-3-662-02931-2. ISBN 9783540564539

Performance Assessment of Optimal Chemotherapy Strategies for Cancer Treatment Planning

Ewa Szlachcic[(✉)] and Ryszard Klempous

Department of Automation, Mechatronics and Control Systems,
Wroclaw University of Science and Technology, Wroclaw, Poland
{ewa.szlachcic,ryszard.klempous}@pwr.edu.pl

Abstract. The paper presents a methodology of using multi-objective differential evolutionary approach to optimize a cancer chemotherapeutic treatment. The structure of optimal and suboptimal solutions will be presented. The constrained Pareto sub-optimal solutions are discussed. The performance assessment of non-dominated and dominated solutions is analyzed to help physicians to choose the most effective solution according to the approximation set of Pareto front.

Keywords: Cancer chemotherapy schedules
Bi-criteria optimization problem · Differential evolution approach
Pareto set · Performance assessment

1 Introduction

A cancer tumor treatment planning consists of using a chemotherapy process, which ought to slow down successively the growth of the tumor and manage the toxic-side effect for the human body of the patient. The very small difference among a treatment process and a survival process leads to determine precisely the set of non-dominated solutions and automatically generate treatment strategies among Pareto solutions. It becomes very important to design treatment strategies for specific tumors and drugs protocols without overdosing the patient in a chemotherapy process.

A chemotherapy is a treatment to eradicate a cancer tumor size using set of toxic, anti-cancer drugs. In a cancer chemotherapy optimization problem schedules of medical treatments were determined based on a mathematical growth model for cancer tumor described by set of differential equations [1–4,6]. The minimization of tumor burden at a fixed period of time and the minimization of drugs concentration with constraints, which described an influence of anticancer drugs for the human body, have been introduced in our earlier papers [11,12]. Mathematical, Gompertz model with a liner cell-loss effect, to define a tumor growth, was used [1,5,7].

© Springer International Publishing AG 2018
R. Moreno-Díaz et al. (Eds.): EUROCAST 2017, Part II, LNCS 10672, pp. 386–393, 2018.
https://doi.org/10.1007/978-3-319-74727-9_46

The non-dominated optimal solutions set is found by the modified differential evolutionary search method for multi-objective optimization problem with the help of constraints normalization procedure [11,12,14,15]. The drug doses should be scheduled to ensure the patients will tolerate its toxic side effects and their survival time will be longer. The search of optimization result may be very time-consuming taking under consideration the whole Pareto set of treatment scenarios and it depends also on patient medical parameters and the experience of physicians. Therefore, in the paper we try to analyze the performance of the optimal bi-criteria strategies calculated as the set of non-dominated solutions, which can explore the wide range of treatment scheduling strategies. The performance measures [8,13,16] were devised, which can gauge the quality and regularity of solutions. Using the drug schedules belonging to the Pareto optimal front with the help of proposed measures, a physician could obtain the effective chemotherapeutic treatment suggestions. On the basis of assumptions concerning oncologist preferences a knowledge about non-dominated front helps a decision maker in choosing the best compromise strategy for a cancer chemotherapy treatment in a decision support system.

2   Chemotherapy Planning

Curative treatments attempts to reduce a cancer tumor and to respect highly toxic drugs and their influence on a patient survival time. The structure of optimal schedule for multi-drugs and drug doses in time intervals to be given determines a parallel process of killing cancer cells and simultaneously killing the healthy cells, what cause damage of the sensitive tissues of the human body. The Gompertz mathematical model was used, the most widely accepted in the practical applications [7–10]. This model based on Gompertz type growth equations takes into account the dynamic relationship between the behavior of the set of drugs and its corresponding concentration level. The model with personalized data for the chosen patient simulates a growth of a cancer tumor and determines a rate of tumor reduction.

The bi-criteria optimization problem for chemotherapy treatment planning is defined as two objective functions described over the strictly defined set of constraints. The first objective function $f_1(x)$ will be to minimize the number of cancer tumor cells at a fixed period of time and the second objective function $f_2(x)$ will be sought to minimize the toxicity of drugs doses. From treatment intervals the optimal vector $x = [x_{ij}]$ is a template for drug doses, where i defined the index of time interval, for $\forall i = \overline{1, n}$ and j means an index of j drug, for $\forall j = \overline{1, D}$ drugs and is determined as below:

$$\min_{x \in X} F(x) = \begin{bmatrix} f_1(x) \\ f_2(x) \end{bmatrix} \tag{1}$$

where: $f_1(x(t)) = n(x(t))$ denotes a number of tumor cells for anti-cancer drug doses x at time t_i, for $\forall i = \overline{1, n}$, and

$$f_2(x) = \int_0^{T_{max}} c_1(x(t))dt \tag{2}$$

The function $c_1(x(t))$ determines the concentration of drugs at time t_i on the set of constraints X:

$$X = \{x : g_j(x) - \overline{g_j} \leq 0\} \qquad for \qquad \forall j = \overline{1, 2D + 2} \tag{3}$$

The set of intricate constraints X secure a patient life before toxic side effect of anti-cancer drugs in the form:

1. The White Blood Cells (WBC) $w(x, t + \tau)$ have be strictly controlled at level higher than a fixed down level W_d:

$$g_1(x) : W_d - w(x, t + \tau) \leq 0 \tag{4}$$

2. The maximum permissible time T_{max} when the patient WBC count remains below a fixed upper lever W_u and over the down level W_d:

$$g_2(x) : t_u(x, T_{max}) - T_u \leq 0 \tag{5}$$

3. The toxicity of j drug $a_j(x, T_{max})$ at T_{max} must not exceed specified limit A_{maxj}:

$$g_3(x) : a_j(x, T_{max}) - A_{maxj} \leq 0 \qquad for \qquad \forall j = \overline{1, D} \tag{6}$$

4. The rate of drug j accumulation $c_{1j}(x)$ cannot exceed maximum dose value C_{maxj}:

$$g_4(x) : c_{1j}(x) - C_{maxj} \leq 0 \qquad for \qquad \forall j = \overline{1, D} \tag{7}$$

The tumor growth according to a Gompertz-type growth equation takes the form [7,11]:

$$\frac{dn(x(t))}{dt} = N_g(x(t)) - N_c(x(t)) \tag{8}$$

where: $N_g(x(t))$ - represents the growth of an untreated tumor, $N_c(x(t))$ - the cell loss term, depending on drugs concentration.

The Pareto optimality definition, based on the concept of a dominance idea [13] was used, as below: A vector $F^a(x)$ of two objective functions is said to dominate a vector $F^b(x)$, if

$$f_i^a(x) \leq f_i^b(x) \qquad for \qquad i \in \{1, 2\} \tag{9}$$

$$\exists j \in \{1, 2\} \qquad for \quad which \qquad f_i^a(x) < f_j^b(x) \tag{10}$$

in the objective functions space.

We already introduced the idea of the non-dominated solutions to calculate the optimal strategies for chemotherapy scheduling and drug doses. The Pareto optimal solutions are found by the Hybrid Differential Evolution Algorithm [11,12] for bi-criteria optimization problem with complex set of constraints. The chemotherapy treatment strategies generates the personalized chemotherapy schedules and dose rates for oncology patients. All these treatment scenarios vary depending on a given patient profile and may change as the chemotherapy process is going on. It leads to an extension of Patient Survival Time (PST) and simultaneously stop an increase of tumor burden and starts its cell-loss if possible. Especially the WBC constraints like (4) and (5) were taken under consideration in constraints normalization scheme. When the WBC constraints would be exceeded it may cause the dangerous deterioration of patients life parameters. It seems very important to choose even a dominated solution, but satisfying the whole set X. Such chemotherapeutic treatment strategy gives the chance of a survival to a treated patient.

In numerical process in some cases it is difficult to calculate the chemotherapy scenarios, which fulfill the whole set of constraints. Despite the selection of personal parameters and different types of differential evolution operators, algorithm could not calculate the feasible and non-dominated solutions. In order to evolve solutions, which fulfilled the subsets of constraints, the different types of domination scheme are proposed according to various set of constraints. In medical treatment procedures two constraints (4) and (5), concerning the WBC strategies, take the significant role.

In the paper the definition of candidate solutions was weakened for the case, when some constraints are not fulfilled. The constrained Pareto dominance procedure [13] was introduced, to respect the situation, when solutions can satisfy only some of the constraints. Then an individual $F^a(x)$ dominates an individual $F^b(x)$ in the objective functions space taking under consideration the partial feasibility of a set of constraints. In discussed problem (1)–(8) the set of constraints (4)–(7) is divided for two subsets:

1. The subset G_1 determines individuals, fulfilling constraints $g_1(x)$ (4) and $g_2(x)$ (5), which concerns the WBC restrictions

$$G_1 = \{x : W_d - w(x, t + \tau) \leq 0, \quad and \quad t_u(x, T_{max}) - T_u \leq 0\} \qquad (11)$$

2. The subset G_2 concerns the constraints $g_3(x)$ (6) and $g_4(x)$ (7) according to the different D drugs.

$$G_2 = \{x : a_j(x) - A_{max,j} \leq 0, c_{1j}(x) - C_{maxj} \leq 0\} \quad for \quad \forall j = \overline{1, D} \quad (12)$$

The objective function $F^a(x)$ is feasible it means the solution x^a fulfills the whole set of constraints, but the $F^b(x)$ is infeasible it means the solution x^b fulfills the subset G_1 but in the subset G_2 some of the constraints can be overloaded for some types of drugs. In this particular case the definition of constrained Pareto dominance scheme when the solution $F^a(x)$ dominates the solution $F^b(x)$ on the given set X is defined as follows:

– The solutions x^a and x^b are feasible and $F^a(x)$ dominates $F^b(x)$ in the objective functions space.
– The case with the violation of G_2 set of constraints. Two individuals are infeasible according to the same constraints and $F^a(x)$ dominates $F^b(x)$ in the same space.
– The case with the violation of some of the G_2 set. Two individuals are infeasible and $F^a(x)$ fulfills more constraints then $F^b(x)$.
– The case when the solution x^a is feasible and x^b is infeasible.

The constraints of G_1 subset have to be always fulfilled. In order to improve the balance between the feasible and infeasible solutions, according to the constrained Pareto dominance scheme the multi-objective differential evolution algorithm was modified and a three stage optimization procedure is proposed:

- the optimization process with Hybrid Differential Evolution Algorithm when the Pareto optimal solutions fulfill all the given constraints,
- in the case of infeasible solutions the optimization procedure, when non-dominated solutions fulfill one of the constrained Pareto dominance case and
- in the case of feasible, but dominated solutions the optimization algorithm calculates the set of dominated individuals.

Using the calculated chemotherapy scenarios belonging to the Pareto optimal front or to the constrained Pareto dominance set of solutions in order to assess the better chemotherapy strategy some measures are proposed in the next section.

3   Performance Treatment

The proposed algorithm calculates the Pareto optimal front if possible on the bi-criteria functions space for the determined treatment period. The results strongly depends on the desired treatment data parameters and considering the patient dynamics. In the whole treatment procedure the WBC count remains all the time among fixed down and up levels.

Initially, the chemotherapy process is given to reduce a tumor burden. However the choice of one of the solutions from the Pareto set constitutes a great difficulty. The cytotoxic chemotherapy reduced the cancer volume, but how to decide about the chemotherapy strategy taking under consideration the toxicity of the anti-cancer drugs. The set of Pareto front is huge and a medical decision belongs to an oncology physician. For small cancer volumes and for the feasible and optimal Pareto chemotherapy strategy can be effective in the chemotherapy treatment process. But for large cancer volumes the Pareto constrained dominance scheme ought to be used. In this case the dynamics of the process was strongly different and the non-dominated but infeasible solution has to be also taken under consideration. It requires to enter a procedure, which can compare optimal solutions and allows to determine the 'good' solution from the physicians point of view. Such strategy ought to improve the quality of patients life by decreasing the strain of the cytotoxic, anticancer chemotherapy process.

The solution vector ought to fulfill one of the constrained Pareto dominance case. The chemotherapy strategy depends on parameters taking in the Gompertz model and on the cancer volume. The side effects, associated with the treatment process for feasible but dominated solution, are considered less toxic than for the Pareto optimal strategy.

The performance assessment of optimal chemotherapy strategy for cancer treatment planning can be analyzed with the help of the quality measures and the diversity measure [8,13,16]. The quality measures helps to assess the quality of Pareto optimal and constrained Pareto dominant solutions. The diversity measure assess the regularity of multi-objective solutions across several sub-optimal feasible but dominated points in the algorithm.

The Pareto-optimal solutions with different quality values create an opportunity to choose the effective solution among the solutions in the approximation set of scenarios. The optimal decisions vary depending on the patients state of health. The oncologist would receive several treatment suggestions that the physician could access according to the some other parameters not considered in the discussed model. The quality measures, used to choose one of the non-dominated or even dominated solutions can be described as below:

1. Total amount of drugs, given to a patient in one treatment period x(t) (TAD)-the set of used cocktail of chosen drugs, calculated as the integral under the optimal decision vector curve x(t) during one treatment period.
2. Toxicity of the treatment schedules (TTS) is the average of the toxicity level according to an approximation set of Pareto optimal or Pareto sub-optimal solutions.

Restricting toxic side-effect the search space is reduced and thus leads to faster search of multi-criteria solutions. We should favor the treatment strategies, which represent the chemotherapy process with low factors: TAD and TTS.

In order to measure the regularity of the Pareto-optimal or sub-optimal chemotherapy strategies, we consider the regularity parameter: Deviation Value from Average Solution (DAS) [8,13]. The DAS parameter is defined as a vector containing the standard deviation of each of the treatment cycle from the average best treatment values of drug doses.

The space under solution curve x(t) (TAD) represents the total amount of drugs, which were used in one period of chemotherapy process. The treatment schedule minimizing the total amount of drugs may be generally preferred from the set of Pareto-optimal solutions. The non-regular and diversified treatment schedule gives the incorrect impact on the patients state of life. Having regular patterns with the smallest regularity metric (DAS) of chemotherapy strategies it could be considered to be more favorable from the patients point of view.

The three scenarios of chemotherapy strategies were described, depending on the feasibility of some set of constraints and the dominance property of objective functions values. Based on quality and regularity measures more regular and less toxic chemotherapy treatments ought to be selected by a physician.

The numerical calculations of Pareto optimal or Pareto sub-optimal fronts show the relative importance of the toxic side effects in the process. The choice of one suitable schedule of drugs and drug doses gives to a physician a proposition of chemotherapy treatment strategy. The personalized chemotherapy treatment schedule and drug doses are determined with the help of performance measures. The performance factors estimate the quality of the solutions: a more regular treatment pattern and significantly weaken the variability of the solutions are proposed. Received results produce the scenarios of possible treatments, leaving final decisions to oncologists.

4   Conclusions

Based on Gompertz mathematical model of a cancer tumor growth and depending on the behavior of the set of anti-cancer drugs and its concentration level the bi-criteria optimization problem allows to find effective chemotherapy strategies as the minimization of a tumor volume at a fixed period of time and the maximization of Patient Survival Time (PST) by minimization of toxic side-effect of prepared anti-cancer drugs. The hybrid Differential Evolution algorithm, expanded by the procedure with constrained Pareto dominance scheme, can assists oncologists in finding chemotherapy treatment scenarios.

The performance measures are proposed to help a decision maker to provide an effective procedure for the chemotherapy process. These parameters device to determine both the quality and regularity of the solutions obtained. The used measures support the design of cancer chemotherapy treatment, helping the practitioner to choose one effective chemotherapy planning for the fixed patient.

References

1. Antipov, A.: Optimal multitherapy strategy in mathematical model of dynamics of the number of nonuniform tumor cells. J. Comput. Syst. Sci. Int. **50**(3), 499–510 (2011)
2. Araujo, R.P., McElwain, D.S.: A history of the study of solid tumour growth: the contribution of mathematical modelling. Bull. Math. Biol. **66**(5), 1039 (2004)
3. Barry, D.W., Underwood, C.S., McCreedy, B.J., Hadden, D.D., Lucas, J.L.: Systems, methods and computer program products for guiding the selection of therapeutic treatment regimens. US Patent 6,081,786, 27 June 2000
4. Bratus, A., Todorov, Y., Yegorov, I., Yurchenko, D.: Solution of the feedback control problem in the mathematical model of leukaemia therapy. J. Optim. Theory Appl. **159**(3), 590–605 (2013)
5. Chumerina, E.: Choice of optimal strategy of tumor chemotherapy in Gompertz model. J. Comput. Syst. Sci. Int. **48**(2), 325–331 (2009)
6. Egorov, I.E.: Optimal feedback control in a mathematical model of malignant tumour treatment with the immune reaction taken into account. Mat. Biol. i Bioinform. **9**(1), 257–272 (2014)
7. Iliadis, A., Barbolosi, D.: Optimizing drug regimens in cancer chemotherapy by an efficacy-toxicity mathematical model. Comput. Biomed. Res. **33**(3), 211–226 (2000)
8. Ochoa, G., Villasana, M., Burke, E.K.: An evolutionary approach to cancer chemotherapy scheduling. Genet. Program Evolvable Mach. **8**(4), 301–318 (2007)
9. Petrovski, A., Shakya, S., McCall, J.: Optimising cancer chemotherapy using an estimation of distribution algorithm and genetic algorithms. In: Proceedings of the 8th Annual Conference on Genetic and Evolutionary Computation, pp. 413–418. ACM (2006)
10. Petrovski, A., Sudha, B., McCall, J.: Optimising cancer chemotherapy using particle swarm optimisation and genetic algorithms. In: Yao, X., et al. (eds.) PPSN 2004. LNCS, vol. 3242, pp. 633–641. Springer, Heidelberg (2004). https://doi.org/10.1007/978-3-540-30217-9_64

11. Szlachcic, E., Porombka, P.: Decision support system for cancer chemotherapy schedules. In: Moreno-Díaz, R., Pichler, F., Quesada-Arencibia, A. (eds.) EURO-CAST 2013. LNCS, vol. 8112, pp. 226–233. Springer, Heidelberg (2013). https:// doi.org/10.1007/978-3-642-53862-9_29
12. Szlachcic, E., Klempous, R.: Differential evolution multi-objective optimisation for chemotherapy treatment planning. In: Moreno-Díaz, R., Pichler, F., Quesada-Arencibia, A. (eds.) EUROCAST 2015. LNCS, vol. 9520, pp. 471–478. Springer, Cham (2015). https://doi.org/10.1007/978-3-319-27340-2_59
13. Tan, K.C., Goh, C.K., Mamun, A., Ei, E.: An evolutionary articial immune system for multi-objective optimization. Eur. J. Oper. Res. **187**(2), 371–392 (2008)
14. Todorov, Y., Fimmel, E., Bratus, S.A., Semenov, Y.S.: An optimal strategy for leukemia therapy: a multi-objective approach, pp. 1–16 (2011)
15. Zaharie, D.: Differential evolution from theoretical analysis to practical insights. In: Proceeding of 19th International Conference on Soft Computing, Brno, Czech Republic, pp. 26–28 (2013)
16. Zitzler, E., Thiele, L., Laumanns, M., Fonseca, C.M., Da Fonseca, V.G.: Performance assessment of multiobjective optimizers: an analysis and review. IEEE Trans. Evol. Comput. **7**(2), 117–132 (2003)

Intelligent Transportation Systems and Smart Mobility

Optimization of Passenger Distribution at Metro Stations Through a Guidance System

Jusuf Çapalar, Aleksander Nemec, Christoph Zahradnik,
and Cristina Olaverri-Monreal$^{(\boxtimes)}$ (iD)

Department of Information Engineering and Security,
University of Applied Sciences Technikum Wien,
Höchstädtplatz 6, 1200 Vienna, Austria
jusuf.capalar@gmail.com, aleksander.nemec@gmail.com,
christoph.zahradnik@gmail.com,
olaverri@technikum-wien.at

Abstract. The steady growth of population in cities demands an efficient subway management system. To alleviate crowding on certain trains and subway lines, especially during rush hours, we propose a system which optimizes passenger distribution. Through visual cues directly displayed on the waiting platform, the passengers are informed about the wagons' occupation rates, so that they can make a decision about which wagon to use before the subway arrives. Experimental results show that the relevance of the implementation is significant for both passenger satisfaction as well as train operators. In combination with a graphical user interface, the advantages of a guidance system could be demonstrated. The acceptance of our system was guaranteed by 75% of the passengers questioned, who stated they would use such a guidance system.

Keywords: Metro stations · Optimal passenger distribution
Guidance system · Surveys · Passenger comfort

1 Introduction

The trend of population growth in cities causes higher dependency on public transportation and therefore a greater need for its optimization. This growth demands consideration of more efficient subway management systems that alleviate overcrowding on popular trains and subway lines, especially during the busiest travel times. The benefits of an optimal passenger distribution at metro stations include more efficient use of existing trains, increased capacity and more average space for every passenger.

In traffic planning road users' needs should be catered to for the sake of efficiency, and safety. We present in this paper an approach for the optimization of passenger distribution that we then evaluate through a model and software platform.

2 Related Work

The behavior of passengers on the station platform in terms of their distribution among train doors has been investigated in several works [1, 2]. These studies showed that the position and number of platform exits and entrances has a significant influence on

© Springer International Publishing AG 2018
R. Moreno-Díaz et al. (Eds.): EUROCAST 2017, Part II, LNCS 10672, pp. 397–404, 2018.
https://doi.org/10.1007/978-3-319-74727-9_47

passenger distribution. For example, as stated in [3], a station with several platform entrances results in a more balanced passenger dissemination. Uneven distribution causes delays for trains when the operator needs to allow extra time for boarding, which results in longer waiting times for passengers. In order to alleviate this situation, passengers on the train platform should be directed to doors where there are less people waiting to board, thereby orienting themselves better before the train arrives.

In [4] an approach was proposed in which the train adjusted its position while stopping depending on passenger distribution on the platform. However, this solution requires longer platforms that are not always available.

Relying on this idea, several field experiments involving adjustments of the train stopping position have been performed at the Schiphol airport train station in Amsterdam, Netherlands, this measure resulting in a 20% decrease of station dwell times during peak demand and a dwell time variation decrease of approximately 50% [5].

The Austrian train service operator ÖBB has also planned measures to lighten congestion at railway platforms. However they do not include any technical implementation, but rather use a system based on a manual count of passengers [6].

In June 2016 Siemens introduced a new system to guide passengers to less occupied areas [7]. However, the guidance system is only implemented within the train. Therefore, there are no benefits that reduce passenger exchange time. Additionally, this method of redistribution of passengers once already inside the train is inefficient if passengers must navigate around luggage, other passengers, or doors at gangway connections in order to get to other sections.

We propose in this paper a technical solution based on a guidance system where the passengers can see from outside the approaching train which wagons are less crowded.

3   System Design

Our proposed concept of passenger distribution aimed to increase the capacity of the existing transport fleet and ensure a smooth passenger exchange, which would result in fewer delays and an overall improved passenger experience through increased interior space and personal comfort.

Using a user-centric approach in order to find out what the specific needs of the passengers were, we deployed a questionnaire among railway users. The functionality of the approach was then tested by a system consisting of a model and software application that simulated a real scenario and several possible variants by means of a further survey and expert analyses.

Our system consisted of a platform that conveyed information related to the occupancy at each carriage. The goal was to direct waiting passengers to the areas of the platform that corresponded to the doors of less crowded wagons on the oncoming train. As the information was provided prior to the train's arrival, passengers could orient themselves ahead of time to the most convenient doors.

To this end, we computed the number of passengers by calculating the weight of each carriage via a control system. This measurement was implemented before the train left the previous station and its result was then broadcast to the next station. The system was developed to be extended to all existing stations.

4 Implementation

As previously mentioned we implemented a model in order to calculate the passenger distribution in one carriage in real time that was then inferred to the remaining set of carriages.

The program graphically simulated a subway station and showed the optimum distribution of passengers through the control system, indicating entrances with less occupancy rate. By determining the distance between the undercarriage of the subway train and its axis, the degree of capacity utilization of the carriage could be inferred.

The measuring time corresponded to the period of time before the departure during which the passengers are aboard and the doors of the train closed. The system read 100 values per second of each distance sensor and calculated the average of the last 20 records. In addition, outliers on the basis of a deviation of plus/minus 20% of the mean value were identified and discarded.

4.1 Requirements Analysis and Results

In order to determine the need for the implementation of a guidance system in Vienna, we measured the movement of passengers inside a train on a regular working day (Wednesday), between 12:00 and 16:00 in 101 trips. We found out that in 45% of the cases the wagons occupancy rate ranged between 0 and 20% moving on average 0.93 people to another place. In 39% of the cases the wagons occupancy rate was 21–69% moving on average 0.79 people to another location inside the wagon. In 16% of the cases the wagons occupancy rate was 70–100% moving on average 0.19 people to another place.

In order for a system to be user friendly, it needs to be effective and as simple as possible, ensuring that it is understood intuitively. To find out which system could provide the carriages occupancy information in the best possible manner, we designed several messages that we then evaluated through an online survey distributed to 121 participants (males = 73, females = 48), that belonged to the following age groups: 15–25: 57 people; 26–45: 50 people; 46–65: 14 people.

Questions related to the color code and text labels, as well as fields for comments were also included in the survey. Figures 1 and 2 show the messages that could be selected within the online survey. The first three displays were designed to be located between two door sections at a station. The last variation consists of light indicators in red or green according to the capacity of the oncoming carriage, located above each entry section in the metro. The latter variation obtained the highest scores and was selected for implementation in an evaluation platform that included a train model with several wagons.

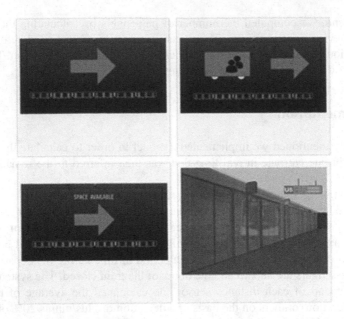

Fig. 1. Design of potential messages to display within a guidance system. (Color figure online)

Fig. 2. Guidance system selected as favorite among the participants in the survey. (Color figure online)

Additionally, we performed a study related to preferences for entering the train. Results from both surveys show that the majority of the passengers entered the metro or train following a precise strategy. Moreover 75% of the participants stated that they would use a guidance system to improve their travel experience with a user interface similar to the one presented in this paper (see Fig. 3).

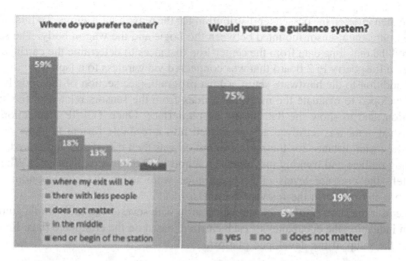

Fig. 3. Results from survey 1 and 2 indicating the preferred entrance location in the metro and the willingness to use a guidance system for find the carriages with less occupancy.

4.2 System Implementation

To compute the passenger distribution in one carriage in real time, we built a model consisting of one carriage for which we calculated the distance of the carriage floor to the ground in real-time. Pressure sensors are already available in trains and are used for regulating the breaking force according to the load of the train and also for compensating the balance immediately as soon people enter the wagons. Relying on this, the air spring was replaced by an ordinary spring that compressed with the weight of a load in the carriage. Figure 4 illustrates the functioning of the developed system.

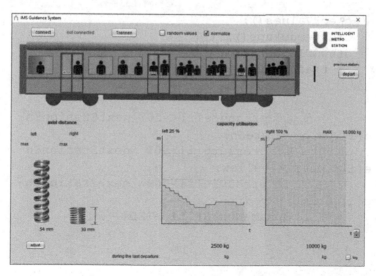

Fig. 4. Model consisting of one carriage for which we calculated the distance to the ground in real-time. The air spring compressed with the weight of a load in the carriage.

We then inferred the passenger distribution for the remaining wagons. To this end we used an ultrasonic sensor mounted between the bogie and the wagon body. The sensor acquired the real-time data from the center-line distances to determine the carriage load through a Raspberry Pi 2 board that was connected via wireless to a laptop.

In addition to the hardware setup, a Java program, (see section of the code below) was developed to evaluate the measured values from the sensors related to the spring deflection, and to show the data in the user interface. Once TCP/IP connection was established, data was retrieved and processed into a readable format for later evaluation. The mean distance for the middle entrance was then calculated and stored. After verifying outliers and backing up values, GUI and LED data were displayed. When the train left, the pressure levels of each entrance were set to the corresponding state through a color code in the relevant station.

The user interface displayed a subway station with several entrances and a guidance system indicating the occupancy of each carriage based on the computed data. Figure 5 depicts the setup of the developed platform.

```
connectTCPIP;
while(true) {
        stringData=readDataLine();
        checkInput();
        pressures=stringData.split(",");               //
split csv; string
        pressure[0]=Integer.parseInt(pressures[0]);   //
assign the parts as integer
        pressure[2]=Integer.parseInt(pressures[1]);
        pressure[1]=(pressure[0]+pressure[2])/2;       //
calculation for the door in the middle
        checkValues();
        createBackups();
        drawGUIGraph();           // for system view
        if trainLeavingNow do for all i {
// send all wagons' data to station as soon the train
leaves; next wagons next 3 i's
                if pressure[i]<=50% nextStationEn-
trance[i].Color = green;
                if pressure[i]>50%  nextStationEn-
trance[i].Color = yellow;
                if pressure[i]>80%  nextStationEn-
trance[i].Color = red;
                updateNextStationLEDs();        // update
them accordingly
        }
}
```

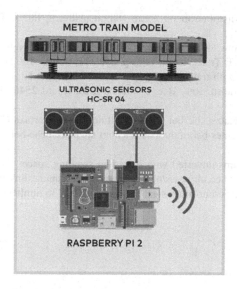 

Fig. 5. System application components: ultrasonic sensors and Raspberry Pi 2, connected by a wireless communication.

5   Conclusion and Future Work

We proposed in this work a system for an even passenger distribution at metro stations. To this end we built a model to evaluate potential technical challenges and test the interactive components. In combination with a graphical user interface, the advantages of a guidance system could be demonstrated.

The acceptance of our system was guaranteed by 75% of the passengers questioned, who stated they would use such a guidance system.

Next steps will include a real life demonstration in a metro station in Vienna that will show the effectiveness of better passenger distribution.

Acknowledgments. This work was partially supported by the "KiTSmart Project – City of Vienna Competence Team for Intelligent Technologies in Smart Cities", project number 18-07 funded by national funds through the MA 23, Urban Administration for Economy, Work and Statistics, Vienna, Austria.

References

1. Szplett, D., Wirasinghe, S.: An investigation of passenger interchange and train standing time at LRT stations: (i) alighting, boarding and platform distribution of passengers. J. Adv. Trans. **18**(1), 1–12 (1984)
2. Wu, Y., Rong, J., Wei, Z., Liu, X.: Modeling passenger distribution on subway station platform prior to the arrival of trains. In: Transportation Research Board 91st Annual Meeting (2012)

3. Krstanoski, N.: Modelling passenger distribution on metro station platform. Int. J. Traffic Transp. Eng. 1(4), 456–465 (2014)
4. Sohn, K.: Optimizing train-stop positions along a platform to distribute the passenger load more evenly across individual cars. IEEE Trans. Intell. Transp. Syst. 14(2), 994–1002 (2013)
5. van den Heuvel, J.: Field experiments with train stopping positions at Schiphol airport train station in Amsterdam, Netherlands. Transp. Res. Rec. J. Transp. Res. Board 2546, 24–32 (2016)
6. Idee eines Bahnfahrers erleichtert Sitzplatzsuche bei ÖBB. M.futurezone.at, German (2016). https://m.futurezone.at/digital-life/idee-eines-bahnfahrers-erleichtert-sitzplatzsuche-bei-oebb/ 218.361.972. Accessed 25 Nov 2016
7. Cecil, N.: New hi-tech Thameslink trains unveiled with increased standing space, Evening Standard (2016). http://www.standard.co.uk/news/transport/new-hitech-thameslink-trains-unveiled-with-fewer-seats-in-each-carriage-but-more-standing-space-a3255236.html#gallery. Accessed 01 Dec 2016

Automatic Vehicle Counting Approach Through Computer Vision for Traffic Management

Arman Allamehzadeh[1], Mohammad S. Aminian[2], Mehran Mostaed[1],
and Cristina Olaverri-Monreal[1(✉)] (iD)

[1] Department of Information Engineering and Security,
University of Applied Sciences Technikum Wien,
Höchstädtplatz 6, 1200 Vienna, Austria
arman4u@gmail.com, mehran.mostaed7@gmail.com,
olaverri@technikum-wien.at
[2] Institute of Transport Planning and Traffic Engineering,
Vienna Technical University, Vienna, Austria
m.s.aminian@hotmail.com

Abstract. Technology based on sensors or cameras that is related to the field of Intelligent Transportation Systems (ITS) can help to alleviate road congestion problems by collecting and evaluating real time traffic data. In this paper, we present an approach to monitor traffic by collecting and processing video streaming information for further analysis in traffic management centers. Results showed a 94% rate of correct vehicle detections in a short period of time with a low rate of false detections.

Keywords: Vehicle counting · Computer vision
Real-time traffic management systems

1 Introduction

Technology based on sensors or cameras that is related to the field of Intelligent Transportation Systems (ITS) can help to alleviate road congestion problems by collecting and evaluating real time traffic data [1]. In traffic management such data is used, for example, to suggest alternative routes to divert traffic from a congested road, or to display other road-related information to drivers to maximize traffic flow and prevent congestions [2].

Traffic monitoring systems supervise traffic volumes and are able to derive other data such as vehicle classification and weight, which is later analyzed through parameters such as speed, density and flow. Speed is defined by the distance traveled per unit of time, density is the number of vehicles per unit length of the roadway and flow is the number of vehicles passing a specific reference point per unit of time [3]. In this paper, we present an approach to traffic counting in roads that are equipped with surveillance cameras. Relying on the EmguCV library for image processing and .NET technologies, traffic information is collected, processed and sent to a server for later statistical analysis.

© Springer International Publishing AG 2018
R. Moreno-Díaz et al. (Eds.): EUROCAST 2017, Part II, LNCS 10672, pp. 405–412, 2018.
https://doi.org/10.1007/978-3-319-74727-9_48

2 Related Work

Current traffic counting techniques mostly rely on inductive loops, piezoelectric, infrared or radar sensors. Although these sensors provide macroscopic information about traffic, they cannot document data for single vehicles without being combined and synced with cameras [4–6].

Computer vision techniques make it possible to exploit traffic cameras solely as a sensor for vehicle detection, counting, tracking and even for speed measurement. Much of road networks are already equipped with surveillance cameras. Cameras are non-intrusive and are used to capture data at high resolution. Like human eyes, cameras capture the scene with details that other sensors like radar, ultrasonic and lasers cannot detect [7]. Hence road operators are very interested in exploiting this technology as tool to derive the real-time traffic information that they need for optimal management of their networks. Data collection technology relying on cameras can have other interesting applications for security tracking, fleet control and electronic toll collection.

Real-time image and video processing to derive traffic data has been applied in many works. A broad review of the literature has been compiled in [8] and cited in this work. For example the robust system Autoscope classified vehicles in real-time and provided traffic analysis reports [9, 10]. A system for counting vehicles and measuring their speed in complex traffic scenarios was presented in [11], and more recently the authors in [12] proposed a hybrid method, based on background subtraction and edge detection for vehicle detection and shadow rejection, to classify and count vehicles in multilane highways. A further approach was proposed through a video analysis method for vehicle counting in [13]. The authors relied on an adaptive bounding box size to detect and track vehicles according to their estimated distance from the camera, given the geometrical setup of the camera. A vehicle counting method based on blob analysis of traffic surveillance video was additionally described in [14] through moving object segmentation, blob analysis, and tracking.

Relying on the techniques of this last work, we present in this paper an approach for vehicle counting in multiple lanes for further analysis in traffic management centers.

3 Technical Implementation

Our approach used the EmguCV vision library in C# to detect the vehicles and process the images. As a platform for developing, running and evaluating the code, we used Visual Studio IDE. All the data was stored in the Microsoft SQL Server. Vehicles are counted individually in the direction that they are travelling.

A login system prevents public access to the records. Black and white are used in order to reduce the size of the stored images. The system works with a live stream in real time using recorded video files. We selected the JSON format for sending the data.

3.1   Video Stream Analysis

A video stream was provided by either a camera or by a recorded video (e.g. by CCTV). After receiving the data, the image processing procedure was initiated and consisted of the following blocks:

- Background building. The background was determined by comparing sequential series of frames through OpenCV functions to create a scene with no vehicle that was then used as background.
- Comparison of two sequential frames. Comparison points for two sequenced frames were determined and a threshold was calculated that enabled the detection of a frame with distinct black/white contrast. The resulting image noise from this operation was then reduced by using Gaussian Blur and morphological image processing noise filters. To reduce the computing process, we first obtained an input frame from the camera, recorded the following frames 2 and 3, and then compared both frames to create a new one that resulted from subtracting them.
- Movement detection. Having subtracted the background from the threshold frame, we obtained a new frame and determined the contours or curve by joining all the continuous points that had the same color or intensity.

3.2   Filtering Process

In this section we provide a more detailed description of the process used to filter the frames sequentially. As stated above, we applied a threshold on the frame to differentiate between black and white. We then applied two common operators to eliminate noise:

- Erosion that slid around by using a slider (kernel) with a size of 3×3 pixels and that delivered white if all the pixels were white, or otherwise remained black.
- Dilation that slid around, and if the entire area was not black, it was converted to white.

Finally we extracted the Binary Large OBject (BLOB) that consisted of a group of connected pixels in a binary image, in order to differentiate between the "large" objects that were relevant to this work (i.e. vehicles) and the other "small" objects that were defined as noise. The image processing procedure is depicted in Fig. 1.

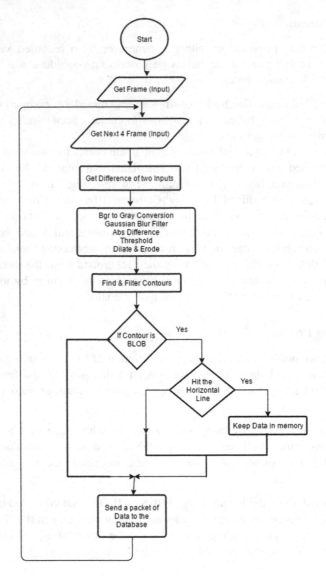

Fig. 1. Image processing procedure for vehicle detection and counting

3.3   Traffic Counting

Relying on the approach in [15] to prevent counting the same vehicle more than once, a horizontal line was inserted in the image that had to be crossed (see Fig. 2). Vehicles were detected and counted only after they cross the line. Afterwards, the data record was sent to the database.

Fig. 2. View of the horizontal line that represents the threshold for counting the vehicles.

3.4 Database Storage

After obtaining the information related to the time stamp, scene image, and plate number, we stored it in a database and repeated all of the above mentioned steps until receiving the stop command. Figure 3 illustrates the process. The administrator is responsible for granting permission to the users to access the data for analysis as well as for configuring new cameras or removing them from the street. Users can be employees who work with a traffic management system.

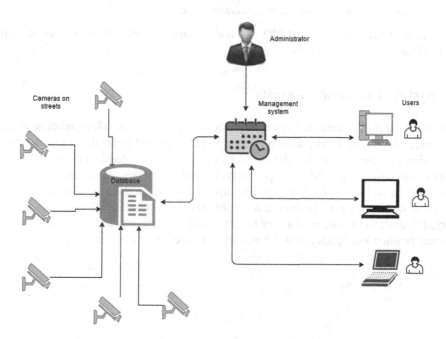

Fig. 3. Traffic data acquisition and storage process

The database contained 4 tables as depicted in Fig. 4 and described below:

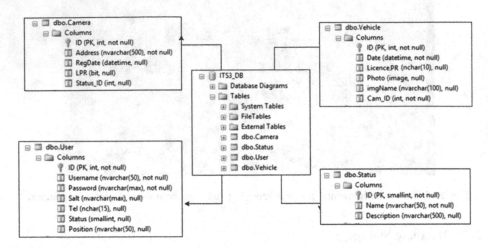

Fig. 4. Database architecture for the storage of the acquired data

– Camera: One or several cameras that can be defined for each street or road as well as been configured, and removed.
– Vehicle: to save the pertinent collected data (e.g. detection time, location id, etc.).
– Status: status of street camera (enabled/disabled).
– User: user data such as username, password, etc.

Licence Plate Recognition (LPR) was additionally defined as a field in the camera and vehicle tables for future use.

4   System Evaluation Results

We evaluated the implemented system by examining a received video stream provided by a video file relying on the work presented in [15] recorded under daylight conditions and using a proper camera installation. The system was evaluated in 6 different streets. Initial evaluations delivered some problems related to the duplicated detection of the same vehicle in two different scenes, but this was solved in the last version.

Final results (see Fig. 5) showed a rate of 94% correct vehicle detection in a short period of time. Results showed a very low rate of false detection. Only in case 6 was the camera position not optimal and the number of false detections increased.

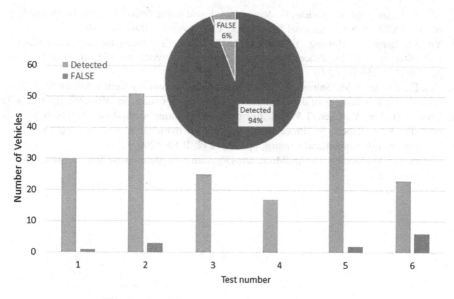

Fig. 5. Results regarding the vehicle detection rate

References

1. Olaverri-Monreal, C.: Automated vehicles and smart mobility related technologies. Infocommun. J. **8**(2), 17–24 (2016)
2. Goncalves, J., Goncalves, J.S., Rossetti, R.J., Olaverri-Monreal, C.: Smartphone sensor platform to study traffic conditions and assess driving performance. In: Proceedings International Conference on Intelligent Transportation Systems (ITSC), Qingdao, China, pp. 2596–2601. IEEE (2014)
3. Hall, F.L.: Traffic stream characteristics, in traffic flow theory. US Federal Highway Administration, pp. 1–32 (1996)
4. Traffic Monitoring Guide: U.S. Department of Transportation (2013). https://www.fhwa.dot.gov/policyinformation/tmguide/tmg_fhwa_pl_13_015.pdf. Accessed 21 Apr 2017
5. Pengjun, Z., McDonad, M.: An investigation on the manual traffic count accuracy. Proc. – Soc. Behav. Sci. **43**, 226–231 (2012)
6. Skszek, S.L.: State-of-the-art report on traffic non-traditional counting methods. Arizona Department of Transportation in Cooperation with U.S. Department of Transportation, Arizona (2001)
7. Ahmadi-Pour, M., Ludwig, T., Olaverri-Monreal, C.: Statistical modelling of multi-sensor data fusion. In: Proceedings International Conference on Vehicular Electronics and Safety (ICVES), Vienna, Austria. IEEE (2017, in press)
8. Loce, R.P., Bernal, E.A., Wencheng, W., Raja, B.: Computer vision in roadway transportation systems: a survey. J. Electron. Imaging **22**(4), 041121, October–December 2013
9. Michalopoulos, P.: Vehicle detection through image processing: the autoscope system. IEEE Trans. Veh. Technol. **40**(1), 21–29 (1991)
10. Carlson, B.: Autoscope clearing the congestion: vision makes traffic control intelligent. Adv. Imaging **12**(2), 54–56 (1997)

11. Ali, A., Bulas-Cruz, J., Dagless, E.: Vision based road traffic data collection. In: Proceedings of the ISATA 26th International Conference, Croydon, pp. 609–616 (1993)
12. Yu, M., Jiang, G., Bokang, Y.: An integrative method for video based traffic parameter extraction in ITS. In: Proceedings Asia Pacific Conference on Circuits and Systems, pp. 136–139. IEEE (2000)
13. Bas, E., Tekalp, A.M., Salman, F.S.: Automatic vehicle counting from video for traffic flow analysis. In: Proceedings of the Intelligent Vehicles Symposium, pp. 392–397. IEEE (2007)
14. Chen, T.H., Lin, Y., Chen, T.Y.: Intelligent vehicle counting method based on blob analysis in traffic surveillance. In: Proceedings Second International Conference on Innovative Computing, Information and Control, pp. 238–242. IEEE (2007)
15. Dahms, C.: MicrocontrollersAndMore. https://github.com/MicrocontrollersAndMore. Accessed 20 Feb 2017

V2V Communication System to Increase Driver Awareness of Emergency Vehicles

Mehran Mostaed, Khaled Aldabas,
and Cristina Olaverri-Monreal[⊠][iD]

Department of Information Engineering and Security,
University of Applied Sciences Technikum Wien,
Höchstädtplatz 6, 1200 Vienna, Austria
mehran.mostaed7@gmail.com, kahled.aldabbas@gmail.com,
olaverri@technikum-wien.at

Abstract. Intelligent Transportation Systems (ITS) applications are supported by systems relying on Dedicated Short-Range Communications (DSRC). This paper presents the implementation steps for a vehicle to vehicle (V2V) communication system intended to increase driver awareness of the surroundings in an emergency situation. Particularly, the system broadcasts basic safety warning messages between an emergency vehicle and other vehicles in the vicinity. The field experiment on communication performance showed that the connection could be established within a range of 20 m. Furthermore, the range could be increased by adding a third communication device and using it to resend the message.

Keywords: Vehicle-to-vehicle communication
Emergency Warning System · DSRC

1 Introduction

According to the World Health Organization (WHO, 2012) fatalities caused by road injuries increased from 1 million in 2000 to more than 1.2 million in 2012. The implementation of road warning messages is intended to increase awareness of dangerous situations, decrease reaction time to them and thereby lower the overall chances of an accident.

Intelligent Transport Systems and Services (C-ITS) enable wireless communication between vehicles and/or traffic infrastructure using a standardized set of messages that are based in the real-time transfer of data. Applications related to Intelligent Transportation Systems (ITS) are supported by systems relying on Dedicated Short-Range Communications (DSRC) that consist of Road Side Units (RSUs) and On Board Units (OBUs) with transceivers and transponders [1]. Vehicle to vehicle (V2V) communication uses IEEE 802.11 or ITS-G5A/B/D standards which operates in 5.9 GHz with a bandwidth on 75 MHz and a range of 1000 m [2].

This paper presents the implementation steps for an Emergency Warning System (EWS) that relies on V2V communication to increase driver awareness of the surroundings in an emergency situation. Particularly, the system broadcasts basic safety

© Springer International Publishing AG 2018
R. Moreno-Díaz et al. (Eds.): EUROCAST 2017, Part II, LNCS 10672, pp. 413–418, 2018.
https://doi.org/10.1007/978-3-319-74727-9_49

warning messages between an emergency vehicle and other vehicles in the vicinity. Figure 1 depicts the communication process.

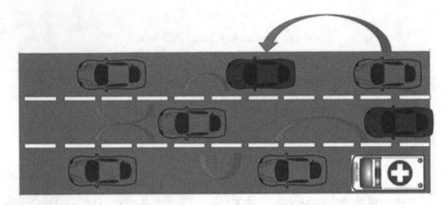

Fig. 1. V2V communication example for an EWS.

2   Related Work

V2V communication is a very extensive area for research and development. For example in [3] a heterogeneous wireless network performance evaluation was performed in vehicle-to-vehicle (V2V) and vehicle-to-infrastructure (V2I) communication. The authors demonstrated the potential of V2V and V2I in heterogeneous networks (HetNet) with Wi-Fi, DSRC and LTE, which guarantees the optimal utilization of available communication options and minimizes the corresponding backhaul communication infrastructure requirements while considering connected vehicle application requirements.

Additional works studied traffic volume estimation by V2V communication [4] or V2X communication for efficient control of fully automated connected vehicles at a freeway merge segment [5].

In [6], authors focused on recently proposed V2V MAC schemes and gave a detailed review of each alongside their strengths and drawbacks. They also discussed pros and cons of V2V MAC, and different ways to achieve communication similarly to [7], in which the radio channel for 5G V2V communications was set at two millimeter wave frequency bands.

In the same line of research, the authors in [8] experimentally characterized V2V millimeter wave radio channel at 38 GHz and 60 GHz frequency bands and related the channel behavior they observed with the measurement environment and setup.

However, in the study presented here, DSRC was used because of the better connection time and signal range, as our focus is on sending a warning message in an emergency situation Utilizing this method, each vehicle functions as an antenna and we can convey our message through a V2V channel so that a clear path can be created for vehicles to take in case of passing an ambulance or police patrol.

3   Technical Implementation

Relying on [3] we implemented a Basic Safety Message (BSM) that is used to exchange safety data regarding vehicle state. The message was broadcast routinely to surrounding vehicles and contained information regarding longitude, latitude and vehicle identity. In order to locate the vehicles, we used an Adafruit Ultimate GPS antenna. Once the BSM was generated, the message was wirelessly transmitted to another vehicle by using the NRF24l01 LNA wireless module. Table 1 summarizes the described components.

Table 1. Communication system components

Hardware	Sender hardware	Receiver hardware
GPS receiver	Adafruit ultimate GPS	
Processor	Raspberry pi	
Antenna	NRF 24L01	
Interface	———————	16 * 2 LCD

In order to make sure that the broadcast BSM was properly obtained, we relied on a receiver with DSRC capabilities. The system also included a computer processing unit that was able to decode the BSM properly and a GPS antenna to verify the relative distance between the sending and the receiving device. In order to convey the message to the driver in an adequate way, we developed an in-vehicle interface to display the warning through a 12 * 2 LCD display.

4   V2V Communication

To establish vehicle communication an accurate and trusted BSM needs to be transmitted and received.

4.1   Message Sending Process

Once its position is known with a GPS sensor, a computer processing unit (Raspberry pi 2) combines the location coordinates with other onboard sensor information (e.g., speed, heading, acceleration) to generate the required BSM data string (see Fig. 2).

Sending

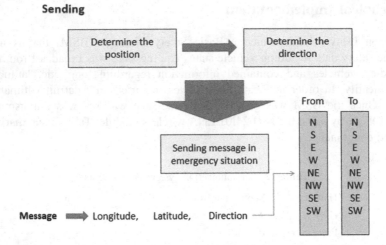

Fig. 2. Message sending process

4.2   Message Receiving Process

In order to establish a connection between 2 vehicles they must be capable of receiving the BSM that is transmitted from a nearby (vehicle) device and it must match the method of BSM transmission (i.e., if the message is transmitted via Dedicated Short Range Radio Communication (DSRC), the receiving device must have a DSRC receiver). It also must have a computer processing unit that can decode the BSM properly. A GPS antenna and receiver is needed to verify the relative distance between the sending device and the receiving device. Figure 3 illustrates the message sending process.

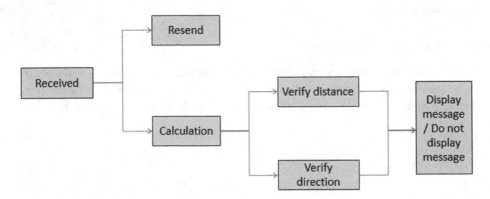

Fig. 3. Message receiving process

5   System Evaluation

The implemented prototype was tested under both lab and real-life conditions. Under lab conditions the devices were able to show the intended emergency message.

Because this system is based on the GPS signal, lab condition tests revolved around sending and receiving the message and how to resend the message to another device. During the field tests, one device was positioned in a vehicle that was marked as an emergency vehicle. The second device was located in a different vehicle. Both devices were capable of sending a message within a range of 20 m. Furthermore, the range could be increased by adding a third device and using it to resend the message.

In order to calculate the maximum possible throughput of a data transfer between the 2 devices, the maximum distance the data could be transferred was calculated for 100 times within 20 days in various situations (i.e., different weather conditions, narrow street with high density, main and secondary street). The results of the evaluation of the developed system are shown on Fig. 4. When the connection was lost, the position of each car was recorded and, based on the recorded position, the point to point distance between cars was calculated. Throughout this test the maximum distance between two cars was 189 m, which was performed in a broadway, and the minimum distance was 85 m. The aforementioned test was performed in some narrow streets of inner-city Vienna.

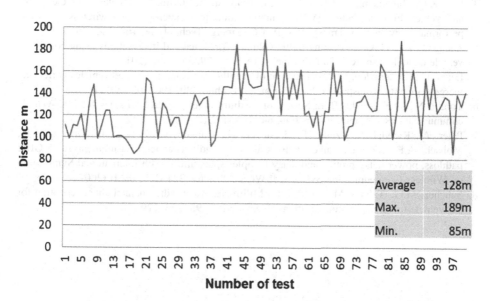

Fig. 4. Results of the prototype evaluation under real-life conditions

6 Conclusion and Future Work

C-ITS have the potential to enhance safety on roads by providing early warnings. However, the success of V2V communication depends on having a fast and reliable network connection, cyber security for V2V and privacy protection. In this paper we presented an Emergency Warning System that relies on V2V communication to increase driver awareness of the surroundings in an emergency situation. Further implementation will determine the direction of movement in cross sections and the position of the vehicle on different roadways.

Acknowledgments. This work was partially supported by the "KiTSmart Project – City of Vienna Competence Team for Intelligent Technologies in Smart Cities", project number 18-07 funded by national funds through the MA 23, Urban Administration for Economy, Work and Statistics, Vienna, Austria.

References

1. Williams, B.: Intelligent Transport Systems Standards. Artech House, Norwood (2008)
2. Adame, T., Bel, A., Bellalta, B., Barcelo, J., Oliver, M.: IEEE 802.11AH: the WiFi approach for M2M communications. IEEE Wirel. Commun. **21**(6), 144–152 (2014)
3. Dey, K.C., Rayamajhi, A., Chowdhury, M., Bhavsar, P., Martin, J.: Vehicle-to-vehicle (V2V) and vehicle-to-infrastructure (V2I) communication in a heterogeneous wireless network– performance evaluation. Transp. Res. Part C: Emerg. Technol. **68**, 168–184 (2016)
4. Zheng, J., Liu, H.X.: Estimating traffic volumes for signalized intersections using connected vehicle data. Transp. Res. Part C: Emerg. Technol. **79**, 347–362 (2017)
5. Letter, C., Elefteriadou, L.: Efficient control of fully automated connected vehicles at freeway merge segments. Transp. Res. Part C: Emerg. Technol. **80**, 190–205 (2017)
6. Torabi, N., Ghahfarokhi, B.S.: Survey of medium access control schemes for inter-vehicle communications. Comput. Electr. Eng. **64**, 450–472 (2017)
7. Paier, A., Karedal, J., Czink, N., Hofstetter, H., Dumard, C., Zemen, T., Tufvesson, F., Molisch, A.F., Mecklenbrauker, C.F.: Car-to-car radio channel measurements at 5 GHz: pathloss, power-delay profile, and delay-Doppler spectrum. In: 4th International Symposium on Wireless Communication Systems, ISWCS 2007, pp. 224–228. IEEE (2007)
8. Sanchez, M.G., Táboas, M.P., Cid, E.L.: Millimeter wave radio channel characterization for 5G vehicle-to-vehicle communications. Measurement **95**, 223–229 (2017)

Cost-Efficient Traffic Sign Detection Relying on Smart Mobile Devices

Mohammad S. Aminian[1], Arman Allamehzadeh[2], Mehran Mostaed[2], and Cristina Olaverri-Monreal[2(✉)] (iD)

[1] Institute of Transport Planning and Traffic Engineering,
Vienna Technical University, Vienna, Austria
m.s.aminian@hotmail.com
[2] Department of Information Engineering and Security,
University of Applied Sciences Technikum Wien,
Höchstädtplatz 6, 1200 Vienna, Austria
arman4u@gmail.com, mehran.mostaed7@gmail.com,
olaverri@technikum-wien.at

Abstract. Neglect of the instructions of road traffic signs is one of the main contributing factors in road accidents. Smartphone traffic sign detection technology can offer significant information about the driving environment and increase driving comfort and traffic safety. It could also have interesting road inventory and maintenance applications. In this paper, we propose a driver assistance system for real-time detection of traffic signs on smartphone platforms using the OpenCV computer vision library. This technology uses the back camera of a smartphone to capture images of the driving environment and then uses advanced image processing functions to detect traffic signs. The field experiment on target traffic signs showed an 85% detection rate. The performance of the application may vary between devices with different processing power and camera quality.

Keywords: Traffic sign recognition · Advanced Driver Assistance Systems
Traffic safety · Computer vision

1 Introduction

Traffic accidents are among the leading causes of fatalities and injuries worldwide. Based on the third global status report on road safety, more than 1.2 million lives are lost annually due to traffic accidents. If no appropriate action is taken, it is estimated that the number will rise to nearly 1.9 million by 2020 [1].

Traffic signs are an important element of the road infrastructure which provides important information about the current state of the road. They are meant to regulate, warn and guide traffic and play a key role in ensuring road safety. Traffic signs are designed to be easily detectable as they sometimes communicate complex information at a glance. The Vienna Convention on Road Signs and Signals has been trying to standardize road signs around the world since 1968. Certain shapes and colors codes have been compiled to categorize different types of traffic signs [2].

© Springer International Publishing AG 2018
R. Moreno-Díaz et al. (Eds.): EUROCAST 2017, Part II, LNCS 10672, pp. 419–426, 2018.
https://doi.org/10.1007/978-3-319-74727-9_50

One approach to minimize distracted driving is the development of Advanced Driver Assistance Systems (ADAS). ADAS are intended to support drivers in performing various driving tasks safely and to minimize the risk of traffic accidents due to driver negligence or erroneous decisions [3]. A subset of ADAS, traffic sign detection and recognition (TSDR) systems offer significant information to drivers about the surrounding environment and road restrictions in real time. Not limited to ADAS, traffic sign detection and recognition systems have also been exploited for other applications such as autonomous driving and sign inventory and maintenance [4, 5].

So far ADAS have been mainly limited to luxury vehicles. Such systems require several built-in sensors, cameras and controls. Unfortunately, a large number of vehicles are still not equipped with such systems. To make up for the lack of ADAS in conventional vehicles, smartphones have attracted the attention of researchers. The prevalence of smartphones and the recent advances in computational power and sensor accuracy of these mobile devices have transformed them into an appealing platform for the development of low-cost driver assistance applications.

We present in this work a low-cost driver assistance system that relies on image processing techniques from the OpenCV library for detecting traffic signs using Android smartphones.

2 Related Literature

OpenCV provides functions for image acquisition and processing which can be exploited for traffic sign detection. The application presented here follows standalone system architecture. It takes in live frames from the back camera of smartphones and uses the smartphone's CPU to process the frames and detect traffic signs. The application does not require supplementary hardware, is user-friendly and cost-efficient.

TSDR has been an important and active research topic in the field of Intelligent Transport Systems (ITS) for the past 3 decades. The earliest study of TSDR is reported in Japan in 1984. Since that time several researchers have analyzed different methods to overcome certain difficulties and improve the performance of TSDR systems. In most real-time experiments, one or two cameras are mounted on the front of vehicles, a PC system is embedded in the car to capture the videos and a display is used to show detected signs. Only a few studies have implemented TSDR systems on smartphone platforms. This is due in part to the immaturity of smartphone camera and processor units in the recent past.

For example the authors in [6] developed a smartphone application to detect and recognize stop signs for inventory and assessment purposes. They utilized color-based detection techniques to identify the areas that differed in brightness and color from surrounding regions, so-called "blob regions". The regions were then tested with trained classifiers to determine if they contained a stop sign.

In [7], still using smartphones, an Android application which received the images and reprocessed them to reduce their resolution and volume was proposed. The frames were then sent to a centralized server for processing. The authors claimed that since the computational power and battery life of smartphones are limited, a client-server approach could better perform the required heavy processing. They used the specific RGB color ratio features to detect the position of the road sign and extract the sign from the original image.

Several studies have compiled the difficulties and challenges of the TSDR process. Factors like variable lighting conditions, air pollution, weather conditions (e.g. fog, rain, sun, and haze), shadows, motion blur, car vibration, sign distortion, fade and partial occlusion can hinder the performance of a TSDR system significantly. Research has led to the development of various methods and strategies to minimize the effects of these influencing factors [8].

Road sign detection involves identifying a certain object from a generally unknown background. To achieve this, there are two main methods for object detection from that can be applied:

- A Haar cascade classifier for traffic sign detection can be trained. This approach requires building a classifier for every traffic sign, which results in a time-consuming detection process as it uses a major part of the processor capacity.
- The second method, color segmentation, offers better performance as it does not require any classifier.

In this work we rely on the second approach and apply color-and shape-based methods to achieve the highest possible precision in detection.

3 Technical Implementation

The developed smart phone application is intended for a smart device with an Android operating system that is equipped with a rear camera, GPS and accelerometer sensors so that the detection and recognition systems work only when the vehicle is moving.

To implement the code for image processing and detection using color-based segmentation method, the following steps were performed: image acquisition, region of interest calculation, color segmentation, blob detection, shape classification and plausible sign detection. We additionally designed a graphical user interface that included check-boxes, push-buttons and a camera display to enable or disable features easily.

In the first step, the mobile application captures a frame using the smartphone's back camera that will be later processed. In order to reduce the detection time, an approximate area is calculated in which the traffic signs are most likely located. This area excludes 15% from the top and 35% from the bottom parts of the original frame. This total 50% of frame reduction significantly reduces the detection time in the search area. Figure 1 shows how this region of interest (ROI) is calculated.

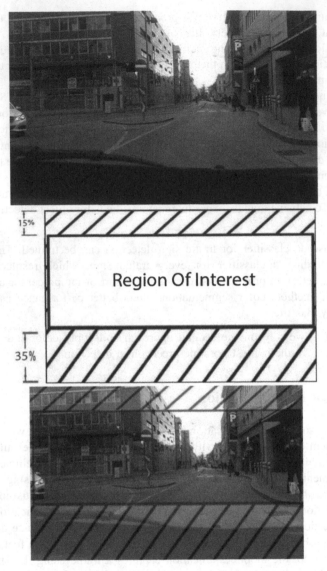

Fig. 1. Calculation of the region of interest (Color figure online)

We targeted in this work those signs which are dominated or bordered by the color red (e.g. stop or entry prohibition signs). To detect them within the ROI, we converted the reduced area from the Red, Green, Blue color model (RGB) to Hue-Saturation-Value (HSV) in order to extract the red color. We used HSV because it is more efficient handling lighting differences, and makes it possible to better discriminate the colors [7].

We then applied Gaussian Blur filters to reduce the noise from the processed frame and detect the red color. As a final step, we applied Binary Large OBject (BLOB) processing to detect the group of connected pixels in the binary image, differentiating between the "large" objects of a certain size that were relevant to this work (i.e. traffic signs) and the other "small" binary objects that were defined as noise. This method was selected due to its effectiveness in finding the targeted objects.

We developed a function to determine the pixels shape and area in order to find the biggest symmetrical target, a process which helps identify falsely-detected objects. Finally, if a red, symmetrical object such as a polygon, circle, triangle, etc. was detected, the application magnified the image and displayed it on the smartphone's screen while triggering an alarm. Figure 2 depicts the implementation steps. Figure 3 illustrates the image processing steps from left to right and top to bottom.

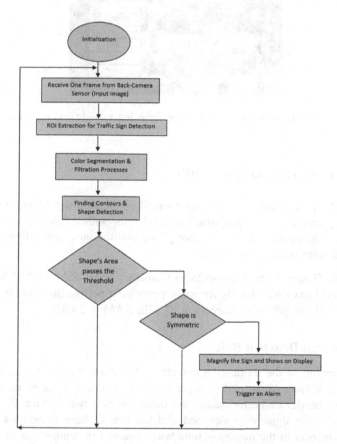

Fig. 2. Overview of the implemented algorithm process.

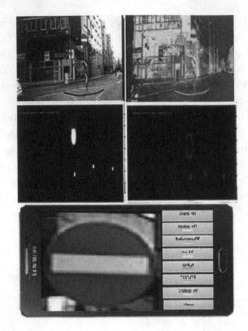

Fig. 3. Overview of the image processing steps from left to right and top to bottom. (Color figure online)

4   Application Evaluation Results

To evaluate the performance of our implemented application regarding sign detection rate, accuracy and time, we performed several tests with a vehicle under daylight conditions and the optimal device position. Two smartphones with different hardware specifications were used in these tests:

(A) Samsung Galaxy Fame (480p@25fps Camera = 5MP, RAM = 512 MB).
(B) Samsung Galaxy A5 with higher image processing performance and better camera resolution (1080p@30fps Camera = 13 MP, RAM = 2 GB).

4.1   Traffic Sign Detection Rate

For the evaluation of the traffic sign detection algorithm, we performed 45 different tests with a minimum of one red sign per test with a detection range of up to 20 m. According to the device performance, the detection rate with device B was higher. 86.6% of all traffic signs were detected and the rate of false detections constituted 31.1%. A reduction in the detection time was observed depending on the RAM and CPU power. The detection time with device B ranged from 211 to 255 ms, the average detection time being 235.6 ms. The detection time with device A ranged from 35 to 45 ms, the average being 40.3 ms. Figures 4 and 5 illustrate the comparative results from both devices.

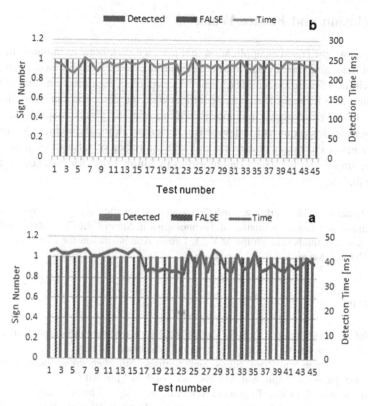

Fig. 4. Detection rates and time results for devices B (above) and A (below). (Color figure online)

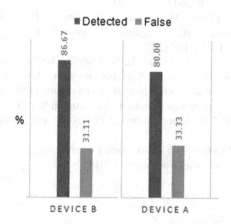

Fig. 5. Comparative results for the detection rates for devices B and A. (Color figure online)

5 Conclusion and Future Work

In this paper, we proposed an application that uses a smartphone to monitor the road for traffic sign detection. The evaluation of the results showed a high object detection rate (86.6%) within a few milliseconds of time. As expected, the detection time differed on smartphones with different hardware specifications, the detection rate being higher for device B. Due to the higher resolution and number of pixels to process, a longer processing time was required for device B. Device A, although having a lower camera quality, detected the target objects as well. The number of false detections could be reduced by improving the detection algorithm in future work, for instance by adding a new filtration stage.

Acknowledgments. This work was partially supported by the "KiTSmart Project – City of Vienna Competence Team for Intelligent Technologies in Smart Cities", project number 18-07 funded by national funds through the MA 23, Urban Administration for Economy, Work and Statistics, Vienna, Austria.

References

1. Road Safety in the South-East Asia Region (2015). www.who.int/violence.../road_safety_status/2015/Road_Safety_SEAR_3_for_web.pdf. Accessed 14 May 2017
2. Vienna Convention on Road Signs and Signals. https://en.wikipedia.org/wiki/Vienna_Convention_on_Road_Signs_and_Signals. Accessed 14 May 2017
3. Olaverri-Monreal, C., Jizba, T.: Human factors in the design of humanmachine interaction an overview emphasizing V2X communication. IEEE Trans. Intell. Veh. **1**(4), 1–12 (2017)
4. Maldonado-Bascon, S., Lafuente-Arroyo, S., Gil-Jimenez, P., Gomez-Moreno, H., Lopez-Ferreras, F.: Road-sign detection and recognition based on support vector machines. IEEE Trans. Intell. Transp. Syst. **8**, 264–278 (2007)
5. García-Garrido, M.A., Ocaña, M., Llorca, D.F., Arroyo, E., Pozuelo, J., Gavilán, M.: Complete vision-based traffic sign recognition supported by an I2V communication system. Sensors (Basel) **12**(2), 1148–1169 (2012)
6. Mertz, C., Kozar, J., Wang, J., Doyle, J., Kaffine, C., Kelkar, A., Nikitha, P., Chan L., Amladi, K.: Smartphone Based Traffic Sign Inventory and Assessment. National Technical Reports Library (2016). utc.ices.cmu.edu/utc/tier-one-reports/Mertz_TSETFinalReport.pdf
7. Xiong, B., Izmirli, O.: A road sign detection and recognition system for mobile devices. In: Proceedings International Workshop on Image Processing and Optical Engineering, Harbin, China (2012)
8. Pritt, C.: Road sign detection on a smartphone for traffic safety. In: IEEE Applied Imagery Pattern Recognition Workshop, Washington (2014)

Stereo Vision-Based Convolutional Networks for Object Detection in Driving Environments

Carlos Guindel[(⊠)], David Martín, and José María Armingol

Intelligent Systems Laboratory, Universidad Carlos III de Madrid, Leganés, Spain
{cguindel,dmgomez,armingol}@ing.uc3m.es

Abstract. Deep learning has become the predominant paradigm in image recognition nowadays. Perception systems in vehicles can also benefit from the improved features provided by modern neural networks to increase the robustness of critical tasks such as obstacle avoidance. This work proposes a vision-based approach for on-road object detection which incorporates depth information from a stereo vision system within the framework of a state-of-art deep learning algorithm. Experiments performed on the KITTI benchmark show that the proposed approach results in significant improvements in the detection accuracy.

Keywords: Object detection · Stereo vision · Deep learning

1 Introduction

Detection of objects from a moving observer is an essential task for a large number of advanced driver assistance systems (ADAS) and virtually every autonomous car. As vehicles are meant to share the road with other users, each with its distinctive behavior, predictions about future traffic situations require an accurate identification of the objects in the surroundings.

While object detection in images is a classic problem in computer vision, traffic scenes are particularly complex due to the diversity of appearances, poses, and occlusions. Additionally, robustness to changes in illumination, weather, and other external factors is an implicit prerequisite for these applications. Challenges posed by driving environments have often been tackled making use of the additional information provided by stereo vision systems [1], which are composed of two nearly-identical cameras displaced horizontally from one another. This setup allows the extraction of depth information about the scene.

On the other hand, deep learning has become ubiquitous in almost every application involving image recognition in the past few years. Convolutional Neural Networks (CNNs) are currently the method of choice after they have demonstrated to be extremely useful in practical applications. Their success stems from their ability to learn hierarchical features which significantly outperform previous hand-crafted features for a variety of computer vision tasks.

In this work, we aim to enhance the performance of a state-of-art object detection framework, Faster R-CNN [2], by incorporating depth information from a stereo camera in a simple, straightforward way.

© Springer International Publishing AG 2018
R. Moreno-Díaz et al. (Eds.): EUROCAST 2017, Part II, LNCS 10672, pp. 427–434, 2018.
https://doi.org/10.1007/978-3-319-74727-9_51

2   Related Work

The standard pipeline for object detection in images entails two main stages: extraction of regions of interest (ROIs) and classification of those proposals. A few years ago, the research interest was focused on hand-crafted features, e.g. HOG [3]. As the complexity of feature extraction schemes was tractable, it was usually possible to perform an exhaustive search over the image using a sliding window approach.

The introduction of CNNs led to a paradigm shift: today, features are learned in a supervised optimization process that makes use of large datasets. The complex hierarchical structures of CNNs involve longer computation times, and sliding-window approaches have become unfeasible due to the huge amount of regions to be classified. For this reason, as well as the large receptive fields featured by the conventional CNN architectures, extraction of ROIs in deep-learning-based object detection schemes remains a very active research area.

Girshick et al. [4] developed the R-CNN paradigm, where CNN features are computed for every candidate ROI and used in a further classification step. The method was further updated in [5] with the introduction of Fast R-CNN.

As a natural evolution, Faster R-CNN [6] extends the CNN approach to the ROI extraction stage, thus resulting in an end-to-end detection framework. The convolutional layers are applied over the image to extract features which are simultaneously used to propose candidate regions and to classify them. The former is performed by a Region Proposal Network (RPN), while the later is carried out with Fast R-CNN, which additionally provides a bounding box refinement. As a consequence, the most time-consuming task, i.e. the computation of the convolutional features, is performed only once.

Despite the impressive performance of Faster R-CNN in generic datasets, e.g. ILSVRC [7], achieved with only a fraction of the cost of more sophisticated models, hypothesis generation remains a substantial limiting factor in performance. As a matter of fact, a significant number of methods in the top positions of the challenging KITTI benchmark [8] are evolutions of the baseline Faster R-CNN approach specifically designed to overcome this limitation, such as the scale-dependent pooling introduced in [9], or the multi-scale CNN presented in [10].

3   Object Detection Approach

We aim to enhance the solid detection baseline provided by Faster R-CNN by leveraging the stereo depth information without significantly altering the original design. For that end, we adapt the setup of the network model to allow the processing of four-channel data structures containing the RGB color channels of the left image and, additionally, a scaled disparity map. Our approach is summarized in Fig. 1.

The disparity map is a data structure which encodes the deviation in horizontal coordinates, d, of corresponding points in both images belonging to the

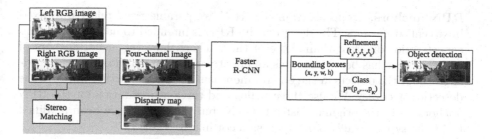

Fig. 1. Proposed object detection method. Our contribution is highlighted in blue. (Color figure online)

stereo pair. Thus, the value of each pixel in our *fourth channel*, $s \cdot d$, is inversely proportional to the scene depth at that location, Z, following the relation:

$$s \cdot d = \frac{f \cdot B}{Z} \qquad (1)$$

where f is the focal length and B the baseline of the binocular pair. Since these values are determined for a particular stereo system, the disparity value is indeed inversely proportional to the actual depth.

The reasoning behind our approach is that region proposal can take advantage of the geometrical information provided by the disparity estimation to segment the foreground objects from the background, thus overcoming the most severe shortcoming of the CNN method. Using disparity values straightforwardly, instead of actual depth values, is expected to benefit the segmentation of objects at closer distances due to the inverse relationship linking both magnitudes.

In summary, our design is intended to preserve the end-to-end nature of the Faster R-CNN detection method while enhancing the performance of the classification, especially for objects represented with a limited number of pixels.

3.1 Parameter Tuning

Before considering the influence of the depth channel in the CNN architecture, we optimized the performance of the baseline Faster R-CNN by tuning its hyperparameters according to the specific requirements of driving environments. The modifications are targeted to the KITTI dataset [8], and include:

1. **Training samples selection.** Samples used in the training procedure are chosen so that their IoU overlap with any ground-truth *DontCare* label, corresponding to distant or unclear objects, is below a certain threshold: 25% for the Fast R-CNN module and 15% for the RPN. On the other hand, only samples eligible to be included in the 'hard' difficulty level are used.
2. **Scale.** Faster R-CNN has been shown [11] to be highly sensitive to the size of the input images. We have found that scaling the original images (with resolutions around 1242×375) to 500 pixels in height, both for training and evaluation, offers a good trade-off between accuracy and computation time.

3. **RPN anchors.** Proposals from the RPN are parametrized relative to fixed boxes called *anchors*. The design of the RPN is intended to handle scales and aspect ratios different than those of the anchors; however, using anchors of multiple sizes has been proven as an effective solution, so the a-priori knowledge about the objects in the environment can be used to further improve the detection accuracy. We use three scales and three aspect ratios for the RPN anchors, as in the original Faster R-CNN; but the values have been modified to fit the typical traffic participants, according to Table 1.

Table 1. Modification in the settings of RPN anchors.

	Original	Proposed
Scales	$\{128^2, 256^2, 512^2\}$	$\{80^2, 112^2, 144^2\}$
Aspect ratios	$\{2:1,\ 1:1,\ 1:2\}$	$\{5:2,\ 5:4,\ 2:5\}$

3.2   Stereo Depth Information

Different alternatives can be adopted to estimate the disparity map from the images of the stereo pair. Henceforth, the following methods are considered:

1. The classical Semiglobal Matching algorithm [12] in its OpenCV implementation [13]; i.e. using block matching and the Birchfield-Tomasi metric.
2. A state-of-art CNN-based algorithm, DispNet [14], currently ranked 9th in the KITTI stereo leaderboard among the published methods[1] and with a reported runtime of 60 ms.

The density of the SGM disparity map is around 90% due to the existence of unmatched pixels. As these *undefined* values could prevent the gradient descent training to converge, we perform a background interpolation to fill the holes, so every pixel (u_0, v) with a undetermined value in the disparity map, $\nexists d(u_0, v)$, is given a value according to:

$$\hat{d}(u_0, v) = \min(d(u_0^-, v), d(u_0^+, v)) \qquad (2)$$

where $d(u_0^-, v)$ and $d(u_0^+, v)$ are the disparities of the contiguous *defined* pixels in the same row. DispNet, on the other hand, provides a 100% dense disparity map.

As mentioned above, values in the disparity map are scaled before entering the CNN, according to the s factor in Eq. 1. This is actually a normalization of the disparity values between 0 and 255/s. Note that the scaling operation must be performed with saturation to prevent overflow. We chose $s = 4$ in order to obtain values close to the pixels in the color channels; this means that only disparities originally in the range between 0 and 64 are distinguishable in the resulting map.

---

[1] http://www.cvlibs.net/datasets/kitti/eval_scene_flow.php?benchmark=stereo.

Given the parameters of the KITTI stereo system, that clipping corresponds to depths from 6 m to the infinite, which is reasonably tailored to the field of view of the camera. Figure 2 depicts an example of the resulting fourth channel (already normalized) for each of the two employed stereo matching approaches.

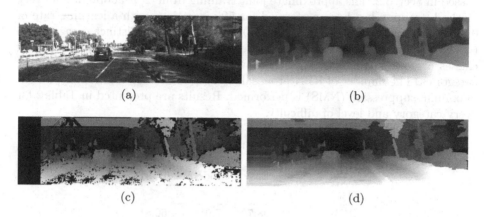

Fig. 2. Example of normalized disparity maps for a frame in the KITTI dataset (a), obtained with the two selected stereo matching approaches: DispNet [14] (b) and SGM [12] with interpolation (d), computed from the original SGM (c).

CNN architectures typically used in image recognition can be applied to our approach with minimal changes: only the filters in the first convolutional layer have to be adapted to accept a four-channel input.

A common practice in training CNN models is to initialize the weights in the convolutional layers using values trained in larger datasets, such as the ILSVRC [7], with the hope that the learned features may still be useful for related applications. As the filters that we use in the first convolutional layer are different from the existing pre-trained models, we initialize the weights in the fourth channel as the mean value of the same weight in the filters corresponding to the preexisting color channels. This approach, which avoids the need to retrain the models from scratch, is based on the assumption that discontinuities in depth are related to discontinuities in intensity. Additionally, we let weights in all the convolutional layers, including the shallower ones, be modified during training to fit the new nature of the data.

4   Results

We compare our approach with the baseline Faster R-CNN to investigate the improvement introduced by the stereo information. We use the already mentioned KITTI object detection benchmark [8] for evaluation. Since the test ground-truth labels are not publicly available, we use the train/validation split by [15] to ensure that images from the same sequence do not exist in both training and validation sets. Following the standard KITTI setup, we use the

Average Precision (AP) metric to evaluate the performance of the object detection pipeline and require IoU overlaps of 70%, for cars, and 50%, for pedestrian and cyclists.

We employ the VGG16 architecture [16], with the minimal changes discussed in Sect. 3.2. The approximate joint training from [2] is adopted. For every method, training has been performed for 50k iterations with a learning rate of 0.001 and then for 30k iterations with 0.0001. The seven distinct categories in the KITTI dataset are considered; however, only *Car*, *Pedestrian* and *Cyclist* classes are evaluated because of the low number of samples in the remaining categories. The number of RPN proposals is limited to 300; additionally, a non-maximum suppression (NMS) is performed. Results are presented in Table 2 for every category and level of difficulty.

Table 2. Detection AP (%) obtained on the KITTI validation set.

Input	Easy	Moderate	Hard
Car			
RGB	88.76	77.01	60.81
RGB+SGM	**89.39**	**77.99**	**66.84**
RGB+DispNet	88.82	77.29	66.56
Pedestrian			
RGB	85.97	68.71	61.41
RGB+SGM	87.39	69.16	63.62
RGB+DispNet	**87.70**	**69.73**	**64.47**
Cyclist			
RGB	65.22	53.67	50.37
RGB+SGM	64.07	52.25	49.68
RGB+DispNet	**66.51**	**55.77**	**52.26**

Detection using the disparity information surpasses the bare RGB approach in almost all cases, with the notable exception of SGM for cyclists. On the other hand, DispNet outperforms the SGM estimation for pedestrians, while SGM shows better results for cars. The improvement introduced by the disparity information is especially noticeable in 'hard' samples, as shown in the summary tabulated in Table 3.

Table 3. Summary of mAP (%) obtained on the KITTI validation set, expressed as the difference in percentage points from the baseline RGB approach.

Input	Easy	Moderate	Hard
RGB	79.98	66.46	57.53
RGB+SGM	+0.30	+0.01	+2.52
RGB+DispNet	**+1.03**	**+1.14**	**+3.57**

The average running time per image of the detection stage is 116 ms using a NVIDIA Titan Xp and Caffe [17]. For preliminary results with fixed weights in the first two convolutional layers during training, please refer to the Extended Abstract of this paper [18].

5 Conclusion and Future Work

We have presented an approach to exploit the spatial information provided by a stereo vision system in order to enhance a well-established object detection method based on Convolutional Neural Networks. Our proposal is particularly suitable for automotive applications, where stereo cameras have frequently been employed to deal with the complexity of the environments without significantly altering the features of the vehicle.

Results have proven the potential of stereo information to enhance the convolutional features produced by the network, leading thus to a significant enhancement of the detection performance. The improvement is especially notable when detecting Vulnerable Road Users (VRU), namely pedestrians and cyclists, frequently identified as the most problematic categories in image recognition.

Further steps might focus on the architecture of the network, adopting either modern architectures, e.g. ResNets, or ad-hoc designs intended to exploit the information extraction from the disparity map. Additionally, some of the developments recently introduced in the literature to overcome the fixed size of the receptive field could be adopted.

This work is intended to be the first step towards a full scene understanding system in our IVVI 2.0 intelligent vehicle [19], an experimental platform for driving assistance systems. This application, along with other critical perception modules, will enable inference about complex traffic situations.

Acknowledgments. Research supported by the Spanish Government through the CICYT projects (TRA2015-63708-R and TRA2016-78886-C3-1-R), and the Comunidad de Madrid through SEGVAUTO-TRIES (S2013/MIT-2713). The Titan Xp used for this research was donated by the NVIDIA Corporation.

References

1. Bernini, N., Bertozzi, M., Castangia, L., Patander, M., Sabbatelli, M.: Real-time obstacle detection using stereo vision for autonomous ground vehicles: a survey. In: Proceedings of the IEEE International Conference on Intelligent Transportation Systems (ITSC), pp. 873–878 (2014)
2. Ren, S., He, K., Girshick, R., Sun, J.: Faster R-CNN: towards real-time object detection with region proposal networks. IEEE Trans. Pattern Anal. Mach. Intell. **39**(6), 1137–1149 (2016)
3. Dalal, N., Triggs, B.: Histograms of oriented gradients for human detection. In: Proceedings of the IEEE Conference on Computer Vision and Pattern Recognition (CVPR), vol. 1, pp. 886–893 (2005)

4. Girshick, R., Donahue, J., Darrell, T., Malik, J.: Rich feature hierarchies for accurate object detection and semantic segmentation. In: Proceedings of the IEEE Conference on Computer Vision and Pattern Recognition (CVPR), pp. 580–587 (2014)
5. Girshick, R.: Fast R-CNN. In: Proceedings of the IEEE International Conference on Computer Vision (ICCV), pp. 1440–1448 (2015)
6. Ren, S., He, K., Girshick, R., Sun, J.: Faster R-CNN: towards real-time object detection with region proposal networks. In: Advances in Neural Information Processing Systems (NIPS) (2015)
7. Russakovsky, O., Deng, J., Su, H., Krause, J., Satheesh, S., Ma, S., Huang, Z., Karpathy, A., Khosla, A., Bernstein, M., Berg, A.C., Fei-Fei, L.: Imagenet large scale visual recognition challenge. Int. J. Comput. Vis. **115**(3), 211–252 (2015)
8. Geiger, A., Lenz, P., Urtasun, R.: Are we ready for autonomous driving? The KITTI vision benchmark suite. In: Proceedings of the IEEE Conference on Computer Vision and Pattern Recognition (CVPR), pp. 3354–3361 (2012)
9. Yang, F., Choi, W., Lin, Y.: Exploit all the layers: fast and accurate CNN object detector with scale dependent pooling and cascaded rejection classifiers. In: Proceedings of the IEEE Conference on Computer Vision and Pattern Recognition (CVPR), pp. 2129–2137 (2016)
10. Cai, Z., Fan, Q., Feris, R.S., Vasconcelos, N.: A unified multi-scale deep convolutional neural network for fast object detection. In: Leibe, B., Matas, J., Sebe, N., Welling, M. (eds.) ECCV 2016. LNCS, vol. 9908, pp. 354–370. Springer, Cham (2016). https://doi.org/10.1007/978-3-319-46493-0_22
11. Fan, Q., Brown, L., Smith, J.: A closer look at faster R-CNN for vehicle detection. In: Proceedings of the IEEE Intelligent Vehicles Symposium (IV), pp. 124–129 (2016)
12. Hirschmüller, H.: Stereo processing by semiglobal matching and mutual information. IEEE Trans. Pattern Anal. Mach. Intell. **30**(2), 328–341 (2008)
13. Kaehler, A., Bradski, G.: Learning OpenCV 3: Computer Vision in C++ with the OpenCV Library. O'Reilly Media, Inc., Sebastopol (2016)
14. Mayer, N., Ilg, E., Häusser, P., Fischer, P., Cremers, D., Dosovitskiy, A., Brox, T.: A large dataset to train convolutional networks for disparity, optical flow, and scene flow estimation. In: Proceedings of the IEEE Conference on Computer Vision and Pattern Recognition (CVPR), pp. 4040–4048 (2015)
15. Chen, X., Zhu, Y.: 3D object proposals for accurate object class detection. In: Proceedings of the Advances in Neural Information Processing Systems (NIPS), pp. 424–432 (2015)
16. Simonyan, K., Zisserman, A.: Very Deep Convolutional Networks for Large-Scale Image Recognition. CoRR abs/1409.1 (2014)
17. Jia, Y., Shelhamer, E., Donahue, J., Karayev, S., Long, J., Girshick, R., Guadarrama, S., Darrell, T.: Caffe: convolutional architecture for fast feature embedding. In: Proceedings of the ACM International Conference on Multimedia, pp. 675–678 (2014)
18. Guindel, C., Martín, D., Armingol, J.M.: Stereo vision-based convolutional networks for object detection in driving environments. In: EUROCAST 2017 - Extended Abstracts, pp. 288–289 (2017)
19. Martín, D., García, F., Musleh, B., Olmeda, D., Peláez, G.A., Marín, P., Ponz, A., Rodríguez Garavito, C.H., Al-Kaff, A., de la Escalera, A., Armingol, J.M.: IVVI 2.0: an intelligent vehicle based on computational perception. Expert Syst. Appl. **41**(17), 7927–7944 (2014)

A Simple Classification Approach to Traffic Flow State Estimation

Aitor del Pino Saavedra Hernández[1], Javier J. Sánchez Medina[1(✉)],
and Luis Moraine-Matias[2]

[1] CICEI – ULPGC, Las Palmas, Spain
adritor8994@gmail.com, javier.sanchez@ulpgc.es
[2] NEC Laboratories Europe, Heidelberg, Germany
luis.matias@neclab.eu

1 Introduction

One of the most important elements in the mobility of the developed cities is the road traffic management. The mobility determines the quality of citizens' living conditions because of many reasons, security, efficiency, and the environmental impact. Focusing on security, according to World Health Organization (WHO), every year two millions of people die as a result of traffic accidents. Moreover between twenty and fifty millions of people suffer non-fatal injuries and a proportion of these people suffer from a disability. These injuries affect both the family economy and the country. For this reason, amongst others, it is required to equip the mobility managers with the proper tools to get a precise idea about the current situation and estimate future state. These tools facilitate the decision-making and the development of mobility.

The aim of this work is at creating a congestion prediction model for the Portuguese city of Porto based on FCD. To obtain such traffic model it has been applied the Knowledge Discovery in Databases (KDD) process, which comprises the following phases: data pre-processing, data mining and interpretation and model evaluation.

2 KDD Process

The goal of the KDD process is to extract knowledge from data in the context of large databases.

The knowledge extracted by the KDD process must have four characteristics: no trivial implicit, unknown and useful.

Fayyad according to the KDD process "is the non-trivial process of identifying valid, novel, potentially useful, and ultimately understandable patterns in data" [1] (Fig. 1).

© Springer International Publishing AG 2018
R. Moreno-Díaz et al. (Eds.): EUROCAST 2017, Part II, LNCS 10672, pp. 435–439, 2018.
https://doi.org/10.1007/978-3-319-74727-9_52

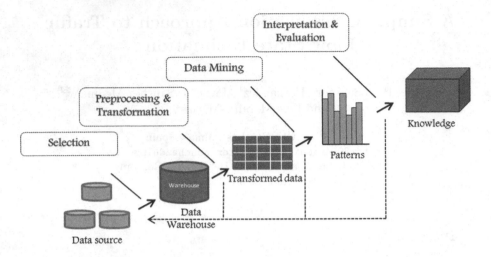

Fig. 1. Stage of KDD process

3   Data Preprocessing

The dataset (this dataset is published here [2]) were obtained by 442 taxis in the city of Porto, in Portugal. These taxis were equipped in order to obtain every 15 s their Global Positioning System (GPS) position and also other variables. Between the different attributes that compose the dataset, it highlights the attribute called Polyline because it contains the sequence of GPS positions of the trip of a taxi.

In the first step of the KDD we applied a selection of attributes in order to reduce the data and so to have the attributes that might be useful. Moreover, two types of data cleaning was applied, first in the attribute called Polyline by the absence of values since we consider inconsistent that the taxi carries out a trip in less than 15 s. The second data cleaning was for erroneous data in the same attribute because there were instances that did not have the complete sequence of GPS positions. After that, we did a transformation of that dataset using a grid in order to discretize the spatial coordinates using a well-known preprocessing tool: grid decomposition Each cell consisted on a maximum and minimum latitude and longitude selection, also sampling by the hour of the day and the day of the week. After that we split the original dataset into N dataset spatio-temporal 3D datasets, assigning all GPS samples to each one of the aforementioned cells. Moreover, a few statistics were obtained for each cell, regarding occupancy, inflow and outflow of vehicles (Fig. 2).

As a final task in this preprocessing stage we tagged each cell. We did that using an analogy to the fundamental diagram of traffic flow which represents the density of vehicles compared to flow of vehicles. The key to this graph is where the highest density of vehicles is obtained. This point is called critical density. The critical density is used to know how it is the traffic flow, because

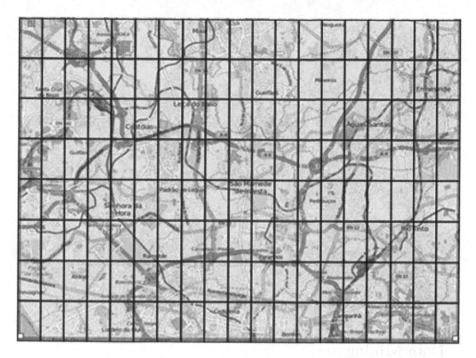

Fig. 2. Grid in Porto

when the density of vehicles is greater than the critical density, the traffic is relatively flowing and stable, while if the density of vehicles is greater than the critical density, it increases the traffic congestion (More information about the fundamental diagram of traffic flow [3,4]).

In our project we have called the critical density as critical N. This parameter indicates the number of vehicles that supports each cell when the outflow of vehicles was maximum. Using this parameter and the number of vehicles of each cell for each day of the week and time of day we have labelled into three categories:

- Free: the number of vehicles is less than 25% of the critical N.
- Synchronized: the number of vehicles is between 25 and 90% of the critical N.
- Congested: the number of vehicles is greater than 90% of the critical.

It is observed that the labelling that we are carrying out is the traffic flow of taxis and it is not general traffic flow. This must be taken into account, since there could be certain hours where there is more traffic of citizens than traffic of taxis. Therefore, we assume that the traffic flow general follows a similar pattern to traffic flow of taxis. This simplification has been made because we have only one dataset with trajectories of taxis and not of the vehicles of citizens of Porto (Fig. 3).

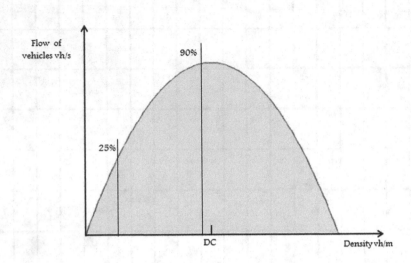

Fig. 3. Fundamental diagram of traffic flow.

4 Data Mining

In this stage of KDD process we decided to make three experiments in order to obtain the best model that predicts the traffic flow. For each of the three experiments we used seven classification algorithms: Bagging + C.4.5, Bagging + Hoeffding Tree, Hoeffding Tree, C.4.5, Random Forest, Part and OneR (information about these classification algorithms [5–8]). Moreover, for each of the three experiments it was used as evaluation method 10-folded cross-validation.

In the first experiment, all the classification algorithms were executed to determine the accuracy rate in which we were. In the second experiment, the most important parameters of each classification algorithms were adjusted in order to improve the accuracy rate and to obtain a more robust model. Finally in the third experiment, we performed a selection of attributes and then we execute classification algorithms.

The best model was obtained in the second experiment, when we adjust the parameters of the classifier Bagging + C.4.5. Moreover, we also evaluated other metrics like the model consistency (variance reduction) and Multi-class Area Under the Curve metric ([9]).

5 Interpretation and Evaluation Patterns

As we already previously commented, the evaluation method used to evaluate classification algorithms was 10-folded cross-validation. Moreover we made a simple interface that represented the set of roads that contained each cell in order to know whether the results were consistent.

6   Conclusion and Future Directions

We think that the prediction of traffic flow is essential to improve the quality of citizens' living since it affects from daily actions like to go to work to environmental pollution. Thus, it is very possible that in a short time there are models that predict of traffic flow with high accuracy.

Using our model we obtained a average accuracy rate of 82.58% which we consider acceptable However we believe that the future could improve the accuracy rate specified of different ways. One way would be to improve the preprocessing data, using the information of neighbouring cells to predict the current cell. Another possibility would be to use the data stream mining methodology (Information about data stream mining methodology [10]), since this would allow us to have a current model of traffic flow situation.

References

1. Fayyad, U., Piatetsky-Shapiro, G., Smyth, P.: From data mining to knowledge discovery in databases. AI Magazine **17**(3), 37–54 (1996)
2. Moreira-Matias, L.: Taxi service trajectory, prediction challenge (2015). https://archive.ics.uci.edu/ml/datasets/Taxi+Service+Trajectory+-+Prediction+Challen ge,+ECML+PKDD+2015
3. Maerivoet, S., Moor, B.D.: Traffic flow theory (2008)
4. Arbaiza, A., Martínez, P.T.: Parámetros fundamentales del tráfico ii. la velocidad. definiciones. percentil 85. velocidad inadecuada y velocidad excesiva. otras variables derivadas. métodos de obtención de datos de los parámetros de tráfico. procedimiento de integración y análisis (2014)
5. Han, J., Kamber, M.: Data Mining: Concepts and Techniques, 2nd edn, chap. 6, pp. 366–367. Morgan Kaufmann Publishers, Los Altos (2006)
6. Data Mining: Concepts and Techniques, 2nd edn, chap. 8, pp. 482–484. Morgan Kaufmann Publishers, Los Altos (2006)
7. Breiman, L.: Machine Learning, pp. 5–32. Kluwer Academic Publishers, Dordrecht (2001)
8. Divya, M., Vijayarani, S.: An efficient algorithm for classification rule hiding. **33**(3) (2011)
9. Tang, K., Wang, R., Chen, T.: Towards maximizing the area under the ROC curve for multi-class classification problems
10. Witten, I.H., Frank, E., Hall, M.A.: Data Mining Practical Machine Learning Tools and Techniques, 3rd edn, chap. 9, pp. 380–383. Morgan Kaufmann Publishers, Los Altos (2011)

SUMO Performance Comparative Analysis of SUMO's Speed Using Different Programming Languages

Samuel Romero Santana[1], Javier J. Sanchez-Medina[1(✉)],
David Sanchez Rodriguez[2], and Itziar Alonso Gonzalez[2]

[1] CICEI – ULPGC, Las Palmas, Spain
sam_r_s@hotmail.com, javier.sanchez@ulpgc.es
[2] IDeTIC – ULPGC, Las Palmas, Spain
{david.sanchez,itziar.alonso}@ulpgc.es

Mobility is an essential part of modern societies. Urban mobility has three important goals:

- Safety
- Efficiency
- Sustainability

Modern cities have a number of challenges ahead. To name a few, population levels are increasing together with increments in pollution, energy demand and environmental impact.

A very important element within urban mobility is Traffic Simulation. There is no doubt about the importance of the development of accurate traffic simulations, in particular for an efficient traffic management and planning. In that context, there is an outstanding tool, SUMO (Simulation Urban Mobility) [1,2], which is not only a complete and customizable microsimulation platform. It is also changing the game rules with its open software approach. Its community of developers is very alive and growing every month, likewise the number of research groups that embrace this tool as main technology for their research plans.

However, SUMO still lacks of something that would be clearly a deal breaker for local administrations regarding other tools and platforms [1,2]. That is performance. Traffic networks are generally very big and computationally expensive to simulate. Real-time performance levels are needed for its online use.

In our lab we are aiming at the parallelization of SUMO [1], but before of that move, we are evaluation the performance improvement through shifting from Python [3] to ANSI C as programming language.

We have studied how SUMO works [1]. SUMO is very important to researchers who work in modelling traffic because it allows to inspect the behaviour of the vehicles, the traffic in general and our infrastructures, therefore being able to detect and correct problems or to improve traffic behaviour before implementing them in real life. For example in platoon driving we can model behaviours, distances between vehicles or improve algorithms for special situations like intersections, before implementing such method in real life with the associated error cost there.

© Springer International Publishing AG 2018
R. Moreno-Díaz et al. (Eds.): EUROCAST 2017, Part II, LNCS 10672, pp. 440–445, 2018.
https://doi.org/10.1007/978-3-319-74727-9_53

```
<configuration>
  <input>
      <net-file value="highwaynet.xml"/>
      <route-files value="newrou.rou.xml"/>
  </input>
</configuration>
```

Fig. 1. Configuration file

To set up all of this, in SUMO [1], we need a simulation file with the extension '.cfg' which is formed by combining of at least other two files.

The file highwaynet.xml 1 determines the map of street over we want to do the simulation. This file has a special structure.

Researchers can get a map using several methods; Google, Openstreetmap, Mapbox, or even a map drawn by ourselves. For that we must know that a street, via, motorway or highway is formed by a set of nodes. The union of a set of nodes form the via.

To define the vehicles ([1] and the route that our vehicles, once we have the map, we need to construct a different file, called ruta.rou.xml 1. This file defines the type of vehicles we want to have in our simulation. To be able to simulate big number of vehicles or traffic jams, a set of vehicles and their routes are automatically created once configured that file automatically. So, researchers can build a battery of vehicles in just a few steps.

For that, it is possible to define repeated vehicle flows, which have the same parameters except for the departure time.

To build a battery of vehicles in an automatic way, we need two files.

– car.flow.xml
– highwaynet.xml

The first one indicates the characteristics and routes for our set of vehicles, and the second one is the map we have used before.

Researchers usually need to control simulations. For example to take the control of an specific car, or to try to extract interesting data like the speed of a particular vehicle, the fuel consumption, the position of a vehicle in the highway or the position of a particular car in a platoon [4].

For that, SUMO allows us to interact with it through TraCI [1]. In a few words we can define TraCI like the short term for "Traffic Control Interface" which gives the access to a running road traffic simulation, it allows to retrieve values of simulated objects and to manipulate their behaviour "online".

TraCI allows to connect SUMO with an script (preferably written in Python) and do what we want in a real time inside of our simulation. To carry out this process, we have to do some changes in our cfg file. We have to put the port where we are going to link our SUMO with the script. This port will be the number 8813.

Now that we know how SUMO works and the numerous calculations that it has to do to show the map of a big city and the thousands of vehicles that circulate through it with their decisions and rules, it is easier to understand that

Fig. 2. Flow: "Set of vehicles put into the simulator in a range of time"

```
<configuration>
    <input>
        <net-file value="highwaynet.xml"/>
        <route-files value="ruta.rou.xml"/>
    <remote-port value="8813"/>
    </input>
</configuration>
```

Fig. 3. Configuration file with remote control

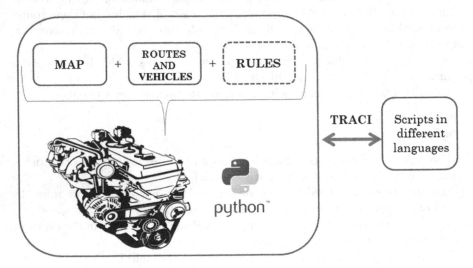

Fig. 4. "SUMO and Python"

to do simulations in real time is very computationally expensive. Simulation speed-up is very important to operate in real-time settings.

In our opinion, to accelerate SUMO simulations it is necessary to study the engine of the same. We realized that SUMO makes its calculations in Phyton [3]. Phyton has a lot of advantages:

– Python is a high level programming language.
– Python is multi-platform.

- Python was developed to work in a high level and it is oriented to portable object.
- Python is easy to read and understand for programmers.
- Python gives rise to quick development by using less code.
- A large number of resources are available for Python.
- Python code is being freely distributed.
- Python allows to scale even the most complex applications with ease.

Python allows people and programmers with little experience to contribute to the community by quickly developing modules. That is one of the reasons explaining SUMO rapid expansion. But Python [3] is an interpreted programming language, which means it is neither performance oriented nor optimized for a particular architecture.

In this paper we propose to use C instead, in order to accelerate SUMO situations. C is a programming language more difficult to write than Python, but it is typically faster, because is a down level programming language, that is compiled (not interpreted).

- C is faster than Python (down level programming language).
- C is a structure programming language.
- C is highly portable language.
- C has the ability to extend itself.
- C language has a rich library which provides a number of built-in functions.
- C offers dynamic memory allocation.

We present a performance comparison study, using two versions of SUMO. The out-of-the-box Python version and another version with some of its functionalities developed in C. We have done our simulations using a net, based in a motorway with two lanes, and a flow of vehicles, between fifty and one hundred cars driving along the motorway during five seconds. That vehicles drive along the motorway in a constant speed without any intersections for this set of experiments.

We use Python because the SUMO engine is implemented in that language, and researchers when want to communicate with SUMO through TraCI use that high level language. However, here we present some promising results using C for several modules.

Results are encouraging. We have evaluated the application speed-up by doing an arithmetic average of thirty simulations in C and other thirty simulations in Python. We have launched a number of different experiments varying the number of vehicles that were in the simulation.

In this graphic we have done 30 tests increasing 100 vehicles in each test, with the following parameters:

- road: highway (GC-1).
- Fixed Steps.
- 100 vehicles.

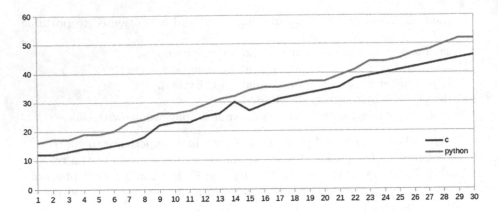

Fig. 5. C vs. Python

In Fig. 5, the x axis represents the different proofs we have done. For each test, we have been increasing the number of vehicles substantially to see the result in milliseconds at Y axis. As we can see, C has a better response. In each proof C is faster than Python, and we suppose that these results will be growing in the time line. These proofs are very simple. If we increase the type of vehicles, number of lanes, streets, highways and start to overburden the simulation we could see better results. For that, we have to implement libraries in C language and do others proofs in the future. In Fig. 6, the axis x represents the different proofs we have done too. In each proof, we have been increasing the number of vehicles substantially to see the result in Mega-Bytes at Y axis. As we can see, C has a better response. In each proof C saves more memory than Python, and we suppose that these results will be growing in the time line. We can assume that C manages memory better than Python in our simulations.

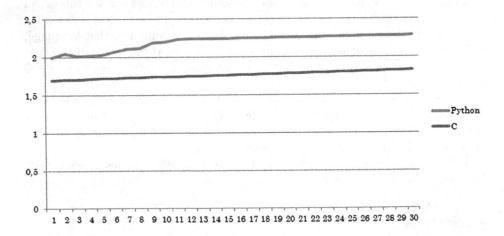

Fig. 6. C vs. Python

As for future plans, before we implement the parallel architecture we have design we will keep on developing some other modules into C, to be sure that we have the maximum from a serial implementation of SUMO.

References

1. Santana, S.R., Sanchez-Medina, J.J., Rubio-Royo, E.: Platoon driving intelligence. A survey. In: Moreno-Díaz, R., Pichler, F., Quesada-Arencibia, A. (eds.) EURO-CAST 2015. LNCS, vol. 9520, pp. 765–772. Springer, Cham (2015). https://doi.org/10.1007/978-3-319-27340-2_94
2. Behrisch, M., Bieker, L., Erdmann, J., Krajzewicz, D.: Sumo-simulation of urban mobility-an overview. In: SIMUL 2011, the 3rd International Conference on Advances in System Simulation, pp. 55–60 (2011)
3. Dobesova, Z.: Programming language python for data processing. In: 2011 International Conference on Electrical and Control Engineering (ICECE) (2011)
4. Santana, S.R., Sanchez-Medina, J.J., Rubio-Royo, E.: How to simulate traffic with SUMO. In: Moreno-Díaz, R., Pichler, F., Quesada-Arencibia, A. (eds.) EUROCAST 2015. LNCS, vol. 9520, pp. 773–778. Springer, Cham (2015). https://doi.org/10.1007/978-3-319-27340-2_95

Overtaking Maneuver for Automated Driving Using Virtual Environments

Ray Lattarulo$^{(\boxtimes)}$, Mauricio Marcano, and Joshué Pérez

Tecnalia Research and Innovation, Parque Científico y Tecnológico de Bizkaia,
Geldo Auzoa, Edif. 700, 48160 Derio, Bizkaia, Spain
{rayalejandro.lattarulo,mauricio.marcano,joshue.perez}@tecnalia.com
http://www.tecnalia.com/en/

Abstract. Among the driving possible scenarios in highways, the overtaking maneuver is one of the most challenging. Its high complexity along with the interest in automated cooperative vehicles make this maneuver one of the most studied topics on the field on last years. It involves a great interaction between both longitudinal (throttle and brake) and lateral (steering) actuators. This work presents a three phases overtaking path planning using Bézier curves, with special interest in the continuity of the curvature. Communication among the vehicles is also considered. Finally, the maneuver will be validated using *Dynacar*, a dynamic model vehicle simulator.

1 Introduction

With the evolution of the Intelligent Transportation Systems (ITS) in the last decade, a great variety of ADAS systems have been tested with successful deployment in commercial vehicles; some examples of these are: lane departure warning and assistance, automatic parking, blind spot monitoring, among others. Most of them are based on on-board sensors such as radars, LiDAR, cameras, ultrasonic sensors and communication V2X.

Great variety of tests related with automated driving have been performed on highways. However, planning constrains and control techniques have not been implemented in an extensive way [1]. One of the biggest and most interesting scenario on highways is the overtaking maneuver. This is defined by a series of constrains and conditions given by the perception systems (detection of the environment and obstacles), and the control stage (speed considerations and communication with nearby vehicles).

The control architecture used in this work was based on [2], with special approach in the behavioral planner, as part of the decision block, which is going to take responsability for the lane change maneuver on highways. We are considering the information obtained from the sensors as frontal LiDAR and cameras to detect obstacles on the road, and communication with other vehicles in a cooperative way.

Three possible stages were established to validate the maneuver. The first is the generation of a parametric (continuous) curve to change the lane, considering

© Springer International Publishing AG 2018
R. Moreno-Díaz et al. (Eds.): EUROCAST 2017, Part II, LNCS 10672, pp. 446–453, 2018.
https://doi.org/10.1007/978-3-319-74727-9_54

the possibilities that a vehicle could come on the opposite way, and the total time to recreate the maneuver. The second stage is to overtake the other vehicle in the opposite lane (adding a secure distance in front of the overtaken vehicle) and the third stage is the return of the vehicle to the lane.

The implementation of these algorithms and the control architecture was performed on MATLAB/Simulink, along with *Dynacar*, a dynamic model vehicle simulator [3].

2   Control Architecture and Decision Module

The architecture used in this work is based on [3]. It is a 6-block generalization to define an automated vehicle control structure; it has been firstly presented in [4]. To accomplish the goals of the current work, two critical blocks are considered: decision and communication (Fig. 1).

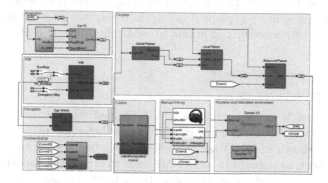

Fig. 1. Control architecture in detail with obstacle avoidance.

The acquisition module gather information from the different on-board sensors on the real platform and on the simulator, as well as, the information from the low level CAN (Ego-vehicle information).

The perception stage considers all the information collected from the acquisition block, and, with the use of different techniques and algorithms, it defines the environment surrounding the vehicle.

Communication stage is relevant for the purposes of this work, because it collects the information coming from other vehicles involved in the maneuver (V2X capabilities). Without this module, the difficulty to accomplish the lane-change maneuver safely, increases.

The control refers to the steering or lateral control and the throttle/brake or longitudinal control. It allows the vehicle to correctly track the trajectory and keep the desire speed.

Actuation is conformed by the actuators of the real and simulated platform; those are: throttle, brake and steering wheel. This block considers the low level control of the actuators.

The decision module has the objective to generate the trajectories that the vehicle has to follow. To accomplish this goal, the global planner reads a basic trajectory file (the assignment of going from point A to point B) that are basic points, for instance, the intersections and roundabouts of the route. After this, a first approach to a single route is done, and then sent to the local planner. It improves the trajectory softness (hard changes in the curvature) using different types of curves, such as Bézier [9]. The behavioral planner is related to the dynamic conditions of the route and how to face them. Some of the maneuvers considered are lane change, obstacle avoidance and overtaking; the latter will be studied in this work.

3   Overtaking Maneuver Planning

This work presents a three phases overtaking path planning, using two 5th order S-Shaped Bézier curves and one straight line. This curve is of special interest due to its capacity to joint with $G2$ continuity [10] with straight segments, both at the beginning and the end of the curve. The design of the path will be presented and the maneuver will be explained taking into account another vehicle in the opposite way. The necessary conditions to make the route free of collisions will be given.

3.1   The Path

In real driving it is common to follow a S-Shaped path in order to make a lane change. Previous works have emulated this behavior using graph search algorithms, as shown in [6]. Other techniques have been presented in [8] where static curves such as sigmoid functions are used to create the S-path. This work use Bézier curves to create a smooth curve to make the transition between lanes.

Parametric Bézier Curve: Given $n + 1$ points in the space, $P_0, P_1, ..., P_n$, a Bézier curve is defined as a combination of these points with the Bernstein Polynomials. Equation 1 shows the general representation of the degree-n Bézier curve [9]. The $P_i's$ are called the control points of $B(t)$, and the parametric curve will lie within their convex hull.

$$B(t) = \sum_{i=0}^{n} \binom{n}{i} t^i (1 - t)^{n-i} P_i \tag{1}$$

In [7] paths for overtaking are designed aiming for $G2$ (curvature) continuity, due to its importance in the reduction of lateral accelerations and improving comfort. With this in mind, the present work shows the design of a S-shaped continuous-curvature path using Bézier curves.

Designing the Path: Due to Bézier curves properties [7] it is possible to build a S-shaped route, placing the control points in specific positions: $\overrightarrow{P_0 P_1}$ must be parallel to the road and placed in the right lane and $\overrightarrow{P_{n-1} P_n}$ parallel to the road and placed in the left lane. This configuration allows a minimum of four control points. Taking into account the curvature in the search of smoothness, it is important to joint the Bézier curve with $G2$ continuity with a straight line, resulting in curvature 0 at both the beginning and the end of the path. Curvature for parametric curve is defined in Eq. 2.

$$k(t) = \frac{\left\| \overrightarrow{B'(t)} \times \overrightarrow{B''(t)} \right\|}{\left\| \overrightarrow{B'(t)} \right\|^3} = \frac{\left\| \overrightarrow{B''(t)} \right\| \, sin(\alpha(t))}{\left\| \overrightarrow{B'(t)} \right\|^2} \tag{2}$$

Where $t \in \mathcal{R}$ such that $0 \leqslant t \leqslant 1$, and $\alpha(t)$ is the angle between vectors $\overrightarrow{B'(t)}$ and $\overrightarrow{B''(t)}$. The intention is to find the proper conditions where $k(0) = k(1) = 0$. Using Eqs. 1 and 2 the following results are obtained:

$$\overrightarrow{B'(0)} = n(P_1 - P_0) = n\overrightarrow{P_0 P_1} \tag{3}$$

$$\overrightarrow{B''(0)} = n(n-1)(P_2 - 2P_1 + P_0) = n(n-1)(\overrightarrow{P_0 P_1} - \overrightarrow{P_1 P_2}) \tag{4}$$

From 2, 3 and 4, $k(0) = 0 \Leftrightarrow \alpha(0) = 0$ or $\overrightarrow{B''(0)} = 0$. This condition is met when P_0, P_1 and P_2 are collinear. By symmetry $k(1) = 0$ if P_{n-2}, P_{n-1} and P_n are collinear as well. Thus, the minimum number of point to achieve a S-shaped Bézier curve with $G2$ continuity with straight lines is six.

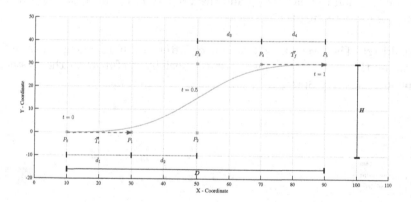

Fig. 2. 5th order S-shaped Bezier curve.

Quintic Bézier Curve: The representation of the 5th order Bézier curve is shown in Fig. 2. It is a degree-5 polynomial with a S-shape graph as previously deduced. To find the corresponding control points the following formulas are given:

$$
\begin{aligned}
P_0 &= (x_i, y_i) & P_5 &= P_0 + D\overrightarrow{p} + H\overrightarrow{q} \\
P_1 &= P_0 + d_1 \overrightarrow{p} & P_4 &= P_5 - d_4 \overrightarrow{p} \\
P_2 &= P_1 + d_2 \overrightarrow{p} & P_3 &= P_4 - d_3 \overrightarrow{p}
\end{aligned}
\tag{5}
$$

Where (x_i, y_i) is the position of the vehicle previous to the overtaking maneuver, $\vec{p}$ is the unit vector in the direction of the road, $\vec{q}$ is the unit vector that is orthogonal to $\vec{p}$, D is the distance between the vehicle and the obstacle, H is the height of the lane change; in general it is equal to the length of the lane. The lengths $d1$, $d2$, $d3$ and $d4$ are design parameters that change features of the curve; specially the curvature is of great interest.

3.2   The Maneuver

This section presents a three phase overtaking maneuver, composed by a lane change, an overtaking and a lane return. This is the normal behavior when avoiding slower cars in highways. The maneuver considers the joint of two Quintic Bézier curves and a straight line. Figure 3 shows the complete scenario with one vehicle as obstacle and a second vehicle in the opposite direction in the left lane.

Lane Change: The first phase is to avoid the obstacle in the road by making a lane change. The path to follow will be a Quintic Bézier curve. The design parameters are the width of the lane H and the critical obstacle distance D_{min} that is the minimum distance to the obstacle to make a safe lane change. This distance is directly proportional to the velocity of the vehicle. [5] define an approximate quadratic function to relate D_{min} and the velocity v. In this case, the LiDAR is responsible for detecting the obstacle.

Overtaking: The overtaking takes place after the lane change, passing the obstacle by the other lane. This phase is specially useful when the obstacle is moving at a constant speed.

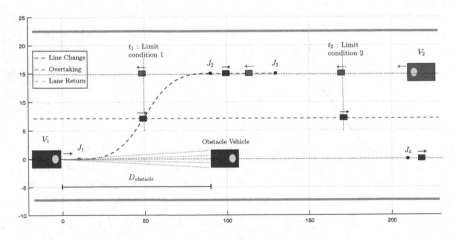

Fig. 3. Overtaking maneuver

Lane Change (Return): When the obstacle has been left behind the vehicle can return to the right lane to continue with the original trajectory. This is done with other Quintic Bézier curve. In order to perform the maneuver without collide with vehicles driving in the opposite direction, it is assumed that exists $V2V$ (Vehicle to Vehicle) communication and its positions are well known. In this case, according to Fig. 3 two critic conditions must be evaluated in $t = t_1$ and $t = t_2$. If after a prediction in time both conditions are hold then the overtaking can take place. In the other case the autonomated vehicle needs to stop until the vehicle in manual mode is no longer a threat.

4   Results and Validation

The simulated scenario was made on Dynacar Simulator and it is composed by one obstacle and two vehicles driving in parallel lanes in opposite direction. The first vehicle $V1$ is driving in automated mode and is able to change the trajectory and take decisions, while $V2$ runs in manual driving. The maneuver consists of $V1$ overtaking an obstacle considering two possible cases:

4.1   Case 1: Overtake Without Stop

This case presents the scenario where the vehicle avoids the obstacle via the overtaking maneuver without stop, after predicting no collision in future states with the overcoming vehicle in the other lane. Figure 4 shows the sequence of the maneuver in the simulator.

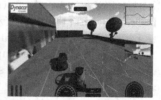

Fig. 4. Overtaking without stop

4.2   Case 2: Overtake with Stop

Other scenario is presented, where the conditions does not allow an overtaking without collision. In this case, the vehicle in automated mode must be in charge of delaying the overtaking until the vehicle in manual mode is no longer a threat.

Figure 5 illustrate this scenario. It can be noticed that after 7 s the automated vehicle starts decreasing the velocity until it stops. It waits for the other vehicle to pass and when the path is safe it activate the overtaking maneuver following the two Quintic Bézier curves previously designed. It can be noticed that $t_1 \neq t_2$, showing that both X and Y coordinate does not coincide at the same time, demonstrating a non-collision maneuver.

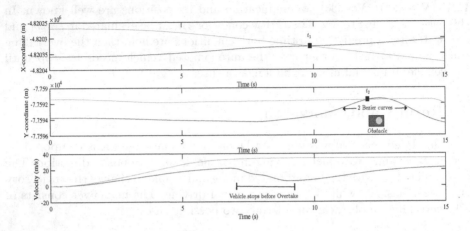

Fig. 5. Overtaking simulation with stop

5   Conclusions and Future Works

In this work, a three phases overtaking maneuver was presented. The path was designed with two S-shaped curves and one straight line, following the common behavior of drivers when overtaking vehicles in highways. The S-shape part was designed with a Quintic Bézier curve, because of its low computational cost and its property of zero curvature at the beginning and end of the curve.

The overtaking was implemented in simulation showing successful results after running two different scenarios without collision. The automated vehicle is capable of both overtaking with and without stop.

As future work, the proposed curve will be designed with additional parameters, such as the time to collision and the kinematics of the vehicle. Non-static obstacles such as vehicles with constant speed will be considered in the near future.

Speed profiles will be taken into account along with a deeper curvature analysis in order to look for comfort when performing an overtaking with higher velocities.

In addition, simulation implementation needs to be tested in a real vehicle, on highways and urban scenarios.

Acknowledgment. This work is partly supported by the H2020 project UnCoVerCPS with grant number 643921.

References

1. Peréz, J., Lattarulo, R., Nashashibi, F.: Dynamic trajectory generation using continuous-curvature algorithms for door to door assistance vehicles. In: IEEE Intelligent Vehicle Symposium (IV), pp. 510–515 (2014)
2. González, D., Peréz, J., Milanés, V., Nashashibi, F.: A review of motion planning techniques for automated vehicles. IEEE Trans. Intell. Transp. Syst. **17**, 1135–1145 (2015)
3. Lattarulo, R., Pérez, J., Dendaluce, M.: A complete modular framework for developing and testing automated driving controllers. In: IFAC World Congress, Toulouse, France (2017)
4. González, D., Pérez, J.: Control architecture for cybernetics transportation systems in urban environments. In: IEEE Intelligent Vehicle Symposium, Gold Coast, Australia, pp. 1119–1124, June 2013
5. Naranjo, J.E., Gonzalez, C., Garcia, R., de Pedro, T.: Lane-change fuzzy control in autonomous vehicles for the overtaking maneuver. IEEE Trans. Intell. Transp. Syst. **9**(3), 438–450 (2008)
6. Katrakazas, C., Quddus, M., Chen, W.-H., Deka, L.: Real-time motion planning methods for autonomous on-road driving: State-of-the-art and future research directions. Transp. Res. Part C **60**, 416–442 (2015)
7. González, D.: Functional Architecture For Automated Vehicles Trajectory Planning in Complex Environments. Doctoral school n432, April 2017
8. Upadhyay, S., Ratnoo, A.: Continuous-curvature path planning with obstacle avoidance using four parameter logistic curves. IEEE Robot. Autom. Lett. **1**(2), 609–616 (2016)
9. Farouki, R.T.: Pythagorean-Hodograph Curves: Algebra and Geometry Inseparable, vol. 1. Springer Science & Business Media (2008)
10. Yang, K., Sukkarieh, S.: An analytical continuous-curvature path-smoothing algorithm. IEEE Trans. Robot. **26**(3), 561–568 (2010)

Effects of Cooperative Lane-Change Behavior on Vehicular Traffic Flow

Christian Backfrieder[1]([⊠]), Manuel Lindorfer[1], Christoph F. Mecklenbräuker[2], and Gerald Ostermayer[1]

[1] Research Group Networks and Mobility, FH Upper Austria, Hagenberg, Austria
{christian.backfrieder,manuel.lindorfer,
gerald.ostermayer}@fh-hagenberg.at
[2] Christian Doppler Lab Wireless Technology for Sustainable Mobility,
Vienna University of Technology, Vienna, Austria
cfm@nt.tuwien.ac.at

Abstract. Modeling of driver behavior is an important issue in microscopic vehicular traffic simulations. Especially for lane-change models, cooperative behavior is in some situations inevitable to ensure a smooth traffic flow and avoid abnormally long waiting times for single vehicles. Especially in case of congestion and unconditionally required lane-changes before intersections, at highway exits or road narrows, cooperation between vehicles can significantly improve traffic flow. We analyze the comprehensive model Cooperative Lane-Change and Longitudinal Behavior Model Extension (CLLxt) from literature with regard to its influence to performance indicators of traffic flow such as average travel time and fuel consumption. To conduct the evaluation, the microscopic traffic simulator TraffSim is used. The simulation study includes evaluations for two representative scenarios - one highway exit scenario and an intersection scenario with lanes including turn restrictions. Additionally, effects of a varying traffic density are analyzed. The results reveal significant improvements for all investigated situations due to the application of CLLxt.

Keywords: Lane-change model · Traffic simulation
Cooperative lane-change

1 Introduction

Beyond doubt, behavior of drivers has a strong impact on vehicular traffic flow. In order to investigate those effects, i.e. gains and losses resulting from more or less cooperative behavior, simulations are an expedient to quantify those effects in numbers. Comprehensive simulations include road networks containing roads with multiple lanes. The discrete decision making of single vehicles whether or not to change their lane is influenced by multiple factors, and often controlled by lane-change models of different types. Obviously, the decision is depending

© Springer International Publishing AG 2018
R. Moreno-Díaz et al. (Eds.): EUROCAST 2017, Part II, LNCS 10672, pp. 454–461, 2018.
https://doi.org/10.1007/978-3-319-74727-9_55

not only on the demand to stick to traffic rules, pass by lane closures or accidents or overtaking maneuvers, but also to follow the own route and gain speed advantage by overtaking others. Additionally, the reaction to other, neighbored vehicles' requirements may trigger the consideration of lane-changes. To do so, influence on the current speed and acceleration may be necessary to execute the lane-change. A cooperative lane-change model considering the discussed issues is presented in [1]. It acts as extension to both lane-change (e.g. MOBIL [2]) and longitudinal models (Intelligent Driver Model (IDM) [3]), and especially provides solutions for situations where cooperation between vehicles is necessary. However, it does not present and quantify potential gains and advantages resulting from cooperative behavior compared to egoistic lane-change models. This paper presents simulation results, which show in which situations cooperation is beneficial and how much resources can be saved as a consequence. It reveals benefits when using cooperative lane-change behavior, especially in terms of average time consumption of vehicles to get from source to destination.

Additionally, another unrealistic effect of non-cooperative models can be counteracted, that is not visible when averaging time and fuel consumption: Single vehicles are convicted to wait for a very long time, because it is impossible to change to the target lane in dense traffic. If the blocking vehicle on the target lane considers also its neighbors, the neighbor vehicle can change lane after a short waiting time and continue its ride, what moves simulation results closer to reality.

2   Related Work

Traffic congestion is indeed a major problem worldwide, especially in urban areas. Also, lane-change models have a significant influence in the field of microscopic traffic simulation [4]. Although microscopic lane-change models are a intensively discussed topic in literature, the analysis of effects and application of such a model in congested situations is a special case which is not considered by many of the basic models [2,5]. A comprehensive driving model including also lane-changes is presented in [6], which is both well configurable and covers many aspects in great detail. Hidas [7] deal with lane-changes in congested situations by analysis of video recordings and try to rebuild the effects in a macroscopic manner. The authors focus on analysis of gaps and classification of lane-changes from observations into defined categories. Other work that mainly deal with characteristic properties of cooperative lane-change models and analysis of lane-change gaps in congested situations can be found in [8,9]. However, the impacts of a cooperative compared to egoistic behavior are not covered by any of the mentioned papers. This is precisely still an important question, and therefore addressed by this paper.

3   Investigation and Setup

The simulations were carried out using different scenarios, configured within the microscopic traffic simulator TraffSim [10]. In order to investigate the effects of cooperative or drawbacks of non-cooperative behavior, situations where cooperation is necessary are simulated correspondingly in order to emphasize the influence of the proposed model extension. This is the case in slight overload of a highway exit or considerable and continuously high traffic emergence for traffic lights.

3.1 Scenarios

Figure 1b shows the analyzed highway scenario, where we simulated 700 vehicles in total with an equally distributed inter arrival time between two consecutive vehicles. The arrival time for the highway was varied from 700 ms to 1100 ms in order to prove the influence of the traffic density on results. The road ends that are marked in green represent the entry road segments for all vehicles. The destinations are randomly distributed between the blue areas.

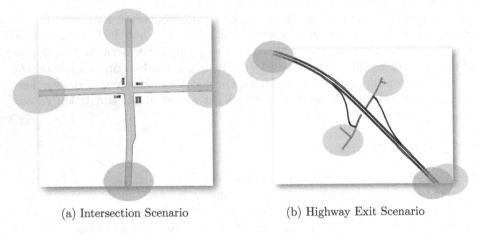

(a) Intersection Scenario   (b) Highway Exit Scenario

Fig. 1. Speed and acceleration graphs for both supplying and invoking vehicles (Color figure online)

In Fig. 1a, the second scenario with an intersection is depicted. The scenario consists of a traffic light with four connected road segments and multiple lanes, including turn restrictions and therefore necessity of lane-changes for some of the vehicles to reach the desired destination. For simplicity, we use only one entry point (green ellipse) for all vehicles and again distribute the destinations between the three remaining arms of the intersection (blue ellipses). Thus, the changes in the results can completely be ascribed to the setup cooperation behaviour. Analogous to the highway scenario, we also vary the vehicle arrival times from 700 to 1300 ms.

3.2 Simulation Configuration

The results are generated by executing each of the scenario configurations twice, once with *CLLxt* disabled and a second time enabled. All vehicles are parametrized with the same longitudinal model IDM [3] with a safe deceleration of $-2\,\text{m/s}^2$, a comfortable acceleration of $0.7\,\text{m/s}^2$. The minimum distance and time between two vehicles is set to 2 m and 1.1 s, respectively, and the δ for the aggressiveness is defined at a medium level of 4.

The lane-change model MOBIL and the cooperative lane-change extension are parametrized as proposed in [1].

4   Simulation Results

The simulation results show that cooperative behaviour leads to significant reduction of average travel time and fuel consumption in both investigated situations. However, the improvements also depends significantly on the traffic density in the highway scenario. Naturally, we have to admit that cooperative behavior is not always beneficial in the same manner, but has the greatest impact for scenarios with a good balance between unproblematic situations with very low traffic emergence and overcrowded situations without any possibility to improve.

4.1   Travel Time and Fuel Consumption

Results reveal slight improvements of around 2.5% average fuel consumption over all vehicles and 5% less average travel time for the intersection scenario, as shown in Fig. 2. These improvements can completely be associated with the cooperative lane-change behaviour, and less waiting times after standstill due to consideration of neighbor vehicles' needs. If these numbers are scaled to large scenarios with many intersections, even better values can be expected.

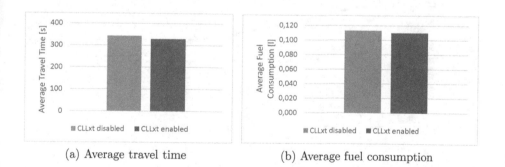

(a) Average travel time   (b) Average fuel consumption

Fig. 2. Simulation results for intersection scenario

The impact of cooperative behavior on the observed values is even more beneficial in the highway scenario. The potential of high waiting times in case of egoistic drivers is higher, which can be ascribed to higher speeds. The effort of slowing down to standstill and accelerating again due to a non-cooperative behavior of

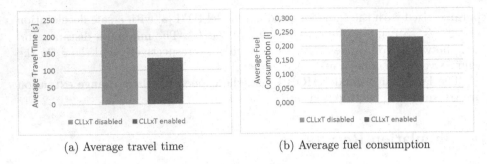

(a) Average travel time   (b) Average fuel consumption

Fig. 3. Simulation results for highway scenario

a neighboring driver is much higher. Figures 2a and 3b reveal improvements of travel time of up to 75% and of fuel consumption of up to 11%.

Our evaluations showed that the grade of improvement is strongly dependent on the vehicle density in the highway scenario, as Fig. 4 shows. Certainly, there is no need for any cooperative lane-change if the density is at such a low level that possible situations are very rare. Consequently, also the difference between the CLLxt 'on' and 'off' curves are low. The same effect is visible for very high densities, where the significant influence on average travel time and fuel consumption is reducible to the congestion, and no improvement is possible due to that fact. This explains the little differences between the curves on the left side of the figure.

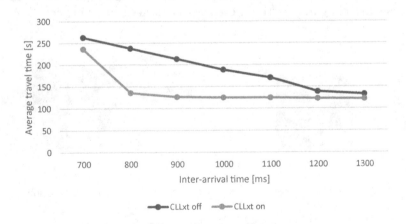

Fig. 4. Travel time vs. density for highway scenario

In contrast to the highway scenario, nearly no correlation between the vehicle density and the percental improvement becomes apparent in the regulated intersection (Fig. 5). The improvement stays constant at the mentioned values. The reason for that is the static configuration of the traffic light cycles, which is independent from the emerging traffic. In other words, this means that although the density increases, the potential for improvements by cooperative lane-changes stays constant, only the queue length before the intersection becomes longer. Nevertheless, the probability of occurrence of situations with necessity to cooperate between vehicles is therefore nearly equivalent, and thus also the improvement curves.

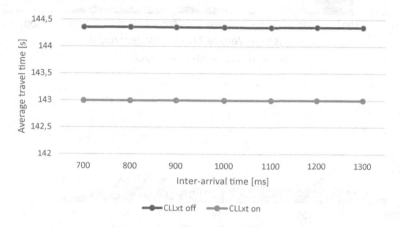

Fig. 5. Travel time vs. density for intersection scenario

4.2 Distribution of Travel Time

Another interesting aspect which reveals significantly in the highway scenario and what one cannot see in the averaged values above is the distribution of travel time among the vehicles, which is depicted in Fig. 6. We picked a representative configuration with the highway scenario and an inter-arrival interval of 800 ms. The upper part shows the histogram over travel time for the non-cooperative behaviour (Fig. 6a). The majority of vehicles has travel times of about 100 to 200 s, as the left columns indicate. However, a considerable amount of vehicles experience travel times of a multiple of the regular time, which of course is disadvantageous for reasons of transport planning for both individuals and operators. The reason for these outliers are the long waiting times for some vehicles with pending lane-changes to the exit. They cannot change to the target lane because it is occupied by upcoming vehicles, which ignore their lane-change demand. In contrast to that, the travel time distribution is more balanced with cooperative behavior enabled (Fig. 6b). The previously occurring high waiting times due to blocked highway exit lanes are avoided for single vehicles, since as soon as an intended lane-change is indicated, the following vehicles take the intention into consideration. If possible, they slightly reduce their speed to precipitate an uncritical lane-change maneuver.

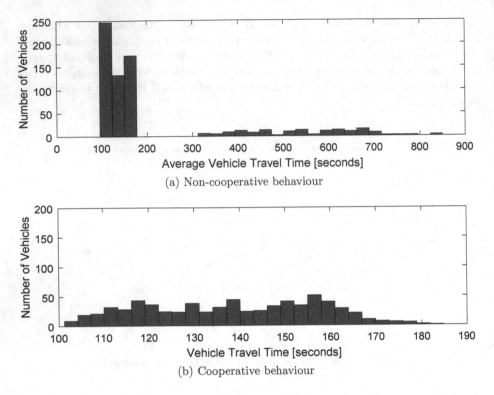

(a) Non-cooperative behaviour

(b) Cooperative behaviour

Fig. 6. Travel time distribution with different behavior settings

5   Conclusion

In this paper we proof the influence of cooperative lane-change on traffic flow by simulations. We analyze the Cooperative Lane-Change and Longitudinal Behavior Model Extension *CLLxt* [1] and compare scenarios with enabled and disabled cooperative behavior regarding the influence of the cooperation between drivers on traffic flow. The simulations are carried out using the microscopic traffic simulator *TraffSim* using two scenarios, one highway scenario and one intersection scenario. Additionally, we vary the traffic density to investigate the influence on the potential of improvement in different situations. Cooperation is essential in all investigated situations, which is demonstrated and quantified in numbers. The results show up to 75% savings in average travel time for the investigated highway scenario, and 5% for the intersection scenario. Fuel consumption decreases by 11 and 5% for the highway and intersection scenario, respectively.

Acknowledgments. This project has been co-financed by the European Union using financial means of the European Regional Development Fund (EFRE). Further information to IWB/EFRE is available at www.efre.gv.at.

Europäische Union Investitionen in Wachstum & Beschäftigung. Österreich.

References

1. Backfrieder, C., Ostermayer, G., Lindorfer, M., Mecklenbräuker, C.F.: Cooperative lane-change and longitudinal behaviour model extension for TraffSim. In: Alba, E., Chicano, F., Luque, G. (eds.) Smart-CT 2016. LNCS, vol. 9704, pp. 52–62. Springer, Cham (2016). https://doi.org/10.1007/978-3-319-39595-1_6
2. Kesting, A., Treiber, M., Helbing, D.: General lane-changing model MOBIL for car-following models. Transp. Res. Rec.: J. Transp. Res. Board, 86–94 (2007)
3. Kesting, A., Treiber, M., Helbing, D.: Enhanced intelligent driver model to access the impact of driving strategies on traffic capacity. Philos. Trans. R. Soc. Ser. A: Math. Phys. Eng. Sci. (1928), 4585–4605 (2010). arXiv: 0912.3613
4. Rahman, M., Chowdhury, M., Xie, Y., He, Y.: Review of microscopic lane-changing models and future research opportunities. IEEE Trans. Intell. Transp. Syst. **14**(4), 1942–1956 (2013)
5. Gipps, P.: A model for the structure of lane-changing decisions. Transp. Res. Part B: Methodol. **20**(5), 403–414 (1986)
6. Toledo, T., Koutsopoulos, H., Ben-Akiva, M.: Integrated driving behavior modeling. Transp. Res. Part C: Emerg. Technol. **15**(2), 96–112 (2007)
7. Hidas, P.: Modelling vehicle interactions in microscopic simulation of merging and weaving. Transp. Res. Part C: Emerg. Technol. **13**(1), 37–62 (2005)
8. Laval, J.A., Leclercq, L.: Microscopic modeling of the relaxation phenomenon using a macroscopic lane-changing model. Transp. Res. Part B: Methodol. **42**(6), 511–522 (2008)
9. Toledo, T., Koutsopoulos, H., Ben-Akiva, M.: Modeling integrated lane-changing behavior. Transp. Res. Rec.: J. Transp. Res. Board (1857), 30–38 (2003). Proceedings of the 82nd Transportation Research Board Annual Meeting, Washington, DC (2003)
10. Backfrieder, C., Ostermayer, G., Mecklenbräuker, C.: TraffSim - a traffic simulator for investigations of congestion minimization through dynamic vehicle rerouting. Int. J. Simul.- Syst. Sci. Technol. IJSSST **15**, 38–47 (2015)

A Comparative Performance Analysis of Variable Speed Limit Systems Control Methods Using Microsimulation: A Case Study on D100 Freeway, Istanbul

Mohd Sadat$^{(\boxtimes)}$, Ismail M. Abuamer, Mehmet Ali Silgu,
and Hilmi Berk Celikoglu

Department of Civil Engineering, Technical University of Istanbul (ITU),
Istanbul, Turkey
{sadatm,ismailmaabuamer,msilgu,celikoglu}@itu.edu.tr

Abstract. In this paper comparative performance analysis of two control methods to determine Variable Speed Limit (VSL) is performed using microsimulation. A case study is performed on D100 Freeway, using both control methods, in a simulated environment of VISSIM. Calibration of VISSIM is sought using Remote Traffic Microwave Sensor data provided by Istanbul Municipality. Integration of MATLAB with VISSIM is realized via COM interface which allows implementation of user defined control methods. Results show that both control methods perform better when compared the no-control case in terms of Total Time Spent in the selected network. The feedback control strategy, based on occupancy, performs relatively better when compared to control strategy using decision tree based on threshold values of volume, occupancy and average speed. Also, the speed profiles achieved with feedback control is more homogenized when compared to resultant speed profiles from control based on volume, occupancy and average speed. It can also be concluded that application area of VSL, acceleration area, the position of signs and detectors affect the performance of VSL control systems.

Keywords: Variable speed limits · Congestion · Microscopic traffic simulation

1 Introduction

Intelligent Transportation System techniques like Ramp Metering, Route Guidance and Variable Speed Limits (VSL) can help in easing out congestion on freeways. While ramp metering and route guidance system have limited scope of application, VSLs can be used for preventing traffic breakdown by regulating flow upstream of a congestion. This is achieved by imposing speed limit before critical density or flow is attained. Fixed control approaches based on occupancy, implemented on I-4 Orlando Florida, and flow, implemented on M-25 England, have been utilized in past. Many studies have reported improved performance with better control strategies, however mostly in hypothetical test scenarios. Aforementioned field implementation also used simple

© Springer International Publishing AG 2018
R. Moreno-Díaz et al. (Eds.): EUROCAST 2017, Part II, LNCS 10672, pp. 462–469, 2018.
https://doi.org/10.1007/978-3-319-74727-9_56

algorithms. Thus this study presents a comparative performance evaluation of two control methods in VSLs implementation: One of the control methods is adapted and modified from [1] which is based on flow occupancy and speed while another one is adapted from [2] based on feedback control based on occupancy to calculate VSLs.

In the following section, relevant literature is summarized. The methodology is explained in the third section. The fourth section presents the simulation results. The final section concludes the paper with possible future research directions.

2 Relevant Literature

Representing vehicular flow mathematically has always been an effective tool in traffic studies. The Mainstream Traffic Flow Control (MTFC) using VSLs to reduce TTS is proposed in [3], while a simple volume, occupancy and speed value based control method is tested in [4]. Using METANET to predict traffic state over a time horizon of 5 min for a mountainous freeway section, where the bottleneck formed due to the lane drop before a tunnel, a model prediction is performed in [5]. They used an objective function to minimize the crash risk utilizing VISSIM for traffic simulation in [5]. MTFC is used in [6] to implement VSLs, where better results with optimal control is reported with the fact that working on continuous speed limits makes it difficult to implement on the field. The safety constraints demand the speed limit to be discrete and sustain for certain period of time before they are changed. The study concluded two key points- Feedback Control was more robust and simple to apply compared to optimal control and it can consider all constraints ensuring safety and practicality. A similar analysis on a hypothetical network using Optimal MTFC and Feedback MTFC with VSLs is performed in [7].

For this case study, two VSL control methods are adopted from [1, 2]. A study to assess the safety and operational impact the proposed VSL algorithm on freeway traffic is conducted in [1]. The algorithm uses a decision tree with threshold values for volume, occupancy, and speed. Simple threshold based VSL controls have been in use in existing freeway. For example, a volume based control is utilized to trigger the VSL on M-25 Motorway. The threshold volumes are sourced from [8] defining the Level of Service (LOS). A feedback control integrated to a microsimulation environment is utilized in [2]. As mentioned above, METANET model for traffic simulation has been extensively used by researchers, but, few points noted in [2] are that METANET being macroscopic divides the stretch into sub-section and speed limit changes are enforced on the whole sub-section but in practice vehicles passing speed limit sign are affected. Further, macroscopic models are not stochastic therefore, driver behavior is better modeled by microscopic models.

A simple feedback control was preferred over optimization because of simplicity and robustness of such controls. VSLs' algorithms-based advanced control methods, such as optimal control and Model Predictive Control (MPC), have not been being tested frequently in the real life with exception of SPECIALIST control method in [9].

3   Case Study Methodology

For traffic simulation, VISSIM is used which is time-step based, stochastic, micro-scopic model that treats vehicles as basic entities. It is based on Wiedemann 99 model for freeway car following behavior of longitudinal movement. For lateral movement, it employs a rule-based algorithm. Control strategies are realized in MATLAB, which is integrated with VISSIM via COM interface. This server-client type setup provides flexibility by allowing user defined controls implementation and access to traffic flow variables in real time. Traffic volume and average speed RTMS data for every two minutes is used to perform calibration of the VISSIM model. Volume is used as input for the model. Freeway traffic parameters, i.e., Headway Time (CC1), Following Variation (CC2) and Oscillation Acceleration (CC7), are modified until the simulated and the real volumes satisfy GEH statistics [10]. Above mentioned parameters are considered since they are reported in [11] and [12] to be the most sensitive parameters having a significant impact on the flow through the network.

Best results are obtained when CC1 = 1 s, CC2 = 5 m and CC7 = 0.30 m/s^2. It is observed that 95.8% of GEH values were within 5 at RTMS 301, 92.5% of GEH values were within 5 at RTMS 533 and 90% of GEH were within 5 at RTMS 534. According to [10] 85% volumes in the simulated model should have GEH less than 5 for accurate representation of real-field traffic flow. Calibration is validated by plotting real and simulated speed profiles as shown in Fig. 1 for RTMS 533. After the calibration simulation is performed, during which value of the volume, occupancy and average speed for every minute is retrieved by MATLAB from VISSIM, desired speed is calculated by the running program code and sending back to VISSIM.

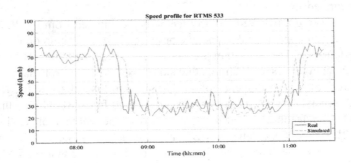

Fig. 1.   Real and simulated speed profile

3.1   Case Study Test Segment

Many segments over D100 freeway in Istanbul suffer from recurrent congestion [13–19]. The section under consideration is 3.5 km westbound 100 from Zincirlikuyu towards Okmeydani. There is a two-lane on-ramp from Barbaros Boulevard followed by an acceleration area, which ends at an off-ramp towards Buyukdere Street. There is another on-ramp from Buyukdere Street followed by a 250 m of an acceleration area. Next, there is an off-ramp followed by another on-ramp. There are three RTMS detectors. RTMS 301 corresponds to first on-ramp while RTMS 533 is stationed at the second on-ramp. RTMS 534 is placed at the third on-ramp, which is towards the end of the section. RTMS 301 and RTMS 533 records point heavy congestion during peak hours. From RTMS 534 it is seen that the flow from on-ramp is relatively low. There is no congestion captured at this point even during peak hours.

3.2   System Control Strategies

As discussed in the previous section, there are various control methods for VSLs. For the present case study, two types of control have been selected. One of the algorithms is adapted from [1], which essentially represents the simple volume, occupancy and speed value based control method. The VSL control based on volume has been implemented on M-25 Motorway, England as well as on E6 motorway, Sweden. Occupancy-based VSL control was implemented on I-4 Orlando, U.S.A. Further, weather based VSL control has been used in past. Thus adoption of the algorithm from [1] is justified as it represents the current methods used in freeway management. The algorithm for control 1 is shown in Fig. 2.

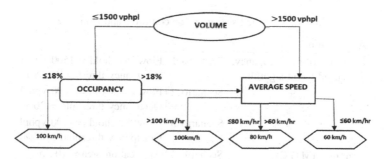

Fig. 2. Algorithm for control 1 adapted from [1]

The other control approach chosen is feedback type that is adapted from [2]. It is a robust and simple control, which is preferred over the optimal control by the authors. It is single-input-single-output type control where the VSL rate $b(\cdot)$ and the occupancy are respectively the control action and the controlled variable. In [2] authors prefer microsimulation environment over macrosimulation due to the practical aspects discussed in Sect. 2. Moreover, feedback control can be readily deployed in the field which makes it preferable. Equation 1 gives MTFC feedback type control:

$$b(k) = b(k-1) + K_I error(k) \tag{1}$$

where; b is the VSL rate (ratio of current speed and speed limit), $error$ = desired occupancy-current occupancy. The desired occupancy was set at 20% for our case as it is determined from the no-control scenario shown in Fig. 3. k is the time step and K_I is the Integral Gain. The value of $K_I = 0.01$ for 0<b> = 0.3, $K_I = 0.008$ for 0.3<b> = 0.6 and $K_I = .011$ for 0.6<b> = 0.9. Table 1 shows all the tested scenarios. The displayed

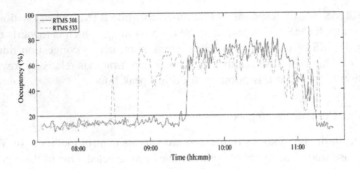

Fig. 3. Occupancy for RTMS 301 and RTMS 533

speed is subjected to two constraints: the difference between displayed speeds of two consecutive VSL signs should not be more than 20 km/h [1] and lower speeds should sustain for at least one minute [20]. These constraints prevent sudden speed changes and allow the traffic to recover before the sign is changed to higher speed values.

Table 1. All tested scenarios for control 1 and control 2

All scenarios		
Control 1 (flow, occupancy, and speed based algorithm	Scenario 1	Flow threshold = 1500 vphpl Occupancy threshold = 18%
	Scenario 2	Flow threshold = 1600 vphpl Occupancy threshold = 20%
	Scenario 3	Flow threshold = 1700 vphpl Occupancy threshold = 22%
Control 2 (MTFC)	Scenario 4	Application area = 100 m Acceleration area = 150 m
	Scenario 5	Application area = 150 m Acceleration area = 200 m
	Scenario 6	Application area = 200 m Acceleration area = 250 m

4   Simulation and Comparative Analysis

The simulation is performed for 4 h with 5 min of warm-up time to saturate the networks with traffic. A Measure of Effectiveness (MOE) is sought in terms of TTS. Table 2 shows the reduction of TTS for control 1 and control along with the no-control case. It can be observed that for control 2 with the scenario no. 4 reduction in the TTS is maximum. Therefore, it is straightforward to conclude that the shorter application area and acceleration results in better performance of feedback type control method for our case. For control 1, scenario no. 3 is the best in terms of TTS reduction, therefore, higher cut-off values for volume and occupancy results in better performance.

Table 2. Summary of TTS reduction for all scenarios

		TTS (veh.h)	% change
No-control	–	1067.63	–
Control 1	Scenario 1	767	−28.15
	Scenario 2	696.45	−34.72
	Scenario 3	679.41	−36.38
Control 2	Scenario 4	586.16	−45.07
	Scenario 5	636.84	−40.31
	Scenario 6	651.54	−38.98

Figure 4 shows the speed profile for no-control, control 1 and control 2 cases, where; the section enclosed by 16 data collection points and the time marked for every two minutes are represented by two axes and the speed is represented by the third axis. It can be observed that onset of congestion starts at point 10, on space axis, corresponding to second on-ramp merging, after 90 min into the simulation and propagates upstream towards point 2, which corresponds to the first on-ramp merging. Congestion only dissipates after point 10 towards the end of the section. The top right plot in Fig. 4 shows the speed profile after control 1 is applied. It shows relatively a better speed profile with reduced congestion over time and space. It is observed, however, that flow becomes unstable towards of the end of the network. The plot at the bottom of Fig. 4 shows the speed profile after control 2 is applied. It can be observed that it has the lowest congestion period, where traffic is relatively stable even after point 10, the 2nd on-ramp merging. One more noticeable difference is that after the application of the VSL control congestion is redistributed in the section and congestion dissipates sooner

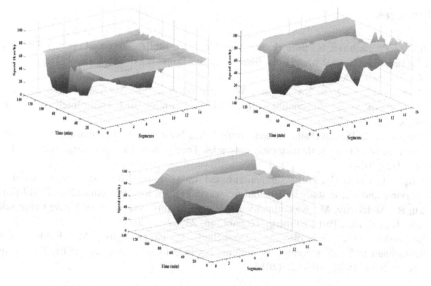

Fig. 4. Speed profiles for no-control, control 1 and control 2

when compared to the no-control case. Furthermore, it can be inferred that VSL systems themselves are unable to obstruct the breakdown in traffic flow when demand is considerably high, however, they can delay the onset of congestion.

5 Conclusions

The aim of the study has been to analyze the performance of two control methods for VSL systems in heavy congestion traffic scenario of D100 freeway. When variable speed limits are applied in very heavy congestion they can successfully delay the onset of the congestion by changing the spatial and temporal distribution of congestion to other areas of the network, which can resolve much earlier compared to the no-control case. When the traffic demand is very high, the performance of the VSL system is not very satisfactory due to the fact that during high traffic demand, traffic breakdown is imminent. VSLs play little to no role in the recovery of traffic, which primarily depends on the reduced demand on main links and ramps. In our case, the feedback type MTFC performed better when compared to flow, occupancy and speed based control. The TTS has been reduced up to 45.07% for the feedback MTFC while it is reduced by up to 36.38% for flow, occupancy and speed based control. It is also observed that placement of detectors and VSL signs also affect the performance of the overall control application. As the future extensions of the study summarized, co-operative methods, including ramp metering as well as lane management, can be tested on the section in interest as there are two bottlenecks due to the presence of the on-ramps. The study will be expanded to include more bottlenecks on D100 and E-80 freeways in order to analyze use of combined control in the network-wide congestion propagation in Istanbul city, as well on the approaches of the Bosphorus strait crossing bridges [21–24].

References

1. Allaby, P., Hellinga, B., Bullock, M.: Variable speed limits: safety and operational impacts of a candidate control strategy for freeway applications. IEEE Trans. Intell. Transp. Syst. **8**(4), 671–680 (2007)
2. Müller, E.R., Carlson, R.C., Kraus, W., Papageorgiou, M.: Microsimulation analysis of practical aspects of traffic control with variable speed limits. IEEE Trans. Intell. Transp. Syst. **16**(1), 512–523 (2015)
3. Carlson, R.C., Papamichail, I., Papageorgiou, M., Messmer, A.: Optimal mainstream traffic flow control of large-scale motorway networks. Transp. Res. Part C: Emerg. Technol. **18**(2), 193–212 (2010)
4. Sadat, M., Celikoglu, H.B.: Simulation-based variable speed limit systems modelling: an overview and a case study on Istanbul freeways. Transp. Res. Procedia **22**, 607–614 (2017)
5. Yu, R., Abdel-Aty, M.: An optimal variable speed limits system to ameliorate traffic safety risk. Transp. Res. Part C: Emerg. Technol. **46**, 235–246 (2014)
6. Iordanidou, G.R., Roncoli, C., Papamichail, I., Papageorgiou, M.: Feedback-based mainstream traffic flow control for multiple bottlenecks on motorways. IEEE Trans. Intell. Transp. Syst. **16**(2), 610–621 (2015)

7. Carlson, R.C., Papamichail, I., Papageorgiou, M.: Local feedback-based mainstream traffic flow control on motorways using variable speed limits. IEEE Trans. Intell. Transp. Syst. **12** (4), 1261–1276 (2011)
8. Highway Capacity Manual: Transportation Research Board of The National Academies (2010)
9. Hegyi, A., Hoogendoorn, S.P.: Dynamic speed limit control to resolve shock waves on freeways-field test results of the SPECIALIST algorithm. In: 13th International IEEE Conference on Intelligent Transportation Systems (ITSC), pp. 519–524. IEEE (2010)
10. UK Highways Agency: UK Design Manual for Roads and Bridges (1996)
11. Gomes, G., May, A., Horowitz, R.: Congested freeway microsimulation model using VISSIM. Transp. Res. Rec.: J. Transp. Res. Board **1876**, 71–81 (2004)
12. PTV AG: VISSIM 5.40 user manual. Karlsruhe, Germany (2011)
13. Celikoglu, H.B.: Flow-based freeway travel-time estimation: a comparative evaluation within dynamic path loading. IEEE Trans. Intell. Transp. Syst. **14**(2), 772–781 (2013)
14. Celikoglu, H.B.: Reconstructing freeway travel times with a simplified network flow model alternating the adopted fundamental diagram. Eur. J. Oper. Res. **228**(2), 457–466 (2013)
15. Celikoglu, H.B.: Dynamic classification of traffic flow patterns simulated by a switching multimode discrete cell transmission model. IEEE Trans. Intell. Transp. Syst. **15**(6), 2539–2550 (2014)
16. Silgu, M.A., Celikoglu, H.B.: Clustering traffic flow patterns by fuzzy c-means method: some preliminary findings. In: Moreno-Díaz, R., Pichler, F., Quesada-Arencibia, A. (eds.) EUROCAST 2015. LNCS, vol. 9520, pp. 756–764. Springer, Cham (2015). https://doi.org/10.1007/978-3-319-27340-2_93
17. Celikoglu, H.B., Silgu, M.A.: Extension of traffic flow pattern dynamic classification by a macroscopic model using multivariate clustering. Transp. Sci. **50**(3), 966–981 (2016)
18. Abuamer, I.M., Celikoglu, H.B.: Local ramp metering strategy ALINEA: microscopic simulation based evaluation study on Istanbul freeways. Transp. Res. Procedia **22**, 598–606 (2017)
19. Abuamer, I.M., Silgu, M.A., Celikoglu, H.B.: Micro-simulation based ramp metering on Istanbul freeways: an evaluation adopting ALINEA. In: 19th International IEEE Conference on Intelligent Transportation Systems (ITSC2016), pp. 695–700 (2016)
20. Khondaker, B., Kattan, L.: Variable speed limit: a microscopic analysis in a connected vehicle environment. Transp. Res. Part C: Emerg. Technol. **58**, 146–159 (2015)
21. Celikoglu, H.B., Dell'Orco, M.: Mesoscopic simulation of a dynamic link loading process. Transp. Res. Part C: Emerg. Technol. **15**(5), 329–344 (2007)
22. Celikoglu, H.B., Dell'Orco, M.: A dynamic model for acceleration behaviour description in congested traffic. In: Proceedings of the 11th International IEEE Conference on Intelligent Transportation Systems (ITSC 2008), pp. 986–991 (2008)
23. Dell'Orco, M., Celikoglu, H.B., Gurcanli, G.E.: Evaluation of traffic pollution through a dynamic link loading model. In: Li, S.C., Wang, Y.J., Cao, F.X., Huang, P., Zhang, Y. (eds.) Proceedings of the International Symposium on Environmental Science and Technology: Progress in Environmental Science and Technology, vol. II, pp. 773–777 (2009)
24. Demiral, C., Celikoglu, H.B.: Application of ALINEA ramp control algorithm to freeway traffic flow on approaches to Bosphorus strait crossing bridges. Procedia Soc. Behav. Sci. **20**, 364–371 (2011)

Bayesian Networks Probabilistic Safety Analysis of Highways and Roads

Elena Mora$^{(\boxtimes)}$, Zacarías Grande, and Enrique Castillo

Department of Applied Mathematics and Computational Sciences,
University of Cantabria, 39005 Santander, Spain
elenamoravil@gmail.com, zgandrade@gmail.com, castie@unican.es

Abstract. A probabilistic safety analysis methodology based on Bayesian networks models for the probabilistic safety assessment (PSA) of highways and roads is presented. The main idea consists of (a) identifying all the elements encountered when travelling the road, (b) reproducing these elements by sets of variables, (c) identifying the direct dependencies among variables, (d) building a directed acyclic graph to reproduce the qualitative structure of the Bayesian network, and (e) building the conditional probability tables for each variable conditioned on its parent nodes. Since human error is the most important cause of accidents, driver's tiredness and attention are used to model how the driver's behaviour evolves with driving and how it is affected by the environment, signs and other factors. A computer program developed in Matlab implements the Bayesian network model from the list of road items and a set of parameter values given by a group of experts. In this way, the most critical elements can be identified and sorted by importance, thus, an improvement of the global safety of the road can be done savings time and money. The proposed methodology is illustrated in real examples of a Spanish highway and a conventional road.

Keywords: Bayesian networks · Road safety
Probabilistic safety analysis

1 Introduction

One of the main concerns of our society today is the great number of accidents that occur on roads. Thus, the improvements on safety traffic analysis on roads are considered very useful. The study of this work focuses on Probabilistic safety analysis (PSA) since it is well-known and the best technique to asses the safety level of a system. It has been widely used for nuclear power plants and in recent years it has been also employed for railway lines (see for example [1,2]).

The analysis presented here is especially based on Bayesian networks (BN) because they have a great power to reproduce multidimensional random variables and present numerous advantages against other existing methods such as fault or event trees. However, there are few analysis of traffic infrastructures based on them. Some of the studies applying BN are for example [3–6].

© Springer International Publishing AG 2018
R. Moreno-Díaz et al. (Eds.): EUROCAST 2017, Part II, LNCS 10672, pp. 470–476, 2018.
https://doi.org/10.1007/978-3-319-74727-9_57

In this work a Bayesian network for probabilistic safety assessment of highways and roads is proposed and described.

2   Bayesian Network Proposed Model

In this section the Bayesian network proposed model is introduced and the steps of its building are detailed. A Bayesian network has two main components: (a) an acyclic graph which define the qualitative structure of the multidimensional variables and (b) the conditional probability tables which quantify the Bayesian network. In order to define the qualitative structure here, is used the subjective method, based on our knowledge of the problem since the existing data are not adequate to provide this information.

The first steps to build the Bayesian network are the selection of the most relevant items that have influence on traffic safety being likely to cause incidents, and the identification of the variables related with these items. In order to identify all possible items, such as curves, stop signals, intersections, roundabouts, tunnels, acceleration or deceleration lines, traffic light signals, and pavement failures, a video is recorded from the start to the end of the road (see [7]). The list of the chosen variables for this model with their definitions and their possible values are shown in Tables 1 and 2 depending on they are related to the driver, infrastructure or the incident respectively.

Table 1. List of variables related to the driver and the infrastructure with their possible definitions and values.

V. Related to the driver	Definition	Values
D: Driver's attention	This variable represents the driver's attention	Distracted, Attentive, Alert
T: Driver fatigue	Measures driver fatigue	A positive value that increases with driving time
Sd: Driver decision on speed	Represents the action of the driver in cases where he must adjust the speed	Correct, Error I, Error II
Dri: Driver type	This variable reflects the quality of the driver	Professional, Experienced, Standard, Bad
AS: Decision of the driver at a traffic light	Represent the decision of the driver at a traffic light	Correct, Error I, Error II
DS: Decision of the driver in a signal	Represents the decision of the driver in a signal, such as Stop, yields or speed limit signals	Correct, Error
V. Related to the infrastructure	Definition	Values
It: Traffic intensity	This variable measures traffic intensity	Light, Medium or Dense
Vis: Visibility	This variable measures the visibility existing at the point considered	Good, Average and Poor
Vt: Type of vehicle	Refers to the type of vehicle	Heavy Vehicle, Automobile and Moto
S: Speed	Is the circulation speed at the point considered	Set of positive values
SS: Signal status	Represents the status of the signal	Green, Yellow, Red
TF: Technical failure	It represents the possible failure of a vehicle, signal, etc.	Yes, No
E: Failure in the environment	It represents the possibility of unwanted events, such as obstructions of the road by stones, trees or other materials, animals as well as defects of the road, clearings, embankments, etc	No fault, Minor fault, Medium fault or Serious fault

Table 2. List of variables related to the incident with their possible definitions and values.

V. Related to the incident	Definition	Values
V: Vehicle failure	Consider the possibility of a vehicle failure	No fault, Minor fault, Medium fault or Serious fault
P: Pavement condition	Represents the state of the pavement	Good condition, Mild failure, Medium failure or Serious fault
Co: Collision	Represents the possibilities of collision with other vehicles that circulate on the road in the same or opposite direction	No Collision, Mild Collision, Medium Collision, or Severe Collision
W: Weather	Represents the type of climate	Fair, Rain/Snow, Wind, Fog, Snow/Ice
CF: Crossover Frequency	Represents the frequency of crossing between two elements of brakes such as vehicles and trains or pedestrians and vehicles at a specific point on the road, such as pedestrian crossings or level crossings	Yes, No
I: Incident	Represents possible incidents that may occur at a particular location on the road or in a no-signal range	Incident, Mild Incident, Medium Incident, and Serious Incident

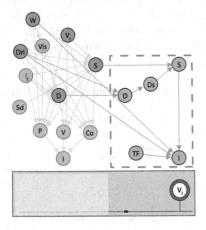

Fig. 1. Sign sub-Bayesian network as example of the different sub-Bayesian networks.

The identification of the direct dependencies among the variables involved is the next step. In order to obtain the acyclic graph different sub-Bayesian networks with a particular structure and associated variables with their corresponding conditional probabilities have been modelled. As a representative example, the sign sub-Bayesian network which is used at locations where some action subject to error is required, is shown in Fig. 1.

Finally, the conditional probability tables are defined using closed formulas.

Once these four steps are covered, all the information needed to calculate the marginal probabilities of the incident types nodes and the ENSI[1] values for each location become available.

---

[1] ENSI refers to the expected number of equivalent severe incidents, where 6.4 medium incidents and 230 light incidents are considered equivalent to one severe incident.

3   Software Development and Outcomes

A computer software written in Matlab with calls to Latex, JavaBayes and BNT software has been developed by our group to obtain the results for this model. As data input it is used a sequential description of all the items encountered when travelling along the highway. The computer program checks the given data for errors, builds the acyclic graph of the Bayesian network, builds the conditional probability tables, calculates the incident probabilities, evaluates the ENSI of incidence nodes, provides a table of the ENSI frequencies for all items, sorts ENSI values by importance, provides the expected number of incidents of each item type, plots the segments Bayesian subnetworks, gives the 'JavaBayes' code, gives the 'BN' code, and provides a report file.

In order to display the results in a useful way, the following elements, shown in Fig. 2 have been included divided by segments: (A) the segment acyclic graph of the Bayesian network, (B) the graphical representation of traffic signs and track elements, (C) the segment characteristics, and (D) a cumulated risk chart where the reader can easily identify the relative importance of the different items by comparing the discontinuities (jumps) of the graph. The whole line representation has been divided into short segments, with the aim of obtaining a more detailed information. In order to obtain a global view of the roads assessment some plant sites, where the riskiest points are allocated and differentiated by severity level, are represented, as well as, different types of tables to facilitate the understanding of the results of the probabilistic safety analysis. In the following Section, representative examples of this graphical information provided are illustrated.

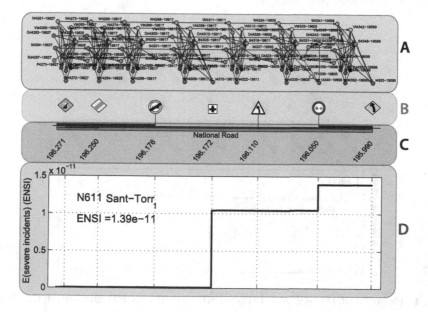

Fig. 2. Example of the information supplied by the computer program.

4   Examples of Application

The usefulness of the model can be appreciated in the next examples. In the first, the National N-634 road from (Kilometer Point, KP) 260.945 to KP 264.870 is considered, with an Annual Average Daily Traffic (AADT) of 2145 vehicles. In Fig. 3 the acyclic graph of the segment between KP 262.180 and KP 262.276 is represented. As commented before, the relative importance of the different

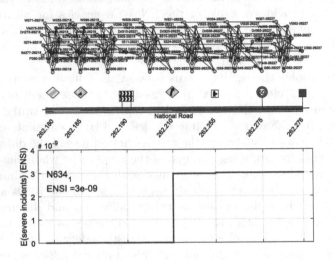

Fig. 3. Example of the segment between KP 260.945 to KP 264.870 of the N-634 national road without improvements.

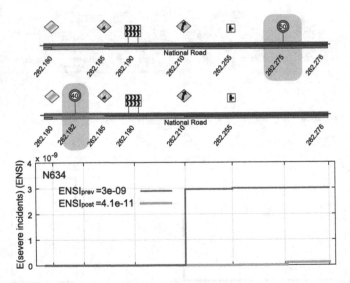

Fig. 4. Example of the segment between KP 260.945 to KP 264.870 of the N-634 national road with proposed improvements.

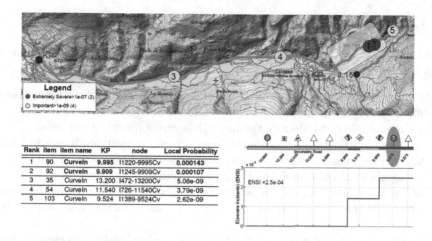

Rank	item	item name	KP	node	Local Probability
1	90	CurveIn	9.995	I1220-9995Cv	0.000143
2	92	CurveIn	9.909	I1245-9909Cv	0.000107
3	35	CurveIn	13.200	I472-13200Cv	5.08e-09
4	54	CurveIn	11.540	I726-11540Cv	3.79e-09
5	103	CurveIn	9.524	I1389-9524Cv	2.62e-09

Fig. 5. Example of the CA-182 secondary road initially.

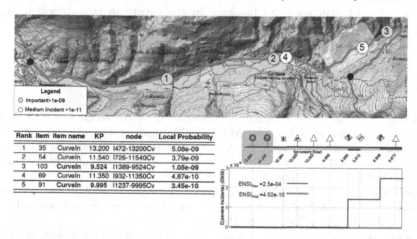

Rank	item	item name	KP	node	Local Probability
1	35	CurveIn	13.200	I472-13200Cv	5.08e-09
2	54	CurveIn	11.540	I726-11540Cv	3.79e-09
3	103	CurveIn	9.524	I1389-9524Cv	1.05e-09
4	69	CurveIn	11.350	I932-11350Cv	4.87e-10
5	91	CurveIn	9.995	I1237-9995Cv	3.45e-10

Fig. 6. Example of the CA-182 secondary road with proposed improvement.

items can be identified by comparing the discontinuities in the graph. It can be easily seen that curves and lateral entries are the most critical items. The worst segment of the line is placed between KP 262.210 and KP 262.416, thus, the speed limit sign (50 Km/h) of KP 262.275 have been reduced to 40 km/h and have been moved to KP 262.184. In Fig. 4 how this change improves significantly the resulting ENSI can be seen.

In the second example, shown in Fig. 5, the CA-182 secondary road trace from KP 9.801 to KP 15.250 with an AADT of 558 vehicles is represented. Here, the solution for the misplacement of a sign (speed limit signal 40 km/h at KP 9.875), which generated two items with a significant high incident probability, was to move the sign from its original location to KP 10.200. In Fig. 6 how the risk decrease significantly and the reduction of the ENSI can be appreciated.

5 Conclusions

Based on the model and outcomes presented in this paper the following conclusions can be drawn:

1. Bayesian network models are very useful to reproduce and to perform a probabilistic safety assessment of highways and roads.
2. This model allows to identify the most dangerous points of any road.
3. It is possible to determine the most frequent events that produce accidents.
4. The real examples commented in this paper show that the method can identify relevant incidents and quantify their probabilities of occurrence.
5. The proposed methodology allows to take corrective measures to the safety problems of the road and to predict accident concentration zones before they occur.
6. Applying this model resources for improving safety and maintenance could be optimized.
7. A lot of work still needs to be done in the future in this direction above all about the parameter estimation. The collaboration of diversity groups of experts would improve the power and the efficiency of the method.

References

1. Castillo, E., Grande, Z., Calviño, A.: Bayesian networks based probabilistic risk analysis for railway lines. Comput. Aided Civil Infrastruct. Eng. **31**, 681–700 (2016)
2. Castillo, E., Grande, Z., Calviño, A.: A Markovian-Bayesian network for risk analysis of high speed and conventional railway lines integrating human errors. Comput. Aided Civil Infrastruct. Eng. **31**, 193–218 (2016)
3. Pawlovich, M., Li, W., Carriquiry, A., Welch, T.: Experience with road diet measures: use of Bayesian approach to assess impacts on crash frequencies and crash rates. TRR: J. Transp. Res. Board **1953**, 163–171 (2006)
4. Persaud, B., Lyon, C.: Empirical Bayes before-after safety studies: lessons learned from two decades of experience and future directions. Accid. Anal. Prev. **39**, 546–555 (2007)
5. Deublein, M.K.: Roadway accident risk prediction based on Bayesian probabilistic networks. ETH Zürich. Ph.D (2013)
6. Deublein, M., Schubert, M., Adey, B.T., de Soto García, B.: A Bayesian network model to predict accidents on Swiss highways. Infrastruct. Asset Manag. **2**, 145–158 (2015)
7. Grande, Z., Castillo, E., Mora, E.: Highway and road probabilistic safety assessment based on Bayesian network model. Computer Aided Civil and Infrastructure Engineering (2017, to appear)

Author Index

Printed in the United States
By Bookmasters